BIOINFORMATICS AND BIOMEDICAL ENGINEERING

PROCEEDINGS OF THE 9TH INTERNATIONAL CONFERENCE ON BIOINFORMATICS AND BIOMEDICAL ENGINEERING, SHANGHAI, CHINA, 18–20 SEPTEMBER 2015

Bioinformatics and Biomedical Engineering

Editors

James J. Chou
Harvard Medical School, USA

Huaibei Zhou
Wuhan University, China

CRC Press
Taylor & Francis Group
Boca Raton London New York Leiden

CRC Press is an imprint of the
Taylor & Francis Group, an **informa** business

A BALKEMA BOOK

Published by: CRC Press/Balkema
P.O. Box 11320, 2301 EH Leiden, The Netherlands
e-mail: Pub.NL@taylorandfrancis.com
www.crcpress.com – www.taylorandfrancis.com

First issued in paperback 2020

ISBN 13: 978-0-367-73767-2 (pbk)
ISBN 13: 978-1-138-02784-8 (hbk)

Visit the Taylor & Francis Web site at
http://www.taylorandfrancis.com

and the CRC Press Web site at
http://www.crcpress.com

Typeset by V Publishing Solutions Pvt Ltd., Chennai, India

Bioinformatics and Biomedical Engineering – Chou & Zhou (Eds)
© *2016 Taylor & Francis Group, London, ISBN 978-1-138-02784-8*

Table of contents

Biomedical materials and products

Biomedical devices and systems

Preface

It is our great pleasure to present the proceedings of *The 9th International Conference on Bioinformatics and Biomedical Engineering (iCBBE 2015)*, held September 18–20, 2015 in Shanghai, China. We would like to take this opportunity to express our sincere gratitude and appreciation to all the authors and participants for their support of this conference.

The research on Bioinformatics and Biomedical Engineering has enormous impacts on science, education, culture and society as well. Actually, the discipline of Bioinformatics and Biomedical Engineering has become a new focus of life science, mathematical science, computer science and electronic information science. More and more scientists all around the world are dedicating themselves to this interdisciplinary area, accumulating a lot of interesting results.

We are proud to see that the previous iCBBE conferences were successful in providing an ideal platform for them to exchange their exciting findings, to stimulate the further development of Bioinformatics and Biomedical Engineering, and to enhance its impacts to various areas of both science and medicine (see, e.g., Medicinal Chemistry, 2015, 11, 218–234). We believe that the 2015 iCBBE will do even better in this regard.

On behalf of the organizing committee, we would like to take this opportunity to express our gratitude to the conference's sponsors: *Wuhan University and The Gordon Life Science Institute*.

Our appreciation and gratitude are also extended to all *the papers' reviewers and the Conference Organization Committee members*. It is impossible to hold such a grand conference without their help and support.

The papers collected in the *"2015 iCBBE Proceedings"* provide the detailed results of some oral presentations that will be of use to the readership.

Editors
Prof. James J. Chou
Harvard Medical School, USA

Prof. Huaibei Zhou
Wuhan University, China
2015

Bioinformatics and Biomedical Engineering – Chou & Zhou (Eds)
© 2016 Taylor & Francis Group, London, ISBN 978-1-138-02784-8

Organization

This volume contains the Proceedings of the 9th International Conference on Bioinformatics and Biomedical Engineering (iCBBE 2015)—held September 18–20, 2015 in Shanghai, China. iCBBE2015 has been organised by Wuhan University and The Gordon Life Science Institute.

INTERNATIONAL PROGRAMME COMMITTEE

Honorary General Chair

Prof. Kuo-Chen Chou, *The Gordon Life Science Institute, USA*

General Chair

Prof. James J. Chou, *Harvard Medical School, USA*

Technical Program Committee

Prof. Shu Q. Liu, *Northwestern University, USA*
Prof. Fengfeng Zhou, *Chinese Academy of Sciences, China*
Prof. Ridha Hambli, *Orleans University, France*
Dr. Fadhl M. Al-Akwaa, *University of Science and Technology, Yemen*
Dr. Yu Chen, *University of Strathclyde, UK*
Prof. Musa Hakan Asyali, *Antalya International University, Turkey*
Prof. Lukasz Kurgan, *University of Alberta, Canada*
Prof. Huabei Jiang, *University of Florida, USA*
Prof. Jerzy Tiuryn, *University of Warsaw, Poland*
Dr. Joseph Chang, *Nanyang Technological University, Singapore*
Dr. Mengxing Tang, *Imperial College London, UK*
Dr. Yanmei Tie, *Harvard Medical School, USA*
Dr. Manuchehr Soleimani, *University of Bath, UK*
Prof. Jinn-Moon Yang, *National Chiao-Tung University, Chinese Taipei*
Dr. Humberto González-Díaz, *University of the Basque Country, Spain*
Dr. Deligianni Despina, *University of Patras, Greece*
Dr. Shuwei Li, *Ambry Genetics, USA*
Dr. Suryani Lukman, *Khalifa University of Science, Technology and Research, UAE*
Prof. Sheng-Xiang Lin, *Laval University Medical Center, Canada*
Dr. Wei-Zhu Zhong, *The Gordon Life Science Institute, USA*
Prof. Rajiv Mahendru, *BPS Government Medical College for Women, India*
Prof. Zodwa Dlamini, *University of South Africa, South Africa*

LOCAL ORGANIZING COMMITTEE

Fang Liu, *Wuhan University, China*
Liang Li, *Wuhan University, China*
Ruoshan Kong, *Wuhan University, China*
Xiaoyan Sheng, *Wuhan University, China*
Yuanyuan Cheng, *Wuhan University, China*
Yujing Zhang, *Wuhan University, China*

Acknowledgements

The Organising Committee members wish to express their sincere gratitude for the financial assistance from the following organisations: Wuhan University, the Gordon Life Science Institute and the 1000 Think Tank.

The technical assistance of all paper peer reviewers and the publisher CRC Press/Balkema is gratefully acknowledged. We are also thankful to the International Programme Committee as well as the members of the Local Organising Committee. Finally, the editors want to acknowledge all peer reviewers for their great efforts and contributions to us.

Editors
Prof. James J. Chou
Harvard Medical School, USA

Prof. Huaibei Zhou
Wuhan University, China

Bioinformatics and Biomedical Engineering – Chou & Zhou (Eds)
© 2016 Taylor & Francis Group, London, ISBN 978-1-138-02784-8

About the editors

Prof. James J. Chou, Harvard Medical School, USA

James J. Chou is a Professor in Biological Chemistry and Molecular Pharmacology at Harvard Medical School. He received his BS in Physics from University of Michigan at Ann Arbor; Ph.D. in Biophysics from Harvard University; and postdoctoral training from NIH. Since 2002, he joined the faculty at Harvard Medical School. Professor Chou has received many prestigious awards such as the Pew Scholar Award in Biomedical Sciences, the Smith Family Award, and the Genzyme Outstanding Achievement in Biomedical Science Award.

Prof. Huaibei Zhou, Wuhan University, China

Prof. Huaibei Zhou received his B.S. in Radio Wave Propagation and Antenna Design in Wuhan University, China in 1984; Ph.D. in Computational Physics from the University of Maryland at College Park and Post-Doc in Biotechnology in the National Institute for Standard and Technology in 1994; and MBA in Engineering Management from George Washington University in 2001. Prof. Huaibei Zhou discovered the chaotic motion of protein molecules by simulating the dynamics of protein molecules in water (the simulation is programmed in C and run in Unix environment), the result has been published as the first paper in this field; more than 30 scientists around the world have cited this paper in their publications.

Biomacromolecular sequence, structure and function

Bioinformatics and Biomedical Engineering – Chou & Zhou (Eds)
© *2016 Taylor & Francis Group, London, ISBN 978-1-138-02784-8*

Relationship between homo sapiens histamine receptors using data mining

S.M. Kim, S.R. Kim & T.S. Yoon
Department of Natural Science, Hankuk Academy of Foreign Studies (HAFS), Yongin, South Korea

ABSTRACT: In this study, we tried to find rules between mRNA sequences of Homo sapiens histamine receptors H1, H2, H3 and H4 with data mining algorithms, namely Apriori and Decision tree. In the Apriori algorithm, we split sequences into 5, 7 and 9 windows. The results showed a strong relationship between the H1 and H4 receptors, and also between the H2 and H3 receptors. The receptors were divided into two groups according to their components. Additionally, we would leave relevancy between the H2 and H3 receptors for another study with a different data mining algorithm. In the case of the H1 and H4 receptors, we found that amino acid "F(phenylalanine)" would be a standard to classify the H1 and H4 receptors. We suggest that H4 could be a mutated form of H1. To support our hypothesis, we conducted an additional experiment with the Decision Tree algorithm, focusing on the existence of amino acid "F". The data showed the difference between the H1 and H4 receptors. In conclusion, the H4 and H1 receptors are related to each other by mutation.

1 INTRODUCTION

Immune system is the system of the animal body that acts as a protection system against the pathogen from the outer environment. However, there are several diseases caused by disruptions in immune system functions, such as allergies. Allergies are a result of exaggerated and hypersensitive responses to certain antigens called allergens. Antigens of the IgE class involves in allergic reaction. IgE antibodies are attached to mast cells in connective tissues by their base. When antigens enter the body, they attach to antigen-binding sites of IgE antibodies. This connects other antigens near IgE antibodies, causing these antigens to band together. Mast cells secrete histamines or other inflammation induction substances from granules. This process is called degranulation. Histamine expands the blood vessel and increases the permeability of capillaries, which causes typical allergy symptoms such as sneezing, runny nose, watery eyes and smooth muscle contractions which may lead to breathing difficulty. Antihistamines blocks histamines from combining to receptor in order to diminish allergy symptoms (Jane. 2011). Histamine is a type of amine that is produced and secreted in the animal immune system. It is used in local immune response in order to cause inflammation. Histamine, which is secreted from mast cells, would bind to its receptor (histamine receptor). There are four kinds of histamine receptor known in the human body, which are named H1, H2, H3 and H4. Previously, it was known that histamine receptor 1 (H1) is highly involved in allergic reaction; however, recent studies have shown that histamine receptor 4 (H4) is also involved. It is supposed that there may be some kind of relationships between these two receptors (Thrumond. 2008) (Fung-Leung. 2004). In this study, we compare the amino acid strand of four types of histamine receptor H1, H2, H3 and H4 by using an Apriori algorithm and a Decision Tree algorithm in order to see relatedness and isoforms between the receptors.

2   MATERIALS AND METHODS

2.1   *Materials*

For the experiment, we collected the mRNA sequence of Homo Sapiens histamine receptors H1 (HRH1)*, H2 (HRH2)**, H3 (HRH3)*** and H4 (HRH4)**** including its transcript variants. The mRNA sequences used can be found in the NCBI database.

2.2   *Methods*

To process the data, we used two kinds of algorithms: Apriori and Decision Tree (Lee. 2014) (Lim. 2014) (Go. 2014).

2.2.1   *Apriori algorithm*
Apriori algorithm is usually used in data processing and in the field of bioinformatics to find rules among the continuative data (Kim. 2014). It shows the volume of common parts existing in the data. By using the Apriori algorithm, we can clearly compare objects with high frequency in certain data with those found in other data. In this study, 5, 7, and 9 windows were used to search for amino acid showing high frequency in each data. Also, broken data, which are rarely found during data processing, were not used in the experiment. This is the sample of the result of the 5 window Apriori algorithm experiment (Kwon. 2014).

Best rules found:

1. amino5=L 42
2. amino3=L 40
3. amino4=L 35
4. amino1=L 33
5. amino4=S 31

In the example, it showed amino5=L 42 as the first rule. This means that amino acid L (Leucine, Leu) is the most frequent amino acid in the 5 window as it showed 42 times repeatedly. Also, statistical compilations focused on the frequency and number of amino acid appearance instead of its location. According to these data, we constructed a graph in order to see the data intuitively.

---

*"Homo sapiens histamine receptor H1 (HRH1), transcript variant 1, mRNA", NCBI Reference Sequence: NM_001098213.1 (4,578 bp linear mRNA).
"Homo sapiens histamine receptor H1 (HRH1), transcript variant 2, mRNA", NCBI Reference Sequence: NM_001098212.1 (4,298 bp linear mRNA).
"Homo sapiens histamine receptor H1 (HRH1), transcript variant 3, mRNA", NCBI Reference Sequence: NM_001098211.1 (4,348 bp linear mRNA).
"Homo sapiens histamine receptor H1 (HRH1), transcript variant 4, mRNA", NCBI Reference Sequence: NM_000861.3 (4,427 bp linear mRNA).
**"Homo sapiens histamine receptor H2 (HRH2), transcript variant 1, mRNA", NCBI Reference Sequence: NM_001131055.1 (2,624 bp linear mRNA).
"Homo sapiens histamine receptor H2 (HRH2), transcript variant 2, mRNA", NCBI Reference Sequence: NM_022304.2 (3,095 bp linear mRNA).
***"Homo sapiens histamine receptor H3 (HRH3), mRNA", NCBI Reference Sequence: NM_007232.2 (2,680 bp linear mRNA).
****"Homo sapiens histamine receptor H4 (HRH4), transcript variant 1, mRNA", NCBI Reference Sequence: NM_021624.3 (3,686 bp linear mRNA).
"Homo sapiens histamine receptor H4 (HRH4), transcript variant 2, mRNA", NCBI Reference Sequence: NM_001143828.1 (3,422 bp linear mRNA).
"Homo sapiens histamine receptor H4 (HRH4), transcript variant 3, mRNA", NCBI Reference Sequence: NM_001160166.1 (3,522 bp linear mRNA).

2.2.2 Decision tree algorithm

While the Apriori algorithm is used to extract the frequency of various amino acids, Decision Tree is an effective alternative method to show the difference between the data clearly or definitely (Lim 2014). With the Decision Tree algorithm, isoform data were gathered from those sequences that were not able to classify with the Apriori algorithm. We determined that the Decision tree Algorithm is the most appropriate data mining algorithm and well-ordered lists, which can perform the function of classifying data and finding classes that represent ability. We compared the histamine receptors H1 and H4, which were expected to give clear data. We divided into 3 window sizes (5, 7 and 9) and split the data up to 7 classes. Class 1 to 4 each responds to the Homo sapiens histamine receptor H1 (HRH1) transcript variant 1 through 4, and class 5 to 7 each responds to the Homo sapiens histamine receptor H4 (HRH4) transcript variant 1 through 3. In this study, we used a 10-fold cross-validation method and rule-based classifiers. To retain accuracy, sets of the rules with a frequency over 0.800 in each class were only used to obtain the conclusion.

3   DISCUSSION

3.1   Apriori algorithm

From the result of the Apriori algorithm, we obtained two significant results. The histamine receptors may be divided into two groups by their amino acid, which would put the H1 and H4 receptors in the same group and the H2 and H3 receptors in the other group. From Table 1, for each histamine receptor, high frequencies of L (Leucine, Leu) and S (Serine, Ser) amino acid were found. As these amino acids exist commonly in all receptors, we defined these two amino acids as a "Basic Group". In contrast, there were amino acids that were only found in a certain type of histamine receptors such as A (Alanine, Ala), F (Phenylalanine, Phe) and G (Glycine, Gly). A and G amino acids were found frequently in the H2 and H3 receptors; however, F amino acid was only found in the H4 receptor. We defined these amino acids as an "Additional Group". We define that "Basic Group" amino acids would carry out a common function of histamine receptor, while "Additional Group" amino acids assigns a distinct function to the receptor. According to this definition, we suggest that the H2 and H3 receptors take a similar position in the immune system such as local inflammation, which differs from the H1 and H4 receptors that takes position in allergy reaction. As the H1 receptor only has peaks of "Basic Group" amino acids, we would name the receptor as a type 0 receptor. Also, the H4 receptor is named as type 0` receptor since it has only one additional amino acid (F) compared with H1 that does not overlap with any other receptors. As this receptor differs only in F amino acid, we believe this single amino

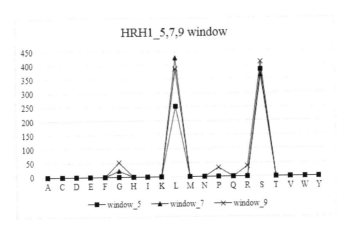

Figure 1.   Graph of HRH1_5,7,9 window.

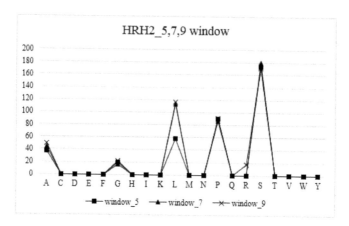

Figure 2.   Graph of HRH2_5,7,9 window.

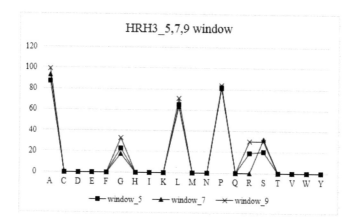

Figure 3.   Graph of HRH3_5,7,9 window.

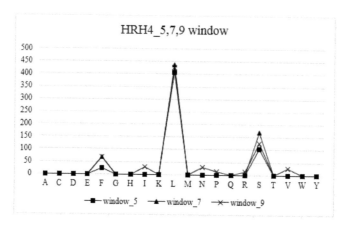

Figure 4.   Graph of HRH4_5,7,9 window.

Table 1. Rule extraction under 5 window.

Histamine receptor	Rule			Frequency
1-1	pos2 = D	pos3 = K		0.800
	pos3 = K	pos5 = P		0.800
	pos1 = C	pos5 = Y		0.800
1-2	pos2 = R	pos5 = D		0.800
	pos1 = Q	pos5 = P		0.800
	pos1 = H	pos5 = A		0.857
1-3	pos2 = G	pos3 = K		0.800
	pos3 = E	pos4 = N		0.800
	pos2 = Q	pos5 = V		0.833
	pos2 = S	pos4 = P	pos5 = L	0.800
1-4	pos2 = S	pos4 = S	pos5 = C	0.800
	pos1 = S	pos2 = L	pos3 = L	0.800
	pos1 = S	pos3 = S	pos5 = W	0.800
	pos3 = C	pos4 = T		0.800
	pos3 = M	pos5 = A		0.800
4-1	Not extracted			
4-2	pos1 = P	pos5 = T		0.800
	pos2 = K	pos4 = F	pos5 = L	0.833
4-3	pos2 = T	pos5 = P		0.800
	pos1 = H	pos4 = V		0.800

Table 2. Rule extraction under 7 window.

Histamine receptor	Rule			Frequency
1-1	pos1 = K	pos2 = A		0.800
	pos2 = M	pos4 = E		0.800
	pos1 = S	pos2 = K		0.800
1-2	pos1 = W	pos2 = A		0.800
	pos1 = C	pos2 = T		0.800
	pos2 = W	pos6 = R		0.800
	pos2 = L	pos6 = Q		0.800
1-3	Not extracted			
1-4	pos1 = M	pos2 = L		0.800
	pos1 = I	pos2 = T		0.800
	pos2 = C	pos3 = T		0.800
	pos2 = G	pos5 = Q		0.800
4-1	pos2 = I	pos7 = I		0.800
	pos2 = S	pos5 = Q	pos7 = L	0.800
	pos2 = N	pos4 = S		0.800
4-2	pos3 = F	pos4 = F		0.800
4-3	Not extracted			

acid makes a small difference between H1 and H4 amino acids, while they still have a more common part. This explains why the H1 and H4 receptors share a common position in the immune system even though they are named differently. Also, the difference between the H2 and H3 receptors would occur by the amount of additional amino acids P and A. Although the H2 and H3 receptors share their additional amino acids A, P and G, the H2 receptor has

Table 3. Rule extraction under 9 window.

Histamine receptor	Rule			Frequency
1-1	pos2 = A	pos9 = I		0.800
	pos5 = R	pos8 = S		0.800
	pos2 = L	pos5 = P		0.857
1-2	pos5 = L	pos7 = G	pos9 = G	0.800
	pos1 = S	pos3 = T		0.857
	pos1 = W	pos3 = T		0.800
	pos1 = S	pos5 = C		0.857
1-3	pos7 = V	pos9 = G		0.800
	pos8 = I	pos9 = W		0.800
	pos4 = T	pos6 = L		0.800
	pos2 = E	pos8 = E		0.800
	pos1 = P	pos8 = H		0.800
	pos3 = L	pos9 = Q		0.800
	pos8 = A	pos9 = P		0.800
	pos2 = E	pos5 = T		0.800
1-4	pos7 = G	pos9 = A		0.800
	pos5 = T	pos9 = T		0.800
	pos4 = E	pos9 = Y		0.800
	pos3 = C	pos5 = C		0.800
	pos4 = N	pos8 = R		0.800
4-1	pos5 = L	pos7=L	pos9 = L	0.800
4-2	pos6 = V	pos8 = N		0.800
4-3	pos6 = T	pos9 = N		0.817
	pos2 = R	pos8 = L		0.857

shown a relatively high peak on P amino acid, while H3 amino acid had a peak on A amino acid. As this difference cannot be defined through the Apriori algorithm, we believe that we have reached the limit of the Apriori algorithm. So, with other data mining algorithm, this difference may be defined.

3.2 *Decision tree algorithm*

Decision Tree algorithm was used in order to find an additional relationship between the H1 and H4 receptors. The purpose of using the Decision Tree algorithm is to find another rule that did not show the Apriori Algorithm. As stated before, data with a frequency over 0.800 were only used. This is because there were too many data with a frequency below 0.800 for the list or sequence analysis and the finding rule. The most important point of this experiment is to find a new rule that we were not able to find using the Apriori algorithm and about F amino acid that belong to the "Additional Group" that we defined in the first experiment. In this way, we obtained an interesting data. It is that F amino acid was found at high frequency in the result of experiment by window 5 and window 7 about class 6, which is an experiment of the H4 receptor. This is an evidence about our suggestion to define the H1 and H4 receptors as type 0 and type 0` receptors by finding similarity with those in the Apriori experiment.

8

4 CONCLUSION

Through two experiments using the Apriori algorithm and the Decision Tree algorithm, we found some important facts. Based on these facts, we made some deduction on the relationship between the H2 and H3 receptors and also on the relationship between the H1 and H4 receptors. The relationship between the H2 and H3 receptor is the first one that we have found. In the experiment with the Apriori algorithm, we found a strong correlation between the H2 and H3 receptors. Both receptors are composed of the same types and frequency of amino acid, which leaves the only difference in the small frequency change of "Additional" amino acid. In other words, this means that another type of data mining algorithm and experiment is required in order to specify the difference. We would leave this part as possibility for the next study and experiment.

The second one is about the H1 and H4 receptors, which is the most important point in our experiment. In the experiment using the Apriori algorithm, we found a large correlation and a small difference between the H1 and H4 receptors, which is whether they did or did not have F amino acid. We decided that an additional experiment is needed in order to find more information. Decision Tree algorithm was chosen for the second experiment. The results did not differ much from the experiment using the Apriori algorithm, which became a big ground for our hypothesis, which is evolutionary variation. This supports our suggestion to be valid. In conclusion, we have analyzed the mRNA sequence of Homo sapiens histamine receptors, which is related to the cause of allergy. The results showed much relevancy between the H1, H2, H3 and H4 receptors. Also, we defined some of receptors to verify our hypothesis. As there are results consistent with the experimental result data, we found that our study was certainly meaningful. Our future task of this study is to develop an integration system for a better environment in studies in the field of bioinformatics.

REFERENCES

Fung-Leung W.P. 2004. Histamine H4 receptor antagonists: the new antihistamines? Current Opinion in Investigational Drugs.
Go E.B 2014. Analysis of Ebolavirus. International Journal of Machine Learning and Computing (IJMLC).
Jane B.R. 2011. BIOLOGY, 9th Edition. California, CA: Pearson Education Inc.
Kim D.Y. 2014. Comparison of Hemagglutinin and Neruaminidase of Influenza A Virus Subtype H1N1, H5N1, H5N2, and H7N9 using Apriori Algorithm. Lecture Note Computer Science (LNCS).
Kwon J.W 2014. Comparison of HTLV and STLV. APCBEE Procedia.
Lee J.H. 2014. Analysis of Malaria Inducing P. Falciparum P. Ovale, and P. Vivax. APCBEE Procedia
Lim S.J. 2014. Analyzing Patterns of Various Avian Influenza Virus by Decision Tree. International Journal of Computer and Electrical Engineering (IJCEE).
Lim S.J. 2014. rRNA of Alphaproteobacteria Rickettsiales and mtDNA Pattern Analyzing with Apriori & SVM. Lecture Note Computer Science (LNCS).
Thrumond R.L. 2008. The role of histamine H1 and H4 receptors in allergic inflammation: the search for new antihistamines. Nat Rev Drug Discov.

Bioinformatics and Biomedical Engineering – Chou & Zhou (Eds)
© *2016 Taylor & Francis Group, London, ISBN 978-1-138-02784-8*

Analysis of Ebolavirus and Marburgvirus using data mining

J. Jeong, B. Kim & T.S. Yoon
Department of International, Hankuk Academy of Foreign Studies, Yongin-si, South Korea

ABSTRACT: Ebolaviruses and marburgviruses, which cause viral haemorrhagic fevers in humans and often prove to be fatal, are part of the Filviridae family. The virus family was first identified in 1967 from green monkeys by laboratory workers in Germany and Yugoslavia and since then, there have been several outbreaks and large epidemics related to it. In this paper, we compared the protein sequences encoded in the RNA genomes of ebolaviruses and marburgviruses particularly focusing on VP24, VP35, and GP. In addition, by examining the position and frequency of each amino acid, we analyzed the similarities between the two members of the filviridae family.

1 INTRODUCTION

Filvirdae is a virus family of RNA viruses belonging to the order *Mononegavirales*. Ascribed to causing hemorrhagic fever in primates including both human and non-humans, members of the family (called filoviruses or filovirids) are identified as hazardous by organizations. The family *Filovirade* contains three virus genera: Cuevavirus, Ebolavirus, Marburgvirus. Ebolavirus can be further divided into five species: Bundibugyo ebolavirus (BEBOV), Reston ebolavirus (REBOV), Sudan ebolavirus (SEBOV), Tai Forest ebolavirus (CIEBOV/TAFV), and Zaire ebolavirus (REBOV) (Henzy, 2014). Ebolavirus and marburgvirus are thought to be zoonotic, passed from animals to humans. Fruit bats of the Pteropodidae family are generally considered to be the host of both viruses; however, apes and chimpanzees are also regarded as possible hosts. More specifically, scientists infer that *rosettus aegypti,* which are fruit bats inside the Pteropodidae family, are the natural hosts of Marburgvirus, transmitting the disease to people and causing it to spread to other humans. The viruses both cause symptoms such as malaise, muscle pain, sudden fever, headache, etc., and at the moment, there is no standardized cure of filovirus diseases (WHO, 2014). The name *Filviridae* comes from the Latin term *filum*, meaning "threadlike" which accurately describes the slender structure of filovirions. Filoviruses contain linear, non-segmented, single-stranded and antisense (often called negative-sense) RNA genomes, ~19-kb long in length. Proteins NP, VP30, VP35, and L form the nucleocapsid; proteins VP24 and VP24 form the viral matrix, and protein GP forms the surface of the particle. In total, filoviruses contain seven proteins that function in its own distinct way.

2 MATERIALS AND METHODS

As mentioned above, *Filovirade* can be classified into three virus genera: Cuevavirus, Ebolavirus, Marburgvirus. This research focuses on the latter two, which are the more commonly known of the three. By using decision tree and apriori algorithm, we tried to obtain the similarities and the differences between the protein sequences of the Ebolavirus and Marburgvirus by analyzing proteins VP24, VP35, and GP from the RNA genome.

2.1 VP24, VP35

VP24 and VP35 target Interferon-Stimulated Genes (ISG) which are proteins produced by interferon, used to combat viruses. Though they attack from different angles, VP24 and VP35 both threaten the host cell's immune system. VP24 block the immune system from transcription, discouraging signaling molecules from reaching the cell nucleus (Xu et al., 2014), and VP35 blocks the path of Type I IFN, keeping them from being produced (Prins et al., 2009).

2.2 GP (Glycoprotein)

The Glycoprotein (GP) is in charge of letting in the virus into target host cells. Located on the surface of the particle, GP mediates the entry.

2.3 Decision tree

A decision tree is a model used to graphically outline the possible consequences of decisions. It consists of three nodes: decision nodes, chance nodes, and end nodes. Drawn from left to right, the branches show the possible actions and the path from root to leaf shows the classification rules. Since a decision tree is a method of displaying algorithm, it is possible to analyze decisions, calculate the probability, and figure out the most optimal strategy; therefore, it is often used in operations research (Rokach, Maimon, 2010). Traditionally, it was drawn manually, but nowadays, graphics program and software are utilized to create it. In this paper, the decision tree was composed of window 5, 7, 9 and experimented with groups VP24, VP35, and GP. We were unable to find the rules of the decision tree when the subjects were too similar to one another.

2.4 Apriori algorithm

Apriori algorithm is the simplest set class theory used to find the frequency of the elements. and the association rules from data. It uses a "bottom up" approach to test candidates against the data, ending when there are no subset extensions left (Shi, 2011). Like the decision tree, the experiment was carried out with windows 5, 7, and 9. The results show the position where certain types of amino acids most frequently appeared.

3 EXPERIMENT

3.1 Decision tree

5 window shows that there is no particular relationship between the GP of Ebola virus and Marburg virus under window 5. Also, according to Table 1, like the Ebola virus, in Marburg virus, position 7 and position 3 occurs most frequently. This implies position 7 and position 3 play important roles in both viruses. Table 2 clearly shows that rules were only extracted in position 8. However, in the case of Ebola virus, pos8 = E was the most frequent, while in the case of Marburg virus, pos8 = S was the most frequent.

According to Table 3, the result of rule extraction of VP24 under 5 window, rules are only extracted at position 4. When looking at the graph as a whole, the frequency of amino acid extracted at position 4 of Ebola virus is higher than that of Marburg virus. No other

Table 1. Rule extraction of GP under 7 window.

Virus	Rule
Ebola	Position 7 = S, Position 3 = H
Marburg	Position 7 = T, Position 1 = I,
	Position 3 = A, and Position 7 = I

Table 2.   Rule extraction of GP under 9 window.

Virus	Rule
Ebola	Position 8 = Q, Position 8 = E, Position 8 = M
Marburg	Position 8 = Y, Position 8 = S, Position 8 = F

Table 3.   Rule extraction of VP24 under 5 window.

Virus	Rule
Ebola	Position 4 = A, Position 4 = P, Position 4 = R
Marburg	Position 4 = I, Position 4 = N

Table 4.   Rule extraction of VP35 under 5 window.

Virus	Rule
Ebola	Position 1 = D, Position 5 = G
Marburg	Position 3 = H, Position 1 = D, Position 2 = K

Table 5.   Rule extraction of VP35 under 7 window.

Virus	Rule
Ebola	Position 1 = Y, Position 1 = N
Marburg	Position 1 = A, Position 1 = E

relationships are proven. No rules can be extracted from VP24 of both viruses under windows 7 and 9.

Table 4 conveys that there are similarities in pos1 = D between VP35 of Ebola virus and Marburg virus. Although their frequency is different from 0.833 to 0.800, sharing of the amino acid in this position might deeply contribute to the similarity of two viruses. According to Table 5, rules are extracted only from the position 1 under 7 window. So, we assume that position 1 will be the crucial factor which makes Ebola and Marburg virus different from each other. Lastly, no rules can be extracted under 9 window.

3.2   *Apriori algorithm*

The results driven from Apriori algorithm are shown below:

Position 4 = T 19, Position 5 = T 18, Position 2 = D 16, Position 1 = T 14

Figure 1 is the result shown by analyzing GP *(Glycoprotein)* of Ebolavirus under 5 windows. Analyzing other proteins of either virus had similar results with this. We chose the most frequent rules of each protein of each virus. As we want to show similar or analogous patterns in each protein of both viruses, we organized the types of amino acid by the frequency of them.

Figure 2, Figure 3, and Figure 4 are the analysis of GP of Ebolavirus and Marburgvirus under windows 5, 7, and 9. 5-window: Comparison between Ebolavirus amino4 Thiamine and Marburgvirus amino1 Thiamine. 7-window: Ebolavirus amino6 Thiamine and Marburgvirus amino4 Thiamine. 9-window: Ebolavirus amino6 Thiamine and Marburgvirus amino5 Thiamine. In all 3 windows, Marburgvirus has higher Thiamine than Evolavirus.

13

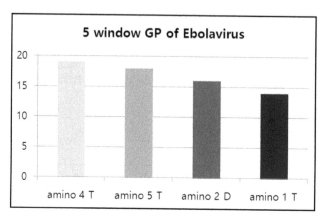

Figure 1.　5 window GP of Ebolavirus amino sequence.

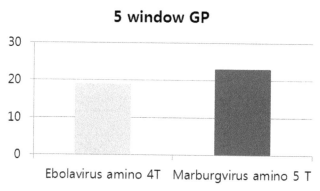

Figure 2.　5 window GP amino sequence of Ebolavirus and Marburgvirus.

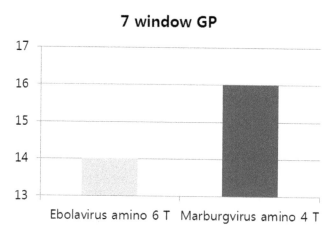

Figure 3.　7 window GP amino sequence of Ebolavirus and Marburgvirus.

Figure 5, Figure 6, and Figure 7 are the analysis of VP24 of both viruses under windows 5, 7, and 9. 5-window: Comparison between Ebolavirus amino1 Leucine and Marburgvirus amino2 Leucine. 7-window: Ebolavirus amino3 Leucine and Marburgvirus amino1 Leucine. 9-window: Ebolavirus amino3 Leucine and Marburgvirus amino3 Leucine. As a result of this analysis, we concluded that Ebolavirus has higher Leucine in VP24 than Marburgvirus has.

Figure 4.　9 window GP amino sequence of Ebolavirus and Marburgvirus.

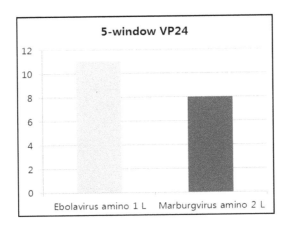

Figure 5.　5 window VP24 amino sequence of Ebolavirus and Marburgvirus.

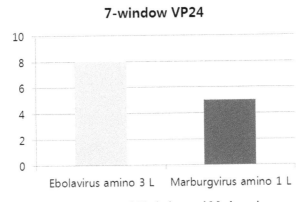

Figure 6.　7 window VP24 amino sequence of Ebolavirus and Marburgvirus.

Figure 8 and Figure 9 are the analysis of VP35 of both viruses under 7 and 9 windows. Under 5-windows, there were no amino acids both viruses share to compare with, so we only chose 7 and 9 windows. 7-window: Comparison between Ebolavirus amino4 Leucine and Marburgvirus amino5 Leucine. 9-window: Ebolavirus amino6 Leucine and Marburgvirus amino3 Leucine. In this case, the figures above show that generally Leucine in VP35 of Marburgvirus is higher than that in Ebolavirus.

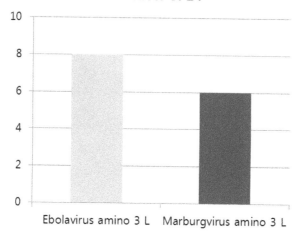

Figure 7.   9 window VP24 amino sequence of Ebolavirus and Marburgvirus.

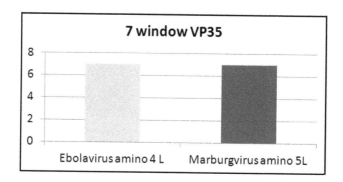

Figure 8.   7 window VP35 amino sequence of Ebolavirus and Marburgvirus.

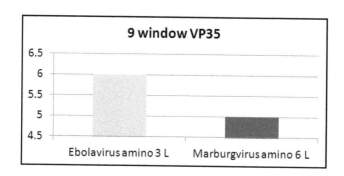

Figure 9.   9 window VP35 amino sequence of Ebolavirus and Marburgvirus.

4   CONCLUSION

As a result of analysis by Apriori algorithm, we found that the higher the windows are, the more detailed and diverse the DNAs were. After using Decision Tree, we were able to figure out similarities between Ebola virus and Marburg virus. Although rules were not extracted from all windows, it was clear that Ebolavirus and Marburgvirus share similar traits for GP,

VP24, and VP35. Since, VP24 and VP35 are related with interferon which is used to combat viruses, we are looking forward to find the methods to cure both viruses. For further research, we want to examine whether having the same hosts leads viruses to possess similar properties and try to find out the most fitting natural host of *Ebolavirus* and *Marburgvirus* from our studies.

REFERENCES

Henzy, Jamie. 7 Sept. 2014. "Five Questions about Filoviruses." *'Small Things Considered'* N.p., Web. 12 Mar. 2015.

Prins, K.C., W.B. Cardenas, and C.F. Basler. "Ebola Virus Protein VP35 Impairs the Function of Interferon Regulatory Factor-Activating Kinases IKK and TBK-1." 2009. *Journal of Virology* 83.7: 3069–077. Web.

Rokach, Lior, and Oded Maimon. 2010. "9. Decision Trees." *Data Mining and Knowledge Discovery Handbook.* N.p.: n.p., n.d. N. pag. Web.

Shi, Zhongzhi. 2011. *Advanced Artificial Intelligence. Singapore:* World Scientific.

World Health Organization. Sept. 2014. Ebola Virus Disease. WHO, Web. 11 Feb. 2015. *Centers for Disease Control and Prevention.* Centers for Disease Control and Prevention, Web. 12 Apr. 2015.

Xu, Wei, Megan R. Edwards, Dominika M. Borek, Alicia R. Feagins, Anuradha Mittal, Joshua B. Alinger, Kayla N. Berry, Benjamin Yen, Jennifer Hamilton, Tom J. Brett, Rohit V. Pappu, Daisy W. Leung, Christopher F. Basler, and Gaya K. Amarasinghe. 2014. "Ebola Virus VP24 Targets a Unique NLS Binding Site on Karyopherin Alpha 5 to Selectively Compete with Nuclear Import of Phosphorylated STAT1." *Cell Host & Microbe* 16.2: 187–200. Web.

Bioinformatics and Biomedical Engineering – Chou & Zhou (Eds)
© 2016 Taylor & Francis Group, London, ISBN 978-1-138-02784-8

Identifying the missing protein in human proteome by structure and function prediction

Q.W. Dong
School of Computer Science, Fudan University, Shanghai, China
Department of Computational Medicine and Bioinformatics, University of Michigan,
Ann Arbor, Michigan, USA
Shanghai Key Laboratory of Intelligent Information Processing, Shanghai, China

K. Wang
College of Animal Science and Technology, Jilin Agricultural University, Changchun, China

ABSTRACT: After the completion of human genome project, the proteome research becomes one of the center problems in post-genomics era. The Human Protein Project aims to identify at least one protein product from each of the human protein-coding genes by using experiment methods. However there are still many proteins without experimental evidence which become one of the major challenges in chromosome-centric human proteome project. Taking into consideration of the complexity of detecting these missing proteins by using proteomics approach, here we provide the structure and function of these missing proteins by bioinformatics methods. 616 "uncertain" missing proteins are extracted from the neXtProt database and the structure and function of these missing proteins are predicted by using state-of-the-art software I-TASSER and COFACTOR respectively. A comprehensive evaluation shows that the results are in good consistent with many manually curated annotations from well-established databases and other mass spectrum datasets. There are 188 foldable proteins (I-TASSER C-Score larger than −1.5) without using any homologous template, which may be native gene-coding proteins. The Gene Ontology function prediction results are in good agreement with the manual annotation from neXtProt database, and also the confidence scores are well correlated with the evaluation metrics with Pearson correlation coefficient of 0.65. The data are deposited into Human Proteome Structure and Function database (HPSF) which can provide valuable references about the missing proteins. The HPSF database is publicly available at http://zhanglab.ccmb.med.umich.edu/HPSF/.

1 INTRODUCTION

Proteins play an important role in biology activities. The successful completion of Human Genome Project (Venter et al., 2001) provides a valuable blueprint about all the genes encoding entire human proteins. However, due to the complexity of proteins and currently underdeveloped proteomics technique, many of the proteins have not been detected and annotated.

Towards exploring the universal space of human proteome, the Human Proteome Organization (HUPO) has recently launched the Human Proteome Project (HPP) (Legrain et al., 2011) including the Chromosome-centric Human Proteome Project (C-HPP) (Paik et al., 2012) and Biology/Disease-Driven HPP (B/DHPP). (Aebersold et al., 2013) The primary goal of the C-HPP is to identify at least one representative protein product and as many post-translational modifications, splice variant isoforms as possible for each of the human genes. This ambitious goal is collaborated by 25-membered international consortium covering 24 chromosomes and mitochondria. (Marko-Varga et al., 2013) The HPP executive committee has established five baseline metrics for C-HPP (Marko-Varga et al., 2013): the Ensembl database (Flicek et al., 2014) provides the number of protein-coding genes; Peptide Atlas

(Farrah et al., 2013) and GPMdb (Craig et al., 2004) provide the highly-confident proteins by mass spectrometry studies; the Human Protein Atlas (Uhlen et al., 2010) provides the corresponding proteins by antibody-based studies. All the human proteins are deposited in neXtProt database (Lane et al., 2012) and assigned a confidence code based on the evidence of the proteins. The proteins from Pe1 level have credible evidence of protein expression and identification by mass spectrometry, immunohistochemistry, 3D structure, or amino acid sequencing. The proteins from PE2 level recognize transcript expression evidence, without evidence of protein expression. The proteins from PE3 level signify the lack of protein or transcript evidence but the presence of protein evidence for a homologous protein in a related species. The proteins from PE4 level hypothesize from gene models, and the proteins from PE5 level come from "dubious" or "uncertain" genes that seemed to have some protein-level evidence in the past but since has been deemed doubtful.

Recent years, the C-HPP has achieved steady progress. Nearly 78 percent of the total protein-coding genes have been identified at credible protein evidence by mass spectrometry, immunohistochemistry, 3D structure, or amino acid sequencing (Lane et al., 2014). However, there are still more than four thousand proteins which have not been identified by any experimental method and are named as "missing proteins" (Lane et al., 2014) which is constituted by the proteins from PE2, PE3, PE4 and PE5 level. Most of those missing proteins are hard to be detected because of the low abundance expression, specific samples, special condition in biology etc. Identification of the missing proteins is one of the challenges in C-HPP. Many efforts have been made towards mining the missing proteins by using large-scale mass spectrum (Wilhelm et al., 2014, Kim et al., 2014) or bioinformatics methods (Ranganathan et al., 2013). On the other hand, one of the recent works by Shidhi et al. (Shidhi et al., 2015) aim to make novel proteins from pseudogenes. By performing a multi-parametric study of the protein equivalents of the 16 pseudogenes from *saccharomyces cerevisiae*, they identified two promising candidates for future protein synthesis in vitro.

This paper will provide comprehensive structure and function analyses of the missing proteins and try to identify the potential native gene-coding proteins from the missing proteins. Because of the limited computational resource, currently we only focus on the "dubious" or "uncertain" (PE5) missing proteins. The structure and function of those missing proteins are predicted by the state-of-the-art software "I-TASSER" (Roy et al., 2010) and "COFACTOR" (Roy et al., 2012) respectively. The results are in consistent with the annotation from HGNC database (Gray et al., 2014) and other mass-spectrum dataset (Kim et al., 2014). All data are publicly available by web-interface, which can provide valuable reference about the "missing proteins".

2 METHODS AND PROCEDURES

2.1 Data source of missing proteins

The neXtProt database (Lane et al., 2012) released at Sep. 19, 2014 is used. The "dubious" or "uncertain" missing proteins with confidence code "PE5" are extracted. Totally, there are 616 proteins in this category with length varying from 21 to 2252. The structures of these proteins are predicted by using I-TASSER software (Roy et al., 2010). The functions including the EC number, the GO terms and the binding sites are predicted by using COFACTOR software (Roy et al., 2012). The subcellular localization is predicted by using Hum-mPLoc (Shen and Chou, 2009). To validate the presented structure and function prediction method, some of the high-confidence proteins from neXtProt are also selected with evidence code as "PE1". We prefer to the proteins whose 3-D structures are known so that the structure prediction method can be accessed. The selected proteins have the similar number (625) and length distribution as the "dubious" missing proteins.

2.2 Pipeline of structure and function prediction for the missing proteins

The procedure for the structure and function prediction is illustrated in Figure 1. For a given protein sequence, the LOMETS program (Wu and Zhang, 2007) is firstly run to get the possible

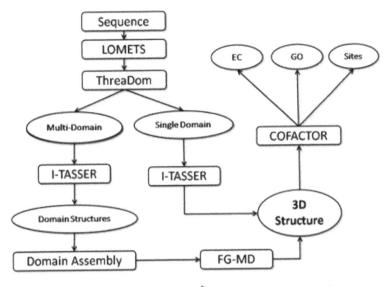

Figure 1. The flowchart of structure and function prediction for missing proteins.

template from **PDB** library, and then the ThreaDom program (Xue et al., 2013) is used to get the domain information. If the result is a single domain, the I-TASSER software (Roy et al., 2010) is used to obtain the final 3-D structure, otherwise the 3-D structure of each domain is obtained by I-TASSER (Roy et al., 2010), and then these structures are assembled to get a single structure, which will be refined with FG-MD program (Zhang et al., 2011) to obtain the final 3-D structure. The COFACTOR program (Roy et al., 2012) is then run to get the function information including ligand-binding site, Gene-Ontology terms, and Enzyme Classification.

All the programs are extensively tested and achieve good assessment on many community-wide experiments. For example, I-TASSER was ranked as the No. 1 server for protein structure prediction in recent CASP7 (Moult et al., 2007), CASP8 (Moult et al., 2009), CASP9 (Moult et al., 2011), and CASP10 (Moult et al., 2014) experiments. The COFACTOR algorithm was ranked as the best method for function prediction in the CASP9 experiments (Moult et al., 2011).

3   RESULTS AND DISCUSSIONS

3.1   *Benchmark test on highly confident protein identification*

To give an unbiased evaluation of the proposed method, the structure and function of the highly confident proteins are predicted in non-homology mode where all the homologous structures identified with sequence identities greater than or equal to 30% are removed. The predicted structures are accessed based on the mapping structures from PDB. TM-score (Xu and Zhang, 2010) is used to quantitatively assess the accuracy of the predicted structure. Additionally, both I-TASSER (Roy et al., 2010) and COFACTOR (Roy et al., 2012) have confidence score to indicate the quality of the prediction. The I-TASSER C-score is defined based on the quality of the threading alignments and the convergence of the I-TASSER's structural assembly refinement simulations. These values usually vary from -5 to 2, where a C-score of higher value signifies a model with a high confidence and vice-versa. An I-TASSER C-score larger than -1.5 indicates a correct model topology. As shown in Figure 2(A), most of the highly confident proteins are well-predicted with very high I-TASSER C-scores, and the average C-score is -0.42. The average TM-score between these highly confident proteins and the corresponding PDB structures is 0.71, which indicates a high quality prediction. The Pearson correlation coefficient between I-TASSER C-scores and the TM-score is 0.74, which means that the I-TASSER C-score is a good indicator of prediction quality.

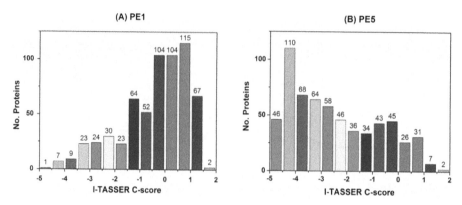

Figure 2.   The distribution of I-TASSER C-score for PE1 (A) and PE5 (B) proteins respectively.

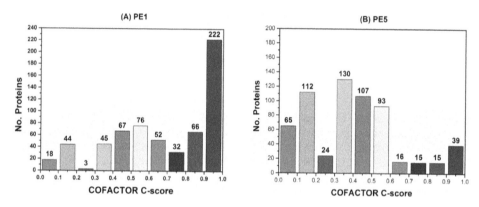

Figure 3.   The distribution of COFACTOR C-score for PE1 (A) and PE5 (B) proteins respectively.

The predicted function is accessed based on the annotation of neXtProt database and the COFACTOR C-score. The COFACTOR C-score is defined based on the C-score of the structure prediction and the global and/or local structural similarity between the predicted models and their structural analogs in the PDB. The COFACTOR C-score has been normal-ized from 0 to 1, where a high value indicates a high confidence prediction. Among the 625 highly confident proteins, 592 proteins have annotation in neXtProt database and one pro-tein has no COFACTOR prediction result, so the evaluation is counted on the 591 proteins. The evaluation metrics used here is the protein-centric metrics as used by Critical Assessment of Function Annotation (Radivojac et al., 2013), where precision is defined as the number of correctly predicted functional terms divided by the total number of prediction, the recall is defined as the number of correctly predicted functional terms divided by the total number of annotation and F-score is a harmonic mean between precision and recall. As shown in Figure 3(A), many of the proteins are predicted with high COFACTOR C-score. The Pear-son correlation coefficient between the COFACTOR C-score and F-score is 0.53, which means the COFACTOR C-score is also a good indicator of the prediction quality.

3.2   Summary of the predicted structure and function of the missing proteins

To provide the most comprehensive information, the structure and function of the missing pro-tein are predicted in homology and non-homology mode. In the homology mode, all the possible templates from PDB library are used no matter they are homologous to the missing proteins or not. In the non-homology mode, only the non-homologous templates are used, where the sequence identities between the target missing protein and the templates are below 30%.

The missing proteins are grouped into three classes based on the result of multi-threading program LOMETS (Wu and Zhang, 2007). For each threading program, the target-template alignment is measured by Z-score, which is defined as the difference between the raw alignment score and the mean in the unit of derivation. A missing protein is classified as "hard" if none of the threading program identifies a template with Z-score larger than the specific threshold, a missing protein is classified as "easy" if on average at least one template per threading program can be detected with Z-score larger than the specific threshold, otherwise the missing protein is classified as "medium". The number of missing proteins in each class is calculated. As expected, the non-homology mode has more medium and hard target than the homology mode. Actually, there isn't a significant difference between the number in homology mode and that in non-homology mode, which means that our threading programs can efficiently detect the remote homologous templates. The predicted structure and function in homology mode may be more accurate than those in non-homology mode, but the results in homology mode cannot guarantee that the proteins are native gene-coding human proteins. Many of the missing proteins are inferred from homologous species or derived from the gene model, so the homologous templates will provide some bias information during prediction. The non-homology mode doesn't use any homologous template. If the structures are well predicted in non-homology mode, the corresponding proteins may be foldable. In this case, the missing proteins may be native gene-coding proteins.

Because missing proteins have not been validated by any proteomics experiment method, the native structure and function of these proteins are unknown. Here the confidence scores of I-TASSER (Roy et al., 2010) and COFACTOR (Roy et al., 2012) are used to access the quality of the prediction. Figure 2(B) shows the distribution of I-TASSER C-score of PE5 proteins in non-homology mode. The C-score is well-distributed in each bin. The foldable proteins (with C-score higher than -1.5) are less than the un-foldable proteins (with C-score lower than -1.5), the reasons may be that the structure of the missing proteins are difficult to be predicted or some of the missing proteins are not native gene-coding proteins. There are 188 proteins with I-TASSER C-score larger than -1.5 in non-homology mode (the best C-score is used for multi-domain proteins). These proteins are foldable without any homologous information, which indicates that they may be native gene-coding proteins.

The distribution of COFACTOR C-score for PE5 proteins in non-homology mode is shown in Figure 3(B). The distribution of COFACTOR C-score is more uneven than that of the I-TASSER C-score. Most values are concentrated between 0.3 and 0.6. There are about 85 proteins with very high COFACTOR. Based on the experience evaluation, the prediction with COFACTOR C-score higher than 0.2 can hit certain correct GO terms. As shown in Figure 3(B), most of the missing proteins have good function prediction.

3.3 *GO function prediction evaluation*

The predicted GO molecular function is mainly enzyme for both missing proteins and highly confident proteins. The two largest groups are "binding" (GO:0005488) and "receptor activity" (GO:0003824), which have the frequencies of about 50% and 30% respectively. The missing proteins are overrepresented with GO items "transporter activity" (GO:0005215) and "receptor activity" (GO:0004872) in comparison with the highly confident proteins, so it can be inferred that there are many membrane proteins in the missing proteins.

Among the proteins of PE5 level, there are 219 missing proteins which have function annotations in neXtProt database (Lane et al., 2012). Although these annotations are not golden standard, they are based on manually curated collection and have high confidence. These annotations are used to evaluate the performance of our function prediction results. In non-homology mode, there are 10 proteins whose GO molecular functions have been perfectly predicted with F-score equal to 1, which means that COFACTOR hits the exact functional terms. There are 71 proteins with recall equal to 1 at non-homology mode, which means that COFACTOR can hit the functional terms but wrongly gives more specific function. By increasing the C-score threshold, COFACTOR can get more accurate prediction as shown in Table 1. The COFACTOR C-score is well correlated with the prediction results as shown.

Table 1. The gene ontology molecular function prediction results on 219 missing proteins.

COFACTOR C-score threshold	Homology mode				Non-homology mode			
	No. proteins	Precision	Recall	F-score	No. proteins	Precision	Recall	F-score
0	219	0.42	0.69	0.46	219	0.40	0.64	0.42
0.2	169	0.50	0.72	0.53	174	0.46	0.65	0.47
0.4	139	0.54	0.67	0.56	127	0.53	0.65	0.53
0.6	78	0.76	0.81	0.76	55	0.81	0.87	0.81
0.8	58	0.86	0.86	0.84	42	0.87	0.85	0.84
0.9	39	0.90	0.90	0.88	27	0.91	0.84	0.84

Table 2. Distribution of missing proteins in different gene loci types after HGNC mapping.

		No. foldable proteins	
Gene loci type	No. missing proteins	Homology mode	Non-homology mode
Gene with protein product	66	31	34
Immunoglobulin gene	2	2	2
Pseudogene	252	144	128
RNA, long non-coding	127	2	2
RNA, ribosomal	1	1	0
Withdrawn	6	0	0
Unknown	47	13	12

In homology mode, the Pearson correlation coefficient between the COFACTOR C-score and F-score is 0.69, where in non-homology mode, the Pearson correlation coefficient is 0.65.

3.4 HGNC mapping analysis

HGNC (Gray et al., 2014) provides a unique name and gene loci type for each of the known gene within human genome. The majority of HGNC data are manually curated (Gray et al., 2014). The gene loci type is valuable information to verify the missing proteins. Within the total 616 missing proteins here, there are 507 proteins which can be mapped to one or more HGNC ID. Table 2 shows the possible gene loci types, the number of missing proteins in each loci and the number of foldable proteins (with I-TASSER C-score larger than −1.5). As expected, the number of foldable proteins in homology mode is more than that in non-homology mode except the gene with protein product class. Actually, the reason to this exception is that the target proteins are medium targets. Half of the proteins in gene with protein product class have been identified as foldable proteins. The others may be hard to be predicted. There are many foldable proteins from the pseudogene class. Pseudogenes are dysfunctional relatives of genes that have lost their protein-coding ability or are otherwise no longer expressed in the cell. Usually, these pseudogenes have homologous protein products during evolution, so I-TASSER can fold many of the proteins in this class.

3.5 Comparison with other mass spectrometry dataset

Mass spectrometry is an effective method to identify peptides. Lots of mass spectrometry data has been deposited in public database, such as PeptideAtlas (Desiere et al., 2006) and GPMDB (Craig et al., 2004). Several groups try to use mass spectrometry to develop the draft of human proteome (Wilhelm et al., 2014, Kim et al., 2014). One of the interesting results reported by Kim et al (Kim et al., 2014) is that they have identified about two-third

(2535/3844) of the "missing proteins" (Lane et al., 2014). Actually the "missing proteins" in that paper refers to the neXtProt proteins whose evidence codes are "PE2", "PE3" or "PE4". This paper focuses on the "PE5" missing proteins. By RefSeq (Pruitt et al., 2014) mapping, 41 "PE5" proteins are also found in Kim's dataset. In non-homology mode, 23 out of the 41 proteins are foldable (I-TASSER C-score>−1.5), which is in good consistent with Kim's results.

3.6   *Web interface*

The predicted structure and function of the 616 missing proteins are publicly available at http://zhanglab.ccmb.med.umich.edu/HPSF/. User can browse or search the specific proteins by clicking the "Browse & Search" link. The proteins can be searched by inputting the neXtProt ID, protein name and corresponding HGNC symbol or name. Both partial and full values are accepted. The input box can automatically provide at most 20 suggestions if any of the record can match the inputted text. The browse or search results page displays general information of the proteins, including the neXtProt ID, the protein name, the gene name, the gene loci type, the ENSEMBL gene, the prediction confidence score (I-TASSER C-score) et al. User can click the link to access the detail information provided by other databases, including neXtProt (clicking neXtProt ID), HGNC (clicking gene symbol) and ENSEMBL (clicking the ensemble gene). To get the detail structure and function information, please click the "HPSF detail" link under the neXtProt ID which will open a new window. The new page will show comprehensive predicted structure and function information about the missing proteins. The structure information includes the predicted secondary structure, the predicted solvent accessibility, the predicted B-factor, the top 5 models and their residue-specific qualities and the domain information. The function information includes the Enzyme commission number (Barrett, 1997) and Gene Ontology terms (Ashburner et al., 2000). The page also give the predicted subcellular localization by Hum-mPLoc (Shen and Chou, 2009). For each missing proteins, the similar proteins from known proteins (PE1 level) are also extracted and displayed. User can also change the results between the homology mode and non-homology mode by simply clicking the corresponding link. The results of each domain can also be accessed by clicking the link of each domain. Finally, the whole database can be downloaded at the download page.

4   CONCLUSION

In this study, 616 missing proteins, which have the lowest confidence (PE5) in neXtProt database, have been investigated by using bioinformatics methods. The structure and function of these proteins have been predicted by using cutting-edge software: I-TASSER and COFACTOR respectively. The prediction is extensively evaluated by extracting well-established evidence about missing proteins, such as neXtProt annotation, HGNC gene loci annotation and mass-spectrometry dataset. The results show that there is good consistency between the prediction and the evidence of the proteins in PE5. There are 188 foldable proteins with high confidence score of I-TASSER structure simulation without using any homologous templates, indicating that these proteins may be native gene-coding proteins. Both structure and function evaluation shows that the missing proteins are over-represented in membrane proteins in comparison with the highly confident proteins. Since the membrane proteins are hard to be separated and purified, detection of the membrane proteins are more difficult than other proteins. The results indicate that there may be more membrane proteins in the missing proteins.

REFERENCES

Aebersold, R., Bader, G.D., Edwards, A.M., et al. 2013. The biology/disease-driven human proteome project (B/D-HPP): enabling protein research for the life sciences community. J Proteome Res, 12, 23–7.
Ashburner, M., Ball, C.A., Blake, J.A., et al. 2000. Gene ontology: tool for the unification of biology. The Gene Ontology Consortium. Nat Genet, 25, 25–9.

Barrett, A.J. 1997. Nomenclature Committee of the International Union of Biochemistry and Molecular Biology (NC-IUBMB). Enzyme Nomenclature. Recommendations 1992. Supplement 4: corrections and additions (1997). Eur J Biochem, 250, 1–6.

Craig, R., Cortens, J.P., Beavis, R.C. 2004. Open source system for analyzing, validating, and storing protein identification data. J Proteome Res, 3, 1234–42.

Desiere, F., Deutsch, E.W., King, N.L., et al. 2006. The PeptideAtlas project. Nucleic Acids Res, 34, D655–8.

Farrah, T., Deutsch, E.W., Hoopmann, M.R., et al. 2013. The state of the human proteome in 2012 as viewed through PeptideAtlas. J Proteome Res, 12, 162–71.

Flicek, P., Amode, M.R., Barrell, D., et al. 2014. Ensembl 2014. Nucleic Acids Res, 42, D749–55.

Gray, K.A., Yates, B., Seal, R.L., et al. 2014. Genenames.org: the HGNC resources in 2015. Nucleic Acids Res.

Kim, M.S., Pinto, S.M., Getnet, D., et al. 2014. A draft map of the human proteome. Nature, 509, 575–81.

Lane, L., Argoud-Puy, G., Britan, A., et al. 2012. neXtProt: a knowledge platform for human proteins. Nucleic Acids Res, 40, D76–83.

Lane, L., Bairoch, A., Beavis, R.C., et al. 2014. Metrics for the Human Proteome Project 2013–2014 and strategies for finding missing proteins. J Proteome Res, 13, 15–20.

Legrain, P., Aebersold, R., Archakov, A., et al. 2011. The human proteome project: current state and future direction. Mol Cell Proteomics, 10, M111 009993.

Marko-Varga, G., Omenn, G.S., Paik, Y.K., et al. 2013. A first step toward completion of a genome-wide characterization of the human proteome. J Proteome Res, 12, 1–5.

Moult, J., Fidelis, K., Kryshtafovych, A., et al. 2007. Critical assessment of methods of protein structure prediction-Round VII. Proteins, 69 Suppl 8, 3–9.

Moult, J., Fidelis, K., Kryshtafovych, A., et al. 2009. Critical assessment of methods of protein structure prediction—Round VIII. Proteins, 77 Suppl 9, 1–4.

Moult, J., Fidelis, K., Kryshtafovych, A., et al. 2014. Critical assessment of methods of protein structure prediction (CASP)—round x. Proteins, 82 Suppl 2, 1–6.

Moult, J., Fidelis, K., Kryshtafovych, A., et al. 2011. Critical assessment of methods of protein structure prediction (CASP)—round IX. Proteins, 79 Suppl 10, 1–5.

Paik, Y.K., Jeong, S.K., Omenn, G.S., et al. 2012. The Chromosome-Centric Human Proteome Project for cataloging proteins encoded in the genome. Nat Biotechnol, 30, 221–3.

Pruitt, K.D., Brown, G.R., Hiatt, S.M., et al. 2014. RefSeq: an update on mammalian reference sequences. Nucleic Acids Res, 42, D756–63.

Radivojac, P., Clark, W.T., Oron, T.R., et al. 2013. A large-scale evaluation of computational protein function prediction. Nat Methods, 10, 221–7.

Ranganathan, S., Khan, J.M., Garg, G., et al. 2013. Functional annotation of the human chromosome 7 "missing" proteins: a bioinformatics approach. J Proteome Res, 12, 2504–10.

Roy, A., Kucukural, A., Zhang, Y. 2010. I-TASSER: a unified platform for automated protein structure and function prediction. Nat Protoc, 5, 725–38.

Roy, A., Yang, J., Zhang, Y. 2012. COFACTOR: an accurate comparative algorithm for structure-based protein function annotation. Nucleic Acids Res, 40, W471–7.

Shen, H.B., Chou, K.C. 2009. A top-down approach to enhance the power of predicting human protein subcellular localization: Hum-mPLoc 2.0. Anal Biochem, 394, 269–74.

Shidhi, P.R., Suravajhala, P., Nayeema, A., et al. 2015. Making novel proteins from pseudogenes. Bioinformatics, 31, 33–9.

Uhlen, M., Oksvold, P., Fagerberg, L., et al. 2010. Towards a knowledge-based Human Protein Atlas. Nat Biotechnol, 28, 1248–50.

Venter, J.C., Adams, M.D., Myers, E.W., et al. 2001. The sequence of the human genome. Science, 291, 1304–51.

Wilhelm, M., Schlegl, J., Hahne, H., et al. 2014. Mass-spectrometry-based draft of the human proteome. Nature, 509, 582–7.

Wu, S., Zhang, Y. 2007. LOMETS: a local meta-threading-server for protein structure prediction. Nucleic Acids Res, 35, 3375–82.

Xu, J., Zhang, Y. 2010. How significant is a protein structure similarity with TM-score = 0.5? Bioinformatics, 26, 889–895.

Xue, Z., Xu, D., Wang, Y., et al. 2013. ThreaDom: extracting protein domain boundary information from multiple threading alignments. Bioinformatics, 29, i247–56.

Zhang, J., Liang, Y., Zhang, Y. 2011. Atomic-level protein structure refinement using fragment-guided molecular dynamics conformation sampling. Structure, 19, 1784–95.

Bioinformatics and Biomedical Engineering – Chou & Zhou (Eds)
© *2016 Taylor & Francis Group, London, ISBN 978-1-138-02784-8*

Prediction of SARS coronavirus main protease by artificial Neural Network

S.P. Jang, S.H. Lee, S.M. Choi, H.S. Choi & T.S. Yoon
Natural Science, Hankuk Academy of Foreign Studies, HAFS, Yongin-si, Republic of Korea

ABSTRACT: SARS (Severe Acute Respiratory Syndrome) corona virus has hugely affected humans for more than ten years. Virus' RNA replicase gene is surrounding the polyprotein la and lab, and the sequence of the polyproteins contains functional proteins, which is an important factor of replication. The experiment was performed based on the "distorted key theory" in order to prevent SARS corona virus from performing replication by inactivating the main protease (also called CoV Mpro) of the virus. After the experiment, Neural Network (NN) was utilized in order to reanalyze the results of polypeptides in the virus. This approach by NN distinguished the fixed patterns in the sequence of cleavage site successfully, and improved the comprehension of the protease structure. The method of preventing the virus replication using competitive inhibitor could be found by analyzing these patterns.

1   INTRODUCTION

Severe Acute Respiratory Syndrome (SARS) is a respiratory disease, which has largely influenced both humans and animals. After the outbreak of SARS in Asia, the WHO declared the coronavirus, which is classified as a single-strand RNA virus of zoonotic origin, as the main cause of SARS. Between November 2002 and July 2003, SARS had a serious effect on 8,273 individuals and caused 775 deaths (9.6% mortality rate) in multiple countries, with most cases in Hong Kong (Chou K.C. 1996). The initial symptom of the sickness was high fever above 38°C (100.4 °F) with an unspecific flu-like symptom, involving breathing difficulties. SARS coronavirus main protease (CoV Mpro) is an enzyme that catalyzes RNA replicase of the virus, which is an essential process for the virus to survive, through replicating essential polyprotein (Marra Marco A. et al. 2003). Perceiving the virus to be disastrous to humans, we decided to perform a study on the cleavage site in CoV Mpro, which is known as the secret of developing drugs against SARS because of its status as an intimidator of SARS existence. Based on the "distorted key theory", we performed an experiment using the NN algorithm to analyze the sequence of the cleavage site and increase its analytic accuracy in order to develop an effective amino acid, which is a key factor to prevent viral replication.

1.1   Neural Network

Deriving the usefulness of the human central nervous system, Neural Network (NN) is used to work out with numerous cases of problems mathematically. Similar to the biological nervous system, there are 'neurons'—simple artificial nodes—which are attached to each other in order to construct a network (Hansen Lars Kai & Peter Salamon 1990). Every artificial neuron works as an individual processor, which enables the total network system not to be affected by errors in some neurons or in other words 'fault tolerance'. Furthermore, NN is able to develop itself through the given situations. This makes it to be used widely in the problem solving of Artificial Intelligence (AI), voice recognition, character recognition and in other diverse fields. As a way to acquire the efficient inhibitor sequence of the protease and assume the structure of protein, we used the NN algorithm. For the input, we inserted amino acid sequences in order to analyze the ones that appear the most. These have high possibility to obtain an effective inhibitor sequence. By using NN as a way of analyzing, we were able to predict more accurately than C5, decision tree and other methods.

1.2 *Distorted key theory*

An useful method of developing inhibitors against SARS is distorted key theory. Effective information for finding inhibitors that are against the key enzyme can be known by protein

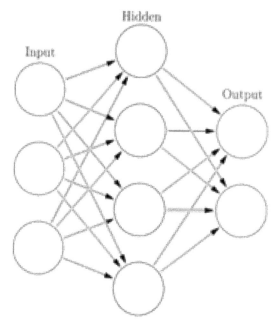

Figure 1. Each circle represents individual nodes that are connected together, arrows represent the output for one side and the input for another side.

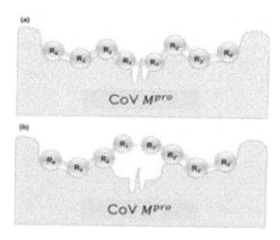

Figure 2. Chou's distorted key theory: (a) the peptide both cleaved by CoV Mpro and effectively bound to the active site of the protease, while the peptide in plate (b) becomes non-cleavable through modification but still bound to the active site. The modified peptide, also called a "distorted key", can become an effective inhibitor against SARS.

cleavage sites. As Koshland's induced fit theory and Fisher's lock-and-key model propose, for the CoV Mpro to cleave a peptide with high possibility, the binding between the active site of the enzyme and the substrate is vital (Zheng Rong Yang 2005). However, once a peptide's scissile bond is modified to a strong hybrid chemical bond, and even though it is still able to bind to the active site, its probability of cleaving would plunge. From this, the peptide can be regarded as the "distorted key", in a way that it can be inserted in the lock but can neither unlock nor come out. As a result of strong binding of the modified peptide, it can operate as an efficient inhibitor against the enzyme (Qi-Shi Du et al. 2007).

2   METHOD AND EXPERIMENT

Neural Network (NN), as the name suggests, is a computer program imitating the information procedure of the biological nervous system. The nervous system consists of basic units called neurons. Neurons are linked to each other, developing a huge network as a whole. The processing of an individual neuron is known as follows: a new electrical signal is received through the dendrites, transferred down the axon, and finally sent to other neighbored neurons. The process of receiving and passing the signal can be carried out only when the signal exceeds the given threshold. Also, a single neuron can be linked with multiple neurons at both ends.

Obviously, NN is a set of nodes in a same topology linked to each other. Each node has input and output links to the others. The fact that these links allow every node to interact with the others and the output value depends on whether the signal surpasses the threshold corresponds to the processing in biological neurons. Reflecting this correspondence, the main function in NN that actually processes the given information is called "neurons."

2.1   Artificial Neural Network

The NN (Neural Network) information procedure can be explained by how perceptrons work (perceptron is a pattern classification device which has learning ability). NN is constructed by several numbers of perceptrons and it works by their cooperative operation. Perception is a one direction, as shown in Figure 1. Several inputs are given and in the next stage each input is multiplied with specific weight factors that range from 0 to 1. After this process, each product value (x*w) is added and the whole sum (Σx*w) will be sent to the neuron. If the sum is higher than the threshold, the signal results in the output, otherwise no signal is sent out.

The experiment output is determined by the NN learning (NN learning is done before the experiment). The learning operation is possible when the input and the output values are designed previously. Let us introduce the progress of the NN learning. First, random weight factors are given to the input value. Second, the result of the first step is compared with the output value that we are trying to analyze. Third, the weight factor multiplied with each of the input value is modified to match the real put with the desired one. After these operations, neurons are able to be classified newly by the NN learning progress.

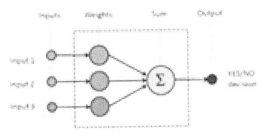

Figure 3.   Perceptron is the principle of NN (Neural Network), calculating the product of the input and weight values and their sum, from which the result can be obtained.

Table 1. Results of 8-mers.

Site	Most frequent	Second-most	Third-most
R4	A: 36.4%	V: 27.3%	P: 18.2%
R3	T: 36.4%	K: 18.2%	R: 18.2%
R2	L: 72.7%	F: 9.1%	M: 9.1%
R1	Q: 100%	–	–

In our experiment, we marked 20 amino acids with numbers ranging from 1 to 20. Next, these numbers were given as inputs to the NN. The input product value and the weight factor were put into the neuron, and the output value is earned by the previously designed NN. Analyzing the frequency of the certain amino acids in a specific spot, it was possible to know the sequence patterns.

The polyprotein, which is related to CoV Mpro, can be parted into two groups: 8-mers and 12-mers. The rules shown between the two groups were considerably different: NC_004718 (TOR2), NC_002645 (HCoV 229E), NC_001846 (MHV), NC_003045 (BCoV), NC_--1451 (IBV), NC_002306 (TGEV), NC_003436 (PEDV), AF208067 (MHVM), AF201929 (MHV2), AF208066 (MHVP), AY278741 (Urbani), AY278488 (BJ01), AY278554 (CUHK-W1), AY282752 (CUHK-su10), and AY291451 (TW1). The experiment was performed with the data of 154 8-mers and 45 12-mers within 14 cleavage sites. Finally, we gained the data from 'Mining SARS-CoV protease cleavage data using non-orthogonal decision trees, a novel method for decisive template selection', researched by Zheng Rong Yang (Zheng Rong Yang 2005).

3 RESULTS

In NN results, we could find a strong rule in the 8-mers experiment, but we could not find a special rule because the 12-mers experiment did not show high accuracy and strong rule. The results for 8-mers are given in Table 1.

3.1 8-mers

The amino acids of performing neural network algorithm with the 8-mers are as follows:

amino1 in {A,I,K,N,P,S,T,V}
amino2 in {A,D,F,G,I,K,L,M,N,Q,R,S,T,V,B}
amino3 in {F,I,L,M,V}
amino4 in {Q}
amino5 in {A,C,G,N,S}
amino6 in {A,C,E,F,G,I,K,L,N,R,S,T,V}
amino7 in {A,D,E,F,G,I,K,L,M,N,P,Q,R,S,T,W,Y}
amino8 in {A,D,E,G,H,I,K,L,M,N,P,Q,R,S,T,V,W}

3.2 12-mers

Unexpectedly, no key rule was found in the performance involving the 12-mers. The presence of different amino acids among all the sites did not have differences that are huge enough to consider as the key factor.

4 CONCLUSIONS

This paper applied the computational program, NN (Neural Network). NN is composed of interaction between perceptrons. NN produces the sequence patterns of cleavage sites by reanalyzing the polypeptides of SARS coronavirus. This outcome leads to representing the effective inhibitor, which induces the inactivation of CoV Mpro based on the "distorted key" theory.

The SARS virus is a very critical and epidemic disease. In order to block the virus from spreading out and killing people, we performed the experiment based on the distorted key theory and the neural network algorithm.

The rule extracted from the 8-mers neural network showed that amino acid Glutamine is the main factor of CoV Mpro. However, we could not find new rules in the 12-mers result.

In our experiment, we found that A, V, T, K, L, Q are key factors that make CoV Mpro to operate. In related work, another experiment that analyzed CoV Mpro performing Support Vector Machine showed the amino sequence A-V-L-Q-S-G-F-R (Shinyoung Lee et al. 2014). In conjunction with the above result, our experiment showed a very accurate result.

REFERENCES

Chou K.C. (1996). "Review: Prediction of human immunodeficiency virus protease cleavage sites in proteins". Analytical Biochemistry 233: 1–14. doi:10.1006/abio.1996.0001. PMID 8789141. (references).

Hansen, Lars Kai, and Peter Salamon. "Neural network ensembles." IEEE transactions on pattern analysis and machine intelligence 12.10 (1990): 993–1001.

Marra, Marco A., et al. "The genome sequence of the SARS-associated corona-virus." Science 300.5624 (2003): 1399–1404.

Qi-Shi Du, Hao Sun and Kuo-Chen Chou, "Inhibitor Design for SARS Coronavirus Main Protease Based on Distorted Key Theory," Medicinal Chemistry, 2007, 3, 1–6.

Shinyoung Lee, Jisue Kang, Jiwoo Oh, Yoonjoo Kim, Jungwon Baek and Taeseon Yoon, "Prediction of SARS Coronavirus Main Protease by Support Vector Machine", IACSIT Press Vol. 59, 2014.

Zheng Rong Yang, "Mining SARS-CoV protease cleavage data using non-orthogonal decision trees; a novel method for decisive template selection," Vol. 21 no. 11 2005, pages 2644–2650. doi:10.1093/bioinformatics/bti404.

Zheng Rong Yang, "Mining SARS-CoV protease cleavage data using non-orthogonal decision trees; a novel method for decisive template selection," Vol. 21 no. 11 2005, pages 2644–2650. doi:10.1093/bioinformatics/bti404.

Algorithms, models and applications

Bioinformatics and Biomedical Engineering – Chou & Zhou (Eds)
© 2016 Taylor & Francis Group, London, ISBN 978-1-138-02784-8

Aquatic biodiversity conservation zoning study based on the fuzzy clustering analysis: A case study of Tieling City, China

H. Liu & L. Xu
*Key Laboratory of Industrial Ecology and Environmental Engineering (China Ministry of Education),
School of Environmental Science and Technology, Dalian University of Technology, Dalian, China*

H. Wang
*Liaoning Province Key Laboratory of Basin Pollution Control,
Liaoning Academy of Environmental Sciences, Shenyang, China*

ABSTRACT: Water ecological conditions of Tieling City were summarized in this paper. An aquatic biodiversity conservation zoning index system was established from four aspects: present situation of aquatic biodiversity, economic value of aquatic biodiversity, pressure and response to the biodiversity, and the importance of the ecosystem. Based on a previous study of the water ecological function zoning of the Liaohe River basin, we selected the functional zone as the evaluation unit. The fuzzy clustering analysis method was used to study the aquatic biodiversity conservation zoning, and the zone protection scheme was put forward.

Keywords: aquatic biodiversity; fuzzy clustering analysis; conservation zoning; Tieling City

1   INTRODUCTION

In China, intensified urbanization has brought great pressure to the water ecosystem in recent years. Farming activities produce a great impact on the water ecosystem, and water pollution and water shortage caused by human activities have also become the major problems of the water environment in China. The water ecosystem is closely tied to the aquatic biodiversity; as a result, research on the aquatic biodiversity and conservation on a large scale is becoming the focus of researchers both at home and abroad.

Because of the easy access to the acquired macrostatistical data, the biodiversity assessment is mostly studied in the unit of administrative regions. The evaluation of the biodiversity difference has been carried out between the province, city, district and county (Wan B T et al. 2007, Zhu W Z et al. 2009), but the research on the evaluation of the unit of functional area is still scarce. Taking the functional area as the unit of evaluation can reflect some information that is excluded by the evaluation of the unit of administrative region, such as the condition of the bottom river. Based on the requirements of the aquatic biodiversity conservation and management of Liaohe River, a set of the aquatic biodiversity conservation zoning index system was established. Remote sensing, literature data and the expert consultation method were used to get the evaluation scores of the unit of ecological function zone, which were chosen as the classification factors for the fuzzy clustering analysis and the following aquatic biodiversity conservation zoning of Tieling City. Suggestions for future aquatic biodiversity management were proposed on the basis of the conservation zoning.

2   STUDY AREA

Tieling City is located in the north of Liaoning Province (Fig. 1). It is one of China's major grain-producing areas, production and processing base of high-quality agricultural products,

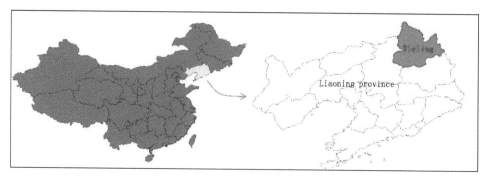

Figure 1.   Location of Tieling City.

and also the city of coal power. The topography of Tieling City is roughly higher from the west to the east (to some extent lower at the center) and lower from the north to the south. Mountains and hills are on both sides of the east and the west, Liaohe Plain is flowing from the north to the south in the central area. Tieling has a temperate zone continental monsoon climate and an annual average rainfall of 700 mm.

The mainstream of Liaohe River flowing in Tieling City includes primary tributary rivers such as East Liaohe River, Zhao Sutai River, Wang River, Qing River, Chai River and Fan River, and secondary tributary rivers such as Er Dao River and Kou River. It has a water resource of 3.141 billion cubic meters. There are 82 reservoirs in the mainstream of Liaohe River flowing in Tieling, four of which have a storage capacity of more than 200 million cubic meters (Qing River reservoir, Chai River reservoir, Nan Chengzi reservoir and Bang Ziling reservoir). The Qing River reservoir is on the mainstream of Qing River, which is mainly used for flood control, irrigation and industrial water supply. It is also used for fish farming, tourism and other comprehensive utilization, and an important water source in the basin. The Chai River reservoir is mainly used for flood control and irrigation, as well as the comprehensive utilization of power generation and fish farming.

River water quality of Liaohe River mainstream flowing in Tieling was stable. Among the nine tributaries monitored, the section water quality of five tributaries (Wang River, Qing River, Chai River, Fan River and Kou River) conformed to the class IV water quality standards, and that of four tributaries (Zhao Sutai River, Chang Gou River, Liang Zi River and Ma Zhong River) conformed to the class V water quality standards. Water quality in the reservoirs (Chai River reservoir and Qing River reservoir) was good overall, and all items conformed to the class II water quality standards in addition to the total nitrogen. The water qualification rate of drinking water source was 100%.

The emission amount of wastewater was 82.41 million tons in 2012; of these, the amount of industrial wastewater emissions was 17.58 million tons and that of urban wastewater emissions was 64.83 million tons. The emission amount of Chemical Oxygen Demand (COD) in the wastewater was 20500 tons; of these, the industrial source was 4700 tons and the source of life was 15800 tons. The emission amount of ammonia nitrogen in the wastewater was 3100 tons; of these, the industrial source was 300 tons and the source of life was 2800 tons.

The total area of the nature reserve was 105231.8 hectares, including one nature reserve at the province level (Tieling Fan River provincial nature reserve) and four nature reserves at the county level (Zengjia Village heron nature reserve in Kaiyuan City, Liaoning Province Ice La mountain raw animal nature reserve in Xifeng County, Kou River wetland and headwater forest nature reserve in Xifeng County, and hazelnut ridge nature reserve). There is one ecosystem function area covering an area of 153433 hectares, namely the Qing River water conservation ecological function protected areas.

3 RESEARCH METHODS

3.1 *Biodiversity conservation zoning index system*

The zoning index system is the basis of the aquatic biodiversity conservation zoning (Liu Y B & Liang C 2009). In this study, an index system of biodiversity conservation division was established from the following four aspects: present situation of the aquatic biodiversity, economic value of the aquatic biodiversity, pressure and response to the biodiversity, and the importance of the ecosystem.

1. Present situation of the aquatic biodiversity

   Aquatic organisms studied in the aquatic biodiversity conservation research mainly include vegetation in the riparian zone, algae, benthic animals and fish. Species diversity refers to the diversity of individual number and the number of species in the ecosystem. A region that contains more species diversity and species number can be credited with rich species diversity. In this study, species diversity was selected for the determination of the aquatic biodiversity, mainly including the following three aspects: number of species, biomass and density of habitat (Wang G R & Du F G 2006, Huang P et al. 2012). Based on the investigation and expert scoring, the aquatic biodiversity can be divided into three levels: high, moderate and low with the assigned points 3, 2 and 1, respectively, corresponding to the degree of protection from high to low (Zhou Z et al. 2009, Huang C X & Wang M Y 2010).

2. Economic value of the aquatic biodiversity

   Biodiversity has a direct use value, an indirect use value and a potential value for humans. The direct value means that organisms supply food, fiber, architecture and furniture material, and other production and life materials for human. The indirect use value means the important ecological functions of biological diversity. Once the wildlife is reduced, the stability of the ecological system will be destroyed and the survival environment of humans will also be affected. The potential use value means the value of many different kinds of wildlife for which there is a lack of research. The use value of a large number of wildlife is unclear, but it is certain that they have a huge potential use value (Ge J W et al. 2006). Based on the investigation and expert scoring, the aquatic biodiversity economic value can be divided into three levels: high, moderate and low with the assigned points 3, 2 and 1, respectively, corresponding to the degree of protection from high to low.

3. Pressure and response to the biodiversity

   This aspect is used to evaluate the agricultural and urbanization pressure as well as the degree of attention to the aquatic biodiversity conservation. Based on the investigation and expert evaluation, taking into account of the vulnerability of the habitat environment, human disturbance, the number of reserve and reserve level, pressure and response to the biodiversity can be divided into three levels: high, moderate and low with the assigned points 3, 2 and 1, respectively, corresponding to the degree of protection from high to low.

4. Importance of the ecosystem

   Wetlands are identified as one of the three major ecosystems (forests and oceans) by the international organization. As aquatic ecosystems, wetland includes the following six categories: (1) rivers, lakes and reservoirs; (2) salt marshes; (3) peat swamp; (4) coast or beach; (5) depression; and (6) swamp. On the one hand, there is considerable importance placed on threatened, endangered or recession species; on the other hand, there is special importance given to the life cycle stage such as wintering grounds and important spawning area of migratory fish. The mangrove swamp, important wetland and reservoir are selected as the important ecosystem area (Zhang L N et al. 2014).

The importance analysis of each kind of ecological system is defined as the comparison of the area and level of each type of ecosystem to study the importance of the ecological system. Using the method of relative comparison, the corresponding score is found to be 0.5, 0.3 and 0.2, with the importance of this ecological system from high to low (Table 1). The score is 0

Table 1. Importance level of the aquatic ecosystem.

Type of the aquatic ecosystem	Importance level		
	High	Moderate	Low
Mangrove swamp	0.5	0.3	0.2
Important wetland	0.5	0.3	0.2
Reservoir	0.5	0.3	0.2

if there is no type under evaluation. This score also corresponds to the degree of protection from high to low. Formula (1) is used for the superposition calculation of the importance point of each ecosystem type in each research unit:

$$IME = \sum_{i=1}^{3} IME_i \qquad (1)$$

where IME refers to the importance score of all ecosystem types within the research unit.

3.2 Protection zoning method—Fuzzy clustering analysis

The fuzzy clustering analysis method can be divided into two parts: (1) calibration, namely establishing a certain affinity–disaffinity relationship between all objects to be classified; (2) classification, i.e. to classify using the fuzzy equivalence relation. There are usually two ways to describe the degree of closeness of things: one is to take each sample as one point in an m-dimensional space and define a certain distance between the points; another one is to use some kind of similarity coefficient to describe the affinity–disaffinity relationship between samples (Huang Y& Cui H R 2011).

1. Establish the fuzzy similar matrix (Qu F H et al. 2011)
   Equipped with n samples, $x_1, x_2, ..., x_n$, each sample has m feature indicators and x_{ij} is used to describe the jth feature indicator of the ith sample. Then, the observation data matrix of the n samples is given by

$$X = \begin{bmatrix} x_{11} & x_{12} & \cdots & x_{1m} \\ \cdots & \cdots & \cdots & \cdots \\ x_{n1} & x_{n2} & \cdots & x_{nm} \end{bmatrix} \qquad (2)$$

where n is the number of samples; m is the number of variables (indicator); and $x_i = (x_{i1}, x_{i2}, ..., x_{im})$.
   In order to establish the fuzzy similar matrix, first we introduce a concept of similar degree r_{ij} to express the similar degree between the xth sample x_i and the jth sample x_j. The ways to determine r_{ij} include the similarity coefficient method and the distance method. The commonly used similarity coefficients are as follows:

$$\text{Included angle cosine}: r_{ij} = \frac{\sum_{k=1}^{m} x_{ik} \cdot x_{jk}}{\sqrt{\sum_{k=1}^{m} x_{ik}^2} \sqrt{\sum_{k=1}^{m} x_{jk}^2}} \qquad (3)$$

$$\text{Minmax method}: r_{ij} = \frac{\sum_{k=1}^{m} (x_{ik} \wedge x_{jk})}{\sum_{k=1}^{m} (x_{ik} \vee x_{jk})} \qquad (4)$$

$$\text{Correlation coefficient}: r_{ij} = \frac{\sum_{k=1}^{m}(x_{ik} - \overline{x}_i)(x_{jk} - \overline{x}_j)}{\sqrt{\sum_{k=1}^{m}(x_{ik} - \overline{x}_i)^2}\sqrt{\sum_{k=}^{m}(x_{jk} - \overline{x}_j)^2}} \tag{5}$$

where $\overline{x}_i$ is the mean value of each indicator of the ith example after the standardized treatment.

As for the distance method, $r_{ij} = 1 - cd(x_i, x_j)$, where c is the proper selected parameter that makes $0 \le r_{ij} \le 1$. The commonly used $d(x_i, x_j)$ to express the distance between the xth sample x_i and the jth sample x_j are as follows:

$$\text{Hamming distance}: d(x_i, x_j) = \sum_{k=1}^{m}|x_{ik} - x_{jk}| \left(\begin{array}{l} i, j = 1, 2, \dots, n \\ k = 1, 2, \dots, m \end{array} \right) \tag{6}$$

$$\text{Euclidean distance}: d(x_i, x_j) = \sqrt{\sum_{k=1}^{m}(x_{ik} - x_{jk})^2} \left(\begin{array}{l} i, j = 1, 2, \dots, n \\ k = 1, 2, \dots, m \end{array} \right) \tag{7}$$

$$\text{Chebyshev distance}: d(x_i, x_j) = \overset{m}{\underset{k=1}{\vee}}|x_{ik} - x_{jk}| \left(\begin{array}{l} i, j = 1, 2, \dots, n \\ k = 1, 2, \dots, m \end{array} \right) \tag{8}$$

In the grading process, a different calculation formula is chosen according to the actual situation. The selection methods will directly affect the classification result.

2. Choose the clustering analysis method

The selection of the clustering method is the most crucial step in clustering analysis. The first step is to determine the sample clustering or the variable clustering, and then choose

Figure 2. Water ecological functional areas of Tieling City.

39

the clustering method to connect between groups, join in the group, nearest-neighbor element, the farthest neighbors and centroid clustering method according to the clustering statistics. The clustering process is performed using SPSS19.0 software.

4 RESULTS AND DISCUSSION

A total of 22 water ecological functional areas of Tieling City are shown in Figure 2, and the evaluation of each functional area is given in Table 2.

Table 2. Evaluation of each functional area.

No.	Code of evaluation unit	Name of the evaluation unit	Status of aquatic biodiversity	Economic value	Pressure and response	Importance of the ecosystem
1	III-2-30	Wan Quan River agricultural development zone	2	1	1	0
2	III-2-35	Chang Gou River agricultural development zone	2	2	1	0
3	III-2-36	Zhao Sutai River downstream agricultural development zone	2	1	1	0
4	III-2-37	Liang Zi River agricultural development zone	1	1	1	0
5	III-3-11	Zhao Sutai River agricultural development zone	2	1	1	0
6	III-3-12	Er Dao River agricultural development zone	2	1	1	0
7	III-3-13	Zhao Sutai River middle stream agricultural development zone	2	1	1	0
8	III-3-14	East Liaohe River agricultural development zone	2	1	1	0
9	IV-5-1	Tieling City development zone	1	1	1	0
10	IV-5-2	Fan River downstream agricultural development zone	1	1	1	0
11	IV-5-3	Fan River middle stream biological habitat zone	3	3	3	0.6
12	IV-5-4	Fan River headwater biological habitat zone	3	3	3	0.6
13	IV-5-5	Chai River downstream hydrology zone	3	3	2	0.6
14	IV-5-6	Chai River upstream water conservation zone	3	3	2	0.6
15	IV-5-7	Zhong Gu River agricultural development zone	2	2	1	0
16	IV-5-8	Qing River downstream agricultural development zone	2	2	1	0
17	IV-5-9	Qing River reservoir hydrologic zone	3	3	2	0.6
18	IV-5-10	Qing River middle-up stream agricultural development zone	2	2	1	0
19	IV-5-11	Qing River headwater water conservation zone	2	2	2	0.6
20	IV-5-12	Kou River headwater water conservation zone	3	3	2	0.6
21	IV-5-13	Kou River middle stream agricultural development	2	2	1	0
22	IV-5-14	Nan Chengzi reservoir hydrologic zone	3	3	2	0.6

Based on the above four protection zoning indicators, we clustered the 22 evaluation samples. Statistical software SPSS19.0 was used to complete the analysis process. In the process of clustering analysis, the standard deviation method was used for the dimensionless of the initial data, namely the Z-score of the standardized data method of SPSS software. The Euclidean distance measure method was used as the measurement level of sample interval. Also, the between-groups linkage method was chosen as the clustering method. Clustering results tree is shown in Figure 3.

The clustering results showed that 11–14, 17, 19, 20 and 22 are classified as class I, 1–3, 5–8, 15, 16, 18 and 21 as class II, and 4, 9 and 10 as class III. According to the aquatic biodiversity conservation zoning index, the class I zones are areas that need important conservation, the class II zones are areas that need less important conservation, and class III zones are areas that need general conservation. The conservation zoning result is shown in Figure 4.

Terrestrial conservation strategies tend to emphasize areas of high habitat quality that can be bounded and protected, which is likely to fail for fresh waters (Boon 2000). This problem of boundary definition impedes the sensible local management of freshwater biodiversity because protection of a particular component of river biota (and often habitat) requires the control over the upstream drainage network, the surrounding land, the riparian zone, and downstream reaches (David Dudgeon et al. 2006). The catchment scale helps resolve the small-scale but damaging conflicts of interest among competing human demands. However, this approach can be problematic in practice, as relatively large areas of land need to be managed in order to protect relatively small water bodies. A promising approach could involve ecological management that integrates the requirements of terrestrial and freshwater environments. This will complicate the process of establishing appropriate boundaries for protected areas. However, from the freshwater perspective, it would have the added advantage of broadening the historic management approach that mainly focuses on biodiversity and habitats within river channels, with dependent floodplains and their inhabitants receiving relatively little attention (Kingsford 2000, Ward et al. 2002).

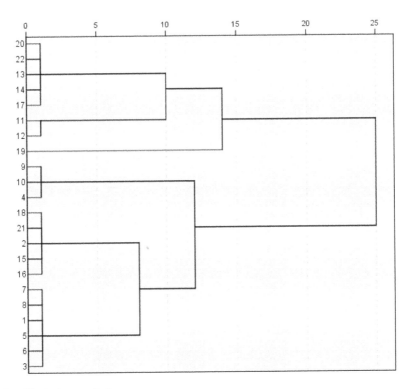

Figure 3.   Clustering results tree.

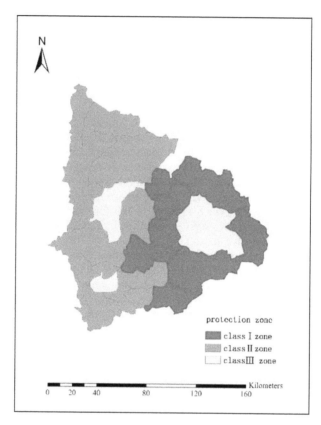

Figure 4.   Conservation zoning result.

Laws and regulations related to the biodiversity conservation are suitable for class I, class II and class III zones. In class I zones, special legislation is also required for nature reserve (Xu S Y & Shen Q J 2009). For areas involved in fishery resources in class I and class II zones, the management of fishery resources should be strengthened (Deng J Y & Jin X S 2000). For areas that consider agricultural development as the main water ecological function in class I, class II and class III zones, the rural environment management should be strengthened. For areas involved in rivers through the city in class III zones, water quality management planning should be established, and the water environment and ecological goal need to be determined. For class III zones, the aquatic habitat should be restored through ecological construction engineering. For areas involved in wetland in class I, class II and class III zones, wetland construction projects should be carried out (Liu B 2013). For urban areas in class I, class II and class III zones, a comprehensive industrial pollution sources improve-ment project should be carried out (Xu S Y & Shen Q J 2009). For rural areas in class III zones, a village environment comprehensive improvement project is needed, as low aquatic biodiversity is mainly caused by non-point source pollution.

5   CONCLUSION

Habitat boundaries division can be used for terrestrial biodiversity protection, but it does not apply to the protection of aquatic biodiversity because of the complexity of the aquatic habitat division. Building the zoning indicators for the division of protection zone is a prom-ising method. As for Tieling, the present situation of the aquatic biodiversity, the economic value of the aquatic biodiversity, the pressure and response to the aquatic biodiversity, and the importance of the ecosystem are chosen as zoning indicators. Other appropriate indicators can be used for other research districts. The hierarchical clustering analysis method can realize the objective classification according to the data characteristics and the degree of similarity of data. The clustering result is realized and can be used for the division of protection zone.

ACKNOWLEDGMENT

The authors acknowledge the support of the National Science and Technology Major Project (Water Pollution Control and Treatment Major Project (No. 2012ZX07505-001-04)).

REFERENCES

Boon, P.J. The development of integrated methods for assessing river conservation value [J]. Hydrobiologia, 2000:413–428.

David Dudgeon, Angela H. Arthington, Mark O. Gessner, et al. Freshwater biodiversity: importance, threats, status and conservation challenges [J]. Freshwater biodiversity, 2006, 81:163–182.

Deng, J.Y, Jin, X.S. Study on fishery biodiversity and its conservation in Laizhou bay and Yellow river estuary [J]. Zoological Research, 2000, 21(1):76–82.

Ge, J.W, Cai, Q.H, Liu, J.K. A new method of economic evaluation of biodiversity in the water ecosystem [J]. Acta Hydrobiologica Sinica, 2006, 30(1):126–128.

Huang, C.X, Wang, M.Y. Condition, Evaluation, and Changing Trend of Aquatics and Microbes on the Section of Liaohe River along Tieling City [J]. Journal of Microbiology, 2010, 30(1):88–92.

Huang, P, Ye, Y.Z, Gao, H.M, et al. Biodiversity survey and evaluation in Henan Province Luo River basin [J]. Journal of Henan normal university (natural science edition), 2012, 40(1):142–145.

Huang, Y, Cui, H.R. Fuzzy cluster analysis method of water quality of lakes [C]. Hubei province: Hubei province water conservancy society, 2011.

Kingsford, R.T. Ecological impacts of dams, waterdiversions and river management on floodplain wetlands in Australia [J]. Austral Ecology, 2000, 25:109–127.

Liu, B. Biodiversity evaluation on Lianhua lake wetland in Tieling [J]. Environmental science and management, 2013, 38(6):88–92.

Liu, Y.B, Liang, C. Water conservation zoning study of the upper Yangtze river [J]. China Rural Water and Hydropower, 2009, (4):10–14.

Qu, F.H, Cui, G.C, Li, Y.F, et al. Fuzzy clustering algorithm and application [M]. Beijing: National defense industry press, 2011.

Wan, B.T, Xu, H.G, Ding, H, et al. Methodology of comprehensive biodiversity assessment [J]. Biodiversity Science, 2007, 15(1):97–106.

Wang, G.R, Du, F.G. Biodiversity assessment of the Songhua river Three Lake wetland [J]. Journal of north China university (natural science edition), 2006, 7(3):278–280.

Ward, J.V, Tockner, K, Arscott, D.B, et al. Riverine landscape diversity [J]. Freshwater Biology, 2002, 47:517–539.

Xu, S.Y, Shen, Q.J. Urban Biodiversity Conservation: The Planning Ideal and the Approach [J]. Modern Urban Research, 2009, (9):12–18.

Zhang, L.N, Li, X.W, Song, X.L, et al. Coupling of Huang-huai-hai wetland ecosystem services and biodiversity protection [J]. Acta Ecologica Sinica, 2014, (14).

Zhou, Z, Li, J, Wang, Z.C. Benthic animal investigation and water quality evaluation of 3 tributaries of Liaohe river tieling section [J]. Environment protection and circular economy, 2009:33–34.

Zhu, W.Z, Fan, J.R, Wang, Y.K, et al. Assessment of biodiversity conservation importance in the upper reaches of the Yangtze River: by taking county area as the basic assessment unit [J]. Acta Ecologica Sinica, 2009, 29(5):2603–2611.

Bioinformatics and Biomedical Engineering – Chou & Zhou (Eds)
© 2016 Taylor & Francis Group, London, ISBN 978-1-138-02784-8

Ecological security assessment of water source based on the GEM-AHP-GMS model

Y. Xin, S.S. Zhang & L. Xu
Key Laboratory of Industrial Ecology and Environmental Engineering (China Ministry of Education),
School of Environmental Science and Technology, Dalian University of Technology, Dalian, China

W. He
Liaoning Province Key Laboratory of Basin Pollution Control,
Liaoning Academy of Environmental Sciences, Shenyang, China

ABSTRACT: Ecological security assessment is the study of identification and judgment of the ecosystem integrity and the ability of maintaining health sustainable under risks. Taking the Liaohe River Basin as the study area, the index system for the water source area of the Liaohe River basin ecological security evaluation and evaluation model was established. Using the combined method of the GEM-AHP (Group Eigenvalue Method-Analytic Hierarchy Process) to calculate the weight of each evaluation indicator, the GMS (Gray system Modeling theory of Software) correlation degree analysis was used to determine the main influence factors. The results showed that the ecological security status of the water source area in Liaohe River Basin was IV (unsafe state) level, the per-capita GDP and population density, which had a great impact on the ecological security. Ecological environment problems in the water source area of the Liaohe River Basin included reduction of resources, low vegetation coverage and insufficient capital investment of pollution control.

Keywords: ecological security assessment; index system; GEM-AHP-GMS; water source area

1 INTRODUCTION

1.1 Research background

Liaohe River Basin is one of the water resource shortage areas in China, with poor water supply conditions, unbalanced distribution in regional water resources and changes in dramatic water resources over time. What's more, industrial and agricultural water consumption has become very concentrated, coupled with poor management, serious pollution, the frequent occurrence of ecological disasters, the problem of ecological security, which have become the focus of attention (Zhang Nan et al. 2009). The Eighteenth National Congress of the Communist Party of China report put the ecological civilization construction in a prominent position (Zheng Hangsheng. 2013). At present, there are many studies on basin ecological security assessment research at home and abroad, such as the ecological security evaluation of Kondratyev (Kondratyev S. et al. 2002) on water resources of Ladoga Lake and its basins. It had a rapid development in health diagnosis and risk assessment system, using conceptual framework models such as the mathematical model, ecological model and exposure response model as effective tools. It made the ecological security evaluation to be a research of in-depth relationship.

1.2 Research status

At home, researchers who studied on the basin ecological security evaluation are Dong Wei (Dong Wei et al. 2010) and Huang Baoqiang (Huang Baoqiang et al. 2012). They used the principal component analysis and analytic hierarchy process to establish the ecological security evaluation model and the index system, selecting the middle and lower reaches of Liaohe River Basin and the upper reaches of the Yangtze River as the study area, respectively. These studies mainly focused on the main cities in the basin, but considered less on ecological security issues in the water source area of the basin. In the process of evaluation, the model is often weighted artificially, with a lack of objectivity to a certain extent. Therefore, this paper introduced the GEM-AHP-GMS model based on the "pressure-state-response" model, evaluated the ecological security status of the water source of the Liaohe River Basin, and analyzed the main influence factors based on the level of security.

2   RESEARCH METHODS

2.1   Establishment of the ecological safety evaluation model and the index system

First, we made research on the evaluation of the ecological security of Liaohe River Basin water source. In this paper, "pressure-state-response" conceptual framework model was used. The main factors affecting ecological security in the water source area were taken into consideration as well as the availability of the actual data of each index (Table 1).

Combined with the evaluation index system, the ecological security comprehensive evaluation model of water source area was established as follows:

$$ESI = \sum_{i=1}^{n} A_i W_i \tag{1}$$

where ESI is the ecological security composite index; A_i is the normalized value of each index; W_i is the weight of each index; and n is the total number of indices, with $i = 1, 2, 3 \dots n$.

The full normalized value is 10. Each indicator of the ecological security evaluation index system is scored by 5 experts in accordance with the present situation of the Liaohe River Basin water source area and the index weight is calculated based on the GEM-AHP model.

Table 1.   Ecological security assessment index system of water source.

Target	Criterion A	Feature B	Index C
Ecological security assessment of water source O	State (A1)	Land (B1)	Land development rate (C1)
		Climate (B2)	The average annual rainfall (C2)
			Average annual temperature (C3)
		Resources (B3)	Per-capita water resources (C4)
			Mineral resources (C5)
		Vegetation (B4)	Vegetation coverage (C6)
		Species (B5)	The diversity index (C7)
			Biological habitat area ratio (C8)
	Pressure (A2)	Population (B6)	The population density (C9)
			The population growth rate (C10)
			The population quality of life (C11)
		Ecological disaster (B7)	Drought area proportion (C12)
			Flood area proportion (C13)
		Economic development (B8)	Per-capita GDP (C14)
			Social stability (C15)
			Emissions intensity (C16)
	Response (A3)	Ecological protection (B9)	Pollution control investment (C17)
			Per GDP water demand (C18)
		Human activities (B10)	Public education expenditure (C19)
			Staff research proportion (C20)

Table 2 presents the state level. Table 3 provides the ecological security evaluation criteria (Li Ruzhong. 2005).

2.2 Configuration of the index weight based on the GEM-AHP model

Scientifically and rationally determining the index weight will be directly related to the validity and reliability of the evaluation results. Professor Qiu Wanhua (Qiu Wanhua et al. 2012) proposed the Group Eigenvalue Method (GEM), which makes up the limitation of the AHP method. The judgment matrix was constructed to describe the construction of the expert scoring matrix, making the decision-making process more objective and accurate.

1. Establishment of the hierarchical structure
   According to the AHP method, the hierarchical structure is established, as shown in Figure 1.
2. The construction of the expert scoring matrix
   Experts are invited to score the importance of the criteria layer relative to the target layer according to their experience, so as to construct the expert scoring matrix of the criterion layer for the target layer.
3. The calculation of relative weights of elements
   The specific algorithm is as follows: the transposing of the score matrix X squaring is denoted as F, namely $F = x^T x$, where the largest eigenvalue of the feature vector is the

Table 2. The state level of water source area evaluation.

Level	Characterization	Index feature
V	Extremely unsafe	Ecosystem services of water sources are near collapse, more than 90% non-compliance, ecological problem becomes an ecological disaster
IV	Unsafe	Ecosystem service function of water source is degenerated, more than 70% non-compliance, more ecological disaster
III	Safe	Water source ecosystem service function has degenerated, more than 50% non-compliance, ecological disasters occasionally occur
II	Very safe	Water source ecosystem services are more comprehensive, more than 70% compliance, the ecological environment is basically undisturbed
I	Extremely safe	Water source ecological system function is strong, more than 90% compliance, less ecological disasters

Table 3. The ecological security evaluation criteria of water source area.

Level	Extremely safe	Very safe	Safe	Unsafe	Extremely unsafe
Value	>9.530	[5.114,9.530)	[3.544,5.114)	[2.047,3.544)	<2.047

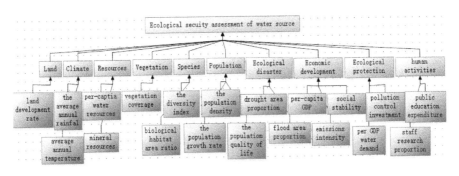

Figure 1. Ecological security assessment index configuration hierarchy model.

optimal decision x*. Under the condition of the precision requirement for ε, x*can be easily obtained using the power method in numerical algebra (Hu Qinghe. 2007).

Based on the establishment of index weight, the ecological security index in Liaohe River Basin water source area is calculated, so as to determine the comprehensive level of ecological security assessment.

2.3   Correlation analysis of ecological security influence factors

According to the level of ecological security assessment in Liaohe River Basin water source area, the main factors affecting the regional ecological security are analyzed and determined, and ecological security factors affecting the correlation degree are analyzed by using the modeling software of gray system theory.

Gray correlation analysis is a method that uses the gray correlation degree sequence to describe the relations of strength, size and order among the factors. Its basic principle is as follows: the closer the geometric shape for the sequence curve, the greater the gray correlation degree among them, vice versa (Guo Wei. 2014).

In this paper, the Deng correlation degree is applied for the gray correlation analysis.

1. Determine the analysis sequence:

$$X_0(k) = (x_0(1),\ x_0(2),\ \dots,\ x_0(n))$$

2. Calculate the difference sequence, range and correlation coefficient:

$$\gamma\big(|X_0(k),\ X_i(k)|\big) = \frac{\min(i)\min(k)\big|X_0(k)-X_i(k)\big| + \zeta\max(i)\max(k)\big|X_0(k)-X_i(k)\big|}{\big|X_0(k)-X_i(k)\big| + \zeta\max(i)\max(k)\big|X_0(k)-X_i(k)\big|} \qquad (2)$$

Generally, satisfactory resolution can be obtained with a resolution factor of 0.5.

3. Calculate the correlation degree and determine the correlation sequence.

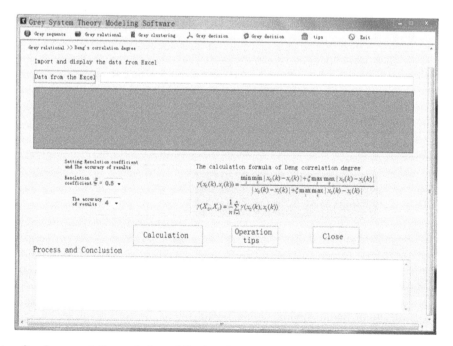

Figure 2.   Gray correlation analysis modeling interface.

3 CASE ANALYSIS

3.1 *The general situation of Liaohe River Basin water source area*

Liaohe River Basin is located between the east longitude of 117°00′~125°30′ and the north latitude of 40°30′~45°10′. The main reservoirs of Liaohe are Qinghe reservoir and Chai River reservoir. The full length of Liaohe River is 523 km, and the watershed area is 41090000 km². Liaohe is the largest natural river in Liaoning Province.

The average annual flow of Liaohe is about 400 m³/s, the average annual runoff is 12.6 billion m³, and the annual average sediment discharge is 20.98 million t. Still, there are chemical plants, paper mills and sewage treatment plants directly or indirectly pouring the industrial wastewater into the river in the Liaohe estuary. Among them, the main pollutant is the Chemical Oxygen Demand (COD), ammonia nitrogen (NH^3-N), Total Nitrogen (TN), Total Phosphorus (TP) and Suspended Solids (SS). The hill area of Liaohe River upstream is faced with serious soil erosion and poor vegetation coverage (about below 30%), and is one of the sand regions with the most severe drought in the Northeast.

3.2 *Determining the weights and assessment results*

Combined with the establishment of the ecological security assessment index system for the water source area of the Liaohe River Basin, the GEM-AHP model is adopted to determine the weight of each index in the system of assessment index. In order to make the results more accurate and reliable, 5 experts were selected to judge, and finally get the weight of each index level of the target layer weights, as given in Table 4.

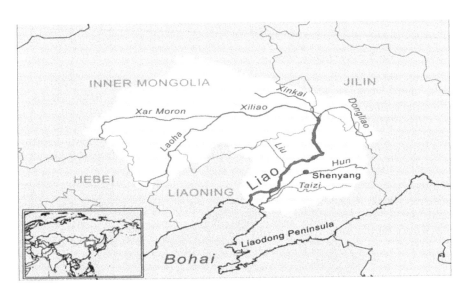

Figure 3.   Liaohe River Basin.

Table 4.   The index weight table of basin.

Index	C1	C2	C3	C4	C5	C6	C7	C8	C9	C10
Weights	0.0359	0.0074	0.0039	0.0241	0.0048	0.0485	0.0258	0.0091	0.1631	0.0307
Index	C11	C12	C13	C14	C15	C16	C17	C18	C19	C20
Weights	0.0867	0.0253	0.1264	0.2065	0.0248	0.0716	0.0130	0.0652	0.0227	0.0045

Table 5. Normalized value of each index in Liaohe River Basin water source.

Index	C1	C2	C3	C4	C5	C6	C7	C8	C9	C10
Normalized value	3	5	4	4	4	4	2	4	3	2
Index	C11	C12	C13	C14	C15	C16	C17	C18	C19	C20
Normalized value	4	3	3	4	4	2	4	4	4	4

Each assessment index is scored by the expert, and the full score of the normalized value is 10 points.

According to the calculation formula of the comprehensive evaluation model of ecological security, the assessment results show that the Ecological Security Index (ESI) in Liaohe River Basin water source was 3.425, the overall ecological security status is at the IV level, and Liaohe River Basin is considered an unsafe area. This suggests that the ecological security status of the Liaohe River Basin water source is bleak. Its environment response is at a very unsafe state, facing greater environment pressure, but there is not enough environmental response and the state of the environment situation is in general. If the adjustment of environmental response is not timely, it may lead to a vicious cycle of watershed ecosystem water sources.

The main problem in the environmental response system is the lack of pollution control capital investment, with the ecological security index being 0.4736. Although there was an establishment of sewage treatment plant to solve the living sewage and agricultural sewage, an inadequate sewage pipe network and other facilities resulted in the slow running of the sewage treatment plant project.

3.3 Gray correlation analysis of factors

The research shows that the Liaohe River Basin water source ecological security status is under an unsafe condition. For a detailed analysis of the main factors on ecological security, the gray relational analysis method is used. We use the weight sequence of each index based on the GEM-AHP model and the normalized value from expert scoring as the dependent variable sequence for gray relational analysis of each index factor. Gray modeling software is applied to obtain the analysis results of the ecological security impact factor for Liaohe River Basin water source, and the calculation process is as follows:

--Start---
Sequence [1] as system feature sequence, the calculation process of the remaining sequence and its Deng correlation degree is as follows:
The [a] step, calculated sequence initial value
The [b] step, calculate the difference sequence
The [c] step, Calculate the range
Range maximum: 942.0756
Range minimum: 0.0000
The [d] step, Calculate the correlation coefficient
The [e] step, the Deng correlation degree of remaining sequence and sequence [1]
--End-------------------------------------

The correlation results are as follows.

It can be seen from Figure 4 that the main indices affecting the ecological security status of Liaohe River Basin are C5, C6 and C17. Vegetation coverage has a great impact on the ecological security. The response to the pressure and environmental protection investment is not enough. The environmental problems are still a matter of concern, and emissions have a greater impact on the degree of security. We need to improve the ecological environment, and reduce exhaust emissions and air pollution. Ecological security situation throughout all

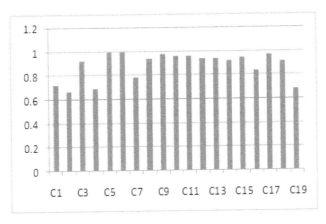

Figure 4. Ecological safety influence factor correlation of Liaohe River Basin.

Liaohe watershed is relatively poor. Though natural conditions are relatively intact, the level of productivity is not enough overall and agriculture is still an important part of the national economy. The response to treatment pressure is poor and the process of urbanization is too fast. In addition, the contradiction between water resources consumption and supply will become the hidden danger of ecological security and stability.

4 CONCLUSION

1. Based on the summary of previous research on the evaluation of ecological security, with improvement and validation of the index system, and according to the respective characteristics of AHP and GEM, this paper put forward the comprehensive algorithm for AHP and GEM to determine the practical index weight.
2. According to the results of ecological security evaluation, the main factors of Liaohe River Basin water source ecological security were identified using the gray theory modeling software on influencing factors for correlation analysis, which is useful for adjusting measures in accordance with local conditions for ecological planning and management.
3. The research showed that the Liaohe River Basin water source ecological security status was under an unsafe condition. The main influencing factors were the vegetation coverage and pollution control investment. It should vigorously strengthen ecological environment construction activities in tree planting and grass. In order to maintain the ecological security in water sources, ecological environment construction activities such as increasing investment in pollution control and increasing vegetation cover can improve the ecological environment, protecting and maintaining the ecological security.

REFERENCES

Dong Wei et al. 2010. Delimitation of water conservation areas in the upper Yangtze River and analysis of influencing factors on eco-security [J]. Journal of University of Science and Technology Beijing, 32(2): 139–144.

Guo Wei. 2014. Grey correlation analysis of the tourist defrauding formed based on factor analysis [J]. China Collective Economy, 4(12): 100–103.

Huang Baoqiang et al. 2012. A Review on Ecological Security Assessment [J]. Resources and Environment in the Yangtze Basin, 21(Z2): 150–156.

Hu Qinghe. 2007. Risk assessment of river basin water resources conflict [J]. Statistics and Decision, (02): 322–331.

Kondratyev S. et al. 2002. Assessment of present state of water resources of Lake Ladoga and its drainage basin using sustainable development indicators. Ecological Indicators, 2:79–92.

Li Ruzhong. 2005. Study on rational allocation of regional water resources based on index system [J]. Systems Engineering-theory and Practice, 128(2): 368–373.

Qiu Wanhua et al. 2012. A neural network financial data classification method based on information entropy [J]. Control and Decision, (02): 286–293.

Zhang Nan et al. 2009. Multi Index Evaluation Method of River Ecosystem Health in Liao River Basin [J]. Research of Environmental Sciences, 22(2): 162–170.

Zheng Hangsheng. 2013. Interpretation of the eighteen major reports on the social construction [J]. Qiushi magazine, 7: 37–39.

Bioinformatics and Biomedical Engineering – Chou & Zhou (Eds)
© *2016 Taylor & Francis Group, London, ISBN 978-1-138-02784-8*

Uncertainty of both eco-corridor and centrality originated from resistant surface: A case study of Liaoning Province, China

M.Q. Yu, G.B. Song, S.L. Liu & S.S. Zhang
Key Laboratory of Industrial Ecology and Environmental Engineering (MOE),
School of Environmental Science and Technology, Dalian University of Technology, Dalian, China

ABSTRACT: Fragmentation poses threats to landscape connectivity, leading to habitat loss and degradation. Resistance surfaces have been increasingly used for modeling eco-corridor to protect biodiversity. However, there is a lack of consideration about the uncertainty originated from the weight values during resistance surface. This study selected Liaoning Province in China as the study area for the uncertainty analysis of eco-corridor construction and centrality calculation. Resistance surfaces and the "least-cost" model were used as theory, and Linkage Mapper as the tool for analysis. A spatial map of different eco-corridor results was achieved. The centrality analysis showed that the significance of multiple-level reserves and eco-corridors declined in the order of national, provincial, municipal and county levels. The relative standard deviation of uncertainty for both reserves and eco-corridors mainly clustered between 0 and 25%.

Keywords: eco-corridor; resistance surface; uncertainty; centrality; Liaoning Province

1 INTRODUCTION

Climate change poses a challenge to biodiversity. Habitat fragmentation resulted from lost connectivity, blocking the movement of animals driven by global warming. In this situation, current reserves will not continue to support all species due to destruction of eco-corridors, which deteriorate the loss of biodiversity (Beier & Brost 2010, Heller & Zavaleta 2009, Thomas et al. 2004). Fragmentation, habitat degradation and habitat loss against landscape connectivity dominate the mechanisms for the reduction and loss of biodiversity (Adriaensen 2003, Rudnick 2012).

China has gone through a rapid economic development since 1978, which has also created challenges to biodiversity. In the past three decades, tremendous changes have occurred, ranging from urbanization, land use change to road network development, leading to habitat loss and landscape fragmentation. Consequently, decreasing species migration and increasing mortality have become a serious issue (Fu et al. 2010, Jiang et al. 2014, Zhai et al. 2014). Urbanization rate in China increased from 18% in 1978 to 51% in 2011 (Tong & Wu 2013), increasingly threatening habitats that are critical for biodiversity (Elmqvist et al. 2013). Similarly, the dietary pattern of Chinese residents updated with more animal products, requiring more land for food production. Food production increased from 1.28 billion to 2.28 billion ha during 1984 to 2012, which occupied more natural protected areas (Zhao et al. 2014). Road development also contributed to the landscape fragmentation, and it is expected that the impact will increase up to 2030 according to the National Road Planning (Li et al. 2005, NDRC 2013).

Eco-corridors are generally linear spaces that facilitate movement between patches, which can enhance landscape connectivity for species migration. Eco-corridor modeling has been used for conservation planning in recent decades (Rudnick 2012). International organizations launched programs to construct large-scale eco-corridors. (1) 1992, the International

Union for Conservation of Nature initiated the European Green Belt Plan, stretching from the Barents Sea to the Black Sea; (2) 2010, the Society for Ecological Restoration launched the North American Wild ways Network, trying to improve the integrity and connectivity of ecosystems in North America; (3) 2002, the Gondwana Link began to protect the biodiversity of southwestern Australia, extending from moist forest area to dry forests and shrubs Nullarbor plain edge (Riecken et al. 2007, Jonson 2010, Bowers & McKnight 2012, Mu 2014).

Resistance surfaces to support the "least-cost" method have been widely used for modeling regional eco-corridors (Zeller 2012). Resistance surfaces against animal mitigation are typically determined spatial factors, such as land use, disturbance density from human activity and road density. Each cell in a resistance map is attributed with a value reflecting the energetic cost, difficulty, or a comprehensive capital of moving across that cell (McRae & Kavanagh 2011). Resistance surfaces have been used to model virtual landscape scenarios (Adriaensen et al. 2003), to identify functional habitat networks (Watts et al. 2010) and to identify local and regional habitat linkages (Van Manen et al. 2007).

However, the knowledge gap on the uncertainty of resistant surfaces by spatial overlaying still exists. In this study, Linkage Mapper (McRae & Kavanagh 2011) was used with supports of GIS and the "least-cost" theory to build eco-corridors in Liaoning Province of China. Eco-corridor uncertainty from resistant surfaces was analyzed. Reserves and eco-corridors at the national, provincial, municipal and county levels were subjected to centrality analysis to calculate the significance for landscape connectivity, respectively. The Relative Standard Deviation (RSD) of centrality results was used for uncertainty analysis.

2 STUDY AREA AND DATA PROCESSING

2.1 Study area

Liaoning lies in northeastern China, with access to the Yellow Sea and the Bohai Gulf (Fig. 1), with a total continental land area of 145 900 km², accounting for 1.5% of China's territory. Landscapes in Liaoning varied greatly, with 59.8% of mountainous regions, 33.4% of plains, and 6.8% of water bodies. Since China made the strategy to implement the revitalization of old industrial bases in Northeast China in 2003, Liaoning has undergone rapid development.

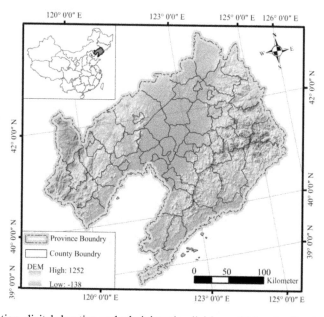

Figure 1.   Location, digital elevation and administrative divisions of Liaoning Province, China.

Table 1. Index system of resistance for eco-corridor in Liaoning Province in China.

First level (index)	Second level (indicator)	Weight
Land use	Urban land	0.6
	Agricultural land	0.4
Human interference intensity	Night-time light intensity	0.4
	Population density	0.3
	GDP	0.3
Linear-feature density	Road-network density	0.5
	River-network density	0.5

Until 2012, the urbanization rate was 65.7%, ranking the second in China and will reach to 75% in 2020. In 2013, the total highway mileage of Liaoning reached 110 072 km, and it is still vigorously developed. The population was 42 million in 2000 and increased to 43.74 million by 2010, attracting a large crowd for tourism and employment. Liaoning is rich in biodiversity and 105 natural reserves were established until 2012, covering 14 national natural reserves, 30 provincial reserves, and 61 municipal and county natural reserves (MEPPRC 2013).

2.2 Index system for mapping resistant surface

Indicators of urban land, agricultural land, gross domestic production, population density, night-time light intensity, road- and river-network density were used to establish the index system. All indicators belonged to Land Use (LU), Human Interference Intensity (HII) and Linear-Feature Density (LED) (Table 1), which were calculated by weighting methods.

2.3 Data and processing

All raster data were converted to the ESRI grid format with a spatial resolution of 1 km², and spatial projections were standardized according to Albers supported by the geographic coordinate system of Krasovsky. Land use, human-interference, and linear-feature density data sets were all converted to a floating format by linear stretching, with values ranging from 0.00 to 100.00.

$$I_s = \left(I - I_{min}\right)/\left(I_{max} - I_{min}\right) \times 100 \tag{1}$$

where I_s is the normalized value of each spatial indicator; I is the pixel value of each original data point; and I_{min} and I_{max} are, respectively, the minimum and maximum of the indicator I image.

3 METHODS

3.1 Resistance surfaces by different weights

Resistance values were generated by the spatial overlaying of different spatial layers in the GIS. Weight variation of different layers directly determined the magnitude of values in pixels. Weight variation led to different resistance surfaces and different eco-corridors with uncertainty. In this part, a total of eleven sets of weights (Table 2) were used to produce different resistance surfaces and different uncertain eco-corridors. The spatial maps of resistance and indicators for the index system are shown in Figure 2.

3.2 Eco-corridor construction and centrality analysis

Linkage Mapper (McRae et al. 2011) was used for extracting eco-corridor and calculating centrality. Each reserve or eco-corridor plays a unique role in protecting biodiversity, but they

Table 2. Weight uncertainty for spatial overlaying of *LU*, *HII* and *LFD* with weights as w_{i1}, w_{i2} and w_{i3}, respectively.

i	w_{i1}	w_{i2}	w_{i3}	i	w_{i1}	w_{i2}	w_{i3}
1	0.4	0.3	0.3	7	0.5	0.1	0.4
2	0.4	0.4	0.2	8	0.3	0.3	0.4
3	0.4	0.2	0.4	9	0.3	0.4	0.3
4	0.5	0.3	0.2	10	0.3	0.5	0.2
5	0.5	0.2	0.3	11	0.3	0.2	0.5
6	0.5	0.4	0.1				

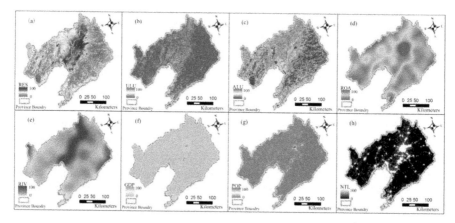

Figure 2. Spatial pattern of influencing factors and resistant surface. (a) Resistance surface (RES), (b) Urban Land (ULU), (c) Agricultural Land (ALU), (d) Road-network density (ROA), (e) River-network density (RIV), (f) Gross Domestic Production (GDP), (g) Population density (POP) and (h) Night-Time Light (NTL).

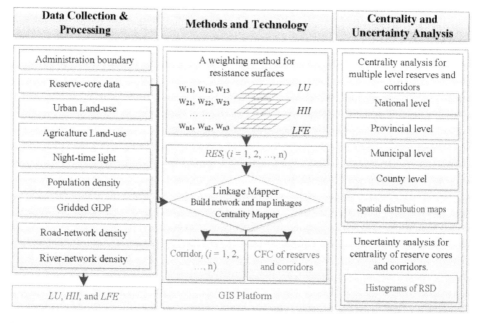

Figure 3. Technical flowchart of regional eco-corridors in Liaoning province under uncertainty covered.

contribute to different degrees for eco-connectivity. Centrality Mapper analyzes the result of link-age networks and calculates Current Flow Centrality (CFC) across the networks (McRae 2012). Reserves were at the national, provincial, municipal, and county levels, and the links were also divided into four levels based on the originating cores. The Relative Standard Deviation (RSD) of CFC for each reserve core or eco-corridor was calculated to analyze centrality uncertainty:

$$RSD = \sqrt{\frac{1}{n-1}\sum_{i=1}^{n}\left(CFC_i - CFC_{i,ave}\right)^2} \Big/ CFC_{i,ave} \times 100\% \qquad (2)$$

where n is the number of CFC for each reserve core or eco-corridor; $CFC_{i,max}$, $CFC_{i,min}$ and $CFC_{i,ave}$ are, respectively, the maximum, minimum and average CFC values of the reserve core or eco-corridor.

4 RESULTS AND ANALYSIS

4.1 *Spatial pattern of eco-corridors*

Eco-corridors were achieved with resistance surfaces. Adding all eco-corridors from different resistance surfaces to one picture is not sensible due to the difficulty of overlaying all results visually. We therefore gave an example (Fig. 4) to illustrate the spatial uncertainty of eco-corridors. Least-Cost Paths (LCPs) were mapped in the local maps to represent corridors. As we can see from Figure 4, there were obvious differences between the 5 LCPs that resulted from the uncertainty of the sharply different weights for resistance surfaces.

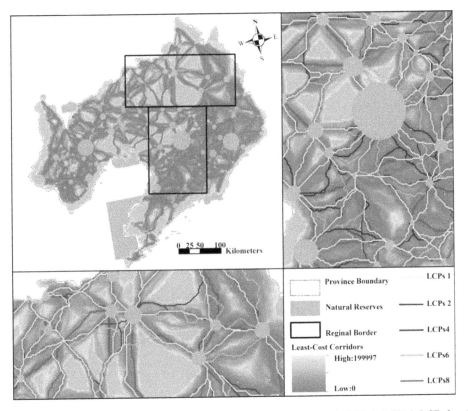

Figure 4. Spatial patterns of eco-corridors in Liaoning province. LCPs1, LCPs2, LCPs4, LCPs6 and LCPs8 represent the uncertainty of eco-corridors originated from different resistance surfaces.

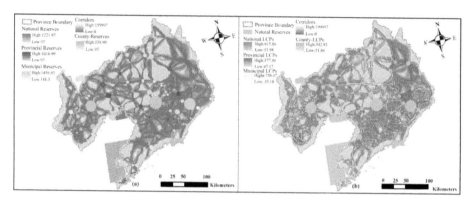

Figure 5. Spatial patterns of natural reserves and eco-corridors' centrality. (a) Current flow of natural reserves at different levels, (b) Current flow of eco-corridors originating from multiple-level reserves.

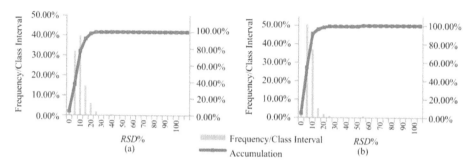

Figure 6. Histograms for the Relative Standard Deviation (RSD) of current flow centrality of (a) reserve cores and (b) eco-corridors.

4.2 Centrality for both natural reserves and eco-corridors at different levels

Each reserve has eleven CFC values as eleven simulations were conducted based on different resistance surfaces. All CFC values were averaged for comparison (Fig. 5a). The centrality presented by the average value and range of CFC for the national, provincial, municipal and county level reserves were 602.79 [97, 1221.97], 548.81 [97, 1016.99], 516.74 [148.5, 1456.83] and 362.06 [97, 538.90], respectively. Although the maximum CFC value of municipal reserves was larger than that of national reserves, the average values declined in the order of national, provincial, municipal and county, indicating that national natural reserves were the most important for Liaoning eco-connectivity compared with other levels. Natural reserves for the Chaoyang fossils group had the maximum CFC value among all the 99 reserves, and should be given a priori protection.

CFC values for the national, provincial, municipal, and county level eco-corridors were spatially expressed in Figure 5(b). Similar to the centrality analysis for the reserve core areas, the results for the eleven simulations were averaged for analysis. The average and ranges of CFC current flow for the national, provincial, municipal and county level reserves were 225.36 [53.98, 615.86], 190.16 [47.17, 377.49], 161.90 [35.18, 758.47] and 132.98 [51.86, 342.95], respectively. Eco-corridors originated from the national reserves had a relatively higher significance, compared with the other three levels. The centrality rank of eco-corridors had the same order as that of national reserves.

4.3 Uncertainty analysis

The centrality analyses of reserve core areas and eco-corridors had uncertainties. The CFC for each reserve core had a range of value based on the eleven results from different resistant surfaces, and it was similar to centrality for each eco-corridor. RSD was used to express the uncertainty of centrality analysis. Histograms (Fig. 6) show the distribution of RSDs for both reserve cores and eco-corridors. As seen from the picture, RSD for the reserve cores and eco-corridors mainly clustered between 0 and 25%, with a proportion of 99.01% for reserve cores and of 99.64% for eco-corridors.

5 CONCLUSION

Uncertainty originating from the weights for resistance surfaces was always ignored. This study focuses on uncertainty and its influence on the centrality of both reserve cores and eco-corridors at multiple levels. The results show that different weights of resistant surfaces lead to uncertainty of eco-corridors, though not very apparent. The significance of both reserve cores and eco-corridors declined in the order of national, provincial, municipal and county levels. Limitations still exist in this study. Owing to the limited time and labor, only eleven sets of weights were simulated, which may be insufficient for the uncertainty analysis. A more detailed study, such as supported by the Monte Carlo simulation, should be further conducted to quantify the uncertainty of eco-corridor construction, which may lead to a deviation of positioning ecological engineers.

REFERENCES

Adriaensen, F., Chardon J.P., De Blust, G. & Matthysen, E. 2003. The application of 'least-cost' modelling as a functional landscape model. Landscape and Urban Planning, 64: 233–247.
Beier, P. & Brost, B. 2010. Use of land facets to plan for climate change: conserving the arenas, not the actors. Conservation Biology, 24: 701–710.
Bowers, K. & McKnight, M. 2012. Reestablishing a healthy and resilient North America: linking ecological restoration with continental habitat connectivity. Ecological Restoration, 30: 267–270.
Elmqvist, T., Fragkias M., Goodness, J. et al. 2013. Urbanization, Biodiversity and Ecosystem Services: Challenges and Opportunities, Chapter 5. Springer Dordrecht Heidelberg New York London.
Fu, W., Liu, S.L., Degloria, S.D., Dong, S.K. & Beazley, R. 2010. Characterizing the "fragmentation-barrier" effect of road networks on landscape connectivity: A case study in Xishuangbanna, Southwest China. Landscape and Urban Planning, 97: 328–328.
Heller, N.E. & Zavaleta, E.S. 2009. Biodiversity management in the face of climate change: A review of 22 years of recommendations. Biological Conservation, 142: 14–32.
Jiang, P.H., Cheng, L., Li, M.C., Zhao, R.F. & Huang, Q.H. 2014. Analysis of landscape fragmentation processes and driving forces in wetlands in arid areas: A case study of the middle reaches of the Heihe River, China. Ecological Indicator, 46(11): 240–252.
Jonson, J. 2010. Ecological restoration of cleared agricultural land in Gondwana Link: lifting the bar at 'Peniup'. Ecological Management & Restoration, 11(1): 16–26.
Li, S.H., Zhou, Q.F. & Wang, L. 2005. Road construction and landscape fragmentation in China. Journal of Geographical Science, 15(1): 123–128.
McRae, B.H. 2012. Centrality Mapper Connectivity Analysis Software. The Nature Conservancy, Seattle WA. Available at: http://www.circuitscape.org/linkagemapper.
McRae, B.H. & Kavanagh, D.M. 2011. Linkage Mapper Connectivity Analysis Software. The Nature Conservancy, Seattle WA. Available at: http://www.circuitscape.org/linkagemapper.
Mu, S.J., Zhou, K.X., Fang, Y. & Zhu, C. 2014. The need and the prospects for developing large-scale green eco-corridors to protect biodiversity. Biodiversity Science, 22(2): 242–249.
Rudnick, D.A., Ryan, S.J., Beier, P. et al. 2012. The role of landscape connectivity in planning and implementing conservation and restoration priorities. Issues in ecology, 16.
Riecken, U., Ullrich, K. & Lang, A. 2007. A vision for the Green Belt in Europe, in: Terry, A., K. Ullrich and U. Riecken: The Green Belt of Europe. From Vision to Reality, IUCN, Gland, Switzerland and Cambridge, UK.
Thomas, C.D., Cameron, A., Green, R.E. et al. 2004. Extinction risk from climate change. Nature, 427(6970): 145–148.
Tong, Y.F. & Wu, Y. 2013. Demographic characteristics and problems during urbanization process of China, Population & Development, 19(4): 37–45.

The list of natural protection area in Liaoning Province. 2013. Ministry of Environmental Protection of the People's Republic of China (MEPPRC). Available at http://websearch.mep.gov.cn.

Van Manen, F.T., Mccollister, F.M., Nicholson, J.M., Thompson, L.M., Kindall, J.L. & Jones, M.D. 2007. Identifying habitat linkages for American black bears in North Carolina, USA. J Wildl Manag, 71: 487–495.

Watts, K., Eycott, A.E., Handley, P., Ray, D., Humphrey, J.W. & Quine, C.P. 2010. Targeting and evaluating biodiversity conservation action within fragmented landscapes: an approach based on generic focal species and least-cost networks. Landscape Ecol., 25: 1305–1318.

Zhai, D.L., Cannon, C.H., Dai, Z.C., Zhang, C.P. & Xu, J.C. 2014. Deforestation and fragmentation of natural forests in the upper Changhua watershed, Hainan, China: implications for biodiversity conservation. Environmental Monitoring and Assessment, 187: 4137.

Zhao, Y.Y., Jiang, L.L. & Wang, L. 2014. Study of the effect of residents' dietary pattern change to the land requirements for food. China Population, Resources and Environment, 24(3): 54–60.

Zeller, K.A., McGarigal, K. & Whiteley, A.R. 2012. Estimating landscape resistance to movement: a review. Landsc. Ecol., 27: 777–797.

Bioinformatics and Biomedical Engineering – Chou & Zhou (Eds)
© *2016 Taylor & Francis Group, London, ISBN 978-1-138-02784-8*

Continuous blood pressure measurement based on carotid artery diameter

H.Y. Li, Y.B. Li, S.L. Li & N. Deng
Institute of Microelectronics, Tsinghua University, Beijing, China

ABSTRACT: A new continuous blood pressure measurement method based on the relationship between blood pressure and artery diameter is proposed. We use ultrasonic sensor to detect carotid artery and get carotid diameter by recognizing the reflected ultrasound from the front and back walls of the carotid artery. Experiment is designed to establish the calculation model of blood pressure and test the accuracy of the proposed method. The results of comparison between the estimated blood pressure and the standard blood pressure show that the maximum deviation of systolic blood pressure is 2.46 ± 3.01 mmHg and the maximum deviation of diastolic blood pressure is 2.33 ± 2.86 mmHg, both are within 5 ± 8 mmHg and up to the standard of the Association for the Advancement of Medical Instrumentation (AAMI).

Keywords: carotid artery diameter; blood pressure; ultrasonic ranging

1   INTRODUCTION

Blood Pressure (BP) is an important physiological parameter that reflects the function of heart and blood vessel. Moreover, continuous blood pressure measurement can record the lesions of heart and body function and plays a great role on the detection and prevention of the cardiovascular diseases.

Current popular studies on non-invasive continuous blood pressure measurement are based on the relationship between Pulse Transit Time (PTT) and blood pressure (Jeong et al. 2005). PTT is usually acquired as the time that pulse wave transit from heart to radial artery, so it needs to detect electrocardiographic (ECG) signal and pulse wave signal on radial artery to get PTT (Gesche et al. 2012). The equipment of PTT measurement are complex, what is worse, it is hard to establish the uniform and accurate calculation model between PTT and blood pressure due to individual differences which leads to relatively large errors.

In this paper, a new continuous blood pressure measurement method based on the relationship between blood pressure and artery diameter is proposed. The measurement system gets carotid artery diameter by ultrasonic ranging and calculates blood pressure by the established blood pressure calculation model. The experimental results show that the method measures blood pressure accurately and meets AAMI standard.

2   EXPERIMENTAL SETUP

The frame of blood pressure measurement system is shown in Figure 1. The A-type ultrasonic probe (5 MHz center frequency, V110 Olympus NDT) transmits ultrasonic signal under the excitation of high-voltage pulse (200 Hz) to detect carotid artery and receives the reflected ultrasound from the front and back walls of the carotid artery (Joseph et al. 2008). The raw ultrasound is preprocessed by the band filter circuit of 1–10 MHz and the amplifying circuit of 20dB. The processed signals are switched to digital signals by A/D module (50 Mbps,

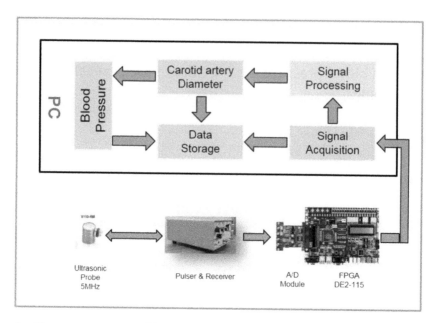

Figure 1. Measurement system architecture.

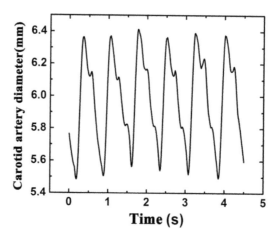

Figure 2.   Carotid artery diameter waveform.

12 bit resolution) and are transmitted to PC. The system of signal acquisition which realizes the real-time signal acquisition and transmission was developed on DE2-115 FPGA (Robson et al. 2006).

The collected signals are processed with MATLAB on PC. Firstly, the envelope $e(t)$ of the ultrasonic signal $x(t)$ is calculated as follows:

$$e(t) = \sqrt{[x(t)]^2 + \{H[x(t)]\}^2}$$

(1)

where $H[x(t)]$ is the Hilbert transform of $x(t)$ given by the following expression:

$$H[x(t)] = -\frac{1}{\pi} \int_{-\infty}^{+\infty} \frac{x(\tau)}{t - \tau} d\tau$$

(2)

The envelop is smoothed using a low pass filter and then detecting the peaks reflected from the front and back walls of the carotid artery to get the time delay ΔT (Matz et al. 2009, Song & Que 2006). Carotid artery diameter D is calculated according to $D = V \times \Delta T/2$, where V is the ultrasonic transit velocity in the blood. The waveform of carotid artery diameter is shown in Figure 2.

3 EXPERIMENT

3.1 Modeling experiment

The study found that the blood pressure variation is closed with the change of arterial diameter during a cardiac cycle (Hardy & Collins 1982, Sass et al. 1998, Sugawara et al. 2000) and proved that blood pressure has a closed relationship with the unit volume of arterial vessel (Zhaorong 1986, Zhaorong & Xixi 1997). The experimental method is as follows:

First, blood pressure of testers are changed by exercise and then we acquire carotid artery diameter of testers by the measurement system and measure standard blood pressure using the criterion sphygmomanometer (OMRON HEM-1020) continuously during the recovery of blood pressure. For getting the synchronous data of blood pressure and carotid diameter, we measure blood pressure and carotid artery diameter synchronously at a fixed time interval of 2 minutes. Systolic Blood Pressure (SBP) and Diastolic Blood Pressure (DBP) are corresponding to the maximum and minimum values of carotid artery diameter respectively in a cardiac cycle.

The result of fitting the experimental data is shown in Figure 3.

The exponential fitting expression is as follows:

$$V = ae^{bP} + c \tag{3}$$

where V is arterial vascular unit volume; P is the blood pressure; a, b, c are undetermined coefficients.

And we have relationship between V and artery diameter R:

$$V = \pi(R/2)^2 \cdot L \tag{4}$$

where L is unit length 1, so we can get the calculation model expression of blood pressure:

$$\pi(R/2)^2 = ae^{bP} + c \tag{5}$$

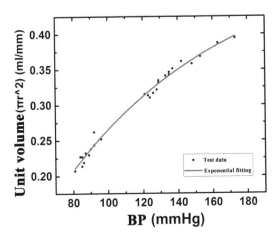

Figure 3. Exponential fitting curve.

3.2 Testing experiment

In order to verify the accuracy of the continuous blood pressure measurement system, we got 6 testers for test. The testers are healthy male volunteers (age: 23–27 years, body mass index: 18–25 kg/m²).

First, we determined the blood pressure calculation expression of each tester in accordance with the above experimental method. Then we measured the standard blood pressure with criterion sphygmomanometer and the calculated blood pressure with the designed system synchronously at the interval of 2 minutes. We got 10 sets of comparison data from each tester.

4 RESULTS AND DISCUSSIONS

4.1 Results

Table 1 shows the comparison results of SBP and Table 2 shows the comparison results of DBP.

The results of comparison between the blood pressure measured by the proposed method and the standard blood pressure show that the maximum Mean Deviation (MD) of Systolic Blood Pressure (SBP) is 2.46 mmHg and the maximum Standard Deviation (SD) is 3.01 mmHg, so the deviation of SBP is 2.46 ± 3.01 mmHg; the maximum MD of Diastolic Blood Pressure (DBP) is 2.33 mmHg and the maximum SD is 2.86 mmHg, so the deviation of DBP is 2.33 ± 2.86 mmHg. The designed continuous blood pressure measurement system meets the AAMI standard that the deviation shall not exceed 5 ± 8 mmHg. The deviations of each tester are illustrated in Figure 4.

The correlation between the standard BP and the calculated BP of all testers is shown in Figure 5. The correlation coefficient between the calculated SBP and the standard SBP is 0.8667 and the correlation coefficient of DBP is 0.7655. It is shown that the BP measured by the designed system are highly correlated with the standard BP.

Table 1. Systolic blood pressure measurement comparison.

Testers	Mean standard SBP/mmHg	Mean calculation SBP/mmHg	Mean deviation/ mmHg	Standard deviation/mmHg
1	111.10	111.37	2.19	2.49
2	120.10	119.51	1.81	2.14
3	126.40	125.30	2.46	3.01
4	113.30	114.56	1.82	2.23
5	122.40	122.84	1.75	1.94
6	125.10	124.44	2.11	2.37

Table 2. Diastolic blood pressure measurement comparison.

Testers	Mean standard DBP/mmHg	Mean calculation DBP/mmHg	Mean deviation/ mmHg	Standard deviation/mmHg
1	77.30	79.12	1.98	2.50
2	80.40	80.84	1.84	2.04
3	87.30	89.58	2.33	2.86
4	75.30	76.54	2.26	2.84
5	82.10	83.19	2.08	2.60
6	85.20	85.82	1.84	2.19

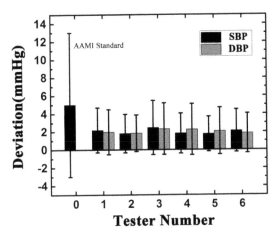

Figure 4.   Comparative deviation of SBP and DBP.

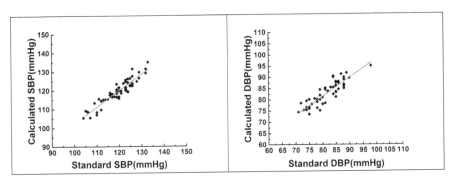

Figure 5.   Correlation between the calculated blood pressure and the standard blood pressure.

4.2   *Discussions*

In our investigation, we found that the carotid diameter can be used for calculating blood pressure continuously. The experimental results were in good agreements with the standard. After 6 human trails, we found that the parameter *b* does not change with individual difference in the calculation model expression (5) and we just need one group of the standard SBP and DBP to determine *a, c* for different individuals. The system is used to measure BP easily because it uses only one ultrasonic probe to detect the carotid artery. However, there is some further work to do, for example, we had better find some hypertensive and the aged people to test the proposed method further and find a way to detect the carotid artery easily and automatically.

5   CONCLUSION

In this paper, a new continuous blood pressure measurement method is proposed. The carotid artery diameter measurement system which is realized by using A-type ultrasonic ranging is designed, and the calculation model of blood pressure based on the relationship between blood pressure and artery diameter is established. Experiments prove that the calculation model is easy to determine and the results of comparison between the standard blood pressure and the calculated blood pressure show that the maximum deviations of SBP and DBP are both satisfied with the standard made by AAMI.

REFERENCES

Gesche, H., Grosskurth, D., Küchler, G., & Patzak, A. 2012. Continuous blood pressure measurement by using the pulse transit time: comparison to a cuff-based method. *European journal of applied physiology*, 112(1), 309–315.

Hardy, H.H., & Collins, R.E. 1982. On the pressure-volume relationship in circulatory elements. *Medical and Biological Engineering and Computing*, 20(5), 565–570.

Jeong, G.Y., Yu, K.H., & Kim, N.G. 2005. Continuous Blood Pressure Monitoring using Pulse Wave Transit Time. *Measurement*, 4, 7.

Joseph, J., & Jayashankar, V. 2008, August. A virtual instrument for real time in vivo measurement of carotid artery compliance. In *Engineering in Medicine and Biology Society, 2008. EMBS 2008. 30th Annual International Conference of the IEEE* (pp. 2281–2284). IEEE.

Matz, V., Smid, R., Starman, S., & Kreidl, M. 2009. Signal-to-noise ratio enhancement based on wavelet filtering in ultrasonic testing. *Ultrasonics*, 49(8), 752–759.

Robson, C.C.W., Bousselham, A., & Bohm, C. 2006. An FPGA-based general-purpose data acquisition controller. *Nuclear Science, IEEE Transactions on*, 53(4), 2092–2096.

Sass, C., Herbeth, B., Chapet, O., Siest, G., Visvikis, S., & Zannad, F. 1998. Intima–media thickness and diameter of carotid and femoral arteries in children, adolescents and adults from the Stanislas cohort: effect of age, sex, anthropometry and blood pressure. *Journal of hypertension*, 16(11), 1593–1602.

Song, S.P., & Que, P.W. 2006. Wavelet based noise suppression technique and its application to ultrasonic flaw detection. *Ultrasonics*, 44(2), 188–193.

Sugawara, M., Niki, K., Furuhata, H., Ohnishi, S., & Suzuki, S. 2000. Relationship between the pressure and diameter of the carotid artery in humans. *Heart and vessels*, 15(1), 49–51.

Zhaorong. 1986. *Cardiovascular Fluid Dynamics*. Fudan University Press.

Zhaorong, & Xixi. 1997. *Hemodynamic Principles and Methods*. Fudan University Press.

Bioinformatics and Biomedical Engineering – Chou & Zhou (Eds)
© 2016 Taylor & Francis Group, London, ISBN 978-1-138-02784-8

A novel saccade signals detection algorithm for EOG-based Human Activity Recognition

Z. Lv
The Key Laboratory of Intelligent Computing and Signal Processing, Anhui University, Hefei, China
Co-Innovation Center for Information Supply and Assurance Technology, Anhui University, Hefei, China

J.N. Guan
School of Computer Science and Technology, Anhui University, Hefei, China

B.Y. Zhou
Co-Innovation Center for Information Supply and Assurance Technology, Anhui University, Hefei, China

X.P. Wu
The Key Laboratory of Intelligent Computing and Signal Processing, Anhui University, Hefei, China

ABSTRACT: A research on the relationship between eye movements and human behavior is a hot topic in the field of Human Activity Recognition (HAR). In this paper, a novel saccade signals detection algorithm for EOG-based HAR, which aims to improve the performance of HAR system, was proposed. In the proposed algorithm, Common Spatial Pattern (CSP) was utilized to build spatial filters, and then use it to process original multi-channel EOG signals. Consequently, feature parameters of different saccade signals can be acquired. To valid the performance of the proposed algorithm, a linear Support Vector Machine (SVM) was chosen. In lab environment, four types of saccade signals corresponding to up, down, left and right were used as analysis objects. Experimental results show that the accuracy recognition ratio is about 97.7%, which reveal that the proposed algorithm has a good classification performance in saccade signals analysis.

Keywords: EOG; saccade signals; Common Spatial Pattern (CSP); joint approximate diagonalization; Support Vector Machine (SVM)

1 INTRODUCTION

HAR (Human Activity Recognition) can be defined as analyzing and identifying systematic of the information such as the types of observation and the patterns of behavior, and the recognition results can be described in a natural language (Aggarwal. 2011). HAR system can perceive the intention of users, so it has broad application prospects in intelligent video surveillance, medical diagnosis, motion analysis and Human-Computer Interface (HCI), which has become a research hot spot in the field of artificial intelligence and pattern recognition (Ni et al. 2013, Yang et al. 2012, Bulling et al. 2008). At present, EOG-based HAR has become a new research spot. In the EOG-HAR system, identifying saccade signals is a critical step, some algorithms of detecting eye movement have been proposed. Among them, Clement proposed a method which use the visual angle of original EOG signals to identify and recognize eye movement signal endpoint (Clement. 1991). Aungsakun utilized the characteristics that the eye movement of EOG signals changed faster to extract eye movement characteristic parameters (Aungsakun et al. 2011). Besides, Antrobus also suggested to use the statistics of eye movement signals and the characteristics of time domain (Antrobus. 1973). The approaches above mainly focus on the time-domain characteristics of EOG signals. Obviously, it is difficult to depict the original

EOG signals correctly in some noise environment (such as the movement of the electrode position, channel distortion, etc.), which cannot be avoided when applied in the real environment, hence feature parameters of saccade signal based on time-domain analysis revealed a poor robustness. On the other hand, it is common to use multi-channel to acquire EOG data in order to obtain more abundant eye movement information (including saccade, fixations, blinks etc.) in the collection of EOG signals. In this case, most of people analyze single channel data specifically according to the position of channels in order to process multi-channels eye signals. Clearly, this method merely considers the variation of single channel signals but ignores the relationship between different channels, which may offer some help to improve the performance of HAR system. In order to solve the above questions, a novel saccade signals detection algorithm for EOG-based HAR was proposed in this paper.

2 GENERATION OF EOG SIGNALS

EOG signals are caused by electric potential difference between the cornea and the retina because of the eye movements and this potential difference is launched by retinal pigment epithelium and light receptor cells, its positive is located in light sensors end, and the cathode is located in the retinal pigment epithelium, the current from the retina flows to the end of the cornea, thereby forming a potential, of which positive is the cornea, and cathode is the retina (Potts & Inoue. 1969). We call this kind of potential EOG signals. Once the eyes move, amplitude of EOG signals will change constantly with the movements of the eyeballs. If we illustrate the change of electric potential on a time axis, we will obtain a curve which is called EOG (Electrooculogram). Figure 1 shows the EOG signal waveform when the eyeballs roll up and down vertically.

3 SACCADE SIGNALS DETECTION ALGORITHM

To realize the feature extraction among four different eye movement signal types which are up, down, left and right respectively, the CSP algorithm (Fukunaga. 1990) was used to establish four spatial filters corresponding to different saccade tasks, which can be utilized to filter the original multi-channels EOG signals. Then the outputs of different spatial filters are feature parameters of eye movements under different saccade tasks; besides, to verify the validity of the EOG features, Support Vector Machine (SVM) proposed by this paper is utilized to identity various kinds of eye movement signals. The block diagram of the proposed algorithm is shown in Figure 2.

It can be shown in Figure 2 that the saccade signal detection algorithm based on CSP mainly contains the following three steps:

1. Data preparation: Multi-channels label EOG data corresponding to up, down, left and right saccades are collected and filtered, and then the preprocessed data are divided into two parts: training data and test data;

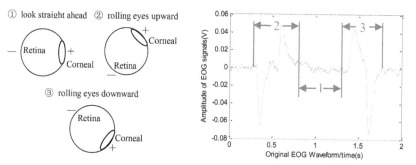

Figure 1.   Eye movements and the corresponding waveform.

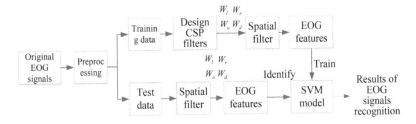

Figure 2. Block diagram of the proposed algorithm.

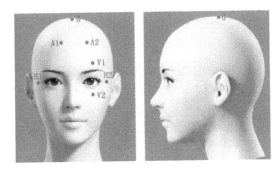

Figure 3. Position of EOG acquiring electrodes.

2. Training: Four spatial filters W_l, W_r, W_u and W_d under different saccade tasks were established by the CSP method, and then these spatial filters were used to process the original multi-channels EOG signals. As a result, the filtered data will be regarded as EOG features parameters and send to the SVM classifier for training;

3. Recognition: Four spatial filters W_l, W_r, W_u and W_d which are established in step 2, were used as spatial filtering and feature extraction for the test data, and then the processed results were output to the SVM classifier to recognize different eye movement signals under different saccade tasks.

3.1 Data acquirement and experimental paradigm design

The collection of EOG is completed by NeroScan acquisition instrument and the Ag/AgCl electrodes are used. In order to obtain the EOG signals from four directions: left, right, up, down and improve the identification accuracy, nine electrodes which placed around the eye are utilized and the position of them are shown in Figure 3.

All subjects have normal or corrected to normal vision, the center point (O) of four observation points locals approximately 2 m in front of eye level and the distance between different observation points and the center point (O) is about 1.5 m. Each trial started with the presentation of a "start" character at the center of the monitor, followed by a 20 ms warning tone ("beep"). After 1 s, a red arrow, which continues for 3 s appears at the center of the monitor for 3 s, pointing to the up, the down, the left or the right respectively. Depending on the direction of the arrow, the subject was instructed to look different observation points and gazed it for a certain time (about 1 second or 2 seconds), then back to the center point (O). It should to be emphasized that the subject can't blink in this three-second process. When the above steps are finished, the subject could blink or relax in order to continue the next trial better. A single experimental paradigm is shown in Figure 4.

3.2 Preprocessing

EOG signal is also a result of a number of factors, not only including eyeball and eyelid movement, but also containing different sources of artifact such as Electroencephalograph

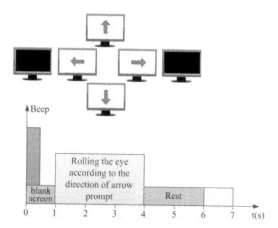

Figure 4. Process of the single experiment.

(EEG), Electrocardiograph (ECG), Electromyogram (EMG), electrodes placement, head movement and influence of the luminance. Therefore, a band-pass filter is used to eliminate the shifting resting potential (mean value) and suppress the noise interrupt.

3.3 Feature extracting of multi-classification eye movements

In the proposed algorithm, we should classify four kinds of eye movements, such as up, down, left and right saccade signals. Therefore, multi-classification problem should be solved in CSP algorithm. In general, Using CSP within the classifier (IN), OVR (One-Versus-Rest) and Simultaneous diagonalization (SIM) was adopted in the realization of multi-classification CSP (Dornhege et al. 2004). Obviously, if the classification becomes more, the former two methods need to generate more spatial filters, which will do harm to the realization of the online function. Consequently, the SIM method is applied in this paper and the process can be described as follows:

Suppose X_l, X_r, X_u and X_d denote 4 classes origin multi-channels eye movement signals corresponding to up, down, left and right saccade, and the normalized covariance matrix is

$$C_l = \frac{X_l X_l^T}{trace(X_l X_l^T)} \quad C_r = \frac{X_r X_r^T}{trace(X_r X_r^T)}$$

$$C_u = \frac{X_u X_u^T}{trace(X_u X_u^T)} \quad C_d = \frac{X_d X_d^T}{trace(X_d X_d^T)} \tag{1}$$

We average multiple sets of training data from the same kind of eye movement modes, obtain the average covariance matrix $\overline{C_l}, \overline{C_r}, \overline{C_u}, \overline{C_d}$ and calculate the sum of the average covariance matrix of all saccade signals $C = \overline{C_l} + \overline{C_r} + \overline{C_u} + \overline{C_d}$, and then decompose the eigenvalue C, that is

$$C = U_0 \sum U_0^T \tag{2}$$

In Equ. 2, U_0 represents the feature vectors, U_0^T represents transposed matrix of U_0, Σ represents diagonal matrix of eigenvalue, then the whitening transformation matrix P can be expressed as:

$$P = \sqrt{\sum{}^1 U_0^T} \tag{3}$$

Transform the average covariance matrices $\overline{C_l}$, $\overline{C_r}$, $\overline{C_u}$ and $\overline{C_d}$ by the whitening matrix P (6×6)

$$S_l = P\overline{C_l}P^T \qquad S_u = P\overline{C_u}P^T$$
$$S_r = P\overline{C_r}P^T \qquad S_d = P\overline{C_d}P^T \tag{4}$$

Use the joint approximate diagonalization method (Cardoso & Souloumiac. 1996) to process S_l, S_r, S_u and S_d, and then an orthogonal matrix U and approximate diagonal matrix Σ_l, Σ_r, Σ_u and Σ_d can be acquired, and satisfy the following relations

$$US_lU^T = \sum{}_l \qquad US_uU^T = \sum{}_u$$
$$US_rU^T = \sum{}_r \qquad US_dU^T = \sum{}_d \tag{5}$$

And

$$\sum{}_l + \sum{}_r + \sum{}_u + \sum{}_d = I \tag{6}$$

In Equ. 6, I represent an identity matrix. In multi-classes case, diagonal matrix Σ_l, Σ_r, Σ_u and Σ_d can be chose by the following equation

$$score(\Sigma) = \max(\Sigma,(1-\Sigma)/(1-\Sigma+(N-1)^2\Sigma)) \tag{7}$$

In Equ. 7, $N = 4$ means four eye movement types, Σ is 4×4 diagonal matrix. The highest score is calculated respectively in Σ_l, Σ_r, Σ_u and Σ_d, then the corresponding row vector U_l, U_r, U_u and U_d will be extracted in the orthogonal matrix U. In this case, different eye movement spatial filters can be computed

$$W_l = U_l^T P \quad W_r = U_r^T P \quad W_u = U_u^T P \quad W_d = U_d^T P \tag{8}$$

Furthermore, source activities related to different eye movement tasks can be acquired

$$Z_l = W_lX \quad Z_r = W_rX \quad Z_u = W_uX \quad Z_d = W_dX \tag{9}$$

In order to classify saccade signals, we take spatial filtering results Z_l, Z_r, Z_u and Z_d from Equ. 9 as feature parameters into the SVM classifier to train the SVM model to classify testing data.

4 EXPERIMENTS AND RESULTS ANALYSIS

All experimental data were collected in our laboratory for the lack of public eye movement database. The HAR-EOG database recorded the 10 subjects (6 males and 4 females, aged between 22–25 years, age mean = 23.6, age sd = 2.7), the experimental environment described in Section 3.1. 4*360(4 types of saccade tasks as well as up, down, left and right, 360 trials in each type saccade) eye movement data were recorded from each subject. The sampling rate was 250 Hz, 6 channels were used for data acquisition.

A band-pass filter (cut frequency: 0.5–12 Hz) was used to remove baseline drift and other noise. Then the spatial filters (W_l, W_r, W_u and W_d) were computed by EOG data label. It should be pointed out that the corresponding relationship between saccade tasks and label is: up-1, down-2, left-3 and right-4. Coefficients of the CSP spatial filter which come from a randomly selected subject data were shown in Figure 5.

Figure 5 shows that the value of H2 electrode is the largest in all channels when the subject rolls to left. Similarly, H1, V1 and V2 rank the first respectively when the subject rolls to right to right, up and down. Furthermore, we can conclude that electrode V1(up rolling), V2(down rolling), H2(left rolling) and H1(right rolling) are respectively corresponding to the 6th, 5th, 1st and 2nd coefficient of CSP spatial filters. The result illustrates the obtained CSP spatial filter achieves the distinction maximization for 4 types of eye movement signals. Using the

71

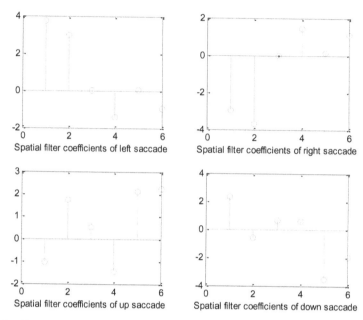

Figure 5.   Coefficients of the CSP spatial filter.

Table 1. 10×5 cross-validation results of 10 subjects.

Saccadic pattern	Subject										Average
	S1	S2	S3	S4	S5	S6	S7	S8	S9	S10	
Up	98.7	99.1	94.2	97.5	98.4	98.6	96.4	99.2	99.0	98.7	**98.0**
Down	96.3	98.4	92.8	95.3	96.7	97.2	95.6	98.7	96.9	98.1	**96.6**
Left	99.1	98.2	95.5	98.4	96.8	97.5	96.3	99.0	98.6	99.2	**97.9**
Right	98.6	99.3	96.7	99.1	98.6	98.1	97.5	98.8	97.7	98.0	**98.2**
Average	**98.2**	**98.8**	**94.8**	**97.6**	**97.6**	**97.9**	**96.5**	**98.9**	**98.1**	**98.5**	**97.7**

above spatial filters to process original multi-channels EOG signals according to Equ. 9, the filtered results (Z_l, Z_r, Z_u and Z_d) are regarded as the feature parameters under different eye movement tasks.

To valid the performance of eye movement feature parameters based on CSP algorithm, we chose a linear support vector machine. Our SVM implementation uses a fast sequential dual method for dealing with multiple classes. This reduces training time considerably while retaining recognition performance. In addition, for acquiring the classification accuracy ratio, we compare recognition results with data labels, if they are the same, the classification is correct, otherwise is wrong. Furthermore, in order to objectively and fairly assess the performance of eye movement feature parameters, a 10*5 cross-validation method was used. The recognition results of 10 subjects (S1–S10) were shown in Table 1.

From Table 1 we can see that the recognition results come from 10 subjects are different, which is due to the different familiarity levels of each subject in experimental environment, experiment equipment and the ability of eye movement controlling. In all subjects, S3 acquires the lowest average recognition rate, and the reason is that this subject sways his body slightly many times unconsciously during 2–3 seconds in the process of data acquiring and it will generate some additional noise interferences, therefore the recognition rate has decreased compared to other subjects. Besides, comparing the down-saccade with other three saccade identification results, we can find that the average recognition accuracy of down-saccade is lower than other three categories, this is because subjects will close their eyelids unconsciously when they perform the down-saccade activity, it would decrease the quality of down-saccade signals. In summary, the overall average accuracy reaches 97.7%, which reveals that the spatial filtering method based on CSP in the process of saccade signal feature extraction is effective.

5 CONCLUSION

In order to improve performance of EOG-HAR system, this paper proposed a novel saccade signals detection algorithm for EOG-based human activity recognition, which construct CSP spatial filter towards four kinds of saccades signals and get four kinds of original activities related to eye movement tasks. In order to test the performance of the proposed eye movement features, we use the SVM algorithm to evaluate them, the average recognition accuracy of which reaches 97.7%. The experimental results validate that the proposed algorithm has a good classification performance in the areas of saccade signals recognition and presents a certain robustness.

ACKNOWLEDGEMENT

The research work described in this paper is supported by National Nature Science Foundation (No. 61271352, No. 61401002), Anhui Province Natural Science Foundation (No. 1408085QF125), Anhui provincial natural science research project of colleges and Universities (No. KJ2014 A011) and Anhui University Academic and Technical Leaders Introduce Engineering Foundation (No. 02303203).

REFERENCES

Aggarwal, J.K. & Ryoo, M.S. 2011. Human activity analysis: A review. *ACM Computing Surveys (CSUR)* 43(3):16.

Antrobus, J.S. 1973. Eye movements and non-visual cognitive tasks. *The oculomotor system and brain functions* 25(6):354–368.

Aungsakun, S. Phinyomark, A. Phukpattaranont, P., et al. 2011. Robust eye movement recognition using EOG signal for human-computer interface. *Software Engineering and Computer Systems* 32(4): 714–723.

Bulling, A. Ward, J.A. Gellersen, H., et al. 2008. *Robust recognition of reading activity in transit using* wearable electrooculography. Berlin: Heidelberg.

Cardoso, J.F. & Souloumiac A. 1996. Jacobi angles for simultaneous diagonalization. *SIAM Journal on Matrix Analysis and Applications* 17(1):161–164.

Clement, R.A. 1991. Natural coordinates for specification of eye movements. *Vision research* 31(11): 2029–2032.

Dornhege, G. Blankertz, B. Curio, G. et al. 2004. Boosting bit rates in noninvasive EEG single-trial classifications by feature combination and multiclass paradigms. *Biomedical Engineering, IEEE Transactions on* 51(6):993–1002.

Fukunaga, K. 1990. Introduction *to statistical pattern recognition*. Netherlands: Elsevier.

Ni, B. Wang, G. & Moulin, P. 2013. Rgbd-hudaact: A color-depth video database for human daily activity recognition. *Consumer Depth Cameras for Computer Vision* 56(7):193–208.

Potts, A.M. & Inoue, J. 1969. The Electrically Evoked Response (EER) of the visual system II. Effect of adaptation and retinitis pigmentosa. *Investigative Ophthalmology & Visual Science* 8(6):605–612.

Yang, Y. Cheng, N. & Zhang, M. 2012. Research on Activity Recognition Method based on Human Motion Trajectory Features. *Journal of Convergence Information Technology* 7(1):103–107.

Bioinformatics and Biomedical Engineering – Chou & Zhou (Eds)
© *2016 Taylor & Francis Group, London, ISBN 978-1-138-02784-8*

Hopf bifurcation analysis of a predator-prey model

D.D. Nie, Z.L. Xiong & W. Wang
Department of Math, Nanchang University, Nanchang, Jiangxi, China

ABSTRACT: A delayed prey-predator system with special predator and generic predator is studied. By analyzing the associated characteristic equation, it is found that Hopf bifurcation occurs when τ crosses some critical value. The stability and direction of the Hopf bifurcation are determined by applying the normal form theory and center manifold theorem. Numerical simulations are performed to illustrate the obtained results.

Keywords: prey-predator system; time delay; stability; Hopf bifurcation; periodic solutions

1 INTRODUCTION AND FORMULATE THE MODEL

In recent years, the properties of periodic solutions are of great interest, which can arise through Hopf bifurcations in delayed systems (J.K. Hale et al.1997). To explain a kind of eco-logical phenomenon, many scholars proposed the following model which includes two kinds of functional response: Holling type-II and Holling type-III. According to those model (Z. Xiong et al. 2008, H. Zhu et al. 2008, H. Wang et al. 2008), we consider the following system

$$
\begin{cases}
\dot{x}(t) = rx(t)(1 - x(t)/K) - \dfrac{cy(t)x(t)}{x(t)+a} - mx^2(t) \\[2mm]
\dot{y}(t) = qy(t)\left(1 - \dfrac{y(t-\tau)}{px(t)}\right)
\end{cases}
\tag{1}
$$

where $x(t), y(t)$ are the densities of rodent and the special predator respectively, m is the den-sities of generic predator. a, c, p, q, r are positive rate constants. The prey grows logitically with carrying capacity K and intrinsic growth rate in the absence of predation. The parameter $\tau \geq 0$ is the feedback time delay of predator species to the growth of species itself.

2 STABILITY OF A POSITIVE EQUILIBRIUM AND THE EXISTENCE OF HOPF BIFURCATIONS

In this section, we discuss the local stability of a positive equilibrium and the existence of Hopf bifurcations in system (1) has a unique positive equilibrium $E^* = (x^*, y^*)$ where

$$
x^* = \frac{(r - cp - ma - ra/K) + \sqrt{(r - cp - ma - ra/K)^2 + 4ar(m + r/K)}}{2(m + r/K)},
$$

$$
y^* = px^*.
$$

$$
f^{(1)} = rx(t)(1 - x(t)/K) - \frac{cy(t)x(t)}{x(t)+a} - mx^2(t), \quad f^{(2)} = qy(t)\left(1 - \frac{y(t-\tau)}{px(t)}\right),
$$

$$
a_{11} = r(1 - 2x^*/K) - \frac{acy^*}{(x^*+a)^2} - 2mx^*, \quad a_{12} = \frac{-cx^*}{a + x^*}, \quad a_{21} = \frac{qy^{*2}}{px^{*2}}, \quad a_{22} = 0, \quad a_{23} = -q. \tag{2}
$$

Linearizing system (1) at $E^* = (x^*, y^*)$, We derive that

$$\begin{cases} \dot{x}(t) = a_{11}x(t) + a_{12}y(t), \\ \dot{y}(t) = a_{21}x(t) + a_{23}y(t - \tau), \end{cases} \tag{3}$$

where $a_{ij}(i = 1, 2; \ j = 1, 2, 3)$ are defined in (2). Hence, the characteristic equation of system (1) at the positive equilibrium E^* takes the form

$$\lambda^2 - a_{11}\lambda - a_{23}(\lambda - a_{11})e^{-\lambda\tau} - a_{12}a_{21} = 0 \tag{4}$$

If $i\omega(\omega > 0)$ is a root of Eq. (4), then

$$\omega^2 + ia_{11}\omega + a_{23}(\lambda - a_{11})e^{-i\omega\tau} + a_{12}a_{21} = 0$$

Separating the real and imaginary parts, we obtain

$$\omega^2 + a_{12}a_{21} = a_{23}(a_{11}\cos\omega\tau - \omega\sin\omega\tau), \ -a_{11}\omega = a_{23}(a_{11}\sin\omega\tau + \omega\cos\omega\tau) \tag{5}$$

It follows from (5) that

$$\omega^4 + (2a_{12}a_{21} + a_{11}^2 - a_{23}^2)\omega^2 + a_{12}^2a_{21}^2 - a_{11}^2a_{23}^2 = 0 \tag{6}$$

If (H1) $a_{23}^2 - a_{11}^2 - 2a_{12}a_{21} > 0$ and $(a_{12}a_{21} + a_{11}a_{23})(a_{12}a_{21} - a_{11}a_{23}) < 0$, then Eq. (6) has only one positive root ω_0 defined by

$$\omega_0 = \left[\frac{1}{2}\left(B + \sqrt{B^2 - 4C}\right)\right]^{\frac{1}{2}} \tag{7}$$

where $B = -(a_{23}^2 - a_{11}^2 - 2a_{12}a_{21})$, $C = a_{12}^2a_{21}^2 - a_{11}^2a_{23}^2$.
Define

$$\tau_j = \frac{1}{\omega_0}\left(\arccos\frac{a_{11}a_{12}a_{21}}{a_{23}(a_{11}^2 + \omega^2)} + 2j\pi\right), \quad j = 0, 1, \cdots \tag{8}$$

Then (τ_j, ω_0) solves Eq. (5). This means that when $\tau = \tau_j$, Eq. (4) has a pair of purely imaginary roots $\pm i\omega_0$. Now let us consider the behavior of roots of Eq. (4) near τ_j.
Denote $\lambda(\tau) = \alpha(\tau) + i\omega(\tau)$ the root of Eq. (4) such that $\alpha(\tau_j) = 0$, $\omega(\tau_j) = \omega_0$. Substituting $\lambda(\tau)$ into Eq. (4) and differentiating both sides of it with respect to τ, we have

$$\left(\frac{d\lambda}{d\tau}\right)^{-1} = \frac{2\lambda - a_{11}}{a_{23}\lambda(a_{11} - \lambda)e^{-\lambda\tau}} - \frac{1}{\lambda(a_{11} - \lambda)} - \frac{\tau}{\lambda}$$

We therefore derive that

$$\mathrm{Re}\left(\frac{d\lambda}{d\tau}\right)^{-1}_{\tau=\tau_j} = \mathrm{Re}\left\{\frac{2\lambda - a_{11}}{a_{23}\lambda(a_{11} - \lambda)e^{-\lambda\tau}}\right\}_{\tau=\tau_j} - \mathrm{Re}\left\{\frac{1}{\lambda(a_{11} - \lambda)} + \frac{\tau}{\lambda}\right\}_{\tau=\tau_j}$$

$$= \frac{1}{\Gamma}\{-2\omega_0^2 a_{23}(a_{11}\cos\omega_0\tau - \omega_0\sin\omega_0\tau) + a_{11}a_{23}\omega_0(a_{11}\sin\omega_0\tau + \omega_0\cos\omega_0\tau) - a_{23}^2\omega_0^2\}$$

$$= \frac{1}{\Gamma}\{2\omega_0^2(\omega_0^2 + a_{12}a_{21}) + a_{11}^2\omega_0^2 - a_{23}^2\omega_0^2\}$$

$$= \frac{\omega_0^2}{\Gamma}\{2(\omega_0^2 + a_{12}a_{21}) + a_{11}^2 - a_{23}^2\} = \frac{\omega_0^2}{\Gamma}\{\sqrt{B^2 - 4C}\}$$

76

where we have used Eqs. (5) and (7), and $\Gamma = a_{23}^2\omega_0^4 + a_{23}^2 a_{11}^2\omega_0^2 > 0$. Hence, it follows that

$$sign\left\{Re\left(\frac{d\lambda}{d\tau}\right)_{\tau=\tau_j}\right\} = sign\left\{Re\left(\frac{d\lambda}{d\tau}\right)^{-1}_{\tau=\tau_j}\right\} = sign\left\{\frac{\omega_0^2}{\Gamma}\sqrt{B^2-4C}\right\} > 0 \qquad (9)$$

Therefore, when the delay τ near τ_j is increased, the root of Eq. (4) crosses the imaginary axis from left to right. In addition, note that when $\tau = 0$, Eq. (4) has roots with negative real parts on if (H2) $a_{11} + a_{23} < 0$ and $a_{11}a_{23} - a_{12}a_{21} > 0$. Thus, by the well-known Rouche theorem we obtain.

Lemma 2.1 Let $\tau_j(j=0,1,\cdots)$ be defined as in (8). Then all roots of Eq. (4) have negative real parts for all $\tau \in [0,\tau_0)$. However, Eq. (4) has at least one root with positive real part when $\tau > \tau_0$, and Eq. (4) has a pair of purely imaginary root $\pm i\omega_0$ when $\tau = \tau_0$. More detail, for $\tau \in (\tau_j, \tau_{j+1})(j=0,1,\cdots)$, Eq. (4) has $2(j+1)$ roots with positive real parts. Moreover, all roots of Eq. (4) with $\tau = \tau_j(j=0,1,\cdots)$ have negative real parts except $\pm i\omega_0$.

Applying Lemma 2.1, the transversality condition (9) and Theorem 11.1 developed in [J. Hale et al.1993], we obtain:

Theorem 2.1 Let (H1) and (H2) hold, let ω_0 and $\tau_j(j=0,1,\cdots)$ be defined as in (7) and (8), respectively.

i. The positive equilibrium E^* of system (1) is asymptotically stable for all $\tau \in [0,\tau_0)$ and unstable for $\tau > \tau_0$.
ii. System (1) undergoes a Hopf Bifurcation at the positive equilibrium E^* when $\tau = \tau_j(j=0,1,...)$.

3 DIRECTION OF HOPF BIFURCATION

In this section, we derive explicit formulate to determine the properties of the Hopf bifurcation at critical values τ_j by using the normal form theory and center manifold reduction (see. For example, B. Hassard et al.1981). Without loss of generality, denote the critical values τ_j by $\tilde{\tau}$, and set $\tau = \tilde{\tau} + \mu$. Then $\mu = 0$ is a Hopf bifurcation value of system (1). Thus, we can work in the phase space $C = C([-\tilde{\tau},0],R^2)$. Let $u_1(t) = x(t) - x^*$, $[u_2(t) = y(t) - y^*]$, Then system (1) is transformed into

$$\begin{cases} \dot{u}_1(t) = a_{11}u_1(t) + a_{12}u_2(t) + \displaystyle\sum_{i+i\geq2}\frac{1}{i!j!}f_{ij}^{(1)}u_1^i(t)u_2^j(t), \\ \dot{u}_2(t) = a_{21}u_1(t) + a_{23}u_2(t-\tilde{\tau}) + \displaystyle\sum_{i+i+l\geq2}\frac{1}{i!j!l!}f_{ijl}^{(2)}u_1^i(t)u_2^j(t)u_2^l(t-\tilde{\tau}) \end{cases} \qquad (10)$$

where

$$f_{ij}^{(1)} = \left.\frac{\partial^{i+j}f^{(1)}}{\partial x^i\partial y^j}\right|_{(x^*,y^*)}, \quad f_{ijl}^{(2)} = \left.\frac{\partial^{i+j+l}f^{(2)}}{\partial x^i\partial y^j\partial y^l(t-\tilde{\tau})}\right|_{(x^*,y^*,y^*)}, \quad i,j,l \geq 0. \qquad (11)$$

here $f^{(1)}$ and $f^{(2)}$ are defined in (2). For the simplicity of notations, we rewrite (10) as $\dot{u}(t) = L_\mu(u_t) + f(\mu, u_t)$, where $u(t) = (u_1(t), u_2(t))^T \in R^2$, $u_t(0) \in C$ is defined by $u_t(0) = u(t+0)$, and $L_\mu : C \rightarrow R$, $f : R \times C \in R$ are given by

$$L_\mu\phi = \begin{pmatrix} a_{11} & a_{12} \\ a_{21} & 0 \end{pmatrix}\begin{pmatrix} \phi_1(0) \\ \phi_2(0) \end{pmatrix} + \begin{pmatrix} 0 & 0 \\ 0 & a_{23} \end{pmatrix}\begin{pmatrix} \phi_1(-\tilde{\tau}) \\ \phi_2(-\tilde{\tau}) \end{pmatrix} \qquad (12)$$

and

$$f(\mu,\phi) = \begin{pmatrix} \sum_{i+j\geq2} \dfrac{1}{i!\,j!} f_{ij}^{(1)}\phi_1^i(0)\phi_2^j(0) \\[4mm] \sum_{i+j+l\geq2} \dfrac{1}{i!\,j!\,l!} f_{ijl}^{(2)}\phi_1^i(0)\phi_2^j(0)\phi_2^l(-\tilde{\tau}) \end{pmatrix} \tag{13}$$

respectively. By Riesz representation theorem, there exists a function $\eta(\theta,\mu)$ of bounded variation for $\theta \in [-\tilde{\tau},0]$ such that

$$L_\mu\phi = \int_{-\tilde{\tau}}^{0} d\eta(\theta,0)\phi(\theta), \quad for\ \phi \in C \tag{14}$$

In fact, we can choose

$$\eta(\theta,\mu) = \begin{pmatrix} a_{11} & a_{12} \\ a_{21} & 0 \end{pmatrix}\delta(\theta) - \begin{pmatrix} 0 & 0 \\ 0 & a_{23} \end{pmatrix}\delta(\theta+\tau) \tag{15}$$

where δ is the Dirac delta function:

$$\delta(\theta) = \begin{cases} 0, \theta \neq 0 \\ 1, \theta = 0 \end{cases}$$

For $\phi \in C^1([-\tilde{\tau},0],R^2)$, define $A(\mu)\phi = \begin{cases} \dfrac{d\phi(\theta)}{d\theta}, & \theta \in [-\tilde{\tau},0), \\[3mm] \displaystyle\int_{-\tilde{\tau}}^{0} d\eta(\mu,s)\phi(s), & \theta = 0. \end{cases}$

and $R(\mu)\phi = \begin{cases} 0, & \theta \in [-\tilde{\tau},0), \\ f(\mu,\phi), & \theta = 0. \end{cases}$

The system (11) is equivalent to

$$\dot{u}(t) = A(\mu)u_t + R(\mu)u_t \tag{16}$$

where $x_t(\theta) = x(t+\theta)$, for $\theta \in [-\tilde{\tau},0)$.

For $\psi \in C^1([0,\tilde{\tau}],(R^2)^*)$, define $A^*\psi(s) = \begin{cases} -\dfrac{d\psi(s)}{ds}, & s \in (0,\tilde{\tau}], \\[3mm] \displaystyle\int_{-\tilde{\tau}}^{0} d\eta^T(t,0)\psi(-t), & s = 0. \end{cases}$

and a bilinear inner product

$$\langle \psi(s),\phi(\theta) \rangle = \bar{\psi}(0)\phi(0) - \int_{-\tilde{\tau}}^{0}\int_{\xi=0}^{\theta} \bar{\psi}(\xi-\theta)d\eta(\theta)\phi(\xi)d\xi \tag{17}$$

where $\eta(\theta) = \eta(\theta,0)$, Then $A(0)$ and $A^*(0)$ are adjoint operators. By discussions in Section 2 and foregoing assumption, we know that $\pm i\omega_0$ are eigenvalues of $A(0)$. Thus, they are also eigenvalues of $A^*(0)$. We first need to compute the eigenvector of $A(0)$ and $A^*(0)$ corresponding to $i\omega_0$ and $-i\omega_0$, respectively. Suppose that $q(\theta) = (1,\rho)^T e^{i\omega_0\theta}$ is the eigenvector of $A(0)$ corresponding to $i\omega_0$. Then $A(0)q(\theta) = i\omega_0 q(\theta)$. It follows from the definition of $A(0)$, (14) and (15) that

$$\begin{pmatrix} a_{11}+i\omega_0 & a_{21} \\ a_{12} & a_{23}e^{i\omega_0\tilde{\tau}}+i\omega_0 \end{pmatrix}(q(0))^T = \begin{pmatrix} 0 \\ 0 \end{pmatrix}.$$

78

We therefore derive that $q(0) = (1, \rho)^T = (1, (i\omega_0 - a_{11}/a_{12}))^T$. On the other hand, suppose that $q^*(s) = D(1, \sigma)e^{i\omega_0 s}$ is the eigenvector of corresponding to $-i\omega_0$. From the definition of $A^*(0)$, (14) and (15) we have

$$\begin{pmatrix} a_{11} + i\omega_0 & a_{21} \\ a_{12} & a_{23}e^{i\omega_0 \tilde{\tau}} + i\omega_0 \end{pmatrix} (q(0))^T = \begin{pmatrix} 0 \\ 0 \end{pmatrix}$$

Which yields $q^*(0) = D(1, \sigma) = D\left(1, -\dfrac{a_{11} + i\omega_0}{a_{12}}\right)$.

In order to assure $\langle q^*(s), q(\theta)\rangle = 1$, we need to determine the value of D. From (17), we have

$$\langle q^*(s), q(\theta)\rangle$$
$$= \bar{D}\left\{ (1, \bar{\sigma})(1, \rho)^T - \int_{-\tilde{\tau}}^{0}\int_{\xi=0}^{\theta}(1, \bar{\sigma})e^{-i\omega_0(\xi-\theta)}d\eta(\theta)(1, \rho)^T e^{i\omega_0 \xi}d\xi \right\}$$
$$= \bar{D}\left\{ 1 + \rho\bar{\sigma} - \int_{-\tilde{\tau}}^{0}(1, \tilde{\sigma})\theta e^{i\omega_0 \theta}d\eta(\theta)(1, \rho)^T \right\} = \bar{D}\left\{ 1 + \rho\bar{\sigma} + \tilde{\tau}a_{23}\rho\bar{\sigma}e^{-i\omega_0 \tilde{\tau}} \right\}$$
$$= \bar{D}\left\{ 1 + \rho\bar{\sigma} + \tilde{\tau}a_{23}\rho\bar{\sigma}e^{-i\omega_0 \tilde{\tau}} \right\}$$

Thus, we can choose $D = \dfrac{1}{1 + \bar{\rho}\sigma + a_{23}\bar{\tau}\rho\sigma e^{i\omega_0 \tilde{\tau}}}$, $\langle q^*(s), q(\theta)\rangle = 1$, such that $\langle q^*(s), q(\theta)\rangle = 1$, $\langle q^*(s), \bar{q}(\theta)\rangle = 0$.

In the remainder of this section, we use the same notations as in B. Hassard et al. 1981. We first compute the coordinates to describe the center manifold C_0 at $\mu = 0$. Let u_t be the solution of Eq. (11) with $\mu = 0$. Defined

$$z(t) = \langle q^*, u_t \rangle, \quad W(t, \theta) = u_t, \theta) - 2\mathrm{Re}\,\{z(t)q(\theta)\} \tag{18}$$

On the center manifold C_0 we have $W(t, \theta) = W\big(z(t), \bar{z}(t), \theta\big)$, where $W(z, \bar{z}, \theta) = W_{20}(\theta)\,z^2/2 + W_{11}(\theta)z\bar{z} + W_{02}\,\bar{z}^2/2 + \cdots$, z and $\bar{z}$ are local coordinates for center manifold C_0 in the direction of q^* and $\bar{q}^*$. Note that W is real if u_t is real. We consider only real solutions. For the solution $u_t \in C_0$ of (11), since $\mu = 0$, we have

$$\dot{z} = i\omega_0 z + \big\langle \bar{q}^*(\theta), f\big(\theta, W(z, \bar{z}, \theta) + 2\mathrm{Re}\{zq(\theta)\}\big)\big\rangle$$
$$= i\omega_0 z + \bar{q}^*(\theta)f\big(0, W(z, \bar{z}, \theta) + 2\mathrm{Re}\{zq(\theta)\}\big)$$
$$= i\omega_0 z + \bar{q}^*(\theta)f\big(0, W(z, \bar{z}, \theta) + 2\mathrm{Re}\{zq(\theta)\}\big) \tag{19}$$
$$\overset{def}{=} i\omega_0 z + \bar{q}^*(0)f_0(z, \bar{z})$$

We rewrite (19) as $\dot{z} = i\omega_0 z + g(z, \bar{z})$. With

$$g(z, \bar{z}) = \bar{q}^*(0)f_0(z, \bar{z}) = g_{20}\frac{z^2}{2} + g_{11}z\bar{z} + g_{02}\frac{\bar{z}^2}{2} + g_{21}\frac{z^2\bar{z}}{2} + \cdots \tag{20}$$

Noting that $u_t(\theta) = (u_{1t}(\theta), u_{2t}(\theta)) = W(t, \theta) + zq(\theta) + \bar{z}\bar{q}(\theta)$ and $q(\theta) = (1, \rho)^T e^{i\omega_0 \theta}$, we have

$$u_{1t}(0) = z + \bar{z} + W_{20}^{(1)}(0)\frac{z^2}{2} + W_{11}^{(1)}(0)z\bar{z} + W_{02}^{(1)}(0)\frac{\bar{z}^2}{2} + \cdots,$$

$$u_{2t}(0) = \rho z + \bar{\rho}\bar{z} + W_{20}^{(2)}(0)\frac{z^2}{2} + W_{11}^{(2)}(0)z\bar{z} + W_{02}^{(2)}(0)\frac{\bar{z}^2}{2} + \cdots,$$

$$u_{2t}(-\tilde{\tau}) = e^{-i\omega_0 \tilde{\tau}}z + e^{i\omega_0 \tilde{\tau}}\bar{z} + W_{20}^{(2)}(-\tilde{\tau})\frac{z^2}{2} + W_{11}^{(2)}(-\tilde{\tau})z\bar{z} + W_{02}^{(2)}(-\tilde{\tau})\frac{\bar{z}^2}{2} + \cdots.$$

79

Thus, it follows from (13) and (20) that

$$g(z,\bar{z}) = \bar{q}^*(0)f_0(z,\bar{z}) \; = \bar{D}(1,\bar{\sigma}) \begin{pmatrix} \sum\limits_{i+j\geq 2} \dfrac{1}{i!\,j!} f_{ij}^{(1)} u_{1t}^i(0) u_{2t}^j(0) \\[2mm] \sum\limits_{i+j+l\geq 2} \dfrac{1}{i!\,j!\,l!} f_{ijl}^{(2)} u_{1t}^i(0) u_{2t}^j(0) u_{2t}^l(-\tilde{\tau}) \end{pmatrix}$$

$$= \bar{D}\left(\frac{A_{20}}{2} z^2 + A_{11} z\bar{z} + \frac{A_{02}}{2} \bar{z}^2 + A_{21} z^2\bar{z} + \cdots\right)$$

where $A_{20} = F_1 + F_2\bar{\sigma}, A_{11} = F_3 + F_4\bar{\sigma},$

$$F_1 = f_{20}^{(1)} + 2f_{11}^{(1)}\rho + f_{02}^{(1)}\rho^2,$$
$$F_2 = 2\rho f_{110}^{(2)} + 2f_{101}^{(2)}e^{-i\omega_0\tilde{\tau}} + \rho^2 f_{020}^{(2)} + f_{200}^{(2)},$$
$$F_3 = f_{20}^{(1)} + 2\mathrm{Re}\{\rho\} f_{11}^{(1)} + \rho\bar{\rho} f_{02}^{(1)},$$
$$F_4 = 2\mathrm{Re}\{\rho\} f_{110}^{(2)} + 2f_{101}^{(2)}\mathrm{Re}\{e^{-i\omega_0\tilde{\tau}}\} + \rho\bar{\rho} f_{020}^{(2)} + f_{200}^{(2)},$$
$$A_{02} = f_{20}^{(1)} + 2\bar{\rho}f_{11}^{(1)} + f_{02}^{(1)}\bar{\rho}^2 + 2\bar{\sigma}\bar{\rho}f_{110}^{(2)} + 2\bar{\sigma}f_{101}^{(2)}e^{i\omega_0\tilde{\tau}} + \bar{\sigma}\bar{\rho}^2 f_{020}^{(2)} + \bar{\sigma}f_{200}^{(2)},$$
$$A_{21} = \left(\frac{1}{2}f_{20}^{(1)}\left(W_{20}^{(1)}(0) + 2W_{11}^{(1)}(0)\right)\right) + f_{11}^{(1)}\left(\frac{1}{2}W_{20}^{(2)}(0) + \frac{1}{2}\bar{\rho}W_{20}^{(1)}(0) + W_{11}^{(2)}(0) + \rho W_{11}^{(1)}(0)\right)$$

$$+ \frac{1}{2}f_{02}^{(1)}\left(2\rho W_{11}^{(2)}(0) + \frac{1}{2}\bar{\rho}W_{20}^{(2)}(0)\right) + \bar{\sigma}f_{110}^{(2)}\left(W_{11}^{(2)}(0) + \rho W_{11}^{(1)}(0) + \frac{1}{2}\bar{\rho}W_{20}^{(1)}(0) + \frac{1}{2}W_{20}^{(2)}(0)\right)$$

$$+ \bar{\sigma}f_{101}^{(2)}(W_{11}^{(2)}(-\tilde{\tau}) + W_{11}^{(1)}(0)e^{-i\omega_0\tilde{\tau}} + W_{20}^{(1)}(0)e^{i\omega_0\tilde{\tau}} + W_{20}^{(2)}(-\tilde{\tau}))$$

$$+ \frac{1}{2}\bar{\sigma}f_{020}^{(2)}(2\rho W_{11}^{(2)}(0) + \bar{\rho}W_{20}^{(2)}(0)) + \frac{1}{2}\bar{\sigma}f_{200}^{(2)}(2W_{11}^{(1)}(0) + W_{11}^{(1)}(0))$$

Comparing the coefficients in (20), we get

$$g_{20} = A_{20}\bar{D}, g_{11} = A_{11}\bar{D}, g_{02} = A_{02}\bar{D}, g_{21} = A_{21}\bar{D} \tag{21}$$

We now compute $W_{20}(\theta)$ and $W_{11}(\theta)$. It follows from (16) and (18) that

$$\dot{W} = \dot{u}_t - \dot{z}q - \dot{\bar{z}}\bar{q}$$
$$= \begin{cases} AW - 2\mathrm{Re}\{\bar{q}^*(0)f_0 q(\theta)\}, \theta \in (0, \tilde{\tau}], \\ AW - 2\mathrm{Re}\{\bar{q}^*(0)f_0 q(\theta)\} + f_0, \theta = 0, \end{cases} \tag{22}$$
$$\overset{def}{=} AW + H(z,\bar{z},\theta)$$

where

$$H(z,\bar{z},\theta) = H_{20}(\theta)\frac{z^2}{2} + H_{11}(\theta)z\bar{z} + H_{02}(\theta)\frac{\bar{z}^2}{2} + \cdots \tag{23}$$

On the other hand, on C 0 near the origin

$$\dot{W} = W_z\dot{z} + W_{\bar{z}}\dot{\bar{z}}. \tag{24}$$

We derive from (22)–(24) that

$$(A - 2i\omega_0)W_{20}(\theta) = -H_{20}(\theta), \; AW_{11}(\theta) = -H_{11}(\theta), \; \cdots \tag{25}$$

80

It follows from (20) and (22) that for $\theta \in [-\tilde{\tau}, 0)$

$$H(z, \tilde{z}, \theta) = -\bar{q}(0) f_0 q(\theta) - q^*(0) \bar{f}_0 \bar{q}(\theta) = -gq(\theta) - \bar{g}\bar{q}(\theta) \tag{26}$$

Comparing the coefficients in (23) gives that for $\theta \in [-\tilde{\tau}, 0)$

$$H_{20}(\theta) = -g_{20}q(\theta) - \bar{g}_{02}\bar{q}(\theta) \tag{27}$$

$$H_{11}(\theta) = -g_{11}q(\theta) - \bar{g}_{11}\bar{q}(\theta) \tag{28}$$

We derive from (25) and (27) and the definition of A that

$$\dot{W}_{20}(\theta) = 2i\omega_0 W_{20}(\theta) + g_{20}q(\theta) + \bar{g}_{02}\bar{q}(\theta)$$

Noting that $q(\theta) = q(0)e^{i\omega_0\theta}$, it follows that

$$W_{20}(\theta) = \frac{ig_{20}}{\omega_0}q(0)e^{i\omega_0\theta} + \frac{i\bar{g}_{02}}{3\omega_0}\bar{q}(0)e^{-i\omega_0\theta} + E_1 e^{2i\omega_0\theta} \tag{29}$$

where $E_1 = (E_1^{(1)}, E_1^{(2)}) \in R^2$ is a constant vector.
Similarly, from (25) and (28), we can obtain

$$W_{11}(\theta) = -\frac{ig_{11}}{\omega_0}q(0)e^{i\omega_0\theta} + \frac{i\bar{g}_{11}}{\omega_0}\bar{q}(0)e^{-i\omega_0\theta} - E_2 \tag{30}$$

where $E_1 = (E_1^{(1)}, E_1^{(2)}) \in R^2$ is also a constant vector.
In what follows, we seek appropriate E_1 and E_2. From the definition of A and (25), we obtain

$$\int_{-\tilde{\tau}}^0 d\eta(\theta) W_{20}(\theta) = 2i\omega_0 W_{20}(0) - H_{20}(0) \tag{31}$$

$$\int_{-\tilde{\tau}}^0 d\eta(\theta) W_{11}(\theta) = -H_{11}(0) \tag{32}$$

where $\eta(\theta) = \eta(0, \theta)$. From (22), it follows that

$$H_{20}(0) = -g_{20}q(0) - \bar{g}_{02}\bar{q}(0) + \begin{pmatrix} F_1 \\ F_2 \end{pmatrix} \tag{33}$$

$$H_{11}(0) = -g_{11}q(0) - \bar{g}_{11}\bar{q}(0) + \begin{pmatrix} F_3 \\ F_4 \end{pmatrix} \tag{34}$$

Substituting (29) and (33) into (31) and noticing that

$$\left(i\omega_0 I - \int_{-\tilde{\tau}}^0 e^{i\omega_0\theta} d\eta(\theta) \right) q(0) = 0,$$

$$\left(i\omega_0 I + \int_{-\tilde{\tau}}^0 e^{-i\omega_0\theta} d\eta(\theta) \right) \bar{q}(0) = 0,$$

we obtain $\left(2i\omega_0 I - \int_{-\tilde{\tau}}^0 e^{2i\omega_0\theta} d\eta(\theta) \right) E_1 = \begin{pmatrix} F_1 \\ F_2 \end{pmatrix}$ which lead to

$$\begin{pmatrix} 2i\omega_0 - a_{11} & -a_{12} \\ -a_{21} & 2i\omega_0 - a_{23}e^{-2i\omega_0\tilde{\tau}} \end{pmatrix} E_1 = \begin{pmatrix} F_1 \\ F_2 \end{pmatrix}$$

81

It follows that

$$E_1^{(1)} = \frac{1}{A}\begin{vmatrix} F_1 & -a_{12} \\ F_2 & 2i\omega_0 - a_{23}e^{-2i\omega_0\tilde{\tau}} \end{vmatrix},$$

$$E_1^{(2)} = \frac{1}{A}\begin{vmatrix} 2i\omega_0 - a_{11} & F_1 \\ -a_{21} & F_2 \end{vmatrix}$$

where $A = \begin{vmatrix} 2i\omega_0 - a_{11} & -a_{12} \\ -a_{21} & 2i\omega_0 - a_{23}e^{-2i\omega_0\tilde{\tau}} \end{vmatrix}$.

Similarly, substituting (30) and (34) into (32), we get

$$\begin{pmatrix} -a_{11} & -a_2 \\ -a_{21} & -a_{23} \end{pmatrix} E_2 = \begin{pmatrix} F_3 \\ F_4 \end{pmatrix}$$

and hence $E_2^{(1)} = \dfrac{1}{B}\begin{vmatrix} F_3 & -a_{12} \\ F_4 & -a_{23} \end{vmatrix}$, $E_2^{(2)} = \dfrac{1}{B}\begin{vmatrix} -a_{11} & F_3 \\ -a_{21} & F_4 \end{vmatrix}$, where $B = \begin{vmatrix} -a_{11} & -a_{12} \\ -a_{21} & -a_{23} \end{vmatrix}$.

Then, g_{21} can be determined by the parameters and delay. Thus, we can compute the following quantities:

$$c_1(0) = \frac{i}{2\omega_0}\left(g_{11}g_{20} - 2|g_{11}|^2 - \frac{|g_{02}|^2}{3} \right) + \frac{g_{21}}{2},$$

$$\mu_2 = -\frac{\mathrm{Re}\{c_1(0)\}}{\mathrm{Re}\{\lambda'(\tilde{\tau})\}}, \quad \beta_2 = 2\mathrm{Re}\{c_1(0)\},$$

$$T_2 = -(\mathrm{Im}\{c_1(0)\} + \mu_2\mathrm{Im}\{\lambda'(\tilde{\tau})\})/\omega_0.$$

Theorem 3.1

i. μ_2 determines the direction of Hopf bifurcation. If $\mu_2 > 0(<0)$, then the Hopf bifurcation is supercritical (subcritical), and the bifurcating periodic solution exist for $\tau > \tau_0(<\tau_0)$;
ii. β_2 determines the stability of bifurcated periodic solutions. If $\beta_2 > 0(<0)$, the bifurcated periodic solutions are unstable (stable);
iii. T_2 determines the period of the bifurcating periodic solutions. If $T_2 > 0(<0)$, the period increase (decrease).

4 NUMERICAL SIMULATIONS

In this section, we present some numerical results of simulating system (1) at $r = 1.5$, $K = 5$, $a = 1$, $c = 0.6$, $p = 0.6$, q 0.5, $m = 0.5$, the system (1) has unique positive equilibrium point $E^* = (1.3365, 0.8019)$. From the results in Section 3, we evaluate that $\omega_0 = 0.5606$, $\tau_0 = 3.1204$, and the positive equilibrium point E^* is asymptotically stable when $\tau \in [0, \tau_0) = [0, 3.1204)$ (see Fig. 1) and unstable. when $\tau > \tau_0$ (see Fig. 2). By the theory of B. Hassard et al.1981, as it is discussed in previous section, we may also determine the direction of the Hopf bifurcation and the other properties of bifurcating periodic solutions. From the formulaes in Section 3 we compute the values of μ_2, β_2 and T_2 as $c_1(0) = 8.9345 - 25.3028\mathrm{i}$, $\mu_2 = -197.5529$, $\beta_2 = 17.8691$, $T_2 = 18.0995$. In the light of Theorem 3.1, since $\mu_2 < 0$, the Hopf bifurcation of system (1) occurring at $\tau_0 = 3.1204$ is subcritical and the bifurcating periodic solution exist when τ crosses τ_0 to the left, also since $\beta_2 > 0$ the bifurcating periodic solution is unstable (see Fig. 2).

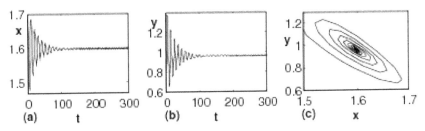

Figure 1. The trajectory graph in t – x (see (a)), t – y (see (b)) and x – y (see (c)) plan of system (1) with r = 1.5, K = 5, a = 1, c = 0.6, p = 0.6, q = 0.5, m = 0.5, τ = 2.8 and initial data x(t) = 1.6, y(t) = 0.5.

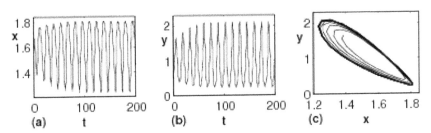

Figure 2. The trajectory graph in t – x (see (a)), t – y (see (b)) and x – y (see (c)) plan of system (1) with τ = 3.8, other parameters are same as Figure 1.

ACKNOWLEDGMENT

This work was supported by the Jiangxi province Natural Science Foundation of China.

REFERENCES

Hailing Wang, Weiming Wang, The dynamical complexity of a Ivlev-type prey-predator system with impulsive effect, Chaos, Solitons and Fractals, vol.38, 2008: 1168–1176.

Hassard B., N. Kazarinoff, Y.H. Wan. Theory and applications of Hopf bifurcation. Cambridge: Cambridge University Press.1981.

Hui Zhu, Zuoliang Xiong, Xin Wang, Analysis of eco-epidemiological model with time delay, Rocky Mountain Journal of Mathematics, vol. 38.2008:1877–1886.

Jack K. Hale, Sjoerd M, Verduyn Lunel. Introduction to functional differential equations. New York:Applied Mathematical Sciences Volume 99.1993.

Jack K. Hale. Theory of Functional Differential Equations. New York: Springer-Verlag.1997.

Zuoliang Xiong, Ying. Xue, Shunyi Li, A food chain system with Holling IV functional responses and impulsive effect. Int. J. Biomath., vol. 1, 2008:361–375.

Bioinformatics and Biomedical Engineering – Chou & Zhou (Eds)
© 2016 Taylor & Francis Group, London, ISBN 978-1-138-02784-8

Finding sorting traces of reversals in the presence of hurdles

B. Tripathi & Amritanjali

Department of Computer Science and Engineering, Birla Institute of Technology, Mesra, Ranchi,
Jharkhand, India

ABSTRACT: Reversal is one of the most commonly observed evolutionary events that affects the order of genes in a single chromosome. By comparing the order of shared genes in two genomes that have evolved by reversals events, their evolutionary scenario is reconstructed. The problem of finding the most plausible sequence of reversals that changed the genomes is complicated by the fact that for most of the cases, several solutions are generated using the computational methods. Much work has been done to reduce the time and space complexities of these approaches. Another issue in the implementation is the problem of hurdles. We modify the existing approaches to enumerate the possible solutions of the sorting by reversals problem in the presence of hurdles and compare their performance.

1   INTRODUCTION

Evolutionary scenario between two species is described by the sequence of rearrangements that changed the arrangement of the shared markers (genes or segment of genes) in their genomes. From the parsimony hypothesis, the most plausible evolutionary scenario involves a minimum number of rearrangements. Reversal is the most common rearrangement operation (McLysaght et al., 2000) that cuts a chromosome at two points and reverses the order and orientation of the genes in the segment between the two points. For two genomes that have evolved by reversals, finding the minimal sequence of reversals that can transform the sequence of shared markers in one genome into that of another is the sorting by reversals problem. Analyzing these mutations can reveal the actual scenario that might have occurred during a period of time. BaobabLuna (Braga, 2009) is a freely available software package that can list all possible sorting sequences of reversals. It can also group the sorting sequences into an equivalence class and represent the class by a unique sequence while counting the total number of sorting sequences in each class. Baudet and Dias (2010) proposed an improved method and generated the solutions in the depth-first manner. Another method to improve the performance is to group the solutions according to the permutations they generate at every step. This reduces the amount of computation as different intermediate solutions resulting in the same permutation are not processed separately (Badr et al., 2011). All these approaches are based on the Hannenhalli and Pevzner (1995) model of the breakpoint graph. However, if bad components are present in the breakpoint graph, the current implementations fail to produce solutions. Braga and Stoye (2010) used a different model employing the Double Cut and Join (DCJ) operation, which is free of bad components. However, finding all possible solutions using this model is still an open problem. In this paper, we modify the existing approaches based on the breakpoint graph model to enumerate the solution space of the sorting by reversals problem in the presence of hurdles and compare their performance.

2 GETTING STARTED

2.1 *Notations*

The order of shared genes in one genome, say target, is represented by an identity permutation π_T of integers 1 to n, where n is the number of shared genes. The relative order and orientation of these genes in the chromosome of the other genome, say source, is represented by a signed permutation of n integers, $\pi = (\pi_1 \dots \pi_n)$. Application of a reversal operation ρ on an interval $[i, j]$ of a permutation π is defined as

$$\pi \cdot \rho = \left(\pi_1 \pi_2 \dots \pi_{i-1} - \pi_j - \pi_{j-1} \dots - \pi_{i+1} - \pi_i \pi_{j+1} \dots \pi_{n-1} \pi_n \right) \tag{1}$$

The reversal ρ is denoted by the unsigned sorted list of elements of π in the interval $[i, j]$. An i-sequence of reversals $\rho_1 \rho_2 \dots \rho_i$ transforms the permutation π into a permutation π' if $\pi \cdot \rho_1 \rho_2 \dots \rho_i = \pi'$. The length of the shortest sequence of reversals that sorts a permutation π into π_T is called the reversal distance (d).

2.2 *Enumerating solution space of sorting by reversals problem*

The BFA method (Braga et al., 2007) enumerates the solution space of the sorting by reversals problem in the breadth-first manner. The set of optimal reversals that can be applied to the input permutation is called as 1-sequences of optimal reversal. On applying them to the source permutation, we get new permutations that are one step closer to the target permutation. For each of these permutations, the next 1-sequences of optimal reversal are computed. When these 1-sequences are concatenated to their corresponding preceding sequence, we get 2-sequences of optimal reversals. Finally, after d-iterations, all the d-sequences of optimal reversals are obtained. Each sequence is one solution in the solution space. The solution space is usually huge and contains many equivalent sorting sequences, so for compact representation, these intermediate sequences are merged into the same equivalence class, using the concept of traces (Diekert and Rozenberg, 1995). The baobabLuna program (Braga, 2009) outputs one trace per class along with the number of sorting sequences in each class. The process of enumerating the sorting traces is similar. At any step, if two i-traces are similar, then they are merged and represented by a single i-trace. However, a running count is maintained to keep track of the number of similar traces merged, which gives the total number of sorting sequences in each trace generated.

In a recent paper, Baudet and Dias (2010) observed that only those i-traces are new in which the next reversal appends at the end of the previous $(i-1)$-trace. All the remaining i-traces can be discarded without checking for similarity with the existing traces. This saves the time spent in comparing a newly formed i-trace with all the existing i-traces. However, by using this concept, it is not possible to count the number of sorting sequences in each trace. To make it more memory-efficient, the d-traces are generated using the Depth-First Approach (DFA) instead of the breadth-first approach. The initial set of 1-sequences is processed in lexicographic order. Starting with first 1-sequence of reversal, corresponding 2-sequences are generated. Continuing like this up to the d-level, the first d-trace is obtained, and it backtracks to the next reversal of the previous level. As all the reversals at each level get processed, it backtracks to the previous level. Finally, it comes back to the first level and proceeds to the next reversal at that level. In this way, all the d-traces are obtained in the depth-first manner. Both the approaches, BFA and DFA, use the breakpoint graph for calculating the reversal distance and identifying the sorting reversals at each step.

2.3 *Breakpoint graph*

The breakpoint graph is constructed from the source and the target permutations. It contains two types of edges. Gray edge connects vertices corresponding to the adjacent genes

86

in the target, and black edge connects vertices corresponding to the adjacent genes in the source. Figure 1(a) describes an example of the breakpoint graph. The arcs are the gray edges, and the straight lines along the axis are the black edges. A cycle is a closed path of alternate gray and black edges, and a component is a set of overlapping cycles. Cycles are said to overlap when their gray edge(s) intersect. A trivial cycle consists of a single pair of gray and black edges. If two adjacent genes in the target have opposite signs, then the corresponding gray edge is called oriented. In addition, the cycle or the component containing one or more oriented gray edge(s) is also oriented. A benign component is either oriented or trivial. Hurdle is an unoriented component that does not separate other unoriented components. If removing the hurdle transforms a non-hurdle into hurdle, then it is called the super hurdle, otherwise simple. Such a non-hurdle is said to be protected. A fortress is present in the graph if all the hurdles are super hurdles and they are odd in number.

Cycles and their interleaving structure give clues to past rearrangement events. By the Hannenhalli and Pevzener duality theorem, the reversal distance is computed as: $d = n + 1 - c + h + f$. Here, n is the size of the permutation, c is the number of cycles, h is the number of hurdles, and the parameter f is 1 if the fortress is present in the graph, otherwise 0. Performing a reversal operation can result in splitting a cycle, merging two cycles or no change in the cycle. Let Δc denote the change in the number of cycles after performing a rearrangement operation, then $\Delta c = \{-1, 0, +1\}$. Based on the value of Δc, a reversal operation is classified as split ($\Delta c = 1$), neutral ($\Delta c = 0$) or joint ($\Delta c = -1$). The reversal operation is valid if $\Delta d = -1$, i.e. $\Delta(h + f - c) = -1$. An optimal (minimal) sorting sequence consists of only valid reversals.

3   HANDLING HURDLES AND FORTRESS

To make the implementation simpler, the existing program fails to produce the solution if the input permutation has a hurdle or fortress. In the proposed work, Siepel's method (2002) is used for finding the possible positions of cut points for applying valid reversals in the presence of hurdles and fortress.

3.1   Hurdle graph

Hurdle graph is constructed for the unoriented components of the breakpoint graph in order to determine their type as simple hurdle or super hurdle or protected non-hurdle. An example of a hurdle graph is given in Figure 1(b).

The algorithm to identify the hurdles and the protected non-hurdles from the hurdle graph is given below:

1. For every vertex in the hurdle graph
   a. If it has degree 2 and belongs to a cycle or has a degree less than 2 and does not belong to a cycle, then the vertex is marked as simple hurdle. Else if its degree is 1, then check

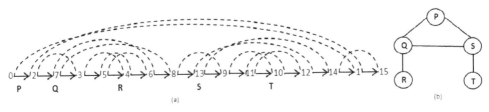

(a)

(b)

Figure 1.   (a) Breakpoint graph for a signed (source) permutation (2, 7, 3, 5, 4, 6, 8, 13, 9, 11, 10, 12, 14, 1, 15). The dotted lines show the gray edges. The arrows show the direction in which the black edges are traversed when traversing the edges in each cycle. The graph has 5 components P, Q, R, S and T. (b) Its hurdle graph, where P is a simple hurdle, Q and S are protected non-hurdles, and R and T are super hurdles.

the degree of its adjacent vertices, if any of them has degree 3 and belongs to a cycle or has degree 2 and does not belong to a cycle, then the vertex is marked as super hurdle.

b. If it has a degree more than 2 and belongs to a cycle or has degree 2 and does not belong to a cycle, then it is not a hurdle and the vertex is marked as protected non-hurdle.

3.2   Sorting reversals to eliminate unoriented components

There are two types of reversals that can overcome a hurdle and fortress, namely the cut reversal and the merge reversal.

Cut Reversal: the extremities of the reversal are in the same cycle of the unoriented component. It is neutral and transforms the unoriented component into an oriented component. It eliminates the hurdle or fortress but does not change the number of cycles, such that $\Delta h + \Delta f = -1$. The possible positions of cut points are determined as follows:

1. Fortress is absent ($f = 0$)
   Perform cut on every pair of black edges of the cycles of the simple hurdle, so that it can be converted into oriented component. The number of simple hurdles must be either more than 1 or can be 1 when the number of super hurdle is even, to avoid the formation of the fortress. Here, Δh is reduced by 1 and Δf remains unchanged, so that ($\Delta h + \Delta f$) is reduced by 1.
2. Fortress is present ($f = 1$)
   Perform cut on all the pairs of black edges of the cycles of the super hurdle or protected non-hurdle, so that the super hurdle can be converted into a simple hurdle component and fortress existing can be resolved. Here, Δf is reduced by 1 and Δh remains unchanged, so that ($\Delta h + \Delta f$) is reduced by 1.

Merge Reversal: the extremities of the reversal are in the cycles of different components, thus rearranging the components, such that the number of cycles and hurdles decrease. This is an example of the joint reversal. Here, $\Delta c = -1$, $\Delta h + \Delta f = -2$, so that $-\Delta c + \Delta h + \Delta f = -1$. The cut points are taken on all pairs of black edges between two components determined as follows:

1. Fortress is absent ($f = 0$)
   a. If the number of simple hurdles is less than or equal to 1, apply reversals between the pairs of:
      i. Super hurdles,
      ii. Super hurdle and benign component with a separating hurdle, and
      iii. Benign component and separating hurdle.
   b. If the number of simple hurdles is 2, apply reversals between the pairs of:
      i. Super hurdles,
      ii. Super hurdle and a benign component with a separating hurdle,
      iii. Benign component and a separating hurdle, and
      iv. Simple hurdles when the number of super hurdles is even.
   c. If the number of simple hurdles is more than 2, apply reversals between the pairs of:
      i. Super hurdles,
      ii. Super hurdle and a benign component with a separating hurdle,
      iii. Benign component and a separating hurdle, and
      iv. Simple hurdles.
2. Fortress is present ($f = 1$)
   a. When a double super hurdle is present, apply reversals between the pairs of:
      i. Super hurdles, which form a double super hurdle, and
      ii. Super hurdle and a benign component with a separating super hurdle, which form a double super hurdle.
   b. When a double super hurdle is not present, apply reversals between the pairs of a super hurdle and a protected non-hurdle.
   c. Apply reversals between the pairs of a super hurdle and a benign component in the same protected non-hurdle.

4   EMPIRICAL RESULTS

The implementation was done using both the breadth-first approach and the depth-first approach. Handling of hurdles was first incorporated into the algorithm proposed by Braga et al. (2007). Then, to improve its efficiency with respect to time and memory, the BFA method was modified. In the modified BFA, the comparison with the existing traces is avoided by using the concept given by Baudet and Dias (see Section 2.3). Next, the DFA method was modified to take care of hurdles, and recursion is used to achieve the depth-first expansion of the solution space (Amritanjali and Sahoo, 2013). A comparative test was carried out with random permutations generated with two test cases, keeping $n = d$ and $n = 2d$. Testing was also done with artificial permutations containing hurdles and fortress. All the tests were carried out on an AMD A-10 2.5 GHz processor with 8 GB RAM running Windows 8.1 64 bit. JVM was allocated with a maximum memory of 1.5 GB (using parameter-Xmx1550 m). The correctness of the implementation has been verified by comparing its output with that of the baobabLuna package for permutations without containing hurdles. The average of the output obtained on these random permutations was taken to get a generalized result. Elapsed time (in seconds) is computed by taking the system's clock value before and after execution by using the currentTimeMillis() method, and the resulting graphs are shown in Figure 2.

The modified BFA and DFA take lesser time than the original BFA. The difference in execution time becomes significantly larger as the size of the permutation and reversal distance increase. The modified BFA and DFA are more or less similar with respect to the time required for computation. Memory analysis was done between the modified BFA and DFA to see the

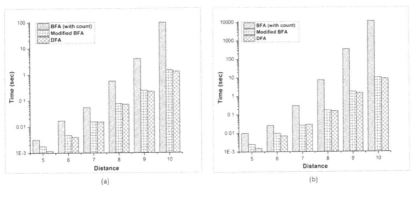

Figure 2.   Average execution time for (a) n = 2d and (b) n = d.

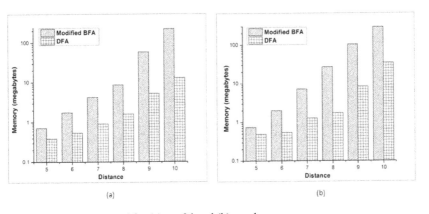

Figure 3.   Average memory used for (a) n = 2d and (b) n = d.

Table 1. Computation results for processing random permutations with hurdles. The time and memory measurements are for DFA.

Input permutation	d	Hurdle	Fortress	Traces	Time (sec)	Memory (bytes)
2, 1, 3, −4, 5, 7, 9, 8, 10, 6, 11, 13, 14, 12	12	Simple, protected non-hurdle	0	172	2.421	903704
2, 7, 3, 5, 4, 6, 8, 13, 9, 11, 10, 12, 14, 1	13	Simple, Super, protected non-hurdle	0	1048	8.790	2893504
17, 1, 3, 8, 4, 6, 5, 7, 9, 11, 13, 12, 14, 10, 15, 2, 16	16	Super, protected non-hurdle	1	1048	120.631	3850136

improvement in memory usage when exploring the solution space in the depth-first manner in comparison with the breadth-first approach. Memory used (in bytes) is computed with the help of a separate thread measuring memory at frequent intervals using the methods totalMemory() and freeMemory(). The graphs for the results obtained are shown in Figure 3.

Memory consumption is more in BFA than in DFA. This is due to the fact that at a time, only one branch is explored and memory is freed before moving to the next branch. Moreover, the plotted graphs show a decrease in the average execution time and used memory as the size of the permutation increases for a given distance, as there is a decrease in the average number of traces. Table 1 gives the output for various permutations with different types of hurdle(s).

5 CONCLUSIONS

Our implementation removes the problem of hurdles and generates the possible evolutionary scenarios that eliminate them. As the implementation is based on Siepel's work, the complete set of sorting traces is obtained. For efficiency, the double super hurdles are not implemented as the chances of their occurrence are negligible. The hurdle version has been implemented using both BFA and DFA approaches. As expected, DFA is more memory efficient than BFA. This version could further be extended to form a parallel version of the algorithm. To reduce the solution space, biological constraints such as the common interval constraint can be applied.

REFERENCES

Amritanjali & Sahoo, G. 2013. Exploring the Solution Space of Sorting By Reversals: A New Approach, International Journal of Information Technology, 3(2): 98–104.
Badr, G., Swenson, K.M. & Sankoff, D. 2011. Listing All Parsimonious Reversal Sequences: New Algorithms and Perspectives. Journal Of Computational Biology, 18(9): 1201–1210.
Baudet, C. & Dias, Z. 2010. An improved algorithm to enumerate all traces that sort a signed permutation by reversals. Proc. ACM Symposium on Applied Computing. 1521–1525.
Braga, M.D.V., Sagot, M., Scornavacca, C. & Tannier, E. 2007. The solution space of sorting by reversals, Proc. Int'l Symp. Bioinformatics Research and Applications (ISBRA 2007). 4463: 293–304.
Braga, M.D.V. 2009. baobabLuna: The solution space of sorting by reversals. Bioinformatics. 25(14): 1833–1835.
Braga M.D.V. & Stoye, J. 2010. The solution space of sorting by DCJ. Journal of Comput. Biol. 17(9): 1145–1165.
Diekert, V. & Rozenberg G. 1995. The Book of Traces. World Scientific: Singapore.
Hannenhalli, S. & Pevzner, P.A. 1995. Transforming cabbage into turnip (polynomial algorithm for sorting signed permutations by reversals). Proc. 27th Ann. ACM Sym on the Theory of Comput., 178–189.
McLysaght, A., Seoighe, C. & Wolfe, K.H. 2000. High frequency of inversions during eukaryote gene order evolution. In Sankoff, D., and Nadeau, J.H., eds., Comparative Genomics, 47–58, Kluwer Academic Press, NY.
Siepel, A.C. 2002. An algorithm to find all sorting reversals. Proc. 6th Ann. Int'l Conf. Comput. Mol. Biol. (RECOMB 2002). 281–290. ACM Press: New York.

Bioinformatics and Biomedical Engineering – Chou & Zhou (Eds)
© 2016 Taylor & Francis Group, London, ISBN 978-1-138-02784-8

Degree distribution of protein-protein interaction networks formed by gene duplication

Z. Ma & H. Xu
Department of Electrical and Computer Engineering, University of Waterloo, Waterloo, ON, Canada

X.G. Wu
Department of Electrical Engineering, Stanford University, Stanford, CA, USA

ABSTRACT: This paper proposes a degree distribution function for protein-protein interaction networks. It is derived from modeling the protein-protein interaction network as a random duplication graph with sparse initial state, where the duplication particularly refers to the gene duplication that produces new proteins and grows the network. This degree distribution reveals some characteristics of protein-protein interaction networks: the majority of nodes are sparsely connected while highly connected proteins also exist; as the growth process continues, more and more highly connected proteins will be produced. Finally, we also show that compared to the widely used scale-free distribution, our degree distribution can fit the experimental data better.

Keywords: protein-protein interaction network; gene duplication; degree distribution function; random duplication graph; scale-free distribution

1   INTRODUCTION

Protein-protein interactions are the physical contacts among proteins due to certain biochemical events. Multiple protein components organized by their protein-protein interactions form up the biological machines that carry out diverse essential biochemical processes. Thus, it is instrumental to understand protein-protein interactions in analyzing cellular functions.

Protein-protein interaction network is the map of protein-protein interactions in a given organism. In the network, proteins are represented as nodes, and an edge exists between two proteins if they can interact with each other. As systems biology advances, development of genome-scale protein-protein interaction networks became possible. To understand how cells and organisms are developed, a comprehensive analysis of the protein-protein interaction networks is of pivotal importance. In this regard, understanding the degree distribution of protein-protein interaction networks has been a major interest for systems biologists, and will be the subject of this paper.

Degree distribution function is defined as $\mathbb{P}(n)$, representing the probability that a randomly chosen protein has n connections, i.e., the percentage of proteins that has n connections in the whole network. Degree distribution characteristic of protein-protein interaction networks comes directly from the networks' growth process. The formation of the networks is a growth process, during which new proteins join the system over a long time period. In this process, new proteins are brought by gene duplication, a major mechanism through which new proteins are generated during molecular evolution (Barabási & Oltvai, 2004). Duplicated genes produce identical proteins that interact with the same protein partners. Therefore, each protein in contact with a duplicated protein gains an extra linkage. Assume gene duplication is purely random. Then highly connected proteins gain a natural advantage—they are more likely to have a link to a duplicated protein than the weakly connected ones. As a result, highly connected proteins are more likely to gain new links, generating proteins with even higher degree.

Since the growth process of the protein-protein interaction networks are similar to the preferential-attachment growth process of scale-free networks (highly connected nodes tend to gain more connections), Barabási and Oltvai claimed that protein-protein interaction networks have a scale-free degree distribution, featuring a polynomial distribution function (Barabási & Oltvai, 2004). This result was well accepted, but it contradicts experimental data. Figure 1(A) is a graphical representation of the DPiM (Drosophila Protein interaction Map) determined by Guruharsha. (Guruharsh et al., 2011), while Figure 1(B) shows the degree distribution of the genome-scale protein-protein interaction networks of Drosophila melanogaster with 7,048 proteins and 20,405 interactions in a log-log plot (Giot et al., 2003).

As shown in Figure 1(B), the degree distribution of DPiM (Drosophila Protein interaction Map) does not follow a scale free distribution which features a straight line in the log-log plot (red line, $\mathbb{P}(n) \sim n^{-1.80}$, the parameter -1.80 is the best fit). Also, Bebek have proved that random duplication graphs will not present a scale-free distribution (Bebek et al., 2006). In a matter of fact, the degree distribution of DPiM shows a concave curve in the log-log plot. Giot proposed that the data can be fitted by a combination of power-law and exponential decay, $\mathbb{P}(n) \sim n^{-\alpha} e^{-\beta n}$ (blue line, in this case $\alpha = 1.20 \pm 0.08$ and $\beta = 0.038 \pm 0.006$, r^2 for the fit is greater than 0.98) (Giot et al., 2003).

However, theoretical result stronger than a fitted curve is needed. We are interested in why protein-protein interaction networks present such a degree distribution pattern, and what the actual degree distribution function is. In this article we model the growth process of protein-protein interaction networks as a random duplication graph, giving a new approach to derive the degree distribution we need. In section 2, we give the definition of a random duplication graph, followed by a theoretical derivation of the degree distribution of random duplication graphs. In section 3, we analyze protein-protein interaction networks as a special case of a random duplication graph where the initial graph is sparse, giving the degree distribution function of protein-protein interaction networks. Also, the properties of the acquired degree distribution function of protein-protein interaction network is analyzed, and a predicted behavior of this growth process is pointed out. Section 4 gives the comparison between our model and experimental data, showing the soundness of our analysis.

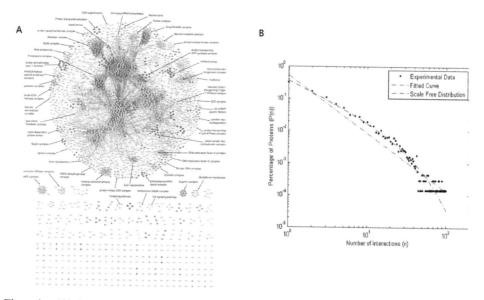

Figure 1. (A) Graphical representation of the DPiM, reused with permission from (Guruharsh et al., 2011). (B) Degree distribution of the genome-scale protein-protein interaction networks of Drosophila melanogaster.

2 DEFINITION AND DEGREE DISTRIBUTION OF RANDOM DUPLICATION GRAPHS

Consider some undirected graph, $G(t_0)$, with t_0 nodes (for simplicity let $t_0 \geq 2$). We start from this initial graph and this initial time t_0, and let time progress in unit of 1. At each time step $t \geq t_0$, choose one node to duplicate uniformly at random, copying all the edges of the original node. For each step of random duplication, one new node is introduced to the graph, so the graph at time t, $G(t)$, would have t nodes. Also, the probability of any node to duplicate at time t is $1/t$. As a result, $G(t)$ is our acquired random duplication graph. A simple illustration of this growth process is shown in Figure 2.

We define $F_t(n)$ as the number of nodes with n connections at time t. Since there are t nodes at time t, the degree distribution function at time t is written as $\mathbb{P}_t(n) = F_t(n)/t$. The information about the initial graph $G(t_0)$ is given as $\{Ft_0(i)$ where $1 \leq i \leq t_0 - 1\}$ (obviously maximum possible degree at time t is $t - 1$). We are interested in the resulting degree distribution function of this growth process. More precisely, we are interested in $\mathbb{E}[\mathbb{P}_t(n)]$, the expectation of resulting degree distribution of $G(t)$, in terms of the information about initial graph, $\{Ft_0(i)$ where $1 \leq i \leq t_0 - 1\}$.

To determine the degree distribution of $G(t)$, we analyzed this growth process. As shown in Figure 3, it is observed that the event {a degree-n node duplicates} will introduce a new node with degree n. Also, the event {a degree-n node's neighbor duplicates} will change this existing degree-n node to a degree-$(n + 1)$ node. And similarly, the event {a degree-$(n - 1)$ node's neighbor duplicates} will change this existing degree-$(n - 1)$ node to a degree-n node.

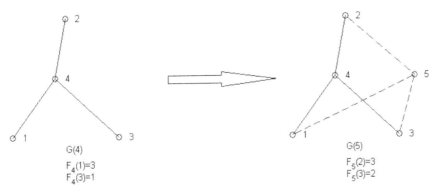

Figure 2. The growth process of random duplication graph. Here node 4 in $G(4)$ is duplicated to produce node 5 in $G(5)$.

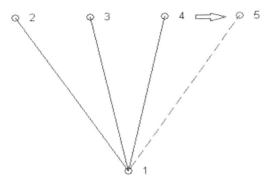

Figure 3. Node 4 duplicates to produce node 5. Node 4 is a degree-1 node itself, as well as a neighbor of a degree-3 node. As a result, duplication of node 4 causes the creation of a new degree-1 node (node 5). Also, node 1 is changed from a degree-3 node to a degree-4 node.

Since every node has a duplication probability of $1/t$, assume no isolated node exists, we obtain the following recursive formula of $\mathbb{E}[F_{t+1}(n)|\{F_t(i)$ where $1 \le i \le t-1\}]$, illustrating the relationship between $G(t+1)$ and $G(t)$.

$$\mathbb{E}[F_{t+1}(n)|\{F_t(i) \text{ where } 1 \le i \le t-1\}] = F_t(n) + \frac{1}{t}F_t(n) - \frac{n}{t}F_t(n) + \frac{n-1}{t}F_t(n-1) \tag{1}$$

The above equation can be solved exactly by writing out each term on the right hand side in terms of earlier time steps.

$$\mathbb{E}[F_t(n)] = \sum_{i=\max\{0, n-t_0+1\}}^{\min\{t-t_0, n-1\}} \frac{[(n-1)(n-2)...(n-i)][(t-n)(t-n-1)...(t_0+i-n+1)]}{(t-1)(t-2)...(t_0+1)(t_0)} \binom{t-t_0}{i} F_{t_0}(n-i) \tag{2}$$

Replacing $n - i$ with a new variable j, we obtain a simpler solution of $\mathbb{E}[F_t(n)]$ in terms of $\{F_{t_0}(i)$ where $1 \le i \le t_0 - 1\}$.

$$\mathbb{E}[F_t(n)] = \sum_{j=\max\{1, n-t+t_0\}}^{\min\{t_0-1, n\}} \frac{\binom{n-1}{j-1}\binom{t-n}{t_0-j}}{\binom{t-1}{t_0-1}} F_{t_0}(j) \tag{3}$$

3 PROTEIN-PROTEIN INTERACTION NETWORKS AS A RANDOM DUPLICATION GRAPH

This model of random duplication graph explicitly depicts the growth process of protein-protein interaction networks. Proteins are represented as nodes, connections between proteins are represented by edges, and gene duplications are represented as random node duplication. Since duplicated genes produce identical proteins that interact with the exact same protein partners, all the edges are copied during node duplication. Consequently, we can model protein-protein interaction network as a special case of random duplication graph.

We take a valid assumption that, at the beginning of the growth process, proteins are sparsely connected. As a result, in $G(t_0)$, we let $F_{t_0}(1) = t_0$ and $F_{t_0}(i) = 0$ for $\forall i \in [2, t_0 - 1]$, i.e., each protein only has one connection. This assumption complies with the situation of the beginning of the biological evolution—a few different proteins gathered together, not a lot of connections were formed. Based on this special initial condition, solution to $\mathbb{E}[F_t(n)]$ from (3) is modified by leaving only the term $j = 1$. The convergence $F_t(n) \to \mathbb{E}[F_t(n)]$ holds if $t_0 \to \infty$, $t - t_0 \to \infty$, and $n = o(t)$.

$$\mathbb{E}[F_t(n)] = \frac{\binom{t-n}{t_0-1}}{\binom{t-1}{t_0-1}} t_0 \quad for 1 \le n \le t - t_0 + 1 \tag{4}$$

To get a simpler expression of the degree distribution function, an upper bound and a lower bound are found for $\mathbb{E}[F_t(n)]$.

$$\left(\frac{t-n-t_0+2}{t-1}\right)^{t_0-2} \cdot t_0 \le \mathbb{E}[F_t(n)] \le \left(\frac{t-n}{t-t_0+1}\right)^{t_0-2} \cdot t_0 \quad for\ t_0 \le n \le t - t_0 + 1 \tag{5}$$

94

These two bounds are asymptotically tight as $t \to \infty$. More precisely,

$$\lim_{t \to \infty}\left[\left(\frac{t-n-t_0+2}{t-1}\right)^{t_0-2}\cdot t_0\right]=\lim_{t \to \infty}\mathbb{E}[F_t(n)]=\lim_{t \to \infty}\left[\left(\frac{t-n}{t-t_0+1}\right)^{t_0-2}\cdot t_0\right] \tag{6}$$

With the help of these bounds, a simple approximation to $\mathbb{E}[F_t(n)]$ is derived.

$$\mathbb{E}[F_t(n)]\approx\left(\frac{t-n}{t-1}\right)^{t_0-2}\cdot t_0 \tag{7}$$

This approximation is sound because this approximation lies within the upper and lower bounds of $\mathbb{E}[F_t(n)]$, and the bounds are asymptotically tight. In mathematical terms, we have the following results between the approximation and the bounds.

$$\left(\frac{t-n-t_0+2}{t-1}\right)^{t_0-2}\cdot t_0 \leq \left(\frac{t-n}{t-1}\right)^{t_0-2}\cdot t_0 \leq \left(\frac{t-n}{t-t_0+1}\right)^{t_0-2}\cdot t_0 \quad for\, t_0 \leq n \leq t-t_0+1 \tag{8}$$

$$\lim_{t \to \infty}\left[\left(\frac{t-n-t_0+2}{t-1}\right)^{t_0-2}\cdot t_0\right]=\lim_{t \to \infty}\left[\left(\frac{t-n}{t-1}\right)^{t_0-2}\cdot t_0\right]=\lim_{t \to \infty}\left[\left(\frac{t-n}{t-t_0+1}\right)^{t_0-2}\cdot t_0\right] \tag{9}$$

Figure 4 shows a comparison between $\mathbb{E}[F_t(n)]$ and the approximation, with $t_0 = 25$ and $t = 200$. It is illustrated that $\mathbb{E}[F_t(n)]$ is well approximated by (7). More importantly, it is shown in Figure 4 that the solution we acquired in (7) presents a concave curve in log-log plot, similar to the degree distribution in DPiM (Drosophila Protein interaction Map).

Consequently, the degree distribution function $\mathbb{E}[P_t(n)]$ is derived from (7).

$$\mathbb{E}[P_t(n)]\sim\left(\frac{t-n}{t-1}\right)^{t_0-2} \tag{10}$$

Replace the $t_0 - 2$ in (10) with a single parameter t_1, we propose a simple degree distribution function for protein-protein interaction networks modeled with random duplication graph.

$$\mathbb{E}[P_t(n)]\sim\left(\frac{t-n}{t-1}\right)^{t_1} \tag{11}$$

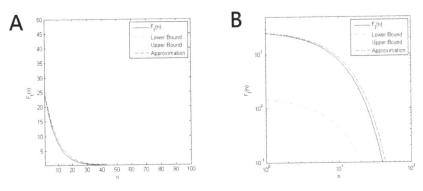

Figure 4. (A) The comparison between the exact solution, the approximation, and the bounds. (B) Log-log plot.

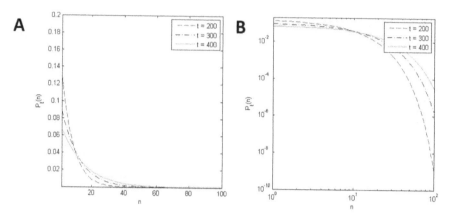

Figure 5. (A) Comparison of our degree distribution at time $t = 200$, $t = 300$, and $t = 400$, with $t_1 = 27$. (B) Degree Distribution in log-log plot.

As a result, we have hereby derived the degree distribution function of protein-protein interaction networks as a special case of random duplication graph where the growth process starts with a sparsely connected initial graph. Two parameters, t_1 and t, stand for the initial scale of the network, and the time steps taken during the growth process separately.

Figure 5 shows the comparison between degree distributions at $t = 200$, $t = 300$, and $t = 400$ (with $t_1 = 27$). We can see that as time t increases from 200 to 400, the portion of highly connected nodes become higher and higher. One important behavior comes from analyzing the acquired degree distribution—as the growth process of random duplication proceeds, more and more highly-connected nodes are formed. This means that this growth process has the ability to produce highly-connected nodes, just like the scale-free distribution, which explains the existence of highly-connected proteins in protein-protein interaction networks. Also, this behavior implies that if a biological organism has gone through longer period of evolution, it is more likely to find highly-connected proteins in the organism.

4   COMPARISON WITH EXPERIMENTAL DATA

We found that the degree distribution of DPiM (Drosophila Protein interaction Map) can be fit by our distribution, with parameters $t_1 = 2600$, and $t = 7048$. Those parameters are chosen to provide the best fit. Our degree distribution function is given as below.

$$\mathbb{E}\left[\mathbb{P}_t(n)\right] \sim \left(\frac{7048 - n}{7048 - 1}\right)^{2600} \tag{12}$$

The r^2 of the fit is greater than 0.96, indicating a good fit. In Figure 6(A), we can see that our distribution provides a good fit for the degree distribution of DPiM. Also, our distribution features a concave curve in the log-log plot in Figure 6(B), which is in consistent with the experimental data.

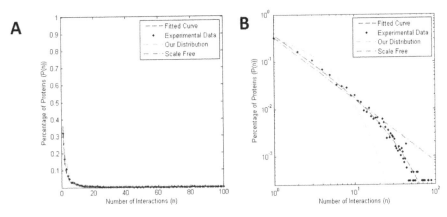

Figure 6. (A) Comparison of our degree distribution function with experimental data, where "fitted curve" stands for $\mathbb{P}(n) \sim n^{-1.20} e^{-0.038n}$, and "scale free" stands for $\mathbb{P}(n) \sim n^{-1.80}$. (B) Degree Distribution in log-log plot.

5 CONCLUSIONS

In this paper we modeled the protein-protein interaction network as a random duplication graph with sparse initial state, and derived a degree distribution under this model. We compared this degree distribution with experimental data, and showed that it can fit the data adequately.

Our model gives a theoretical analysis of the growth process of protein-protein interaction network. Moreover, we have shown that it is the growth process that leads to the specific characteristics in protein-protein interaction networks, in which the majority of proteins are sparsely connected. However, highly connected proteins also exist.

We can make a further prediction based on our analysis—as the growth process continues, more and more high degree proteins will be produced.

REFERENCES

Barabási A.L. & Oltvai, Z.N. 2004, "Network Biology: Understanding the Cell's Functional Organization," *Nature Reviews* 5: 101–113.\

Bebek G. et al., 2006, "The degree distribution of the generalized duplication model," *Theoretical Computer Science* 369: 234–249.

Giot L. et al., 2003, "A Protein Interaction Map of Drosophila melanogaster," *Science* 302: 1727–1736.

Guruharsh K.G. et al., 2011, "A Protein Complex Network of Drosophila melanogaster," *Cell* 147(3): 690–703.

Computational evolutionary biology

The application of *COI* and *ITS* genes in the molecular phylogeny of water beetles

C. Guo
University of Chinese Academy of Sciences, Beijing, China
State Key Laboratory of Forest and Soil Ecology, Institute of Applied Ecology,
Chinese Academy of Sciences, Shenyang, China

D.J. Bian & L.Z. Ji
State Key Laboratory of Forest and Soil Ecology, Institute of Applied Ecology,
Chinese Academy of Sciences, Shenyang, China

ABSTRACT: A 825 base pair region of the mitochondrial Cytochrome Oxidase I (*COI*) gene and a 450 base pair region of the Internal Transcribed Spacer (*ITS*) gene were sequenced for six samples of water beetles. All sequences were aligned using Bio-Edit software, and molecular phylogenetic trees were constructed using Mega 3.1 software. The phylogenetic relationships of the six samples were discussed in this research from the molecular level, hoping that it can present some useful information for the use and control of water beetles.

1 INTRODUCTION

Mitochondrial DNA sequence and nuclear ribosomal DNA sequence data have been extensively applied to the study of the population structure and phylogenetic relationships of animal species, including insects (Caterino, 2000). The mitochondrial Cytochrome Oxidase I (*COI*) gene and the Internal Transcribed Spacer (*ITS*) sequence provide a wealth of variation that may be particularly useful for generating phylogenies in taxa and identifying species whose morphological differences are subtle (So, 2005). The *COI* gene has two important advantages: first, the universal primers for this gene are very robust, enabling recovery of its 5′ end from representatives of most, if not all, animal phyla (Folmer, 1994); second, *COI* appears to possess a greater range of phylogenetic signals than any other mitochondrial genes. *ITS* is located between 5.8s rDNA and 28S rDNA. 5.8S rDNA and 28s rDNA are very conservative sequences, showing that different insects are always similar. Hence, many taxonomists identify sibling species, egg, larva and pupae with *ITS* (Ratcliffe, 2003). In fact, the evolution of this gene is rapid enough to allow the discrimination of not only closely allied species, but also phylogeographic groups within a single species (Cox, 2001).

Water beetle belongs to a main group of Coleoptera, for which molecular biology has been applied in recent research (Michael, 2004). Molecular data may provide a valuable complement to morphological evidence, particularly in situations in which taxa are weakly supported or unstable in traditional analyses (Miller, 1997). In this paper, we have two aims: (1) to examine whether the specimens were new species; (2) to test the utility of mitochondrial and nuclear DNA markers whether they provide information that the two species whose morphological differences are subtle can be identified, which can provide useful information on the use and control of water beetles.

2   MATERIALS AND METHODS

2.1   *Samples*

Wild adult specimens were collected from various locations by net. Water beetles were immediately killed and dropped directly into 95%–100% EtOH and frozen until extraction. Prior to DNA extraction, the specimens were tentatively identified by phenotype, based on morphological characteristics and locality. Six samples are *Elmomorphus tongi* sp. n.(female), *Elmomorphus tongi* sp.n.(male), *Elmomorphus ganzhouensis* sp. n., *Elmomorphus lichuanensis* sp. n., *Elmomorphus longnanensis* sp.n., and *Stenomystax jiangjinensis* sp. n.

2.2   *DNA extraction, amplification and sequencing*

DNA was extracted from individual beetles, a male and a female per species. Specimens soaked in absolute ethyl alcohol were first dabbed with filter paper or volatilization to eliminate most of the alcohol. Beetle was snap-frozen in liquid nitrogen and ground separately with a mortar. These powders was dissolved in 800 µL extraction buffer (2.5% CTAB, 1.4 M NaCl, 0.1 M Tris-HCl, 0.05 M EDTA, 0.2% β-mercaptoethanol (BME) (V/V)), incubated for 1.5 h at 65°C, and then added 4 µL proteinase K to incubate for 2.5 h at 45°C. Phenol-chloroform was used for DNA extraction once, chloroform for DNA extraction twice, two times volume analytical pure ethanol and 1/10 times volume 3 M NaAc for sedimentation, and dried naturally, 20 µL sterilized ddH$_2$O for solution, and stored at 4°C.

Sequences of *COI* were amplified as a single fragment of 825 bp, using primers co-R-(5′CAACATTTATTTTGATTTTTGG3′) and co-F-(5′TCCAATGCACTAATCTGCCTTA3′). A single fragment of 450 bp of *ITS* was amplified using primers 5S1-(5′TGCGCG-TCAACTTGTGATGC3′) and 28S1-(5′AGCGGGTAATCCCACCTGATCTGAG3′). All new sequences generated in the study were deposited in the GenBank database under accession nos. EF119775-EF119780 and EF189141-EF189146.

The following PCR cycling conditions were used for *COI* amplification: 4′ at 94°C; 30″ at 94°C, 45″ at 51°C, 90″ at 72°C, repeated for 35 cycles; 10′ at 72°C. When *ITS* were amplified, its annealing temperature was at 56°C and repeated for 30 cycles. Amplification products were purified using an EZ-10 Spin Column PCR Products Purification Kit. Then DNA fragments were cloned into pUCm-T using a pUCm-T Vector Cloning Kit. Sequences obtained from clones were compared with sequences in the GenBank database using the BLASTn and BLASTx programs.

2.3   *Sequence alignment and phylogenetic analysis*

Raw sequences were examined and corrected using BioEdit. These sequences were then aligned using CLUSTAL W.

Neighbor Joining (NJ) analyses were performed using MEGA 3.1. The Kimura two-parameter model of nucleotide substitution was selected for the analyses based on Nei's empirical guidelines for choosing an appropriate distance measure for phylogenetic inference. When the overall sequence divergence was not high (d < 0.3), simpler models, such as Jukes-Cantor or Kimura two-parameter distances, were recommended. The reliability of clustering patterns in trees was determined by the interior branch test (= standard error test) and bootstrap test (1000 replications).

3   RESULTS

3.1   *DNA extraction and amplification*

After 0.8% agarose gel electrophoresis, the extracted DNA was used in the next step. *COI* and *ITS* gene sequence fragments were obtained by PCR amplification (Fig. 1). Compared with sequences in the GenBank database, these fragments were target fragments.

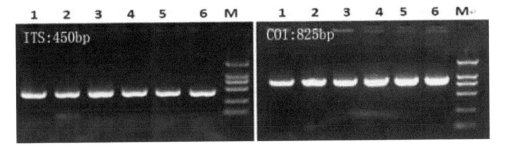

Figure 1. Electrophoretic pattern of *ITS* and *COI* sequence amplification. 1–6: *E. tongi* sp. n. (female), *E. tongi* sp.n. (male), *E. ganzhouensis* sp. n., *E. lichuanensis* sp. n., *E. longnanensis* sp.n., *S. jiangjinensis* sp. n. M: Marker DL2 000 (2000 bp, 1000 bp, 750 bp, 500 bp, 250 bp and 100 bp).

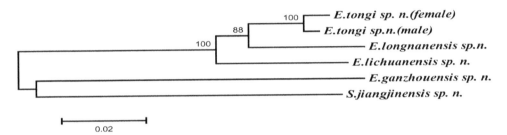

Figure 2. Phylogenetic tree using *COI* sequences data.

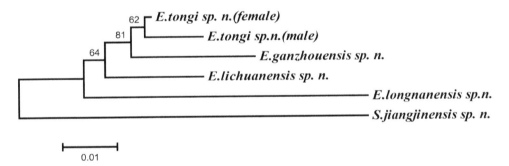

Figure 3. Phylogenetic tree using *ITS* sequence data.

3.2 *COI and ITS sequencing analysis*

The amplified *ITS* fragment in *E. tongi* was 450 bp, their base composition was 21.5%T, 29.0%C, 20.0%A, 29.5%G, and showed a GC bias. The intra-specific alignment of *ITS* showed no intraspecific variation. The *COI* sequence in *E. tongi* was 825 bp, coding for 276 amino acids. The mean base composition was 31.4%T, 14.0%C, 39.2%A (39.2%–39.3%), 15.3%G (15.2%-15.4%) and showed an AT bias. The alignment of *COI* showed intra-specific variation on 5 bases or 4 amino acids.

3.3 *Molecular phylogeny analysis*

The phylogenetic tree with NJ analyses using *COI* sequences data showed that the same species was on one clade, and that the clade of *E. tongi* and the outgroup clade of *E. longnanensis* belonged to the same group (Fig. 2). The tree constructed by *ITS*2 sequences data showed

103

that *E. ganzhouensis* sp. n. was closer to the *E. tongi* sp. n. (Fig. 3). In other words, we suggest that the different species of water beetles can be classified by *COI* and *ITS* genes.

4 DISCUSSION

The identification of new species was a difficult task, and we may meet the convergent evolution and the morphological characters variation under different geographical environments. Although DNA sequence data was a powerful tool in aiding the classification of organisms, the alignment of *COI* sequences was straightforward, as indels were uncommon, reinforcing the results of earlier work showing the rarity of indels in this gene (Ros, 2007; Ben, 2007; Jeyaprakash, 2009). *ITS*2 was a non-coding transcribed portion of DNA and hence was shown to be evolving at a higher rate than the coding regions of the genes on either side (Sanderson, 2008; Lytynoja, 2008). The phylogenetic analysis using *COI* and *ITS*2 provided evidence for the new water beetles being a new species.

Molecular methods can overcome the deficits of morphological approaches to species discrimination: the bounds of intraspecific diversity will be quantifiable; sibling species will be recognizable; taxonomic decisions will be objective and all life stages will be identifiable. We suggest the further study of related species carefully, in a combined morphological molecular framework when necessary.

ACKNOWLEDGEMENTS

This study was supported by the National Natural Science Foundation of China "Taxonomic study on Chinese Elmidae (31201742)".

REFERENCES

Ben D.T., Melamed S., Gerson U., et al. *ITS*2 sequences as barcodes for identifying and analyzing spider mites (Acari: Tetranychidae) [J]. Experimental and Applied Acarology, 2007, 41(3): 169–181.

Caterino M.S., Cho S., Sperling F. The current state of insect molecular systematics: a thriving Tower of Babel [J]. Annul Review Entomology, 2000, 45: 1–15.

Cox A.J., Hebert P. Colonization, extinction and phylogeographic patterning in a freshwater crustacean [J]. *Mol. Ecol.*, 2001, 10: 371–386.

Folmer O., Black M., Hoeh W., et al. DNA primers for amplification of mitochondrial cytochrome *c* oxidase subunit I from diverse metazoan invertebrates [J]. *Mol. Mar. Biol. Biotechnol.*, 1994, 3: 294–299.

Jeyaprakash A., Hoy M.A. First divergence time estimate of spiders, scorpions, mites and ticks (subphylum: Chelicerata) inferred from mitochondrial phylogeny [J]. Experimental and Applied Acarology, 2009, 47(1): 1–18.

Lytynoja A., Goldman N. Phylogeny-aware gap placement prevents errors in sequence alignment and evolutionary analysis [J]. Science, 2008, 320(5883): 1632–1635.

Michael B., Chris H., Steven J., et al. A highly modified stygobiont diving beetle of the genus *Copelatus* (Coleoptera, Dytiscidae): taxonomy and cladistic analysis based on mitochondrial DNA sequences [J]. *Systematic Entomology,* 2004, 29: 59–67.

Miller J., Brower A., DeSalle R. Phylogeny of the neotrophical moth tribe Josiini (Notodontidae: Dioptinae): comparing and combining evidence from DNA sequences and morphology [J]. *Biological Journal of the Linnean Society,* 1997, 60: 297–316.

Ratcliffe S.T., Webb D.W., Weinzievr R.A., et al. PCR-RFLP identification of Diptera (Calliphoridae, Muscidae and Sarcophagidae)-a generally applicable method [J]. *Journal of Forensic Science,* 2003, 48(4): 783–785.

Ros V.I.D., Breeuwer J.A.J. Spider mite (Acari: Tetranychidae) mitochondrial *COI* phylogeny reviewed: host plant relationships, phylogeography, reproductive parasites and barcoding [J]. Experimental and Applied Acarology, 2007, 42(4): 239–262.

Sanderson M.J. Pylogenetic signal in the eukaryotic tree of life [J]. Science, 2008, 321(5885): 121–123.

So Y.L., Hyungjin P., Kyung S.B., et al. Molecular identification of *Adoxophyes honmai* (Yasuda) (Lepidoptera: Tortricidae) based in mitochondrial *COI* gene [J]. *Molecules and Cell,* 2005, 19(3): 391–397.

Bioinformatics and Biomedical Engineering – Chou & Zhou (Eds)
© *2016 Taylor & Francis Group, London, ISBN 978-1-138-02784-8*

Genome-wide identification and phylogenetic analysis of the Ethylene-Insensitive3 (EIN3) and EIN3-Like (EILs) gene family in melons

Y. Ma, Y. Tu, X.L. Chen & X.G. Liu
Department of Biological Science and Technology, Baotou Teacher's College, Baotou, China

ABSTRACT: Four Ethylene-Insensitive3 (EIN3) and EIN3-Like (EILs) genes were identified from the genome-wide database of melon. Phylogenetic analysis, numbers of intron analysis and alignment of sequence similarity was performed in EIN3/EILs gene family in melon. By means of analyzing ESTs which matched with EIN3/EILs proteins, the result show that EIN3/EILs genes were expressed in different tissues, part of genes specific expressed with high abundance in fruit and callus, and different members of EIN3/EILs gene family exhibits difference in patterns of expression. The results of protein interaction network showed that CmEIL02 integrate with multiple ERF proteins.

1 INTRODUCTION

Ethylene-Insensitive3 (EIN3) and EIN3-Like (EILs) is a relatively small gene family in higher plants. *Arabidopsis*, tobacco, tomato, rice each contain 6 (Guo and Ecker, 2004; Chao et al., 1997), 5 (Kosugi and Ohashi, 2000; Rieu et al., 2003), 4 (Tieman et al., 2001; Yokotani et al., 2003) and 6 (Mao et al., 2006; Hiragaet al., 2009) EIN3/EILs transcription factors, respectively. Three EIN3-like genes were cloned in tomato (Tieman et al., 2001), knockouting any one of three genes did not show insensitive to ethylene, the experimental results indicated that these three genes are redundant in function, and play the role of synergistic complementary (Tieman et al., 2001). *LeEIL1-3* sustained expression during fruit development in tomato, *LeEIL4* continuously up-regulated during fruit ripening (Tieman et al., 2001; Yokotani et al., 2003). *MA-EIL2* up-regulated the expression during fruit development in banana, but the expression level of other *MA-EILs* did not change significantly (Mbeguie-A-Mbeguie et al., 2008).

Ethylene is an important hormone in plant growth and development, which process a wide range of regulation of plant development, including seed germination, seedling growth, lateral root development, leaf and flower falling, fruit ripening, tissue aging, biotic and abiotic stress responses (Brown, 1997; Lelievre et al., 1997; Morgan and Drew, 1997; Smalle and Van Der Straeten, 1997). EIN3/EILs proteins are important nuclear transcription factors in ethylene signal transduction pathways in higher plants. To date, EIN3/EILs which belong to a small plant-specific family of transcription factors have been isolated from a wide variety of plant species. The transcription factors contain a highly conserved N-terminal amino acid, including several DNA-binding regions such as acidic domain, proline-rich and basic domains. They regulate expression of related genes by binding directly to a Primary Ethylene Response Element (PERE). EIN3/EILs plays as a pivotal nuclear transcription factor in ethylene signal transduction pathway, expression of EIN3/EILs is capable of activating known ethylene transduction (Guo and Ecker, 2004; Lin et al., 2009; Chen et al., 2005). It acts on the ethylene response genes such as *ERF1* (Ethylene response factor), and launch a series expression of ethylene regulating target gene, which was the positive regulator in ethylene response. Not only PERE sequence is presence in *ERF* promoter downstream region, but also in promoter region of ripening, senescence-related genes, such as *ACO* (1-aminocyclopropane-1-carboxylate oxidase) (Blume et al., 1997; Lasserre et al., 1997).

Melon (*Cucumis melo* L.) is the model plant of Cucurbitaceae in studying fruit development and ripening. The sequencing results of whole-genome of melon have been completed and released in June 2012. The results of whole-genome sequencing of melon provide data support for genome-wide analysis of gene family identification and phylogeny.

These genes have been identified in a number of plants, but they have not yet been identified in melons. This study performed predictive computer analysis for EIN3/EILs proteins in melons. The data generated from this study will contribute to studies on the selection of appropriate candidate genes from the EIN3/EILs family in melon for further functional characterization and understanding of the precise regulatory checkpoints that operate during developmental and stress responses.

2   MATERIALS AND METHODS

2.1   *Searching of database*

An Arabidopsis protein with a typical structure of the EIN3 domain was selected as the probe sequence in the National Center for Biotechnology Information (NCBI) database (ACCESSION: NM_106032), and local Basic Local Alignment Search Tool (BLAST) (Altschul et al., 1990) of the Bioedit 7.0.9 tool (Hall, 1999) was used to search the similar protein sequences in the CM_protein_v3.5 database (https://melonomics.net/files/Genome) which subordinates to MELONOMICS (https://melonomics.net), with the aim of identifying all the candidate proteins that contain the EIN3 domain in the melon genome. Sequences with an E-value $\leq 10^{-3}$ were regarded as candidate proteins with an EIN3 domain, while other parameters were the default values.

To further confirm the postulated EIN3/EILs family genes, the amino acid sequences were then searched for the EIN3/EILs domain using the Simple Modular Architecture Research Tool (SMART) (Schultz et al., 1998; Letunic et al., 2009).

2.2   *Sequence alignment and phylogenetic analysis*

To compare the evolutionary relationships of EIN3/EILs family members, multiple sequence alignment was applied, by way of Clustal W (Larkin et al., 2007), on already obtained EIN3/EILs protein sequences.

A phylogenetic tree was constructed with aligned EIN3/EILs protein sequences of the domains using MEGA5.10 (Tamura et al., 2007) and the Neighbor-Joining (NJ) method (Saitou and Nei, 1987) with poisson correction, pairwise deletion and bootstrap (1000 replicates; random seeds) as parameters. Simultaneously, the Maximum Parsimony (MP) method of MEGA5.10 software (Tamura et al., 2007) was employed to create a second phylogenetic tree with a bootstrap of 1000 replicates to validate the results from the NJ method.

2.3   *Expression analysis of EIN3/EILs genes*

The expression characteristics of the EIN3/EILs gene family were detected using ESTs. We used the nucleotide sequences of the EIN3/EILs, which are available from Cucurbit Genomics Database (CuGenDB), as probing sequences for the proteins in melons. A nucleotide BLAST analysis in the CuGenDB EST database was completed. As a result, the sequences that had a maximum identity of >90 and <100% were screened (Zhang et al., 2012). Redundant sequences were removed from the data set. The EST annotation information of the obtained sequences was used to analyze the expression patterns in different tissues.

2.4   *Analysis of EIN3/EILs protein regulatory networks*

Amino acid sequence of EIN3/EILs was used as the query sequence, interaction network of candidate protein is constructed in STRING (Search Tool for the Retrieval of Interacting Genes/Proteins, http: string.embl.de). Analysis of EIN3/EILs protein family members is performed for regulatory relationships in interaction network.

3 RESULTS

3.1 *Identification of EIN3/EILs gene in melon*

Four non-redundant protein sequences which contains typical domain is obtained via searching of local BLASTp in CM_protein_v3.5 (https://melonomics.net/files/Genome) which is the genome annotation of protein libraries in MELONOMICS database (https://melonomics.net) (Table 1). Four candidate proteins all contain typical EIN3 domain which filtered by analysis of Simple Modular Architecture Research Tool (SMART) (Schultz et al., 1998; Letunic et al., 2009) online. The amino acid sequence of EIN3 in 4 family proteins was extracting via SMART (Letunic et al., 2004). The resulting of constructed phylogenetic tree is that EIN3/EILs gene family includes a duplicate gene pair in melon (Fig. 1).

3.2 *Phylogenetic analysis of EIN3/EILs gene family in melon*

Four EIN3/EILs genes were renamed *CmEIL01~CmEIL04*. In order to investigate the evolutionary relationships of EIN3/EILs protein, no roots phylogenetic tree of Neighbor-joining method was constructed, parameters are as follows: poisson correction, delete pairs and the rest default. For 1000 bootstrap analysis of EIN3/EILs protein after multiple analysis of sequence alignment, it showed expectations of EIN3/EILs protein family in each node of phylogenetic tree are high, this is most likely due to the relatively large EIN3 domain, the amino acid sequence is highly conserved between family members. The reliability of phylogenetic trees in Neighbor-Joining method has been verified by another phylogenetic tree of Maximum-parsimony method.

The resulting of phylogenetic tree shows all members of the family contain an intron and two exons (Fig. 1, Table 1), which speculate this gene family is high conservative in

Table 1. The characteristics of EIN3/EILs family genes in melon.

Gene name	Gene ID	Scaffold distribution	Length of ORF	Introns	Exons
CmEIL01	MELO3C019931	Scaffold00040	1848	1	2
CmEIL02	MELO3C015633	Scaffold00025	1878	1	2
CmEIL03	MELO3C017592	Scaffold00031	1257	1	2
CmEIL04	MELO3C015210	Scaffold00024	1812	1	2

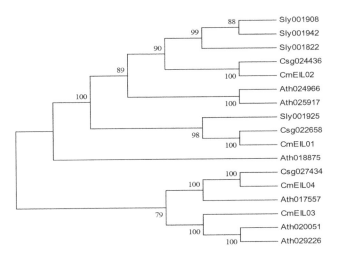

Figure 1. Phylogenetic analysis of EIN3/EILs family genes in Arabidopsis thaliana, *Solanum lycopersicum, Cucumis sativus* and *Cucumis melo* (Ath. *Arabidopsis thaliana*, Csg. *Cucumis cucumber*, Sly. *Solanum lypersicon*).

the differentiation in the species. We found one pair of duplicate genes that *CmEIL01* and *CmEIL02* in analysis of phylogeny. These two genes is likely to be co-exist before differentiation of monocots and dicots in the same ancestral genomes with two different types of EIN3/EILs gene which form via inserting different number of introns and gene duplication.

3.3 Expression analysis of EIN3/EILs gene in melon

In melon EST database Cucurbit Genomics Database (CuGenDB), searching of matched high score annotated unigene with four EIN3/EILs genes, indicating that four EIN3/EILs genes are provided with transcriptional activity. Expression analysis in different tissues shows that, *CmEIL01* specifically expressed in the mature fruits of melon (Fig. 2), *CmEIL02* expressed in fruits, in addition, expressed in the leaves with relatively high expression abundance. *CmEIL03* expressed in callus-specific, relating with abiotic stress responses. *CmEIL04* expressed in melon with tissue-specific pattern, *CmEIL01* and *CmEIL04* are likely to play a crucial role in the fruit ripening process of the melon.

3.4 Regulatory network analysis of EIN3/EILs family protein

STRING is a interactive database of protein interactions in *Arabidopsis*. Protein sequence of CmEIL03 was submitted to the database to get the highest similarity of the target sequence. The mapping protein sequences of submitted sequence were constructed with interaction network in STRING database. The central figure of the red protein in the interaction network is composed of maps (Fig. 3).

CmEIL03 interplay with 6 ERF family proteins which contain a single AP2 domain, respectively (Fig. 3). These ERF family proteins probably involved in stress response and

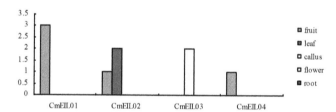

Figure 2.   Expression analysis of EIN3/EILs gene in melon.

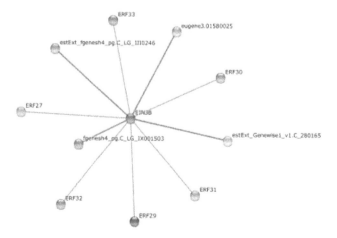

Figure 3.   Interaction network of CmEIL03 protein with other proteins of *Arabidopsis thaliana* Thickness of the line represented the strength of the interaction(ERF27, ERF29~33, eugene3.01580025, estExt_Genewise1_v1.C_280165,   fgenesh4_pg.C_LG_IX001503,   estExt_fgenesh4_pg.C_LG_III0246   are proteins in *Arabidopsis thaliana*).

plant hormone signal transduction processes which play a role in regulation of melon fruit development and ripening.

4 DISCUSSION

Results of this study indicate that four EIN3/EILs proteins were identified in melon which possess two introns. The length of coding region in EIN3/EILs gene family is very close to each, which indicated this gene family was extreme conservative in the evolution and connoted that the gene family has important function in higher plants. Replication and differentiation of gene lead to differentiation gene function and emergence of new genes.

Zhang (2003) supports that tandem and partial replication of gene lead to the initial homologous genes is fully redundant in functions. Along with reducing the selection pressure, two homologous genes can accumulate on their different genetic variants to form new genes with different structures and functions.

According to Moore and Purugganan's theory (2005), duplication of gene may have three different results, firstly, copies of one gene retain the function of the ancestral gene, and the other one copy does not have expression of activity which become pseudogene; Second, wherein a copy retain the functionality of the ancestral gene, but due to the accumulation of genetic variation and other factors so as to have an ancestral gene that does not have a new function; Again, two copies contain partially function of ancestral gene, which cooperate to complete the original ancestor all function. Based on the results of these genes, we speculate generation of these genes is not tandem duplication, but genome duplication generating all members of EIN3/EILs protein family.

In addition, analysis of expression pattern showed that EIN3/EILs gene family is high abundance of expression in melon fruit, such as *CmEIL01*, *CmEIL02*, *CmEIL04*; *CmEIL02* was detected with expression in leaves; *CmEIL03* was detected with high abundance of expression in callus, which is consistent with the view EIN3/EILs protein mainly play crucial role on stress response and regulation of development. *CmEIL01* and *CmEIL02* is a pair of gene duplication which possess highly homology, the presence of these two genes may be functionally redundant in stage of development and ripening in melon fruit, which co-regulated synergistically functions of signal transduction during ripening of fruit.

CmEIL02 and *CmEIL03* interact with multiple ERF family proteins, most of the proteins involved in interactions with a similar function, therefore, there may be functional redundancy between *CmEIL02* and *CmEIL03*, both of which possibly co-regulated functional genes which regulated by *ERF* gene, and thus indirectly regulated expression of fruit development and ripening-related functional genes.

This study would be of interest to plant molecular biologists, particularly those interested in the EIN3/EILs gene family. They could use this premise of this study to predict proteins in other types of plants.

ACKNOWLEDGEMENTS

This work was supported by Universities of Science and Technology Research Project in Inner Mongolia (NJZC14251) and Youth Science Foundation of Baotou Teachers College (BSYKJ2013-28).

REFERENCES

Altschul, S.F., Gish, W., Miller, W., Myers, E.W., Lipman, D.J. 1990. Basic local alignment search tool. *Journal of Molecular Biology* 215: 403–410.
Blume, B., Barry, C.S., Hamilton, A.J., Bouzayen, M., Grierson D. 1997. Identification of transposon-like elements in non-coding regions of tomato ACC oxidase genes. *Molecular and General Genetics* 254: 297–303.

Brown, K.M. 1997. Ethylene and abscission. *Physiologia Plantarum* 100: 567–576.

Chao, Q., Rothenberg, M., Solano, R., Roman, G., Terzaghi, W., Ecker, J.R. 1997. Activation of the ethylene gas response pathway in Arabidopsis by the nuclear protein Ethylene-Insensitive3 and related proteins. *Cell* 89(7): 1133–1144.

Chen, Y.F., Etheridge, N., Schaller, G.E. 2005. Ethylene signal transduction. *Annals of Botany* 95(6): 901–915.

Guo, H. & Ecker, J.R. 2004. The ethylene signaling pathway: new insights. *Current Opinion in Plant Biology* 7(1): 40–49.

Hall, T.A. 1999. BioEdit: a user-friendly biological sequence alignment editor and analysis program for Windows 95/98/NT. *Nucleic Acids Symposium Series* 41: 95–98.

Hiraga, S., Sasaki, K., Hibi, T., Yoshida, H., Uchida, E., Kosugi, S., Kato, T., Mie, T., Ito, H., Katou, S., Seo, S., Matsui, H., Ohashi, Y., Mitsuhara, I. 2009. Involvement of two rice Ethylene Insensitive3-Like genes in wound signaling. *Molecular Genetics and Genomics* 282(5): 517–529.

Kosugi, S. & Ohashi, Y. 2000. Cloning and DNA-binding properties of a tobacco Ethylene-Insensitive3 (EIN3) homolog. *Nucleic Acids Research* 28(4): 960–967.

Larkin, M.A., Blackshields, G., Brown, N.P., Chenna, R., McGettigan, P.A., McWilliam, H., Valentin, F., Wallace, I.M., Wilm, A., Lopez, R., Thompson, J.D., Gibson, T.J., Higgins, D.G. 2007. Clustal W and Clustal X version 2.0. *Bioinformatics* 23: 2947–2948.

Lasserre, E., Godard, F., Bouquin, T., Hernandez, J.A., Pech, J.C., Roby, D., Balague C. 1997. Differential activation of two ACC oxidase gene promoters from melon during plant development and in response to pathogen attack. *Molecular and General Genetics* 254: 211–222.

Lelievre, J.M., Latche, A., Jones, B., Bouzayen, M., Pech, J.C. 1997. Ethylene and fruit ripening. *Physiologia Plantarum* 101: 727–739.

Letunic, I., Doerks, T., Bork, P. 2009. SMART 6: recent updates and new developments. *Nucleic Acids Res* 37: 229–232.

Letunic, I., Copley, R.R., Schmidt, S., Ciccarelli, F.D., Doerks, T., Schultz, J., Ponting, C.P., Bork P. 2004. SMART 4.0: towards genomic data integration. *Nucleic Acids Res* 32: 142–144.

Lin, Z.F., Zhong, S.L., Grierson D. 2009. Recent advances in ethylene research. *Journal of Experimental Botany* 60(12): 3311–3336.

Mao, C.Z., Wang, S.M., Jia, Q.J., Wu, P. 2006. *OsEIL1*, a rice homolog of the Arabidopsis EIN3 regulates the ethylene response as a positive component. *Plant Molecular Biology* 61(1–2): 141–152.

Mbeguie-A-Mbeguie, D., Hubert, O., Fils-Lycaon, B., Chillet, M., Baurens, F.C. 2008. EIN3-like gene expression during fruit ripening of Cavendish banana (*Musa acuminate* cv. *Grande naine*). *Physiologia Plantarum* 133(2): 435–448.

Moore, R.C. & Purugganan, M.D. 2005. The evolutionary dynamics of plant duplicate genes. *Current Opinion in Plant Biology* 8(2): 122–128.

Morgan, P.W. & Drew, M.C. 1997. Ethylene and plant responses to stress. *Physiologia Plantarum* 100: 620–530.

Rieu, I., Mariani, C., Wetering, K. 2003. Expression analysis of five tobacco EIN3 family members in relation to tissue-specific ethylene responses. *Journal of Experimental Botany* 54(391): 2239–2244.

Saitou, N. & Nei, M. 1987. The neighbor-joining method: a new method for reconstructing phylogenetic trees. *Molecular biology and evolution* 4(4): 406–425.

Schultz, J., Milpetz, F., Bork, P., Ponting, C.P. 1998. SMART, a simple modular architecture research tool: identification of signaling domains. *Proc Natl Acad Sci USA* 95: 5857–5864.

Smalle, J. & Van Der Straeten, D. 1997. Ethylene and vegetative development. *Physiologia Plantarum* 100: 593–605.

Tamura, K., Dudley, J., Nei, M., Kumar, S. 2007. MEGA4: Molecular Evolutionary Genetics Analysis (MEGA) software version 4.0. *Mol Biol Evol* 24: 1596–1599.

Tieman, D.M., Ciardi, J.A., Taylor, M.G., Klee, H.J. 2001. Members of the tomato LeEIL (EIN3-like) gene family are functionally redundant and regulate ethylene responses throughout plant development. *Plant Journal* 26(1): 47–58.

Yokotani, N., Tamura, S., Nakano, R., Inaba, A., Kubo, Y. 2003. Characterization of a novel tomato EIN3-like gene (LeEIL4). *Journal of Experimental Botany* 54(393): 2775–2776.

Zhang, J.Z. 2003. Evolution by gene duplication: An update, Trends in Ecology and Evolution. 18(6): 292–298.

Zhang, C.H., Shangguan, L.F., Ma, R.J., Sun, X., Tao, R., Guo, L., Korir, N.K., Yu, M.L. 2012. Genome-wide analysis of the AP2/ERF superfamily in peach (*Prunus persica*). *Genet Mol Res.* 11: 4789–4809.

Biomechanics

Bioinformatics and Biomedical Engineering – Chou & Zhou (Eds)
© *2016 Taylor & Francis Group, London, ISBN 978-1-138-02784-8*

In vitro and *in vivo* biomechanical research on cervical arthroplasty and fusion

Z.H. Liao
Department of Mechanical Engineering, Tsinghua University, Beijing, China

W.Q. Liu
Biomechanics and Biotechnology Lab, Research Institute of Tsinghua University in Shenzhen, China

ABSTRACT: This study aims to evaluate the biomechanical and clinical evidence available and provide a systematic summary of cervical arthroplasty and fusion. The development of cervical cadaver specimens and well-defined test protocols for cervical biomechanical test are introduced. The major progress of biomechanical evaluation about cervical fusion, cervical arthroplasty, multilevel hybrid surgery and cervical coupled motion are analyzed. Concluding all these papers, the well-defined biomechanical testing method should be improved unceasingly. More studies about multilevel hybrid surgery and coupled motion characters are needed to reach a more reliable conclusion.

Keywords: cervical spine; biomechanics; arthroplasty; fusion; hybrid

1   INTRODUCTION

As the golden standard, Anterior Cervical Discectomy and Fusion (ACDF) has been the most accepted procedure for the surgical therapy of cervical spondylosis for decades. However, ACDF altered the normal biomechanical characteristics of the cervical spine because the mobility decreased dramatically at the instrumented levels and the motion increased at the adjacent levels. Although ACDF often gets satisfactory effects in the short-term, the biomechanical stresses at adjacent levels increases according to fusion and this may result in the acceleration of Adjacent Segment Degeneration (ASD). As an alternative choice to fusion, Artificial Cervical Disc Replacement (ACDR) has been applied for cervical degenerative disc disease. Recent studies report that short-term results after ACDR were at least as good as those for ACDF (Ryu W.H.A et al. 2013).

On the other hand, only 43% of patients could meet the rigorous requirement for ACDF and became candidates for arthroplasty. ACDF will still be the golden standard for cervical surgery if the contraindications for ACDR exit. However, various complications related to anterior plate have been verified such as dysphagia and dysphonia. Although a lot of changes have been made and the profile of current anterior plates has turned to more thinner in comparison to earlier designs, the sizes of plates are still bigger (Beutler W.J. et al. 2012). Based on these results, Spacer with Integrated Plate system (SIP) has been exploited to provide stability as reliable as anterior plates by preventing complications such as dysphagia. The new design of SIP has zero-profile which could avoid new implant contact with the soft tissue and esophagus in front of the cervical spine. Without mechanical stimulation, the dysphagia rate may decrease.

As the advance of cervical fusion and non-fusion surgery, many controversies in clinic emerge. Can SIP avoid dysphagia effectively? How about ACDR used for multilevel replacement? Is there biomechanical and clinical evidence on hybrid surgery for multilevel cervical degenerative disc diseases? So the objective of this systematic review is to identify the *in vitro* biomechanical and *in vivo* clinical studies on cervical surgeries, summarize the current concepts and provide a foundation of evidence on the safety and efficacy of several surgeries in the treatment of single-level and multilevel cervical degenerative disc diseases.

2 CERVICAL CADAVER SPECIMENS FOR BIOMECHANICAL TEST

In spinal research, biomechanical *in vitro* testing is necessary for pre-clinical evaluation of new surgical procedures and implants. Beyond question, human cadaver specimens would be the best option for a biomechanical test, but it is quite difficult to obtain human spine specimens. Porcine, sheep and calf spinal columns are often used as a compromise for human spine specimens in biomechanical tests, because these animal spine columns have similarities compared to adult human spines in terms of anatomy, size and material characteristics. Wilke et al. revealed that more similarities than other breeds exit while comparing certain regions of the porcine spine to human spine (Wilke H.J. et al. 2011). The most similarities exit in the cervical region of C1–C2 and the upper and middle thoracic segments. Only qualitatively similarities exit when comparing the lower thoracic and the lumbar area to the human spine. It is not suitable as biomechanical testing model for the residual cervical segments from C3 to C7. DeVries et al. reported that because of large range of motion and neutral zone the sheep cervical spine is too flexible (DeVries N.A. et al. 2012). The large neutral zone may result in the coupled motion while axial rotation and lateral bending. Sheng et al. showed that the C2–C3 and C6–C7 of porcine can be used in nearly all biomechanical experiments as the replacement of the human cervical spine (Sheng S. et al. 2011). Although the ROM of calf is bigger than human cervical spine, some similarities in biomechanics also exit in the C2–C3 and C3–C4 of calf in comparison to human. In conclusion, it should be noted that each animal model compared with human specimens always only represents a compromise.

3 WELL-DEFINED TEST METHOD FOR CERVICAL BIOMECHANICAL TEST

Flexibility and stiffness have been carried out for quite a long time as two kinds of well-defined test protocols (Fig. 1). The flexibility test method is also called pure moment method (Panjabi M.M. et al. 2007) and this method has two significant advantages: the pure moment is always equally whether it is applied to the end-vertebrae or it is applied to any segment of the specimen; the pure moment stays the same regardless of the deformation of the spine during the whole testing.

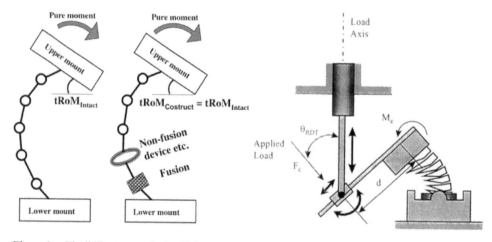

Figure 1.   Flexibility test method utilizing pure moment (left) (Panjabi M.M. et al. 2007) and stiffness test method which is based on displacement-input protocol (right) (Foley K.T. et al. 1999).

The stiffness test method is based on the displacement-input protocol (Foley K.T. et al. 1999). A linear bearing and splined shaft assembly are often needed with a rotating joint fixed to the actuator. The flexion–extension axis of the spine is placed eccentric to the load axis of the actuator, so flexion–extension bending moments and a compressive load could be applied to the upper pot.

Panjabi (Panjabi M.M. et al. 2007) considers that the stiffness method is seldom used compared to the flexibility method. Because the ideal rotation axis is difficult to be determined and is easy to change, the defined rotation axis in the stiffness method could not be the ideal one all the time. Since only the ideal rotation axis can produce natural physiological spinal motions, the defined rotation axis may result in constrained spinal motions and may cause the specimen to injury.

However, when it is needed to simulate the motion-redistribution in a biomechanical test, the displacement-input stiffness protocol is ideally suited for this purpose. The total ROM in the stiffness method keeps the same before and after the surgery, and it is easy to observe the comparative changes at the adjacent spinal levels.

In recent years, an applicable well-defined test method was developed and it was known as the multidirectional hybrid method (Panjabi M.M. et al. 2007). Using this new method, adjacent-level effects can be observed reliably in various conditions. The unconstrained pure moment was used to produce well-defined rotation-moment in the hybrid method. On the other hand, the shortcomings of displacement-input methods can be avoided in this new method.

In vitro conditions, follower load is usually used during experimental test which can simulate muscle strength. However, previous studies (Paxinos O. et al. 2009) suggest that the application of a cervical follower load would decrease motion, bending moments and shear forces of functional spinal units. If a follower load is used, the biomechanical differences between ACDF and ACDR would likely decrease.

In general, there is no a test protocol which can be called as a generally acknowledged test method up to now.

4 BIOMECHANICAL EVALUATION PROGRESS IN CERVICAL SURGERY

4.1 *Biomechanical research on new cervical implant*

It has been reported that more than 90% rate of complaints has been alleviated for the patients following ACDF. However, complication such as dysphagia has been increasing after ACDF. It was reported that rates of dysphagia varied from 2% to 67%. The incidence of chronic dysphagia may be as high as 13.6% with continuous symptoms for more than two years. In recent research cervical plate profile and volume have been identified as the major factor causing this problem (Majid K. et al. 2012). So the Spacer with Integrated Plate system (SIP) has been developed to reduce the probability of dysphagia as it lowers the impact of instrument operation because of its smaller anterior profile and less operative time (Scholz M. et al. 2011) (Fig. 2). Stein et al. reported similar biomechanical stability between SIP with three integrated screws and ACP in cervical spine fusion protocol by *in vitro* study (Stein M.I. et al. 2014). Paik et al. revealed similar feasibility result of the SIP device for single-level fixation (Paik H. et al. 2014). However, SIP devices should be used carefully for the need of multilevel cervical fusion unless posterior fixation was supplemented with them. Nevertheless, SIP devices were reported by Clavenna et al. to show comparable stability to traditional ACP for two-level and three-level fusion (Clavenna A.L. et al. 2012). More research on the SIP devices used for multilevel surgery should be developed. Wojewnik et al. presented in flexion-distraction injury simulation that the locked screw spacer showed significantly reduced motion while the stabilizing effect of the variable angle screw was less satisfying (Wojewnik B. et al. 2013). Beutler et al. reported that adding SIP to adjacent level of a two-level anterior plate shows equivalent stability to a three-level anterior plate (Beutler W.J. et al. 2012). So the traditional protocol by replacing the original plate with a new longer plate can be changed with adding SIP and the complications associated with the anterior plate may overcome.

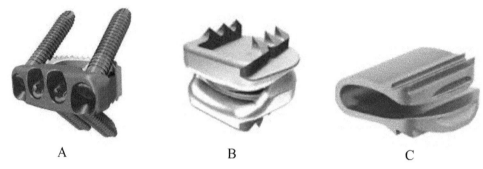

A   B   C

Figure 2. New-type implants for cervical spine. (A) Spacer with integrated plate (SIP). (B) A metal on metal cervical disc with a saddle shaped design (CerviCore disc). (C) The standalone U-shaped implant (DCI).

In clinical research with a mean follow-up time of 13.9 months, Hofstetter et al. presented similar fusion rates between ACDF with anterior plating (96.0%) and SIP (95.2%) (Hofstetter C.C. et al. 2013). However, SIP may reduce risk of dysphagia after fusion surgery and all patients which have moderate or severe swallowing difficulties had accepted ACDF with anterior plate. Miao et al. revealed that in Chinese population Zero-P showed satisfactory fusion rates and lower incidence of postoperative dysphagia (Miao J. et al. 2013). Different from previous studies, Vanek et al. found no significant difference in the incidence of postoperative dysphagia between patients with SIP implantation and a dynamic plate (Vanek P. et al. 2013).

According to non-fusion technology, the cervical disc prosthesis retains the preoperative kinematics at adjacent levels at the 5-year follow-up and was proved a viable option for spinal operation (Ryu W.H.A et al. 2013). In recent years, non-fusion technology has gained new progress in multilevel ACDR, new model ACDR and other type of non-fusion implant. Faizan et al. compared the adjacent level biomechanics of two-level disc replacement, two-level fusion and a hybrid construct which can adjoin disc replacement level to fusion system (Faizan A. et al. 2012). In the two-level fusion model, adjacent level motions, facet loads and endplate stresses increased markedly. On the other hand, adjacent level motions, facet loads and endplate stresses show similarities to intact in the two-level disc replacement model. Colle et al. investigated the biomechanical characteristics of CerviCore total disc replacement (Fig. 2), and found that ROM, LZ, and coupling were relatively maintained after ACDR in a cadaver model (Colle K.O. et al. 2013).

Despite ACDR is the most common non-fusion implant, there are some other non-fusion implants such as the standalone U-shaped implant (DCI). Mo et al. revealed that DCI could retain the spinal kinematics and showed minimum influence on the adjacent soft tissues, but the shortcomings of high stress distribution also exits in terms of the standalone configuration (Mo Z.J. et al. 2014) (Fig. 2).

On the whole, new treatment concepts and new type of cervical implants were developed continuously either in fusion or in non-fusion field. The relevant biomechanical research is reasonably important for clinic.

4.2 Biomechanical research on hybrid surgery

Hybrid Surgery (HS) incorporates ACDF and ACDR at different levels. The purpose is to combine the advantages in both fusion and replacement techniques. ACDR has proven to be beneficial in preserving motion of cervical spine and slowing adjacent level degeneration. These benefits are more significant according to multilevel surgeries. However, indications for ACDR are more severe than ACDF and maybe not all surgery levels can meet the indications. So HS is proposed to tailor ACDF or ACDR to the selected levels. On the one hand, HS can partly preserve segmental motion of the instrumented cervical spine, avoid long-level fusion and may prevent adjacent segment degeneration in the future. On the other hand, hyper-mobility can also be avoided (Jia Z. et al. 2014).

HS is an emerging procedure with need to evaluation and the relevant research is increasing. Cunningham et al. reported no significant difference in adjacent-level motion between two-level HS and intact groups (Cunningham B.W. et al. 2010). Barrey et al. [23] revealed that hybrid and two-level ACDR constructs showed better biomechanical result than two-level arthrodesis, as stress contribution occurred only at lower adjacent level while two-level arthrodesis caused ROM increase at both upper and lower adjacent levels (Barrey C. et al. 2012). Faizan et al. showed that the spinal stiffness after HS was close to intact construct in all bending motions except extension (Faizan A. et al. 2012). Compared to the two-level fusion model, HS showed closer adjacent level motions, facet loads and endplate stress to that in intact construct. Lee et al. reported that a hybrid constructs were biomechanically advantageous than two-level fusion in reducing compensatory adjacent-level hyper-mobility (Lee M.J. et al. 2011). Martin et al. found that ACDR placed adjacent to a two-level ACDF faced increased biomechanical challenge compared to a stand-alone ACDR (Martin S. et al. 2011). Therefore an artificial disc used in HS construct should have to bear the increased loads and wear and this may influence the expected life of the prosthesis.

From a single surgeon series, Cardoso et al. studied thirty-one HS patients (twenty-four patients received a two-level and seven patients received a three-level hybrid construct) with 18 month (mean duration) clinical and radiological follow-up, and identified HS as a safe and effective alternative to multilevel fusion (Cardoso M.J. et al. 2011). Hey et al. provided a direct comparison of HS, ACDF and ACDR patients (7 in each group) with a minimum follow-up of 2 years (Hey H.W.D. et al. 2013). They found that HS shows comparable to ACDF and ACDR in consideration of feasibility and safety, and appears superior according to less time for returning to work, HS seems an attractive option in carefully selected patients.

But so far, is hybrid surgery of the cervical spine a good balance between fusion and arthroplasty? The answer isn't clear. Recent studies have demonstrated that the biomechanical characters of ACDR adjacent to a single-level fusion change little when compared to ACDR alone (Liu W.Q. et al. 2014). However, the biomechanical environment becomes more rigorous when ACDR is placed adjacent to a two-level fusion (Sekhon L.H.S et al. 2005). Because device failure or dislocation has not been found in the included studies, the more studies with large sample size and long-term follow-up are necessary to confirm the doubt.

4.3 *Biomechanical research on coupled motion*

The phenomenon of coupled motions has been well known. *In vitro*, it occurs in the thoracic and lumbar spines, but it is most remarkable in the cervical spine. Panjabi et al. reported that in flexion and extension the coupled motions were slight only with some translations in the sagittal plane (Panjabi M.M. et al. 2001). But as the axial rotation was applied the coupled lateral bending in the same direction could be observed at all levels. And as the lateral bending was applied, the coupled axial rotation in the same direction also presented. Pu et al. also revealed that the normal motion of cervical spines was a 3D coupled motion, especially in LB/RB direction, where a 35% rotation around X- axis existed (Pu T. et al. 2014). *In vivo*, Nagamoto et al. compared the control group to the cervical spondylosis group and found that mean axial rotation and mean coupled lateral bending observably decreased at C5–C6 and C6–C7 while mean coupled lateral bending markedly increased at C2–C3 and C3–C4 for the cervical spondylosis group. Despite both two groups showed the similar coupled motions (Nagamoto Y. et al. 2011).

On the other hand, through a series of complex, multi-planar motions, using hybrid method testing, Daniels et al. compared motions with implanted ACDR and ACDF to motions without implanted ACDR and ACDF (Daniels A.H. et al. 2012). In test seven biomimetic modes were used: Axial Rotation (AR); Flexion/Extension (FE); Lateral Bending (LB); combined FE and LB; combined FE and AR; combined LB and AR; and combined FE, LB, and AR. They found a trend about increased motion at adjacent levels in ACDF specimens in comparison to ACDR specimens.

5 CONCLUSION

In general, ACDF is still the most accepted procedure with satisfactory clinical outcome to date. As an alternative technique for ACDF, ACDR which can preserve the motion of instrumental level and prevent overload of adjacent levels has gained more and more popularity. However, the results of ACDR are less established than that of ACDF and need longer time follow-up.

In the future, the well-defined biomechanical testing method should be improved unceasingly so that the new devices, e.g., artificial discs, nucleus replacement, and flexible stabilizers, can be adequately tested *in vitro* and contrasted with fusion simulations. More studies about multilevel hybrid surgery and coupled motion characters are needed to reach a more reliable conclusion.

ACKNOWLEDGMENT

This project is supported by the National Key Technology R&D Program of China (Grant no. 2012BAI18B05) and Technology Innovation Program of Development Special Fund of Shenzhen Strategic Emerging Industry (Grant no. JSGG20130624162316425).

REFERENCES

Barrey, C. 2012. Cervical disc prosthesis versus arthrodesis using one-level, hybrid and two-level constructs: an in vitro investigation. European Spine Journal 21(3): 432–442.

Beutler, W.J. 2012. A biomechanical evaluation of a spacer with integrated plate for treating adjacent-level disease in the subaxial cervical spine. The Spine Journal 12(7): 585–589.

Cardoso, M.J. 2011. Cervical hybrid arthroplasty with 2 unique fusion techniques: Technical note. Journal of Neurosurgery: Spine 15(1): 48–54.

Clavenna, A.L. 2012. The biomechanical stability of a novel spacer with integrated plate in contiguous two-level and three-level ACDF models: an in vitro cadaveric study. The Spine Journal 12(2): 157–163.

Colle, K.O. 2013. Biomechanical evaluation of a metal-on-metal cervical intervertebral disc prosthesis. The Spine Journal 13(11): 1640–1649.

Cunningham, B.W. 2010. Biomechanical comparison of single-and two-level cervical arthroplasty versus arthrodesis: effect on adjacent-level spinal kinematics. The Spine Journal 10(4): 341–349.

Daniels, A.H. 2012. Examination of cervical spine kinematics in complex, multi-planar motions after anterior cervical discectomy and fusion and total disc replacement. The International Journal of Spine Surgery 6(1): 190–194.

DeVries, N.A. 2012. Biomechanical analysis of the intact and destabilized sheep cervical spine. Spine 37(16): E957–E963.

Faizan, A. 2012. Adjacent level effects of bi level disc replacement, bi level fusion and disc replacement plus fusion in cervical spine-a finite element based study. Clinical Biomechanics 27(3): 226–233.

Foley, K.T. 1999. The in vitro effects of instrumentation on multilevel cervical strut-graft mechanics. Spine 24(22): 2366.

Hey, H.W.D. 2013. Is hybrid surgery of the cervical spine a good balance between fusion and arthroplasty? Pilot results from a single surgeon series. European Spine Journal 22(1): 116–122.

Hofstetter, C.C. 2013. Zero-Profile anchored spacer reduces rate of dysphagia compared to ACDF with anterior plating. J Spinal Disord Tech 17(1): 636–642.

Jia, Z. 2014. Hybrid surgery for multilevel cervical degenerative disc diseases: a systematic review of biomechanical and clinical evidence. European Spine Journal 23(8): 1619–1632.

Lee, M.J. 2011. Disc replacement adjacent to cervical fusion: a biomechanical comparison of hybrid construct versus two-level fusion. Spine 36(23): 1932–1939.

Liu, W.Q. 2014. Biomechanical comparison of three-level cervical hybrid contract: an in vitro investigation. J Tsinghua Univ (Sci&Technol) 54(5): 685–689.

Majid, K. 2012. A comparative biomechanical study of a novel integrated plate spacer for stabilization of cervical spine: an in vitro human cadaveric model. Clinical Biomechanics 27(6): 532–536.

Martin, S. 2011. Kinematics of cervical total disc replacement adjacent to a two-level, straight versus lordotic fusion. Spine 36(17): 1359–1366.

Miao, J. 2013. Early follow-up outcomes of a new zero-profile implant used in anterior cervical discectomy and fusion. Journal of spinal disorders & techniques 26(5): E193–E197.

Mo, Z.J. 2014. Biomechanical effects of cervical arthroplasty with U-shaped disc implant on segmental range of motion and loading of surrounding soft tissue. European Spine Journal 23(3): 613–621.

Nagamoto, Y. 2011. In vivo three-dimensional kinematics of the cervical spine during head rotation in patients with cervical spondylosis. Spine 36(10): 778–783.

Paik, H. 2014. Do stand-alone inter-body spacers with integrated screws provide adequate segmental stability for multilevel cervical arthrodesis? The Spine Journal 14(8): 1740–1747.

Panjabi, M.M. 2007. Hybrid multidirectional test method to evaluate spinal adjacent-level effects. Clinical Biomechanics 22(3): 257–265.

Paxinos, O. 2009. Anterior cervical discectomy and fusion with a locked plate and wedged graft effectively stabilizes flexion-distraction stage-3 injury in the lower cervical spine: a biomechanical study. Spine 34(1): E9–E15.

Panjabi, M.M. 2001. Mechanical properties of the human cervical spine as shown by three-dimensional load–displacement curves. Spine 26(24): 2692–2700.

Pu, T. 2014. In vitro study on biomechanical comparison between cervical arthroplasty and fusion. Journal of Medical Biomechanics 29(2): 105–112.

Ryu, W.H.A. 2013. Long-term kinematic analysis of cervical spine after single-level implantation of Bryan cervical disc prosthesis. The Spine Journal 13(6): 628–634.

Scholz, M. 2011. A new zero-profile implant for stand-alone anterior cervical inter-body fusion. Clinical Orthopaedics and Related Research® 469(3): 666–673.

Stein, M.I. 2014. Biomechanics of an integrated inter-body device versus ACDF anterior locking plate in a single-level cervical spine fusion construct. The Spine Journal 14(1): 128–136.

Sekhon, L.H.S. 2005. Cervical arthroplasty after previous surgery: results of treating 24 discs in 15 patients. Journal of Neurosurgery: Spine 3(5): 335–341.

Sheng, S. 2011. Biomechanical comparison with cervical spines of porcin, calf and human. Journal of Medical Biomechanics (5): 380–384.

Vanek, P. 2013. Anterior Inter-body Fusion of the Cervical Spine with Zero-P Spacer: Prospective Comparative Study—Clinical and Radiological Results at a Minimum 2 Years after Surgery. Spine 38(13): E792–E797.

Wilke, H.J. 2011. Biomechanical in vitro evaluation of the complete porcine spine in comparison with data of the human spine. European Spine Journal 20(11): 1859–1868.

Wojewnik, B. 2013. Biomechanical evaluation of a low profile, anchored cervical inter-body spacer device in the setting of progressive flexion-distraction injury of the cervical spine. European Spine Journal 22(1): 135–141.

Bioinformatics and Biomedical Engineering – Chou & Zhou (Eds)
© 2016 Taylor & Francis Group, London, ISBN 978-1-138-02784-8

Mechanical conditions affect intervertebral disc degeneration concerning its water retention

W. Weina
Department of Biomedical Engineering, Tsinghua University, Beijing, China

W.Q. Liu
Department of Biomedical Engineering, Tsinghua University, Beijing, China
Department of Mechanical Engineering, Tsinghua University, Beijing, China

Z.H. Liao
Department of Mechanical Engineering, Tsinghua University, Beijing, China

ABSTRACT: Intervertebral disc plays an important role in conducting pressure and conferring flexibility of the spine. The disc degeneration occurs gradually with an increase in age. However, according to clinical data, specific occupations and lifestyles will largely aggravate this process. Surgery is an effective treatment for severe intervertebral disc disease, but spinal mechanical change caused by operation scheme and implants may accelerate the adjacent segments' lesion. This paper reviews related research for a deep insight into the relationship between mechanical conditions and intervertebral disc degeneration focusing on its water retention in mechanical conditions. In conclusion, moderate load and movement is recommended to maintain the health of the intervertebral disc. Therefore, biomechanical properties of different implants and operation schemes could be used as an evaluation criteria to determine its safety.

Keywords: load; water response; metabolism; remodeling; operative assessment

1 INTRODUCTION

1.1 *Change in composition with the intervertebral disc degeneration*

Intervertebral Disc (IVD) is a heterogeneous structure, which comprises a central gelatinous Nucleus Pulposus (NP) surrounded circumferentially by the Annulus Fibrosus (AF). The AF, a multiple-lamellae structure, is generally divided into the inner Layer Fibrosus (IA) and the outer layer fibrosus (OA). Type I collagen, the main component of OA, provides the basic frame and tensile of IVD. Type II collagen is abundant in IA and constitutes a network that is encompassed and interspersed by Proteoglycan (PG). High concentration and hydrophilic characteristic of PG produce enough osmotic pressure, ensuring a high water content of NP to resist compression load (Lai *et al.*, 1991). With increasing age, first, the content of type II collagen and its capture ability towards PG decrease. Then, PG and water molecule of NP and IA, and to a lesser degree IA, leak out. Simultaneously, type I collagen begins to appear in NP and increases the development of fibrosis. The composition of regional difference, which is large in young people, decreases with the deepening degeneration and becomes almost negligible in severe degenerated disc (Antoniou *et al.*, 1996).

1.2 *The mechanical environment of IVD*

Degeneration of IVD is a natural and inevitable process that mostly arises during twenties. Some specific activities or certain habits can exacerbate this process, such as long time staying in a vibration environment (e.g. truck driving), prolonged bending and twisting, the fatigue work, heavy manual labour, sedentariness or acute violent stress (Farfan, 1984, Adams and Hutton, 1983a, Elfering *et al.*, 2002). Anterior Cervical Discectomy Fusion (ACDF) is a golden standard to relieve the cervical pain caused by particular disc degeneration, but the degeneration subsequently occur in the adjacent segment of the surgical section. One theory is that the adjacent intervertebral disc compensates the lost motion of fusion segment and changes the original mechanical environment, which finally leads to the degeneration of the adjacent intervertebral disc. There are other theories affirming that the degeneration of the adjacent intervertebral disc probably is still a natural process based on the patient's habits (Helgeson *et al.*, 2013). Thus, to study the relationship between mechanical conditions and degeneration of IVD may help us to recognize ACDF better.

2 WATER RESPONSE WITH DEGENERATION

2.1 *Resistance model of IVD*

In the equilibrium state of IVD, the swelling pressure (Ps1), due to high osmotic pressure in NP, equals the hydrostatic pressure (Ph1) produced by tensile of collagen network and muscle. If there is an external load that increases the hydrostatic pressure, Ph2 > Ps1, the equilibrium will be disturbed. Also, the fluid will be squeezed out, with the concentration of proteoglycan and osmotic pressure increasing. The hydrostatic pressure will drop with the fluid discharging, the new equilibrium will be established when Ph2 = Ps2, and Vice versa (Urban *et al.*, 1979).

2.2 *The property of water response*

For the normal disc, most of the loads concentrate in NP and IA, where water changes drastically. The flow rate of the fluid under pressure is nonlinear, which slows over time. The experimental results show that to restore the water lost in the past 16 h of activities requires 8 h in which 70% of lost water is regained in the first hour. This is determined by the creep characteristic of the tissue of the intervertebral disc (Masuoka *et al.*, 2007, Ayotte *et al.*, 2001, Malko *et al.*, 2002, Tyrrell *et al.*, 1985). Besides, the flow volume is determined by the magnitude and frequency of loads, and almost negligible when the frequency is too high or the magnitude is too weak (Malko *et al.*, 2002). With increasing age and degeneration deepening, OA gradually bears more loads (Adams *et al.*, 1996, Adams, 2006). Meanwhile, IVD sacrifices more height with relatively less water discharge under pressure, and needs more time to recuperate (Urban and McMullin, 1988).

3 REGULATION EFFECT OF MECHANICAL CONDITIONS

3.1 *Load regulation: Gene expression*

3.1.1 *Static load*

Load affects the water content, oxygen level and PH in IVD. Sustained load may change the metabolism mode of the cell and remodel the matrix. A certain intensity of sustained load is recommended to keep the IVD healthy. It has been proved that the hydrostatic pressure of IVD when humans sprawl is about 0.1~1 Mpa, which can accelerate the synthesis of aggrecan and type II collagen in NP and IA (Ohshima *et al.*, 1995). When the hydrostatic pressure is too low, the proteoglycan in NP is released and the anabolism and renovation are inhibited (Urban and Maroudas, 1981).

However, if the load exceeds a certain threshold, catabolism becomes a predominant metabolism rather than anabolism. In this process, the synthesis of type II collagen, the most magnitude sensitive substance, is inhibited first and then its content decreases significantly in IA. Subsequently, the synthesis and content of aggrecan in NP decline and cell apoptosis increases significantly (Ohshima *et al.*, 1995).

3.1.2 Immobilization

To deprive the implication of immobilization from static load, researchers designed a static bending experiment. They found a synthesis reduction of PG and collagen on the concave side (compression) instead of the convex side (uncompression) (Court *et al.*, 2007). Animal (mouse) experiments have further demonstrated that increased loads on IVD can expedite the immobilization effect (Iatridis *et al.*, 1999). Though it is difficult to deprive the implication of immobilization from static load experiments, it suggests that increasing load is able to amplify the influence of immobilization and accelerate IVD degeneration.

3.1.3 Dynamic load

Dynamic load is the other factor that regulates the metabolism of the extracellular matrix of IVD depending on two parameters: frequency and strength. The sensitivity of different parts of IVD to dynamic loads is distinct (MacLean *et al.*, 2005). In NP, the anabolism genes of type I and II collagen and aggrecan mRNA can be up-regulated at 0.01 Hz under 1 Mp, while catabolism genes prefer 1 Hz. Conversely, in AF, gene expression mainly depends on the magnitude instead of the frequency (Maclean *et al.*, 2004, Walsh and Lotz, 2004, Wuertz *et al.*, 2009). If the frequency is higher than 1 Hz, catabolism of IVD increases remarkably and peaks at 5 Hz, which is the resonance frequency of IVD. Unfortunately, most of the vibration frequencies of vehicles in operation are between 3 and 8 Hz. This might be an explanation for the high IVD degeneration morbidity in vehicle drivers (Kasra *et al.*, 2006).

Similar to the static load, the overloading vibration strength is harmful as well. In 1998, Hotton fixed the spring between two bony vertebras of dogs to simulate high intension in movement. In this experiment, the content of PG and type II collagen decreases. Type I collagen increases in NP but decreases in AF, which could be considered as a sign of early degeneration, since the content and synthesis of type I collagen should be inconspicuous in the NP of normal IVD (Hutton *et al.*, 1998).

3.2 Load regulation: Substance transport

The metabolism of mature IVD is via fluid exchange: diffusion and convection. Micromolecules, such as oxygen and glucose, are interchanged by diffusion, while macromolecules, such as cytokines that are necessary to regulate metabolism, are transported by convection (Ferguson *et al.*, 2004). Sustained overloading diminishes the space between cells and prevents water diffusion. This results in oxygen inadequacy that generates the pH increasing in IVD and finally inhibits the synthesis of PG and collagen. Besides, the decreasing glucose content in this process increases cell apoptosis, which accelerates disc degeneration (Huang and Gu, 2008).

However, moderate intensity and frequency of load are necessary to maintain the disc's health. The speed and volume of liquid in convection under certain levels of dynamic load are higher than static load, which is helpful to stimulate the anabolic process. It has been proved that the water of IVD lost during 1 h sprint is greater than that lost during 7.5 h relative static activities (Adams and Hutton, 1983b, White and Malone, 1990).

3.3 Load affects the structure of IVD

The overloading strength can cause trauma of AF tissue and change the pressure model. Trauma increases the sagittal coefficient and circular coefficient, so the water content and diffusion effect decrease in AF (Gu *et al.*, 1999). The metabolism is inhibited subsequently and the solidification of tissue makes IVD become more vulnerable (Lotz *et al.*, 1998).

Long time immobilization is harmful to structure of IVD as well. Similar to the nature degeneration process, the disorganization of AF and decrease in tissue axial compliance, angular laxity and thickness of IVD are observed in a long time immobilization. Moreover, in the peripheral tissue of the joint, the contents of GAG and water decline as well. It may affect the elasticity and tension of the tissue, and give rise to exogenous tissue damage (Akeson *et al.*, 1973).

3.4 *Accumulation of mechanical effect*

The mechanical effect on matrix remodeling can be accumulated with the daily loading duration and sustained days (Wuertz *et al.*, 2009, Adams and Dolan, 1997). The accumulated anabolic products, such as PG and type II collagen in NP and IA area as well as type I collagen in AF, increase significantly in a short duration when moderate loads are applied to IVD until a new balance is achieved (MacLean *et al.*, 2005). However, a long period of insufficient rest will never recover the lost water and will accelerate IVD degeneration as well (Lotz and Chin, 2000).

Furthermore, the collagen cells turnover at a very slow speed for its high cell density and cross-linking structure of AF. So, the injured collagen molecules persist in the fibril structure and become more and more extensive with increasing age (Roughley, 2004). Such damages eventually result in tissue loss and destroy the mechanical strength of the tissue eventually. More so, the trauma may produce pain that limits the motion of the spine and generates the immobilization effect (Yong-Hing and Kirkaldy-Willis, 1983). This immobilization may exacerbate the degeneration, though the effect on matrix remolding can recover when the activity returns to normal (Muller *et al.*, 1994).

4 CONCLUSION

The degeneration of IVD is a complex process, which is affected by many factors. However, it is no doubt that mechanical conditions can regulate the gene expression, structure and substance transport of IVD.

Some clinical results suggest that there is a high risk of adjacent segment degeneration after fusion surgery. The reason has been speculated as the compensation action of adjacent segments. However, according to the *in vitro* experiment, we can find that the intradiscal pressure of adjacent segment after fusion surgery increases by 50%, but it is still in the range of 0.1–1 Mpa, which is definitely in the "safe window" as we described above (Eck *et al.*, 2002, Chang *et al.*, 2007) (Chang *et al.*, 2007). Admittedly, the data obtained from the *in vitro* experiment is not accurate for the change in water, elastic and strength of ligaments and muscles (Dekutoski *et al.*, 1994). However, it is still vulnerable to say that ACDF might generate the degeneration of adjacent segment only according to the pressure augment.

Furthermore, different degeneration degrees of IVD generate a distinct response to the same mechanical condition. If there is already a sign of degeneration of adjacent segments, it presents less compressive ability and more vulnerable to get trauma (Adams *et al.*, 2000, Chosa *et al.*, 2004). This may account for an abnormal clinical phenomena that multiple-level fusion presents less adjacent segment degeneration accidents compared with single-level fusion, since the adjacent segments of single-level fusion are more likely to suffer degeneration already, though adjacent segments of multiple-level fusion bear more movement and load.

Thus, to deeply know the reason of adjacent segment degeneration after fusion surgery needs to comprehensively consider the original degeneration degree itself and the mechanical condition including vibration frequency, strength, daily duration, and rest time. Moreover, the proper rehabilitation exercise is a key part to retard the adjacent segment degeneration process after fusion surgery.

REFERENCES

Adams, M.A. & Roughley, P.J. (2006), "What is intervertebral disc degeneration, and what causes it?", *Spine,* Vol. 31 No. 18, pp. 2151–2161.

Adams, M.A. & Dolan, P. (1997), "Could sudden increases in physical activity cause degeneration of intervertebral discs?", *Lancet,* Vol. 350 No. 9079, pp. 734–5.

Adams, M.A. & Hutton, W.C. (1983a), "The effect of fatigue on the lumbar intervertebral disc", *J Bone Joint Surg Br,* Vol. 65 No. 2, pp. 199–203.

Adams, M.A. & Hutton, W.C. (1983b), "The effect of posture on the fluid content of lumbar intervertebral discs", *Spine (Phila Pa 1976),* Vol. 8 No. 6, pp. 665–71.

Adams, M.A., Freeman, B.J., Morrison, H.P., Nelson, I.W. & Dolan, P. (2000), "Mechanical initiation of intervertebral disc degeneration", *Spine (Phila Pa 1976)*, Vol. 25 No. 13, pp. 1625–36.

Adams, M.A., Mcmillan, D.W., Green, T.P. & Dolan, P. (1996), "Sustained loading generates stress concentrations in lumbar intervertebral discs", *Spine (Phila Pa 1976)*, Vol. 21 No. 4, pp. 434–8.

Akeson, W.H., Woo, S.L., Amiel, D., Coutts, R.D. & Daniel, D. (1973), "The connective tissue response to immobility: biochemical changes in periarticular connective tissue of the immobilized rabbit knee", *Clin Orthop Relat Res*, No. 93, pp. 356–62.

Antoniou, J., Steffen, T., Nelson, F., WinterbottOM, N., Hollander, A.P., Poole, R.A., Aebi, M. & Alini, M. (1996), "The human lumbar intervertebral disc: evidence for changes in the biosynthesis and denaturation of the extracellular matrix with growth, maturation, ageing, and degeneration", *J Clin Invest*, Vol. 98 No. 4, pp. 996–1003.

Ayotte, D.C., Ito, K. & Tepic, S. (2001), "Direction-dependent resistance to flow in the endplate of the intervertebral disc: an ex vivo study", *J Orthop Res*, Vol. 19 No. 6, pp. 1073–7.

Chang, U.K., Kim, D.H., Lee, M.C., Willenberg, R., Kim, S.H. & Lim, J. (2007), "Changes in adjacent-level disc pressure and facet joint force after cervical arthroplasty compared with cervical discectomy and fusion", *J Neurosurg Spine*, Vol. 7 No. 1, pp. 33–9.

Chosa, E., Goto, K., Totoribe, K. & Tajima, N. (2004), "Analysis of the effect of lumbar spine fusion on the superior adjacent intervertebral disk in the presence of disk degeneration, using the three-dimensional finite element method", *J Spinal Disord Tech*, Vol. 17 No. 2, pp. 134–9.

Court, C., Chin, J.R., Liebenberg, E., Colliou, O.K. & Lotz, J.C. (2007), "Biological and mechanical consequences of transient intervertebral disc bending", *Eur Spine J*, Vol. 16 No. 11, pp. 1899–906.

Dekutoski, M.B., Schendel, M.J., Ogilvie, J.W., Olsewski, J.M., Wallace, L.J. & Lewis, J.L. (1994), "Comparison of in vivo and in vitro adjacent segment motion after lumbar fusion", *Spine (Phila Pa 1976)*, Vol. 19 No. 15, pp. 1745–51.

Eck, J.C., Humphreys, S.C., Lim, T.H., Jeong, S.T., Kim, J.G., Hodges, S.D. & An, H.S. (2002), "Biomechanical study on the effect of cervical spine fusion on adjacent-level intradiscal pressure and segmental motion", *Spine (Phila Pa 1976)*, Vol. 27 No. 22, pp. 2431–4.

Elfering, A., Semmer, N., Birkhofer, D., Zanetti, M., Hodler, J. & Boos, N. (2002), "Risk factors for lumbar disc degeneration: a 5-year prospective MRI study in asymptomatic individuals", *Spine (Phila Pa 1976)*, Vol. 27 No. 2, pp. 125–34.

Farfan, H.F. (1984), "The torsional injury of the lumbar spine", *Spine (Phila Pa 1976)*, Vol. 9 No. 1, pp. 53.

Ferguson, S.J., Ito, K. & Nolte, L.P. (2004), "Fluid flow and convective transport of solutes within the intervertebral disc", *J Biomech*, Vol. 37 No. 2, pp. 213–21.

Gu, W.Y., Mao, X.G., Foster, R.J., Weidenbaum, M., Mow, V.C. & Rawlins, B.A. (1999), "The anisotropic hydraulic permeability of human lumbar anulus fibrosus. Influence of age, degeneration, direction, and water content", *Spine (Phila Pa 1976)*, Vol. 24 No. 23, pp. 2449–55.

Handa, T., Ishihara, H., Ohshima, H., Osada, R., Tsuji, H., & Obata, K.I. (1997). "Effects of hydrostatic pressure on matrix synthesis and matrix metalloproteinase production in the human lumbar intervertebral disc", Spine, Vol. 22 No. 10, pp. 1085–1091.

Helgeson, M.D., Bevevino, A.J. & Hilibrand, A.S. (2013), "Update on the evidence for adjacent segment degeneration and disease", *Spine J*, Vol. 13 No. 3, pp. 342–51.

Huang, C.Y. & Gu, W.Y. (2008), "Effects of mechanical compression on metabolism and distribution of oxygen and lactate in intervertebral disc", *J Biomech*, Vol. 41 No. 6, pp. 1184–96.

Hutton, W.C., Toribatake, Y., Elmer, W.A., Ganey, T.M., Tomita, K. & Whitesides, T.E. (1998), "The effect of compressive force applied to the intervertebral disc in vivo. A study of proteoglycans and collagen", *Spine (Phila Pa 1976)*, Vol. 23 No. 23, pp. 2524–37.

Iatridis, J.C., Mente, P.L., Stokes, I.A., Aronsson, D.D. & Alini, M. (1999), "Compression-induced changes in intervertebral disc properties in a rat tail model", *Spine (Phila Pa 1976)*, Vol. 24 No. 10, pp. 996–1002.

Kasra, M., Merryman, W.D., Loveless, K.N., Goel, V.K., Martin, J.D. & Buckwalter, J.A. (2006), "Frequency response of pig intervertebral disc cells subjected to dynamic hydrostatic pressure", *J Orthop Res*, Vol. 24 No. 10, pp. 1967–73.

Lai, W.M., Hou, J.S. & Mow, V.C. (1991), "A triphasic theory for the swelling and deformation behaviors of articular cartilage", *J Biomech Eng*, Vol. 113 No. 3, pp. 245–58.

Lotz, J.C. & Chin, J.R. (2000), "Intervertebral disc cell death is dependent on the magnitude and duration of spinal loading", *Spine (Phila Pa 1976)*, Vol. 25 No. 12, pp. 1477–83.

Lotz, J.C., Colliou, O.K., Chin, J.R., Duncan, N.A. & Liebenberg, E. (1998), "Compression-induced degeneration of the intervertebral disc: an in vivo mouse model and finite-element study", *Spine (Phila Pa 1976)*, Vol. 23 No. 23, pp. 2493–506.

Maclean, J.J., Lee, C.R., Alini, M. & Iatridis, J.C. (2004), "Anabolic and catabolic mRna levels of the intervertebral disc vary with the magnitude and frequency of in vivo dynamic compression", *J Orthop Res*, Vol. 22 No. 6, pp. 1193–200.

Maclean, J.J., Lee, C.R., Alini, M. & Iatridis, J.C. (2005), "The effects of short-term load duration on anabolic and catabolic gene expression in the rat tail intervertebral disc", *J Orthop Res*, Vol. 23 No. 5, pp. 1120–7.

Malko, J.A., Hutton, W.C. & Fajman, W.A. (2002), "An in vivo MRI study of the changes in volume (and fluid content) of the lumbar intervertebral disc after overnight bed rest and during an 8-hour walking protocol", *J Spinal Disord Tech*, Vol. 15 No. 2, pp. 157–63.

Masuoka, K., Michalek, A.J., Maclean, J.J., Stokes, I.A. & Iatridis, J.C. (2007), "Different effects of static versus cyclic compressive loading on rat intervertebral disc height and water loss in vitro", *Spine (Phila Pa 1976)*, Vol. 32 No. 18, pp. 1974–9.

Muller, F.J., Setton, L.A., Manicourt, D.H., Mow, V.C., Howell, D.S. & Pita, J.C. (1994), "Centrifugal and biochemical comparison of proteoglycan aggregates from articular cartilage in experimental joint disuse and joint instability", *J Orthop Res*, Vol. 12 No. 4, pp. 498–508.

Ohshima, H., Urban, J.P. & Bergel, D.H. (1995), "Effect of static load on matrix synthesis rates in the intervertebral disc measured in vitro by a new perfusion technique", *J Orthop Res*, Vol. 13 No. 1, pp. 22–9.

Roughley, P.J. (2004), "Biology of intervertebral disc aging and degeneration: involvement of the extracellular matrix", *Spine (Phila Pa 1976)*, Vol. 29 No. 23, pp. 2691–9.

Tyrrell, A.R., Reilly, T. & Troup, J.D. (1985), "Circadian variation in stature and the effects of spinal loading", *Spine (Phila Pa 1976)*, Vol. 10 No. 2, pp. 161–4.

Urban, J.P. & Maroudas, A. (1981), "Swelling of the intervertebral disc in vitro", *Connect Tissue Res*, Vol. 9 No. 1, pp. 1–10.

Urban, J.P. & Mcmullin, J.F. (1988), "Swelling pressure of the lumbar intervertebral discs: influence of age, spinal level, composition, and degeneration", *Spine (Phila Pa 1976)*, Vol. 13 No. 2, pp. 179–87.

Urban, J.P., Maroudas, A., Bayliss, M.T. & Dillon, J. (1979), "Swelling pressures of proteoglycans at the concentrations found in cartilaginous tissues", *Biorheology*, Vol. 16 No. 6, pp. 447.

Walsh, A.J. & Lotz, J.C. (2004), "Biological response of the intervertebral disc to dynamic loading", *J Biomech*, Vol. 37 No. 3, pp. 329–37.

White, T.L. & Malone, T.R. (1990), "Effects of running on intervertebral disc height", *J Orthop Sports Phys Ther*, Vol. 12 No. 4, pp. 139–46.

Wuertz, K., Godburn, K., Maclean, J.J., Barbir, A., Donnelly, J.S., Roughley, P.J., Alini, M. & Iatridis, J.C. (2009), "In vivo remodeling of intervertebral discs in response to short- and long-term dynamic compression", *J Orthop Res*, Vol. 27 No. 9, pp. 1235–42.

Yong-Hing, K. & Kirkaldy-Willis, W.H. (1983), "The pathophysiology of degenerative disease of the lumbar spine", *Orthop Clin North Am*, Vol. 14 No. 3, pp. 491–504.

Viscoelasticity of the intervertebral segments after fusion under continuous compression

B.Q. Pei, Z.Y. Liu & H. Li
School of Biological Science and Medical Engineering, Beihang University, Beijing, China

Y.Y. Pei
Senior High School Grade Two of the High School Affiliated to Beihang University, Beijing, China

ABSTRACT: Cushioning is the essential function of the spine to protect itself, and the major form for the spine to resist the external load. Viscoelasticity is one of the most important indicator of the cushioning capacity of the spine. By comparing the creep characteristics before and after the fusion, this paper examines the influence of fusion on the cushioning capacity of the spine. Ten ovine lumbar two-level segments were continuously compressed under an axial pressure of 0.4 Mpa for 30 minutes. Each sample was used twice in intact and fused groups respectively. Fused groups were internally fixed rigidly without curvature changes. The deformation of the adjacent segment were recorded using an optics automatic extensometer built in the Instron mechanical testing system. The creep curves of the whole samples were also recorded and a 5-parameter rheological model was used for curve fitting. The equilibrium displacement and time were calculated using the fitting functions. There were no significant differences between axial and lateral creep deformations, while the total creep displacement decreased significantly. L/S_E, L/S_2 and τ_2 were significantly smaller, but there was no significant difference on L/S_1 and τ_1. S_1 and S_2 increased significantly, and the equilibrium displacement and time reduced significantly. Based on these results, we can draw the following conclusions. The fusion with only stiffness increased had no effect on creep displacement of the adjacent segment. The viscosity of the whole spine decreased but the stiffness increased after fusion. The shortened equilibrium time illustrates that the whole cushioning capacity decreased after fusion. Fast and slow responses correspond bone tissue and Intervertebral Disc (IVD) tissue respectively, and fusion mainly changed the viscoelasticity properties of IVD tissue, while no significant change was found on viscoelasticity properties of bone tissue.

Keywords: creep; fusion; adjacent segment; viscoelasticity; cushioning capacity; continuous compression

1 INTRODUCTION

Fusion has been widely applied to clinical treatment of spinal diseases; intervertebral fusion removes the movement of the diseased IVD through the method of internally fixed fusion auxiliary with bone graft, which changes the original movable connection into a rigid connection between the vertebrae. Therefore, the increase of stiffness on spine after operation is the most significant change. Spine with increased stiffness requires more load to complete the original movement. Many patients are afraid of the pain on the surgical site, and it is rare to move the same scope before operation. However, spine must remain upright and patients walk in short time is the common load type after the surgical operation.

Cushioning is the essential function of the spine to protect itself, and the major form for the spine to resist the external load. When there is external load, the IVD will extend the load time by its creep deformation, thus to reduce the speed of the load for spine protection. Cushion capacity should be reduced after fusion but how much it reduced is not so clear now. Viscoelasticity is the major index to measure the cushion capacity of spine.

It is required to contain liquid phase and solid phase composition the same time when described creep. Therefore, the rheological model is a good method for description of the time dependence of IVD (*Cassidy et al. 1990; Li et al., 1995*). The rheological model describes viscoelastic properties using mathematics method through the combination of spring and damper. It can be used to predict mechanical state of IVD in the expected time (*Pollintine et al., 2010*). Some studies explore displacement-time behavior on viscoelastic body in the creep by using a spring and a damper in parallel (Voigt model). Then describe the initial elastic response followed by a spring installed in series. There are many factors related to time dependent behavior, including geometric structure and composition.

This paper adopts two-level segments model of ovine lumbar vertebra, to obtain the creep property changes on whole and adjacent segments after fusion. Simulate the lumbar compression load when human stand under daily load, continuously action for 30 minutes. 5-parameter equation was used to fitting its creep displacement, find out the corresponding tissue change and its cause.

2 MATERIAL AND METHOD

2.1 Sample processing

In total, Ten two-level motion segment specimens of ovine were tested in this research. Each sample was used in intact group first and in fused group subsequently. The recovery time between the first intact experiment and the second fused experiment was long enough. The ovine (about one year old) lumbar vertebra samples were obtained from the market nearby within a few hours death. The sample was cut at IVD between L3 and L4, giving two motion segment L1-L3 and L4-L6, keep the superior and inferior surface parallel to each endplate. The muscles, fats surrounding by the specimens were removed left only vertebras, cartilages and ligaments. The specimens were then wrapped with gauze soaked in saline and sealed with cling films. After the processing, all the samples were stored in a refrigerator temperature in −20 degree until they were prepared to experimentation.

The inferior surface has to embed using denture based resin denture powder and denture liquid making the specimens satisfy the apparatus before the experiment. 40 mg denture powder was poured into a container, then 20 ml denture liquid was added and mixture with it. In dough-stage, the mixture change into a dough-like plastic substance and stickiness was disappeared, it was the best stage to insert the specimen, specimen was inserted into the middle of the container. Thirty minutes later, the embedding was finished when the specimen fully integrated with the denture powder.

A fusion was performed after the intact experiment. Four screws were attached to the steel across the IVD among the two bottom segments. No additional bone graft or a cage was attached as rigidity change was the main factor in our experiment. Curvature didn't change after fusion.

2.2 Experiment methods

The specimens were unfrozen at room temperature before the experiment. Specimen was fixed on INSTRON electric servo hydraulic test frame shown in Figure (1). A preload consist of a triangular cyclic wave compressive loading was applied, which trough value was 0.1 MPa, wave peak was 0.4 MPa, frequency of 1/3 Hz. The cyclic compressive loading was disappearing 60 seconds later. The pressure approach to 0.4 MPa in 1.5 s, holds for 30 minutes (simulate stand) and then unload it. Total displacement and stress were recorded by INSTRON electric servo hydraulic test frame. Meanwhile, an optics automatic extensometer built in the Instron mechanical testing system was used to capture and calculate the displacement of the axial and lateral. Four marker points were marked on the adjacent IVD convenient for camera capture and calculate the displacement.

Figure 1.   INSTRON electric servo hydraulic test frame and intact and fused specimens.

The displacement (d, mm) was a function of time (τ, s), applied load (L, N), and stiffness (S, N/mm) were fit to a 5-parameter theological model (Eq. (1)) is composed of two Voigt solids and a spring in series (*Keller et al., 1998; Johannessen et al., 2006*). This model was applied to fitting the creep curves before and after fusion. Thus the viscoelasticity parameters were obtained.

$$d(t = t_i \rightarrow t_{i+1}) = L \times \left[\left(\frac{1}{S_1}(1 - e^{-\frac{t}{\tau_1}}) \right) + \left(\frac{1}{S_2}(1 - e^{-\frac{t}{\tau_2}}) \right) + \frac{1}{S_E} \right] \qquad (1)$$

Expand the brackets and simplify it and get an equation (Eq. (2)) below.

$$d(t = t_i \rightarrow t_{i+1}) = \frac{L}{S_1} + \frac{L}{S_2} + \frac{L}{S_E} - \frac{L}{S_1}e^{-\frac{t}{\tau_1}} - \frac{L}{S_2}e^{-\frac{t}{\tau_2}} \qquad (2)$$

Some simple letter represent the constant parameter in Eq. (2) were used.

$$a = \frac{L}{S_1} + \frac{L}{S_2} + \frac{L}{S_E}, \, b = \frac{L}{S_1}, \, c = \frac{1}{\tau_1}, \, d = \frac{L}{S_2}, \, e = \frac{1}{\tau_2}$$

Then Eq. (2) could represent like below:

$$d(t = t_i \rightarrow t_{i+1}) = a - be^{-ct} - de^{-et} \qquad (3)$$

Five parameters in Eq. (1) were corresponds to different meanings. L was the creep load, S_E represents for elastic response, τ_1 and S_1 corresponds to fast response while τ_2 and S_2 corresponds to slow response, $i \rightarrow i+1$ was related to the creep and recovery. Elastic stiffness coefficient S (N/mm) and the time constant τ(s) of each process were calculated through fitting. Elastic response represents the 1.5 s when loads applied and the first 30 seconds of the creep. The equilibrium displacement was calculated at $t = \infty$ when the viscous effect was disappearing completely. But it was simplified as 99% of the equilibrium displacement in our experiment. The cushion capacity was proportional to the equilibrium displacement when the external applied load was a constant value.

Displacement-time cures were calculated by substituting the parameters in Eq. (3) with the average data obtained from the experiment using the MATLAB fitting curves toolbox.

The statistical difference was analyzed using SPSS statistical analysis software v.19.0 (SPSSInc, Chicago, IL, USA). Lateral and axial creep before and after fusion were tested. The total displacement before and after fusion was tested. Five parameters in Eq. (3), the equilibrium displacement and the equilibrium time before and after fusion were tested. The statistical research belongs to the one-way repeated measures MANOVA, T test were performed to determine the significance, significance was set at $p = 0.05$.

3 RESULTS

Axial displacement in intact group was 0.31 (± 0.13) mm, it is 0.36 (± 0.13) mm in fused group. Lateral displacement in intact group was 0.07 (± 0.05) mm, it is 0.06 (± 0.06) mm in fused group. Both of them showing no significant difference. Total displacement in intact group was 0.73 (± 0.11) mm, it is 0.47 (± 0.13) mm in fused group. Total displacement significantly smaller after fusion, mainly caused by the increase of stiffness.

MATLAB Curve fitting tool was used for curve fitting. A 5-parameters rheological model was used for fitting the displacement-time curves. The model consists of elastic response (S_E), fast response (τ_1, S_1, L/S_1) and slow response (τ_2, S_2, L/S_2). The result of the parameters were shown in Table (1).

Elastic response including the displacement in 1.5 s when initial load applied and the first 30 s of the creep. L/S_E decreased significantly after fusion indicate the instantaneous resistance to load increased when the high rigidity equipment was fixed in it. Thus could protect the operation segment from a clinical perspective.

Fast response had no significant change except S_1. The physical meaning of fast response is a tissue consists of a small viscosity in parallel with a high rigidity spring. The displacement won't change significantly as the rigidity of this tissue changed significantly while viscosity didn't change after fusion. So we can judge that the fast response corresponding to the vertebra. The internal fixed equipment in parallel with the vertebral and then the rigidity of the specimens increased, so S_1 increased significantly. Viscosity parameter didn't change significantly because nothing that would change viscosity was added. The creep displacement of vertebra didn't change significantly when rigidity increased because it's original displacement was small under 0.4 Mpa.

All parameters in slow response changed significantly after fusion. τ_2 and L/S_2 decreased while S_2 increased significantly. The physical meaning of slow response is a tissue consists of a high viscosity in parallel with a small rigidity spring. We could judge that the slow response was corresponding to IVD tissue as the viscosity and rigidity of this part changed

Table 1. Parameters before and after fusion. (*: $p < 0.05$, **: $p < 0.001$).

	Intact group	Fused group
τ_1 (s)	73.85 ± 8.18	73.25 ± 8.60
L/S_1 (mm)	0.16 ± 0.06	0.120 ± 0.06
S_1 (N/mm)*	958.16 ± 265.66	1392.88 ± 467.15
τ_2 (s)**	2236.29 ± 208.23	1770.47 ± 174.09
L/S_2 (mm)**	1.00 ± 0.15	0.51 ± 0.13
S_2 (N/mm)**	142.35 ± 27.15	284.80 ± 52.09
L/S_E (mm)*	0.74 ± 0.34	0.48 ± 0.11
S_E (N/mm)	225.66 ± 96.84	302.48 ± 72.10
Equilibrium displacement*	1.91 ± 0.45	1.11 ± 0.24
Equilibrium time*	8944.775 ± 1188.635	6782.04 ± 835.2192

Table 2. The proportion of each parameters contribute to total equilibrium displacement.

	Intact (%)	Fused (%)
L/S1	8.4	10.8
L/S2	52.4	45.9
L/SE	38.7	43.2

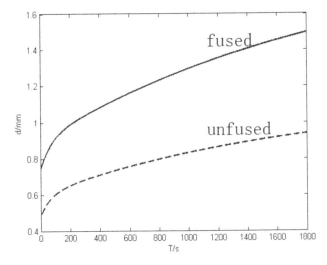

Figure 2. Displacement-time curves before and after fusion.

after fusion. Viscosity property was more important than rigidity property in IVD tissue, and dominate the creep displacement, is the main embodiment of the cushion capacity. The internal fixed equipment will suffer more load though we didn't cut the IVD in fusion, the fusion IVD almost didn't suffer any load. Thus two-level segment left only one IVD after fusion, leading to the viscosity property τ_2 decreased significantly. IVD contributes the most to the total displacement, it will lead to a significant decrease on L/S$_2$. A high rigidity equipment added to the segment will result in an increase of rigidity, so S$_2$ increased significantly.

The percentage of each component contribute to equilibrium displacement was calculated using data in Table (1). From Table (2) we can know the proportion of slow response reduced after fusion, the proportion of fast response and elastic response increased accordingly. Means the displacement of vertebra will be larger. 52% of equilibrium displacement undertake by two vertebra before turns to 45.9% of equilibrium displacement taken by adjacent IVD after fusion. We can see that adjacent IVD suffer more even though the axial and lateral displacement didn't change significantly.

Equilibrium displacement and equilibrium time was associated with the cushion capacity. A large equilibrium displacement when the load was a constant denotes each displacement suffers a smaller load, it was same to equilibrium time. The curve fitting result was shown in Figure (2). Equilibrium displacement and equilibrium time in intact group was significantly higher than fused group indicates the cushion capacity was decreased after fusion. Means the load speed was increased, so the spine has an increased probability to break down.

4 DISCUSSIONS

Intact two-level ovine lumbar sample and the two-level ovine lumbar sample with increased stiffness are applied with continuous compressive load for 30 minutes separately, their creep viscoelastic properties are compared and analyzed, and it is found that the cushion capacity of the fusion group sample with increased stiffness decreases significantly. Meanwhile, adjacent segments of fusion group and spine bear a greater cushion burden.

Loading time of 30 minutes is relatively short compared with many researches on the viscoelastic properties of IVD (usually the loading time ranges from 4 hours to 8 hours (Adams et al., 1996)), but for daily load, it is not realistic to continuously sit or stand for 4–8 hours. 30 minutes of continuous load is reasonable, and many experiments adopt loading time of 30 minutes (Wang H. 2014). As for the selection of sample, samples of sheep spine are very important model in many studies (Goldschlager et al., 2011). Many researches using ovine vertebra as they sample (Wang et al., 2014). According to the regeneration treatment strategy, the IVD tissue of human and sheep has similar biochemical composition and biomechanical situation, so it is very appropriate to use ovine as an animal model. Although there are some differences in the appearance and size, the sheep's average intervertebral pressure (0.75 Mpa) (Reitmaier et al., 2012) is very similar to previous results (Wilke et al., 1999) about human's intervertebral pressure in activities. According to the requirements of this research, the research on the impact of changing stiffness on viscoelastic properties before and after change spinal fusion is studied; finally the impact of fusion stiffness on cushion function is studied, it is appropriate to take sheep sample as study subject.

5   CONCLUSIONS

1. Fusion with stiffness increased only has no effect on creep displacement of adjacent segments.
2. The whole viscosity decreased and stiffness increased of the whole spine after fusion.
3. The shortened equilibrium time illustrates that the whole cushioning capacity decreased after fusion.
4. Fast and slow response correspond bone tissue and Intervertebral Disc (IVD) tissue respectively, and fusion mainly changed the viscoelasticity properties of IVD tissue, while no significant change on viscoelasticity properties of bone tissue.

REFERENCES

Adams M.A., McMillan D.W., Green T.P., et al. Sustained loading generates stress concentrations in lumbar intervertebral discs [J]. Spine, 1996, 21(4): 434–438.
Cassidy J.J., Silverstein M.S., Hiltner A., et al. A water transport model for the creep response of the intervertebral disc [J]. Journal of Materials Science: Materials in Medicine, 1990, 1(2): 81–89.
Goldschlager T., Rosenfeld J.V., Ghosh P., et al. Cervical interbody fusion is enhanced by allogeneic mesenchymal precursor cells in an ovine model [J]. Spine, 2011, 36(8): 615–623.
Johannessen W., Cloyd J.M., et al. Trans-endplate nucleotomy increases deformation and creep response in axail loading. Ann Biomed Eng. 2006, 34(4): 687–696.
Keller T.S., Hansson T.H., et al. In vivo creep behavior of the normal and degenerated porcine intervertebral disk: a preliminary report. J. Spinal Disord. 1998, 1(4): 267–278.
Li S., Patwardhan A.G., Amirouche F.M.L., et al. Limitations of the standard linear solid model of intervertebral discs subject to prolonged loading and low-frequency vibration in axial compression [J]. Journal of biomechanics, 1995, 28(7): 779–790.
Pollintine P., van Tunen M.S.L.M., Luo J., et al. Time-dependent compressive deformation of the ageing spine: relevance to spinal stenosis [J]. Spine, 2010, 35(4): 386–394.
Reitmaier S., Wolfram U., Ignatius A., et al. Hydrogels for nucleus replacement—facing the biomechanical challenge [J]. Journal of the mechanical behavior of biomedical materials, 2012, 14: 67–77.
Wang H., Weiss K.J., Haggerty M.C., et al. The effect of active sitting on trunk motion [J]. Journal of Sport and Health Science, 2014, 3(4): 333–337.
Wilke H.J., Neef P., Caimi M., et al. New in vivo measurements of pressures in the intervertebral disc in daily life [J]. Spine, 1999, 24(8): 755–762.

Biosignal/image processing and analysis

Bioinformatics and Biomedical Engineering – Chou & Zhou (Eds)
© 2016 Taylor & Francis Group, London, ISBN 978-1-138-02784-8

Recognition of sequential upper limb movements based on surface Electromyography (sEMG) signals

B.Y. Zhang, Z.T. Zhou, E.W. Yin, J. Jiang & D.W. Hu
College of Mechatronics and Automation, National University of Defense Technology, Changsha, Hunan, China

ABSTRACT: Recently, surface Electromyography (sEMG) signals have been widely used for the detection of limb actions and applied as control signals in a Human-Computer Interaction (HCI) system. As previous studies indicated, typical natural human actions are often composed of several sequential movements, whereas different human actions are composed of different sequential movements. These observations could be addressed and taken advantage of to improve the performance of the HCI system, if the sequential movements could somehow be efficiently detected and correctly recognized. In this paper, an approach to recognize human action by detecting the sequential movements from the sEMG signals of upper limb muscles, was proposed and realized. The sEMG signals of Anterior Deltoid muscle (AD), Biceps Brachii muscle (BB) and Flexor Digitorum Superficialis muscle (FDS) were acquired, and the features of Mean Absolute Value (MAV) and Waveform Length (WL) were then extracted. Seven-class actions were recognized by means of decision-tree, and the classification accuracy of MAV and WL reached to 95.24% and 94.05%, respectively. In addition, the features of MAV and WL were fused on both feature-level and decision-level, and the classification accuracy increased to 96.43% and 98.81%, respectively. The results indicated that the sequential feature of sEMG signals could be exploited effectively and multi-feature fusion method could increase the classification accuracy further. The approach of this paper contributed to realize a more practical HCI system by recognizing human upper limb movements sequentially.

Keywords: sEMG signals; upper limb; sequential movements; decision-tree; multi-feature fusion

1 INTRODUCTION

Surface electromyography (sEMG) signals are the summation of each motor unit action potential exposed to the electrode (Bida 2005). In micro view, sEMG signals contain the information of the number and spatial pattern of the motoneurons, as well as the conduction velocity of the neuropotentials. In macro view, sEMG signals contain the information of the accurate motor direction and torque of the joints, which is related to the motor pattern of the muscles (Fukuda 2006).

To date, sEMG signals have been widely and effectively applied for the detection and recognition of limb actions. Li et al. acquired the elbow joint angle and the sEMG signals of biceps brachii muscle and triceps brachii muscle, then mapped the non-linear relationship between the arm movement status and the sEMG signals (Li 2005). Phinyomark et al. recognized six-class wrist movements by using 37 time domain and frequency domain features of sEMG signals, and evaluated the performance and application scope of each feature (Phinyomark 2012). Building on this foundational research, sEMG signals have been widely used as control signals in a Human-Computer Interaction (HCI) system (Sankai 2010). Moving towards functional application, a lower limb rehabilitation exoskeleton system based on sEMG signals was recently designed and developed by Zhejiang University in China (Li 2009). Rosen et al. realized an upper limb motion control with the sEMG signals from the muscles near shoulder and elbow joints (Rosen 2007).

Lately, the HCI system based on natural human actions has been studied further. Previous studies have shown that the human complex actions are composed of sequential discrete movements. Human motor nervous system could analyze the requirements of tasks, and generate proper sequential movements in real time (Gross 2002, Rucci 2007). To catch a glimpse of natural human actions, sequential movements need to be well employed in the interaction paradigm design. Then, the HCI system based on natural human actions could be realized.

In this paper, we proposed and realized an approach to recognize human action by detecting the sequential movements from the sEMG signals of upper limb muscles. First, we acquired the sEMG signals from Anterior Deltoid muscle (AD), Biceps Brachii muscle (BB) and Flexor Digitorum Superficialis muscle (FDS). Second, the features of Mean Absolute Value (MAV) and Waveform Length (WL) were extracted. Third, decision-tree and multi-feature fusion method were used to recognize seven-class sequential movements. The experimental results indicated that the classification accuracy of multi-feature fusion method increased comparing to single-feature method. The approach of this paper could be used for online control and considered as an idealized model for natural human actions.

2 EXPERIMENTS

2.1 Task description and experimental procedure

There were three simple movements: shoulder joint forward spin 90 degrees (S), elbow joint inward spin 90 degrees (E) and five-finger grasp (H). Combining the three movements in order, there were six-class actions: 'S-E-H', 'S-H-E', 'E-S-H', 'E-H-S', 'H-S-E' and 'H-E-S'. The different actions had the same initial position and final position, and only had different order relation of sequential movements. In a run, there were seven trials: six above actions and a 'Relax' state. Four healthy male subjects with the average age of 24.25 participated in the experiments in good state. Moreover, the seven trials generated randomly so that subjects could not prejudge the actions coming soon. Each subject executed three runs and there were 84 trials totally.

Before the experiments started, subjects familiarized the actions in ten minutes. Then, subjects stood in comfortable condition and wait the prompt messages. When the prompt message, such as 'S-E-H', appeared on the screen, the subject reacted to the prompt message in four seconds. After the prompt message disappeared, the subject executed the corresponding actions in three seconds. After a trial was executed, the subject had a rest in five seconds so that the next trail would be well executed. When the seven trials were executed, the prompt message 'End' appeared and the run was over. To eliminate the error of muscle fatigue, there was two minutes between two adjacent runs. Thus, the experimental process was finished.

2.2 Signal acquisition and preprocessing

The non-invasive method was used to acquire the sEMG signals, which was secure and convenient. The sEMG signals of AD, BB and FDS were recorded by the MediTrace Ag/AgCl electrodes and amplified by the BrainAMP MR Plus signal amplifier made by Brain Products in Germany. The movements of S, E and H were mainly actuated by AD, BB and FDS, respectively. As shown in Figure 1, the three muscle positions were: AD (CH.1 & 2), BB (CH.3 & 4) and FDS (CH.5 & 6). The sEMG signals were mainly recorded in the CH.1, 3 & 5, and the CH.2, 4 & 6 were regarded as reference.

The sEMG signals were sampled at 250 Hz frequency. The spatial filtering of pairwise subtraction from the six channels was used and the raw data turned into three channels. Then, the electrocardiogram and common noise were eliminated as well as the SNR enhanced. To eliminate the power frequency, the sEMG signals were trapped at 50 Hz and the 10–40 Hz band-pass filter was chosen. The sEMG signals were divided into representative segments, which was convenient to analyze the characteristics of the movements in the sEMG signals.

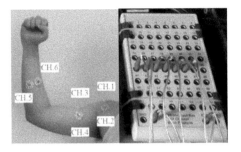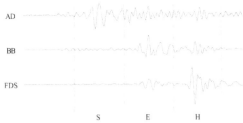

Figure 1. The electrode positions (CH.1~CH.6) and the sEMG signals after preprocessing. AD (Anterior Deltoid muscle), BB (Biceps Brachii muscle) and FDS (Flexor Digitorum Superficialis muscle).

The overlapped segmentation method was employed, and it was the key point to set the window length and the increment well.

3 SIGNAL PROCESSING

3.1 Feature extraction

The inherent characteristics of sEMG signals are effectively illustrated by means of feature extraction method. In this way, the correctness and instantaneity are ensured as well as the redundant information is furthest reduced. Comparing to frequency domain features, time domain features could be good and easy in sEMG signal classification (Phinyomark 2012). Therefore, two time domain features, MAV and WL, were extracted.

MAV is one of the most popular features used for sEMG signal analysis. MAV is an average of absolute value in the segment, which relatively accurately illustrates the change trend of the absolute value in sEMG signals. MAV could be defined as Equation 1:

$$MAV = \frac{1}{N}\sum_{i=1}^{N}|x_i| \tag{1}$$

WL is the cumulative length of sEMG signal waveform over the time segment, which is a complexity measure of sEMG signals. Different from MAV, the data between two near time points are mainly illustrated by WL. WL could be defined as Equation 2:

$$WL = \sum_{i=1}^{N-1}|x_{i+1} - x_i| \tag{2}$$

3.2 Recognition method

3.2.1 Recognition method of single-feature

Since the six-class actions were composed of three different simple movements in order, MAV and WL data segments in the stimulus duration were divided into three intervals so as to detect the three movements sequentially. As shown in Figure 2, the three channels were divided into nine intervals, in which P11~P33 was the peak value of the corresponding interval.

When the joint is still, the relevant muscles are resting-state. Once the joint begins to move, the relevant muscles will become action-state. At this time, the absolute value of the relevant muscle sEMG signals will apparently increase as well as the signal waveform will oscillate more violently, so MAV and WL will be also observed the corresponding peak value. Therefore, the effective peak value was the maximum in this interval. For instance, the effective peak values of 'S-E-H' were P11, P22 and P33 (underlined in Fig. 2). The judgment method was to eliminate the errors and omissions of nonstandard movements effectively.

137

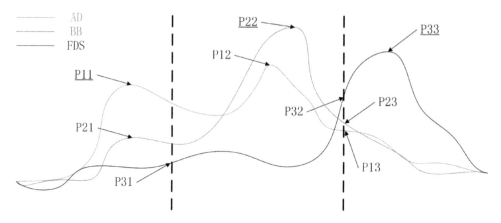

Figure 2. The schematic diagram of interval segmentation. P11~P33 was the peak value of the corresponding interval.

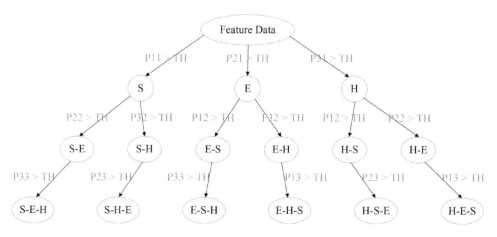

Figure 3. The decision-tree for recognition. P11, P21, P31, etc. were the effective peak values.

According to previous works and experiences, MAV and WL of '*Relax*' are not observed the apparent peak value and the absolute values are very low, so a threshold (TH) was set to recognize the action '*Relax*'. As shown in Figure 3, decision-tree method was used to recognize the other six-class actions. P11, P21, P31, etc. were the effective peak values, and if there was no effective peak value greater than the TH, the action would be regarded as '*Relax*'.

3.2.2 Recognition method of multi-feature fusion

Multi-feature fusion method is employed to increase the classification accuracy further, which is divided into three levels: data-level, feature-level and decision-level fusion (Qu 2002). Since the features were directly acquired from the same raw data, data-level fusion method could not be employed. The methods of feature-level and decision-level fusion are illustrated below:

Feature-level fusion is the middle level fusion, which is to analyze and process the features to a new fused feature. The superiority of feature-level fusion is to reduce the redundant information and retaining the advantages of original features. Weighted average method was employed to fuse MAV and WL. It could be defined as Equation 3:

$$NEW = k \cdot MAV + (1-k) \cdot WL \qquad (3)$$

138

The value range of k is (0, 1) and we chose $k = 0.6$. Eventually, according to the method in chapter 3.2.1, six-class actions were recognized with the new fused feature and the classification accuracy was calculated.

Decision-level fusion is the high level fusion, which is to process the same raw data with different features. Preprocessing, feature extraction and recognition are independently completed, as well as the basic judgments are also established by means of chapter 3.2.1. We chose logic disjunction, which is an operation that two logical values have a false value if and only if both of its operands are false. It could be defined as Equation 4:

$$JudgeFINAL = JudgeMAV \vee JudgeWL \qquad (4)$$

The final associated inference consequence was acquired action by action. Finally, the classification accuracy was obtained.

4   RESULTS AND DISCUSSION

4.1   *Results of single-feature recognition*

The classification accuracy of MAV reached to 95.24% and WL was 94.05%. In general, the result of MAV was slightly superior to WL. As shown in Table 1, the classification accuracy of each subject was greater than 90%. The difference of each subject was quite small, which meant that single-feature method was well used for sequential movement recognition no matter who the subject was. As a consequence, the results of single-feature recognition were considerable and significative.

4.2   *Results of multi-feature fusion recognition*

The features of MAV and WL were fused on both feature-level and decision-level. The classification accuracy increased to 96.43% and 98.81%, respectively. As shown in Table 2, the classification accuracy of each subject was greater than 95%. As a consequence, the results of sequential movement recognition reached to a higher level by means of multi-feature fusion.

Table 1.   The results of single-feature recognition.

	MAV			WL		
Subjects	All	Correct	Accuracy (%)	All	Correct	Accuracy (%)
Subject 1	21	19	90.48	21	20	95.24
Subject 2	21	20	95.24	21	20	95.24
Subject 3	21	20	95.24	21	19	90.48
Subject 4	21	21	100.00	21	20	95.24
Sum	84	80	95.24	84	79	94.05

Table 2.   The results of multi-feature fusion recognition.

	Feature-level			Decision-level		
Subjects	All	Correct	Accuracy (%)	All	Correct	Accuracy (%)
Subject 1	21	20	95.24	21	21	100.00
Subject 2	21	20	95.24	21	20	95.24
Subject 3	21	20	95.24	21	21	100.00
Subject 4	21	21	100.00	21	21	100.00
Sum	84	81	96.43	84	83	98.81

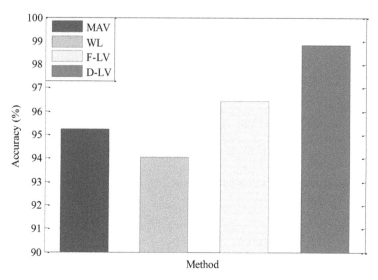

Figure 4. The comparison between multi-feature fusion and single-feature recognition.

In addition, a better multi-feature fusion method could be designed to increase the classification accuracy further.

4.3 *Method comparison and result analysis*

In general, as shown in Figure 4, the results of multi-feature fusion recognition were superior to single-feature recognition. The classification accuracy of feature-level fusion and decision-level fusion increased 1.19% and 3.57% comparing to MAV, respectively; and also increased 2.38% and 4.76% comparing to WL, respectively. From the point of each subject, almost all the sequential movements were recognized by means of multi-feature fusion, and the error of each subject was only one at most. Therefore, multi-feature fusion method could be more suitable for sequential movement recognition than single-feature recognition.

5 CONCLUSION AND FUTURE WORK

In this paper, the principal objective was to recognize seven-class actions composed of sequential movements. We extracted the features of MAV and WL as well as employed decision-tree and multi-feature fusion method. The results of single-feature recognition have met the basic requirements. Then, the classification accuracy has increased by means of multi-feature fusion on both feature-level and decision-level. The approach of this study could be widely and effectively used in many clinical and engineering applications.

The future research directions are discussed below:

First, the recognition approach needs to be optimized to increase the classification accuracy further. For instance, we could extract more effective features and design a better classifier or a better algorithm for multi-feature fusion recognition method.

Second, real-time online control by using sequential movements needs to be conducted. We suppose to improve the performances, such as accuracy, instantaneity and stability, which are important for human-computer coordinated control.

Third, more natural actions need to be studied instead of specified sequential movements to discover the order relation of natural human actions. We supposed to realize a natural HCI system, which appears to be an interesting direction of future work.

ACKNOWLEDGMENT

This work was partly supported by the National Natural Science Foundation of China (Project 91320202 and Project 61375117).

REFERENCES

Bida, O. 2005. Influence of electromyogram (EMG) amplitude processing in EMG-torque estimation. Worcester Polytechnic Institute, USA.

Fukuda, O. & Tsuji, T. 2006. A human-assisting manipulator teleoperated by EMG signals and arm motions. *IEEE Transactions on Robotics and Automation* 19(2): 210–222.

Gross, J. & Timmermann, L. 2002. The neural basis of intermittent motor control in humans. *Proc. Natl. Acad. Sci.* 99(1): 2299–2302.

Li, Q.L. 2009. Study on sEMG based exoskeletal robot for upper limbs rehabilitation. Harbin Institute of Technology, China.

Li, X.F. & Yang, J.J. 2005. Motion state identification of human elbow joint based on EMGs. *Chinese Journal of Biomedical Engineering* 24(4): 416–420.

Phinyomark, A. & Phukpattaranont, P. 2012. Feature reduction and selection for EMG signal classification. *Expert Systems with Applications* 39(1): 7420–7431.

Qu, J.S. & Wang, Z.Z. 2002. Data fusion based technology for remote sensing image processing. *Journal of Image and Graphics* 7(10): 985–993.

Rosen, J. & Perry, J.C. 2007. Upper limb powered exoskeleton. *International Journal of Humanoid Robotics* 4(3): 529–548.

Rucci, M. & Bullock, D. 2007. Integrating robotics and neuroscience: brains for robots, bodied for brains. *Advanced Robotics* 21(1): 1115–1129.

Sankai, Y. 2010. HAL: Hybrid assistive limb based on cybernics. *International Journal of Robotics Research* 66(1): 25–34.

Bioinformatics and Biomedical Engineering – Chou & Zhou (Eds)
© 2016 Taylor & Francis Group, London, ISBN 978-1-138-02784-8

Contrastive study on the enhancement of pitch adjustable Electrolarynx speech based on spectral subtraction technology

Y. Li, L. Wang, Y.J. Feng & H.J. Niu
School of Biological Science and Medical Engineering, Beihang University, Beijing, China

ABSTRACT: Spectral subtraction method is the most common way to eliminate the radiated noise of Electrolarynx (EL) whose pitch is constant. However, for pitch adjustable EL speech, it is not mentioned whether this method is effective. Thus, this paper conducts a study on the enhancement of pitch adjustable EL speech based on spectral subtraction technology. Several spectral subtraction algorithms including the Classical Spectral Subtraction (CSS), Multi-Band Spectral Subtraction (MBSS), Perceptual Weighting Spectral Subtraction (PWSS) and Weighting Function Spectral Subtraction (WFSS) had been performed and subjective and objective methods were used to evaluate and compare the performance. A native male speaker of mandarin Chinese who had been trained to be familiar with using EL was instructed to read twenty daily mandarin sentences in a soundproof room, which was recorded as the raw data of the EL speech. The subjective and objective results showed that the enhanced EL speech by CSS still remained noise obviously though part of radiated noise was eliminated. While the other three methods including MBSS, PWSS and WFSS could remove the noise effectively and improved the speech acceptability, but had little influence on speech Intelligibility. Therefore, the spectral subtraction method can upgrade acceptability of pitch adjustable EL speech and improve speech quality.

Keywords: Electrolarynx; pitch adjustable; spectral subtraction; speech enhancement

1 INTRODUCTION

Esophageal speech, tracheo-esophageal speech, and Electrolarynx (EL) speech are three main ways of voice rehabilitation for laryngectomees. Due to easy learning and phonation, more than half of laryngectomees use the EL as their primary way of daily communication (Liu et al., 2004). Although the EL is very popular, some serious drawbacks still exist that limit the understanding of EL speech, such as the unnaturalness and the unpleasant radiated noise. To improve the speech quality by eliminating the radiated noise becomes one of the goals for researchers.

Barneyet et al. reported that the intensity of the radiated noise was about 20~25 dB when the mouth was closed, and the concentration of noise energy was between 400~1 kHz and between 2~4 kHz (Barney et al., 1959). Weiss et al. found that as the difference of characteristics of neck tissues, the intensity of the radiated noise varied over 4~15 dB across the subjects for the same device. In addition, the perceptual test results indicated that the intelligibility of EL speech obviously decreased when the Signal To Noise Ratio (SNR) was less than 4 dB (Weiss et al., 1979). In order to reduce the radiated noise effectively, an increasing number of techniques have been developed including the acoustic shielding technology and signal processing methods (Norton and Bernstein, 1993, Espy-Wilson et al., 1998). Since the signal processing methods can eliminate the noise during the sound transmission, they have attracted great attention. There are two main signal processing methods which have been put forward to improve the intelligibility and naturalness of EL speech. One is adaptive noise canceling (Niu et al., 2003), which additionally needs a reference input and has a high computational load. The other is spectral subtractive-type algorithm (Liu et al., 2006a, Pandey et al., 2002), which subtracts the estimated noise spectrum from the noisy speech spectrum to obtain the clean speech. Due to the simplicity of implementation and low computational load, the spectral subtraction method is widely used and has become the primary choice for real time applications. However, a serious disadvantage of this classical method is that the enhanced speech is accompanied by unpleasant musical noise which is characterized by tones with random frequencies, which decreases the speech quality. For the sake of solving this problem, several solutions have been presented. Li et al. proposed a multi-band spectral subtraction algorithm by taking into account the non-uniform effect of radiated noise on the spectrum of EL speech (Li et al., 2009), which adopted different spectral subtraction coefficients at different frequency band and obtained a better noise reduction effect. Liu et al. applied the perceptual weighting technique on the spectral subtraction based on frequency-domain masking properties of the human auditory system (Liu et al., 2006b), which also reduced the musical noise. In addition, Verteletskaya et al. introduced a weighting function based on Linear Predictive Coding (LPC) to reshape the enhanced speech spectrum and attenuate the noise spectrum components lying outside identified formants regions (Verteletskaya and Simak, 2011), which effected a substantial reduction of the musical noise without significantly distorting the speech.

Above all, various spectral subtractive-type algorithms have been developed to enhance the pitch constant EL speech and improve the speech quality in some extent. However, for the pitch adjustable EL speech which has developed in recent years, whether these methods above are effective is not concerned. Because of the tunable pitch of EL, the frequency and intensity of the radiated noise are certainly unstable. Therefore, in this study, a variety of spectral subtractive-type techniques will be applied to improve the quality of pitch adjustable EL speech. Combined with subjective and objective methods, the noise reduction performance will be compared.

2 METHOD

2.1 Spectral subtraction method

The spectral subtraction is based on the principle that the enhanced speech can be obtained by subtracting the estimated spectral components of the noise from the spectrum of the noisy speech. Assuming that the additive noise $d(n)$ will be stationary and uncorrelated with the clean speech signal $s(n)$, the noisy speech $y(n)$ can be written as (Boll, 1979):

$$y(n) = s(n) + d(n) \tag{1}$$

For the Classical Spectral Subtraction (CSS) method, the subtraction of the noise estimate is expressed as follows:

$$|S(\omega)|^2 = \begin{cases} |Y(\omega)|^2 - |D(\omega)|^2 & if \ |Y(\omega)|^2 > |D(\omega)|^2 \\ 0 & else \end{cases} \tag{2}$$

where $|S(\omega)|$, $|Y(\omega)|$ and $|D(\omega)|$ represent the enhanced speech spectrum, the noisy speech spectrum and the noise spectrum estimate, respectively. After the subtraction is computed in the frequency domain, the clean speech estimate in the time-domain can be directly reconstructed by an Inverse Fast Fourier Transform (IFFT) and overlap add.

Due to the spectral subtraction parameters of the classical spectral subtraction algorithm used for speech enhancement are fixed and cannot be adapted frame by frame, the enhanced speech spectrum would have some false peaks when the estimated noise spectrum is smaller than the real noise spectrum, resulting that the enhanced speech is accompanied by unpleasant musical noise. In order to minimize the musical noise, Li et al. proposed a multi-band spectral subtraction algorithm by taking into account the non-uniform effect of radiated noise on the spectrum of EL speech that divided the entire spectrum into five non-overlapping bands: [0~300 Hz (Band 1), 300~1 KHz (Band 2), 1 K~2 K (Band 3), 2 K~3 K (Band 4), 3 K~5 K (Band 5)], and spectral subtraction was performed independently in each band. Hence the estimate of the clean speech spectrum in the ith band is obtained by (Li et al., 2009):

$$|S(\omega)|^2 = \begin{cases} |Y(\omega)|^2 - \alpha_i\delta_i \cdot |D(\omega)|^2 & b_i \le \omega \le e_i \\ \beta \cdot |D(\omega)|^2 & else \end{cases} \quad (3)$$

where α_i is the over-subtraction factor of the ith frequency band, which is a function of the segmental SNR. The value of scaling factor α_i higher than 1 results in high SNR level of the enhanced speech, but too high value may cause distortion in perceived speech quality. And δ_i is a tweaking factor that can be individually set for each frequency band to customize the noise removal properties. b_i and e_i are the beginning and ending frequency of the ith frequency band. β is the spectral floor, which is introduced to prevent the spectral components of the enhanced speech spectrum to descend below the lower bound β. $|D(\omega)|^2$ and reduce the spectral excursions of noise peaks (as compared to when the negative components are set to zero) so as to suppress the amount of musical noise. Here we set $\beta = 0.001$, the values of δ_i are empirically determined and set to 1.2, 1.5, 1.8, 1.6, 1.2.

To reduce the musical noise caused by the fixed spectral subtraction parameters, Liu et al. applied the perceptual weighting technique to adapt the subtraction parameters frame by frame in the enhancement process based on frequency-domain masking properties of the human auditory system. The perceptual weighting filter is calculated as follows (Liu et al., 2006b):

$$P(z) = \frac{A\left(\dfrac{z}{\sigma_1}\right)}{A\left(\dfrac{z}{\sigma_2}\right)} = \frac{1 - \sum_{k=1}^{p} a_k \sigma_1^k z^{-k}}{1 - \sum_{k=1}^{p} a_k \sigma_2^k z^{-k}} \quad (4)$$

where $A(z)$ is LPC polynomial, a_k are the short-term linear prediction coefficients, which can be calculated from the noisy speech, σ_1 and σ_2 ($0 \le \sigma_2 \le \sigma_1 \le 1$) are parameters that control the energy of the error in the formant regions and p is the prediction order. Here $p = 15$, $\sigma_1 = 1$, $\sigma_2 = 0.8$. The frequency response of the perceptual filter, $T(\omega)$, can be calculated directly from the noisy speech. Thus the subtraction parameters α and β can be adapted in time and frequency based on $T(\omega)$ towards to different environment.

Besides, Verteletskaya et al. introduced a weighting function based on LPC to reshape the enhanced speech spectrum and attenuate the noise spectrum components lying outside identified formants regions, which effects a substantial reduction of the musical noise without significantly distorting the speech. The weighting function $W(\omega)$ is derived from the spectral envelope $L(\omega)$, which is obtained by means of linear prediction analysis. The computation rule is below (Verteletskaya and Simak, 2011):

$$W(\omega) = \begin{cases} \left[\dfrac{L(\omega)}{\tau}\right]^\gamma & if\ L(\omega) < \tau \\ 1 & else \end{cases} \quad (5)$$

where the threshold value τ is a constant for all frequencies and for all speech segments. In a strongly voiced segment of speech, only small portions of the spectrum will be attenuated, where as in quiet segments most of the spectrum may be attenuated. A value of the threshold about 5% of peak amplitude of the speech is found to work well in this study. The power term $\gamma = 2$, and larger value of γ will make the attenuation harsher.

2.2 Experiments

A native male speaker of mandarin Chinese who had been trained to be familiar with using EL device (the range of the pitch is from 60 Hz to 200 Hz) participated in the experiment. The speaker was instructed to read twenty daily mandarin sentences. The EL was opened to produce the radiated noise for 1~2 seconds before the speaker read as the silence estimated noise. The raw data of EL speech were collected by using a microphone mounted at a distance of about 10 cm from the mouth of the speaker at a sampling frequency of 48 kHz with 16-bits per sample. The recording procedure was carried out in a soundproof room.

The Classical Spectral Subtraction (CSS), the Multi-Band Spectral Subtraction (MBSS), the Perceptual Weighting Spectral Subtraction (PWSS) and the Weighting Function Spectral Subtraction (WFSS) were used to remove the radiated noise of EL speech by computer simulation in MATLAB2010 environment. The frame length was 2048, each frame was 25% overlapped. Frames were windowed with Hamming window and 2048 points Fast Fourier Transformation (FFT) was applied to each frame. The Praat software was used to analyze and compare the quality of the original and enhanced EL speech. And the Energy Ratio (ER) of the original and enhanced EL speech was also computed as follows:

$$ER = \frac{\sum x^2(n)}{\sum y^2(n)} \tag{6}$$

where $x(n)$ represents the enhanced EL speech and $y(n)$ represents the original speech.

The Mean Opinion Score (MOS) method was selected to evaluate the quality of the enhanced EL speech, which included two aspects: acceptability (excellent: 5, good: 4, common: 3, poor: 2, bad: 1) and intelligibility (understanding: 1, not understanding: 0). The perceptual experiments were individually carried out in a soundproof room with eight listeners who were native speakers of Mandarin Chinese. All of the listeners had no reported history of hearing problems and were unfamiliar with EL speech so that they could be safely regarded as naïve listeners. The listening test data which had 100 sentences including the original and enhanced EL speech by various spectral subtractive-type algorithms were randomized for the listening experiments. During the experiments listeners were instructed to write down the scores and what they heard. The intelligibility score was calculated as the mean percentage of correct responses to sentences for all listeners, and the acceptability score was calculated as the mean value of the opinion scores for all listeners.

3 RESULTS

3.1 Acoustic analyses

The radiated noise of the original EL speech can be seen clearly from the waveform and spectrogram of Figure 1(a), especially in non-speech region, and the noise affects the whole spectrum. Compared with the original speech, Figure 1(b) waveform and the ER values of Figure 2 show that the enhanced EL speech by CSS still obviously remains noise in non-speech section though part of radiated noise whose energy are almost equal to 7% of the original speech energy is reduced. Equally, the remaining noise can be also seen form the spectrogram of Figure 1(b). The waveforms and spectrograms of Figure 1(c), (d), (e) indicate that MBSS, PWSS and WFSS algorithms appear to be much better in reducing radiated noise, especially in non-speech section. Nevertheless, it is found that some high frequency component of the normal speech may be eliminated by WFSS, which causes the ER minimum of WFSS method. In addition, the pitch graphs show that there is no difference between the pitch of original EL speech and that of the enhanced speech.

Figure 2 suggests that the energy of the enhanced EL speeches were lower than that of the original EL speech. The ER value of CSS method was 0.93, which was the largest. And the others decreased to 0.84, 0.82, and 0.79. Obviously, the WFSS method obtained the lowest value.

146

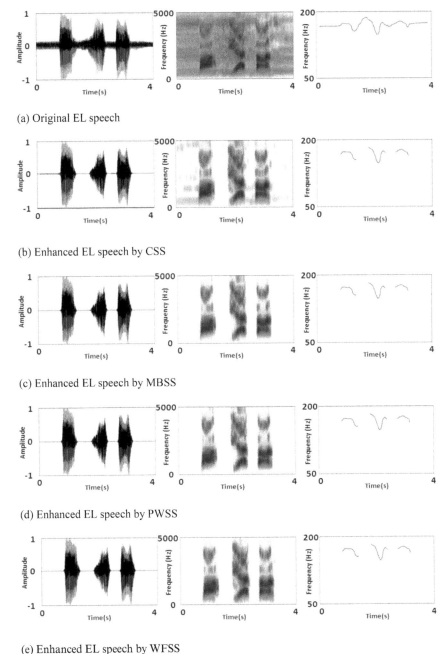

(a) Original EL speech

(b) Enhanced EL speech by CSS

(c) Enhanced EL speech by MBSS

(d) Enhanced EL speech by PWSS

(e) Enhanced EL speech by WFSS

Figure 1.   Waveform, spectrograms and pitch of EL speech "ba dian ban".

3.2 *Perceptual analyses*

Figure 3 shows the perceptual scores of the original speech and the enhanced speech. The acceptability and intelligibility of the original EL speech were 1.70 ± 1.08 and $(40.63 \pm 14.9)\%$, respectively. After the enhancement using CSS, MBSS, PWSS and WFSS methods, the scores were 2.39 ± 1.01 and $(43.12 \pm 15.11)\%$, 2.73 ± 0.99 and $(45.01 \pm 15.22)\%$, 2.94 ± 1.03 and $(46.25 \pm 14.57)\%$, 3.14 ± 1.04 and $(49.37 \pm 13.99)\%$, respectively.

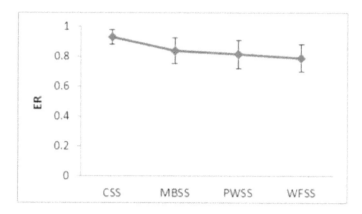

Figure 2.   ER values of the enhanced and original EL speech.

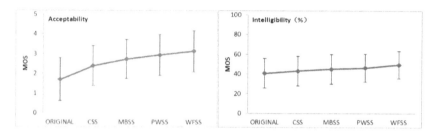

Figure 3.   Perceptual results of the original and enhanced EL speech.

4   DISCUSSION

In order to investigate whether spectral subtraction method is effective to the pitch adjustable EL speech, several spectral subtraction algorithms including CSS, MBSS, PWSS and WFSS are performed and compared in this study. The results of waveform and spectrogram suggest that the MBSS, PWSS and WFSS methods all have a better noise reduction than CSS method. Similarly, in the experiments of Li and Liu, for the pitch constant EL speech, the reduction effect of CSS is respectively worse than that of MBSS and PWSS methods. Nevertheless, the remaining noise in the non-speech section of the enhanced speech of CSS method is less than that of CSS method in this study, which indicates that maybe it is more difficult to eliminate the radiated noise of adjustable pitch than the pitch fixed radiated noise. Moreover, the diagrams of fundamental frequency show that the spectral subtractive-type methods have no influence on EL pitch.

Generally, the ER value of the enhanced EL speech cannot represent the speech quality, but it can reflect the noise reduction effect objectively. In this study, the ER value of CSS method is lowest, which can be also found from the spectrogram that the enhanced speech of CSS method remains much noise. And the other three algorithms obtain the approximated values, among which the value of WFSS method is lower than others. These results are consistent with the spectrogram analysis. The experiment results of Li showed that the SNR of the speech enhanced by MBSS method gave about 6dB improvement at the lower SNR condition increasing to about 12dB improvement at the higher SNR condition than CSS method. Meanwhile, according to the experiment of Verteletskay, the SNR of the enhanced speech by WFSS method gave about 3dB improvement at the lower SNR condition increasing to about 0.2dB improvement at the higher SNR condition than CSS method. All results above reflect that the MBSS and WFSS methods have a better noise reduction performance than CSS method.

Perceptual analyses indicate that the acceptability of the enhanced EL speech by various spectral subtractive-type methods increased than the original speech, which is because the noise reduction making the enhanced speech to be sounded more pleasant. Among them, the CSS method obtained lowest score, and the scores of MBSS and PWSS methods were lower than WFSS method because of the little remaining noise in the enhanced speech, which is consistent with the acoustic analyses. While the scores of intelligibility of the enhanced EL speech were approximated, the reason may be that the reducing noise has little influence on meaning of speech. Toward to the experiment of Li, the original EL speech was produced by the laryngectomized speaker using the pitch constant EL. The score of acceptability was 2.1, and the scores of the enhanced speech by CSS and MBSS methods increased to 2.5 and 3.25, which are larger than that of this study. There are two possible reasons: one is that the radiated noise of pitch adjustable EL is variable, which decreases the noise reduction effect. The other is that the original EL speech produced by the laryngectomized speaker is better than that produced by the normal speaker. This is believed to be related to the fact that the neck tissue of the normal speaker is hard and noncompliant so that more radiated noises are produced during speech. This seems to be similar to laryngectomized patients with firm and fibrotic neck tissue. And in the experiment of Liu, the original EL speech was produced by the normal speaker using the pitch constant EL. The score of acceptability was 1.25, and the scores of the enhanced speech by CSS and PWSS methods increased to 1.6 and 2.05, which were smaller than that of this study, that is may because the pitch of the EL using in this study is adjustable, which makes the EL speech more natural, especially for short words and increases the intelligibility of the EL speech so that affects the acceptability in some extent.

In a conclusion, with regard to the selected EL speech whose pitch is adjustable, the enhancement methods based on spectral subtraction can effectively reduce the radiated noise of EL speech and improve the acceptability, which is greatly helpful for improving speech quality, nevertheless, have little influence on intelligibility.

ACKNOWLEDGEMENT

This work was supported by the National Key Technology R&D Program in the 12th Five Year Plan of China (2012BAI33B03), the Program for New Century Excellent Talents in University (NCET-11-0772) and State Key Laboratory of Software Development Environment (SKLSDE-2014ZX-12).

REFERENCES

Barney, H., Haworth, F. & Dunn, H. 1959. An experimental transistorized artificial larynx. *Bell system technical Journal,* 38, 1337–1356.

Boll, S. 1979. Suppression of acoustic noise in speech using spectral subtraction. *Acoustics, Speech and Signal Processing, IEEE Transactions on,* 27, 113–120.

Espy-Wilson, C.Y., Chari, V.R., Macauslan, J.M., Huang, C.B. & Walsh, M.J. 1998. Enhancement of electrolaryngeal speech by adaptive filtering. *Journal of Speech, Language, and Hearing Research,* 41, 1253–1264.

Li, S., Wan, M. & Wang, S. 2009. Multi-Band Spectral Subtraction Method for Electrolarynx Speech Enhancement. *Algorithms,* 2, 550–564.

Liu, H., Wan, M., Wang, S. & Niu, H. 2004. Aerodynamic characteristics of laryngectomees breathing quietly and speaking with the electrolarynx. *Journal of Voice,* 18, 567–577.

Liu, H., Zhao, Q., Wan, M. & Wang, S. 2006a. Application of spectral subtraction method on enhancement of electrolarynx speech. *The Journal of the Acoustical Society of America,* 120, 398–406.

Liu, H., Zhao, Q., Wan, M. & Wang, S. 2006b. Enhancement of electrolarynx speech based on auditory masking. *Biomedical Engineering, IEEE Transactions on,* 53, 865–874.

Niu, H.J., Wan, M.X., Wang, S.P. & Liu, H.J. 2003. Enhancement of electrolarynx speech using adaptive noise cancelling based on independent component analysis. *Medical & Biological Engineering & Computing,* 41, 670–678.

Norton, R.L. & Bernstein, R.S. 1993. Improved laboratory prototype electrolarynx (LAPEL): using inverse filtering of the frequency response function of the human throat. *Annals of biomedical engineering,* 21, 163–174.

Pandey, P.C., Bhandarkar, S.M., Bachher, G.K. & Lehana, P.K. Enhancement of alaryngeal speech using spectral subtraction. Digital Signal Processing, 2002. DSP 2002. 2002 14th International Conference on, 2002. IEEE, 591–594.

Verteletskaya, E. & Simak, B. 2011. Noise reduction based on modified spectral subtraction method. *IAENG International journal of computer science,* 38, 82–88.

Weiss, M.S., Yeni-Komshian, G.H. & Heinz, J.M. 1979. Acoustical and perceptual characteristics of speech produced with an electronic artificial larynx. *The Journal of the Acoustical Society of America,* 65, 1298–1308.

Bioinformatics and Biomedical Engineering – Chou & Zhou (Eds)
© 2016 Taylor & Francis Group, London, ISBN 978-1-138-02784-8

Comparison between neural network or neural network with genetic algorithm and analysis of EEG signal

Yasmin Abdul Rahim & Magdi M. BkrSudan
University of Science and Technology, Khartoum, Sudan

ABSTRACT: Electroencephalogram (EEG) refers to the recording of the brain's spontaneous electrical activity. It consists of 5 sub-band signals that can be traced and analyzed to detect many diseases. The use of manual prediction cases is a difficult method to get an accurate classification of the EEG signal. This problem increases the number of misdiagnosis that commonly plagues all classification systems. This research aims to analyze Electroencephalogram (EEG) signal parameters using two ways: first, by using only the neural network; second, by using the Genetic Algorithm (GA) and Neural network and then comparing the two ways. A total of 80 well-known reference cases were used in this study. We had four cases of normal open eye, normal close eye, epilepsy free seizures and epilepsy seizures. Signals were classified based on statistical features and total power spectrum extracted from the signals. The feed-forward neural network was applied to classify the case of the signals.

Keywords: electroencephalogram; genetic algorithm; neural network

1 INTRODUCTION

Electroencephalography (EEG) is the recording of electrical activity along the scalp. EEG measures voltage fluctuations resulting from ionic current flows within the neurons of the brain. In the clinical context, EEG refers to the recording of the brain's spontaneous electrical activity over a short period of time, usually 20–40 minutes, as recorded from multiple electrodes placed on the scalp. An EEG signal is a measurement of current that flows during synaptic excitations of the dendrites of many pyramidal neurons in the cerebral cortex. When brain cells (neurons) are activated, the synaptic current is produced within the dendrites. This current generates a magnetic field measured by the Electromyogram (EMG) and a secondary electrical field over the scalp measured by EEG systems (Attwood, H.L., & MacKay, W.A., 1989).

In healthy adults, the amplitudes and frequencies of such signals change from one state of a human to another, such as waking and sleeping. The characteristics of the waves also change with age. There are five major brain waves distinguished by their different frequency ranges. These frequency bands from low to high frequencies, respectively, are called delta (δ), theta (θ), alpha (α), beta (β) and gamma (γ). Delta waves lie within the range of 0.5–4 Hz. These waves are primarily associated with deep sleep and may be present in the waking state. Theta waves lie within the range of 4–7.5 Hz. Theta waves appear as consciousness slips towards drowsiness. Alpha waves appear in the posterior half of the head. For alpha waves, the frequency lies within the range of 8–13 Hz, and commonly appears as a round or sinusoidal-shaped signal. Alpha waves have been thought to indicate both a relaxed awareness without any attention or concentration. A beta wave is the electrical activity of the brain varying within the range of 14–26 Hz (though in some literature, no upper bound is given). A beta wave is the usual waking rhythm of the brain associated with active thinking, attention, focus on the outside world, or solving concrete problems, and is found in normal adults. The frequencies above 30 Hz (mainly up to 45 Hz) correspond to the gamma range (Ashwal, S., & Rust, R. 2003).

The present work has two goals. First, it aims to extract features from EEG signals and classify using the neural network; second, it aims to extract features from EEG signals and select the optimum features using the genetic algorithm and then classify by using the neural network.

2   MATERIALS AND METHODS

In order to analyze the EEG signal, we need a row data to import it in Matlab. Then, we need to perform preprocessing to remove noise from them. We then extract features from this signal. These three steps are similar in both cases, for a comparison between them. Therefore, we first explain these three cases.

2.1   *EEG row data*

For data selection and recording techniques, four sets denoted as A, B, D and E, with each containing 20 single channel EEG segments of 23.6 sec duration, were considered in the study. Sets A and B consisted of segments taken from surface EEG recordings that were carried out on five healthy volunteers using a standardized electrode placement scheme. Volunteers relaxed in an awake state with eyes open ~A! and eyes closed ~B!, respectively.

Segments in set D were recorded from within the epileptogenic zone, while set D contained only activity measured during seizure free intervals, and set E only contained seizure activity. All EEG signals were recorded with the same 128-channel amplifier system. The data were written continuously onto the disk of a data acquisition computer system at a sampling rate of 173.61 Hz. Band-pass filter settings were 0.53–40 Hz (G. Andrzejak,1,2,* Klaus Lehnertz,1,† Florian Mormann,1,2 Christoph Rieke,1,2 Peter David,2 & Christian E. Elger1 2001). Thus, a total of 80 signals were used, with 20 of them representing a signal divided for A, B, D and E.

2.2   *Signal preprocessing*

Band Pass filter was used with a low cut-off frequency of 0.53 Hz and a high cut-off frequency of 40 Hz. Two notch filters were used to remove the EOG and EMG artifact. The 4th order Butterworth were used.

The objective of filtering was to improve signal quality.

We then used the band pass filter to extract the 5 sub-signals from the EEG signal as follows:

Delta wave had a band up to 4 Hz;
Theta wave had a band from 4 to 7 Hz;
Alpha wave had a band from 7 to 14 Hz;
Beta wave had a band from 14 to 30 Hz; and
Gamma wave had a band from approximately 30 to 100 Hz (Nandish M., Stafford Michahial, Hemanth Kumar P., Faizan Ahmed., 2012).

3   FEATURE EXTRACTION

In pattern recognition, feature extraction is a special form of dimensionality reduction. When the input data to an algorithm is too large to be processed and it is suspected to be notoriously redundant (much data, but not much information), then the input data will be transformed into a reduced representation set of features (also named feature vector). Transforming the input data into the set of features is called feature extraction. We depend on nine statistical features and total power spectrum for sub-signals to get accurate results.

152

The statistical features used include: Mean, Median, Standard deviation, Entropy, Root mean square, Variance, Skewness, Kurtosis and Zero cross rate.

These features with total power spectrum are extracted from Delta, Theta, Alpha, Beta and Gamma. At the end of this stage, we have 50 useful features.

These features have two tracks:

- To enter the neural network to classify them.
- The use of a genetic algorithm is to reduce the number of features and then enter the neural network to classify it.

3.1 First track

The 50 features are entered to the neural network in order to classify them.

3.1.1 Neural network classification
We used the feed-forward back propagation neural network with 4 hiding layers, trainlm training function and sigmoid transfer function.

The data were divided into 75%, 15% and 15% for training, validation and testing, respectively.

3.2 Second track

3.2.1 Genetic algorithm
The 50 features were moved to the stage of the selection using the genetic algorithm, and selected the features based on their fitness function.

$$\text{Fitness function} = bsxfun(@minus, A(i), b(i)) \geq 2 \qquad (1)$$

where A = first row of features

b = second row of features

$bsxfun$ = apply element-by-element binary operation (minus in this case) to two arrays with singleton expansion.

Based on fitness function, useful features were selected. From the 50 features, 24 useful features were selected. These 24 features were moved to the classification stage.

They were dependent on the single-point crossover type and order changing mutation type.

3.2.2 Neural network
As in the first case, we had used the same characteristics of the artificial neural network (feed-forward back propagation with 4 hiding layers, trainlm training function and sigmoid transfer function).

The data were divided into 75%, 15% and 15% for training, validation and testing, respectively.

4 RESULTS

Based on our study, both methods gave a 91.67% correct classification: the first method provided a 0.08 mean square error, while the second method gave only 0.05. Figures 1, 2 and 3 show the results of the first method, while Figures 4, 5 and 6 show the results of the second method.

To evaluate these results and ratios for the two methods and compare between the two, we used accuracy, sensitivity, specificity, positive predictive values and negative predictive values of 64 test signals subject to the same previous steps.

153

4.1 Result of the first method

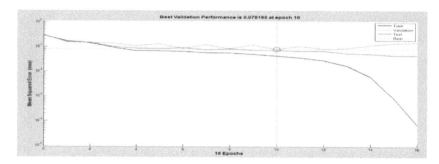

Figure 1. Neural network performance.

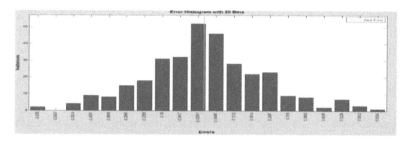

Figure 2. Neural network error histogram.

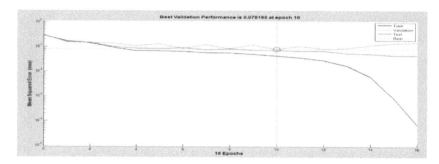

Figure 3. Confusion matrix.

Table 1. Evaluation of performance for neural network classification.

Parameter	Case 1	Case 2	Case 3	Case 4	Mean
TP	11	5	12	16	
FN	5	11	4	0	
TN	41	47	41	43	
FP	7	1	7	5	
ACC	81.25	81.25	82.8	92.2	84.375
SEN	68.75	31.25	75	100	73.25
SPE	85.4	97.9	85.4	89.6	89.58
PPV	61.1	83.3	63.15	76.2	70.94
NPV	89.1	81.03	91.1	100	90.31

154

4.2   Result of the second method

The genetic algorithm with the neural network gave 0.05 mean square error.

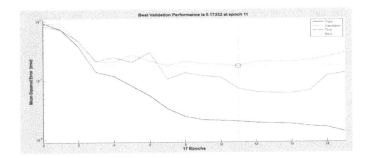

Figure 4.   Neural network performance.

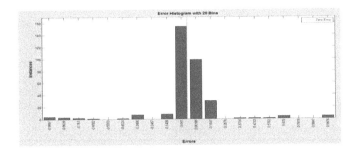

Figure 5.   Neural network error histogram.

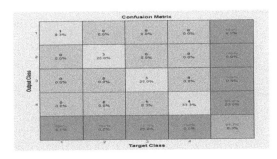

Figure 6.   Confusion matrix.

Table 2.   Genetic algorithm and neural network.

Parameter	Case 1	Case 2	Case 3	Case 4	Mean
TP	16	9	13	15	
FN	0	7	3	1	
TN	40	47	47	47	
FP	8	1	1	1	
ACC	87.5	87.5	93.75	96.88	91.41
SEN	100	56.25	81.25	93.75	82.81
SPE	83.3	97.9	97.9	97.9	94.25
PPV	66.67	90	92.86	93.75	85.82
NPV	100	87.04	94	97.96	94.75

5   CONCLUSION

From this study, we conclude that it is possible to detect and analysis brain signal via signal processing software, which is a powerful and accurate way. Based on the previous experimental results, we can draw the following conclusions. The use of the genetic algorithm with neural network to classify the signals is more accurate than the use of the neural network alone. The use of the genetic algorithm to select features reduces the error rate. The proposed method is of low cost as it can be implemented in computer.

ACKNOWLEDGEMENT

The author thanks Dr Magdi M Bkr for help and guidance. The author also thanks her family and her husband for their constant love and support.

REFERENCES

Attwood, H.L., & MacKay, W.A., 1989. Essentials of Neurophysiology, B.C. Decker, Hamilton, Canada.

Ashwal, S., & Rust, R., 'Child neurology in the 20th century', Pedia. Res., 53, 2003, 345–361.

Feature Extraction and Classification of EEG Signal Using Neural Network Based Techniques.

G. Andrzejak,1,2,* Klaus Lehnertz,1,† Florian Mormann,1,2 Christoph Rieke,1,2 Peter David,2 & Christian E. Elger1, Received 14 May 2001, Indications of nonlinear deterministic and finite-dimensional structures in time series of brain electrical activity: Dependence on recording region and brain state Ralph.

Nandish M., Stafford Michahial, Hemanth Kumar P., Faizan Ahmed., 2012.

www.wikibedia.com.

Bioinformatics and Biomedical Engineering – Chou & Zhou (Eds)
© *2016 Taylor & Francis Group, London, ISBN 978-1-138-02784-8*

Feature optimization for pathological voice based on BP neural network

Y. Zeng, W.P. Hu & D.D. Liang
GuangXi Normal University, Guilin, Guangxi, China

ABSTRACT: An exploration of the contribution of traditional features and nonlinear features to pathological voice recognition is addressed in this paper. A total of 20 kinds of features that have previously been shown to be effective for pathological voice recognition problem are used in this experiment. A saliency measure by the BP neural network is then used to evaluate the contribution of these 20 kinds of features to the recognition problem. Then, the best set of features for pathological voice recognition is selected to perform the identify problem. With the Support Vector Machine (SVM), we get the highest recognition rate of 98.67%, with the average recognition rate of 88.66%.

Keywords: pathological voices; BP neural network; feature ranking; feature optimization

1 INTRODUCTION

The diagnosis of pathological voice is an important subject in the field of clinical medicine. Clinical measurement of pathology detection typically involves laryngoscopy, stroboscopy, glottography, electromyography and video kimography (Henríquez et al. 2009). All these methods may cause discomfort to patients, which may lead to incorrect diagnosis (Henríquez et al. 2009). Compared with the above methods, the objective acoustics method is a non-invasive, quick and automatic technique (Henríquez et al. 2009). The severity of the voice pathology detection is still an open problem.

Usually when processing the pathology detection, the traditional features or the nonlinear features are used. The traditional features represent the voice's linear characteristics, while the nonlinear features represent the voice's non-linear characteristics. Thus, it becomes imperative to explore the contribution of these two characteristics to the pathological voice identification, and to gain the ability to fully characterize the pathological voice. The saliency measure will be used for the ranking of the traditional features and nonlinear features. Gan et al. performed the pathology detection with a single traditional feature and a single nonlinear feature (Gan et al. 2014). The result shows that the nonlinear features can characterize the voice pathology better. Then, our result will be compared with it.

In Section 2, the database used for the pathological voice recognition is provided. In Section 3, a brief introduction of the 20 kinds of features used for the experiment is provided. In Section 4, the technique for feature selection including the BP neural network and saliency measure is provided. Section 5 applies the saliency measure to the comparison of the two classes of features. Section 6 discusses the contribution between the traditional features and the nonlinear features.

2 DATABASE

The database (Zhao et al. 2013 & Gan et al. 2014) consisted of 151 speakers, 78 healthy speakers and 73 pathological speakers. The voice was recorded in a quiet and small room, where the surrounding noise was below 45 dB. The recordings of the normal quality of voice were obtained from the speakers without any laryngeal disease and the recordings of the abnormal quality of voice were obtained from the speakers with disordered speech. For our experiment, the vowel /a/ was sampled from the speakers, and each speaker phonated lasting from 3 to 4 seconds, sampled 3 times, and the sampled frequency was 16 kHz.

3 FEATURE EXTRACTION

3.1 Traditional features

Features based on traditional linear techniques are extracted on the basis of short-time stability. Due to the good difference between the pathological and normal voice, the techniques have been widely used. Ten kinds of traditional features were usually used for the recognition of pathological voice, which include fundamental frequency (F0), Mel Cepstral Coefficients (MFCC), Linear Prediction Cepstral Coefficient (LPCC), formant, absolute frequency jitter (Jita), frequency jitter percentage (Jitt), frequency perturbation entropy (PPQ), amplitude jitter percentage (Shim), amplitude jitter (ShdB) and shimmer entropy (APQ). Methods for evaluating all of the traditional features can be found elsewhere (Gan et al. 2014).

3.2 Nonlinear features

3.2.1 The largest Lyapunov exponent λ_{max}
The largest Lyapunov exponent represents the average divergence rate of neighboring trajectories in the state space (Wolf 1986), whose estimation has been proposed by Rosenstein et al (Rosenstein et al. 1993).

3.2.2 Value of the first minimum of mutual information function
Attractor in the phase space represents the long-term behavior of the system (Henríquez et al. 2009). Takens proved that the delay method can reconstruct a phase space from a time series of the system (Takens et al. 1981). The time delay (T) can be estimated by the First Minimum of Mutual Information function (FMMI), which is evaluated by the following equation (more details can be found elsewhere (Henríquez et al. 2009)):

$$I(\tau) = \sum_{i,j} p_{ij}(\tau) \ln \left[\frac{p_{ij}(\tau)}{p_j p_i(\tau)} \right] \tag{1}$$

where p_{ij} = joint probability; p_i = probability of x; and p_j = probability of y.

3.2.3 Box dimension and intercepts
Aerodynamic and acoustic theory indicates that a chaotic mechanism exists in the voice signal, which is a complex nonlinear process. Fractal theory can effectively describe chaotic signals. In this paper, box dimension and intercepts are used to quantitatively describe the fractal characteristics of pathological voice signals. The calculation method has been described elsewhere (Takeo 1996).

3.2.4 Entropies
Entropy describes the quantity of disorder or complexity of a system (Henríquez et al. 2009). Features of Shannon entropy, sample entropy, fuzzy entropy, multi-scale entropy and second-order Rényi entropy were used to characterize the pathological and normal voice in this paper. Gan et al. provided methods to calculate all of the entropies, the details of which can be found in their paper (Gan et al. 2014).

3.3 Features

Features used for ranking are provided in Table 1. The index is their identifier.

Table 1. Features' index definition.

Index	Feature	Index	Feature
T1	F0	F11	Lyapunov
T2	MFCC	F12	FMMI
T3	LPCC	F13	Box dimension
T4	Formant	F14	Intercepts
T5	Jita	F15	Hurst
T6	Jitt	F16	Rényi entropy
T7	PPQ	F17	Shannon entropy
T8	Shim	F18	Sample entropy
T9	ShdB	F19	Fuzzy entropy
T10	APQ	F20	Multi-scale entropy

4 METHODS

In this section, the Back-Propagation (BP) neural network for ranking the input features' usefulness will be introduced. First, the structure of the network will be described, and then the sensitivity of the network outputs to its inputs is used to rank the input features.

4.1 BP neural network

The BP can be used to perform feature selection. The architecture of the BP network is shown in Figure 1. Each node in the network uses the transfer function, as shown in Figure 2. The network shown uses one hidden layer. Cybenko showed that at most one hidden layer is required for approximating functions (Cybenko 1989 & Barron 1989).

The performance of the network is directly related to the number of the hidden nodes. For the experiment, we consider the number of the hidden nodes ranging from 1 to 50, and then for each number of the hidden nodes, we train 10 times and get the average recognition rate. Figure 3 shows the classification rate in different numbers of the hidden nodes. It can also be seen that the choice of the number of the hidden nodes should be "15".

4.2 Algorithm for feature ranking

We consider the network in Figure 1. Ruck et al. proved that the saliency of the input (Ruck et al. 2009) can be formulated as follows:

$$\Lambda_j = \sum_{x \in S} \sum_i \sum_{x_j \in D_j} \left| \frac{\partial z_i}{\partial x_j}(\mathbf{x}, \mathbf{w}) \right| \tag{2}$$

where $x = j$ feature's m network inputs; S = training set; i = output layer; and D_j = R points for input.

We then calculate the desired derivative of the output with respect to the input. To compute the desired derivative, the chain rule is required. As the transfer function of the node is sigmoidal, we can get the following equation:

$$\frac{\partial z_i}{\partial x_j} = z_i(1 - z_i) \frac{\partial}{\partial x_j} \left(\sum_m w_{mi}^2 x_m^1 + \theta_i^2 \right) \tag{3}$$

where z_i = output of the nod i in the output layer; w_{mi}^2 = weight that connects the node m in the output layer to the node i in the hidden layer; x_m^1 = output of the node m in the hidden layer; and θ_i^2 = threshold of the node i in the output layer.

Continuing the process through the input layer yields

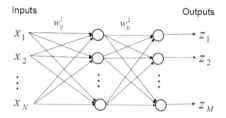

Figure 1. BP neural network.

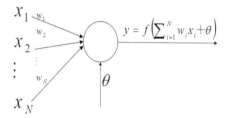

Figure 2. Function of the network node.

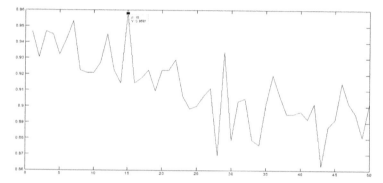

Figure 3. Average recognition rate of 1 to 50 hidden nodes.

$$\frac{\partial z_i}{\partial x_j} = z_i\left(1-z_i\right)\sum_m w_{mi}^2 x_m^1 \left(1-x_m^1\right)\frac{\partial}{\partial x_j}\left(\sum_n w_{nm}^1 x_n^0 + \theta_n\right) \qquad (4)$$

Finally,

$$\frac{\partial z_i}{\partial x_j} = z_i\left(1-z_i\right)\sum_m w_{mi}^2 x_m^1 \left(1-x_m^1\right)w_{jm}^1 \qquad (5)$$

From Eq. 5, we can see that the derivative of the outputs with respect to the input depends on the current input to the network, the weights and the output of each node. Also, Ruck et al. proved that the input should be independently sampled (Ruck et al. 2009); otherwise, it will cause a "dimension curse".

5 EXPERIMENT AND RESULTS

In this section, the saliency measure introduced in the previous section will be used to rank the input features. Then, the best subset of features will be selected to identify the pathological voice.

5.1 *Feature ranking*

The classification problem consisted of a set of 10 traditional feature parameters and 10 non-linear feature parameters. These features were chosen for the recognition task because they have previously been shown to be effective for pathological voice recognition (Gan et al. 2014). The experiment consisted of 100 training networks. Each network was started with a different set of random weight. The order of each training network was different. After each network was trained, the saliency values for each input, Λ_j, were calculated. For each network, the input with the lowest Λ_j was ranked number 1 and the input with the highest was ranked number 20. For each input, we take 10 training network's saliency values on average. Then, we obtain 10 ranks for the 20 features, which are listed in Table 2.

From Table 2, we can draw several conclusions: (1) the method of feature ranking based on the BP neural network appears to be robust; (2) the saliency of nonlinear feature parameters (index F11–F20) is almost higher than that of the traditional features (index T1–T10); and (3) the F0 (index T1) from the traditional feature set made a high contribution to the pathological voice recognition.

5.2 Recognition based on SVM

Based on the above results, five kinds of feature parameters were selected into the subset of features, namely the Hurst parameter, FMMI, fuzzy entropy, second-order Rényi entropy and F0.

The Support Vector Machine (SVM) is better than the BP neural network in classification, so we choose the SVM as the recognition pattern. A detailed description of SVM classifiers can be found elsewhere (Vapnik et al. 1995). Recognition based on the SVM obtains identification results, as given in Table 3.

As shown in Table 3, the subset of features consists of Hurst parameter, attractor and fuzzy entropy, with the highest recognition rate of 98.67% and the average rate of 88.66%. Then, by continuously adding the features into the subset, both the average rate and the maximum rate are decreased. Compared with the traditional features, the F0 is added into the subset. As shown in Table 3, the average recognition rate is improved while the highest recognition rate declines.

6 DISCUSSION

In this paper, the saliency measure used for ranking the importance of features by a BP neural network has been evaluated. The result from Table 2 indicates that the nonlinear features can characterize voice pathology better than the traditional features. Then, the recognition result from Table 3 indicates that the saliency measure is robust for ranking the importance of features. Moreover, when the subset of features consists of the three features, the Hurst parameter, FMMI and fuzzy entropy, the best maximum identification rate (98.67%)

Table 2. 10 ranks of the 20 features.

Order	Order of the ten times' feature ranking										Final feature ranking	Average saliency measure
	1	2	3	4	5	6	7	8	9	10		
1	F15	F15	F15	F15	F15	F15	F12	F12	F12	F16	F15	126.388
2	F19	F16	F12	F12	F12	F12	F15	F15	F15	F15	F12	124.669
3	F16	F19	F19	T1	F19	F19	F19	F19	F19	F19	F19	110.8523
4	F12	F12	T1	F19	F16	F16	F16	T1	F16	F12	F16	105.850
5	F17	F17	F16	F16	T1	T1	T1	F16	T1	F17	T1	101.917
6	F20	F20	F17	F17	F17	F17	F17	F17	F17	T4	F17	92.849
7	T1	F11	F18	F18	F18	F13	F18	F18	F13	T1	F18	86.021
8	T9	T1	F20	F20	F13	T4	F13	F13	F18	F20	F13	85.426
9	F13	T4	T4	F13	T4	F18	T4	F14	T4	F13	F20	81.425
10	F11	F13	F13	T4	F20	F20	F20	T4	F20	F11	T4	83.004
11	T6	F14	F11	T6	F14	F14	F14	F20	F14	F18	F14	75.807
12	T3	T9	T6	F11	T9	T6	T9	T9	T6	T9	T6	71.434
13	T8	F18	F14	T2	T6	T9	T2	T2	T2	F14	T9	71.129
14	T2	T5	T2	F14	T2	T2	F11	T6	T9	T2	F11	69.931
15	T4	T2	T9	T9	F11	F11	T6	F11	T8	T6	T2	69.528
16	F14	T3	T8	T3	T5	T8	T5	T5	T3	T3	T3	64.821
17	T5	T6	T3	F10	T3	T5	T3	T3	F11	T5	T5	63.774
18	F18	T10	T5	T8	T7	T3	T7	T8	T7	T10	T8	63.069
19	T10	T7	T10	T5	T8	T7	T8	T7	T5	T8	T7	61.319
20	T7	T8	T7	T7	T10	T10	T10	T5	T10	T7	T10	60.454

Table 3. Recognition rates.

Features	Average rate %	Max rate %
Hurst+FMMI	87.33	96.00
Hurst+FMMI+fuzzy entropy	88.66	98.67
Hurst+FMMI+fuzzy entropy++Rényi entropy	88.03	96.00
Hurst+FMMI+fuzzy entropy+Rényi entropy+F0	89.12	97.33
Hurst+FMMI+F0	88.11	97.33
Hurst+FMMI+fuzzy entropy+F0	88.99	96.00
Hurst+FMMI+Rényi entropy+F0	89.03	97.33

was obtained. When the subset of features consists of the five features, the Hurst parameter, FMMI, fuzzy entropy, second-order Rényi entropy and F0, the best average recognition rate (88.66%) was obtained. From the above results, it is suggested that the nonlinear acoustic characteristic can more fully characterize the pathological voice.

The saliency measure based on the BP neural network can perfectly reduce the feature dimension and prove the recognition rate. Also, it effectively reflects the individual characteristic's contribution to the voice pathology detection. Although there is not much contribution made by the traditional features, they cannot be rule out as they contain the vocal characterization information. So, in the voice pathology identification problem, we should consider the effectiveness of the two types of feature integration, and not a single class of features alone. This is also one of the extensions.

ACKNOWLEDGMENTS

This work was supported by the National Natural Science Foundation of China (No. 61062011, No. 61362003) and the GuangXi Key Lab of Multi-source Information Mining & Security, Electronic Engineering College, Guangxi Normal University.

REFERENCES

Barron, A.R. (1989, December). *Statistical properties of artificial neural networks*. In Decision and Control, 1989., Proceedings of the 28th IEEE Conference on (pp. 280–285). IEEE.

Cortes, C., & Vapnik, V. (1995). *Support-vector networks. Machine learning, 20(3), 273–297.*

Cybenko, G. (1989). Approximation by superpositions of a sigmoidal function. *Mathematics of control, signals and systems, 2(4), 303–314.*

Gan, D., Hu, W., & Zhao, B. (2014). A comparative study of pathological voice based on traditional acoustic characteristics and nonlinear features. *Journal of biomedical engineering, 31(5), 1149–1154.*

Henríquez, P., Alonso, J.B., Ferrer, M., Travieso, C.M., Godino-Llorente, J., & Díaz-de-María, F. (2009). Characterization of healthy and pathological voice through measures based on nonlinear dynamics. *Audio, Speech, and Language Processing, IEEE Transactions on, 17(6), 1186–1195.*

Rosenstein, M.T., Collins, J.J., & De Luca, C.J. (1993). A practical method for calculating largest Lyapunov exponents from small data sets. *Physica D: Nonlinear Phenomena, 65(1), 117–134.*

Ruck, D.W., Rogers, S.K., & Kabrisky, M. (1990). Feature selection using a multilayer perceptron. *Journal of Neural Network Computing, 2(2), 40–48.*

Salhi, L., & Cherif, A. (2013, April). Selection of pertinent acoustic features for detection of pathological voices. *In Modeling, Simulation and Applied Optimization (ICMSAO), 2013 5th International Conference on (pp. 1–6). IEEE.*

Takens, F. (1981). Detecting strange attractors in turbulence (pp. 366–381). *Springer Berlin Heidelberg.*

Takeo, F. (1996). Box-counting dimension of graphs of generalized Takagi series. *Japan Journal of Industrial and Applied Mathematics, 13(2), 187–194.*

Wolf, A. (1986). Quantifying chaos with Lyapunov exponents. *Chaos, 273–290.*

Zhao, B.X., & Hu, W.P. (2013). Recognition of Pathological Voice Based on Entropy and Support Vector Machine. *Chinese Journal of Biomedical Engineering, 5, 006.*

Bioinformatics and Biomedical Engineering – Chou & Zhou (Eds)
© 2016 Taylor & Francis Group, London, ISBN 978-1-138-02784-8

A graph-based method for blood vessel segmentation of retinal images

J.D. Zhang
Department of Electronics and Communication, Shenzhen Institute of Information Technology, Shenzhen, China

W.H. Jiang
Longgang District, Shenzhen, China

C.X. Zhang & Y.J. Cui
Department of Electronics and Communication, Shenzhen Institute of Information Technology, Shenzhen, China

ABSTRACT: Segmentation of blood vessels from retinal images plays a crucial role for diagnosing eye diseases. In this paper, we propose an automatic segmentation method for blood vessels in retinal images. Firstly, we preprocess the retinal images to enhance vessels and overcome the lighting variation. Then, a graph-based algorithm is exploited to partition the retinal image into multiple regions. Finally, we classify each region as retinal region or non-vessel region with K-mean algorithm. The proposed method is validated on the publicly available DRIVE database, and compared with the state-of-the-art algorithms.

Keywords: medical image segmentation; retinal images; graph-based algorithm; K-mean algorithm

1 INTRODUCTION

In the analysis of retinal images, segmentation of blood vessels plays a crucial role for diagnosing complications due to hypertension, diabetes, arteriosclerosis, cardiovascular disease and stroke (Kanski, 1989). Automatic blood vessel segmentation system could reduce the doctors' workload, and increase the effectiveness of preventive protocols and early therapeutic treatments. However, the retinal images have low contrast, and large variability is presented in the image acquisition process (Roychowdhury et al., 2014), which deteriorates automatic blood vessel segmentation results.

Many approaches have been reported in the area of blood vessel segmentation of retinal image, including rule-based method (Marinez-Perez et al., 2007), model-based method (Jiang et al., 2003; Vermeer et al., 2004; Chaudhuri et al., 1989; Al-Diri et al., 2009), matched filtering (Odstrcilik et al., 2013; Cinsdikici et al, 2009), and supervised method (Roychowdhury et al., 2014; Marin et al., 2011; Niemeijer et al., 2004; Staal et al., 2004; Kande et al., 2007).

Martinez et al. (2007) proposed a rule-based method based upon multi-scale feature extraction and a multiple pass region growing procedure. As model-based method, a general framework based on a verification-based multi-threshold probing scheme was presented by Jiang et al. (2003). In the matched filtering method (Odstrcilik et al., 2013; Cinsdikici et al, 2009), a 2-D linear structuring element is exploited to extract a Gaussian intensity profile of the retinal blood vessels, using Gaussian and their derivatives, for retinal blood vessel enhancement. Neimeijer et al. (2004) implemented a K-nearest neighbor classifier. A 31 component pixel feature vector was constructed with the Gaussian and its derivatives up to order 2 at 5 different scales, augmented with the gray-level from the green channel of the original image.

In this paper, we partition the retinal images into two types: vessel and non-vessel. Firstly, we preprocess the retinal images to enhance vessels and overcome the lighting variation. Then, a graph-based method is exploited to segment the retinal images into multiple vessel regions and no-vessel regions. Finally, we label the segmented regions with K-mean algorithm.

The rest of this paper is organized as follow. Section 2 describes our proposed method. In Section 3, experimental results are presented, followed by the conclusion in Section 4.

2 METHODS

A retinal image is viewed as a graph $G = (V, E)$, where each node $v_i \in V$ corresponds to a pixel in the image and the edges in E connect certain pairs of neighboring pixels. Therefore, we may apply the graph-based method in order to segment the image into multiple regions (vessel regions and non-vessel regions). For improving the segmentation results, we preprocess the retinal image for vessel enhancement.

2.1 Preprocessing

Retinal images often show important lighting variations, poor contrast and noise (Roychowdhury et al., 2014). To overcome these problems and extract the pixel feature accurately, we preprocess the images as the following steps.

1. RGB to green conversion: In original RGB retinal images, the green channel shows the best vessel-background contrast, while the red and blue channels show low contrast and are noisy. So, we select the green channel from the RGB retinal image, and the green channel intensity of each pixel is taken as the intensity feature. Figure 1 (a) is the original RGB retinal image from DRIVE database, and the green channel image is shown in Figure 1 (b).
2. Background homogenization: Retinal images often contain background intensity variation because of uniform illumination. In the present work, the shade-correction method mentioned in (Niemeijer et al., 2005; Zhang et al., 2010) is used to remove the background lightening variations. The shade-correction image of Figure 1 (b) is presented in Figure 1 (c).

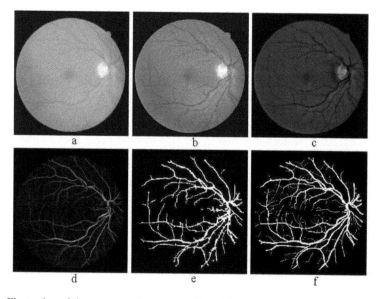

Figure 1.   Illustration of the preprocessing process. (a) Original RGB retinal image. (b) The green channel of the original image. (c) Shade-corrected image. (d) Vessel enhanced image. (e) The segmentation result with our proposed method. (f) The manual segmentation result by the first specialist.

3. Vessel enhancement: As the contrast between the blood vessels and the background is generally poor in the retinal images, vessel enhancement is utilized for estimating the complementary image of the homogenized image, and subsequently applying the morphological top-hat transformation with a disc of eight pixels in radius. Figure 1 (d) is the vessel enhancement image of Figure 1 (c).

2.2 Feature extraction

Hence, for each pixel p, a multi-dimensional feature vector is constructed with the intensity feature of the green channel image I_{gp} and the intensity feature of the image after vessel enhancement I_{ep}, denoted by $f_p = [I_{gp}, I_{ep}]$.

2.3 Graph-based segmentation method

Felzenszwalb (2004) proposed an efficient graph-based images segmentation method. An important characteristic of this method is its ability to preserve detail in low variability image regions which ignoring details in high variability regions. Retinal images have lighting variations and poor contrast. Therefore, we exploit Felzenszwalb's method for blood vessel segmentation could overcome the problem of lighting variations and poor contrast in retinal images. Moreover, this method is highly efficient, running in time nearly linear in the number of image pixels. So, this method can be used in the medical processing application.

Let $G = (V, E)$ be an undirected graph with vertices $v_i \in V$, the set of elements to be segmented, and edges $(v_i, v_j) \in E$ corresponding to pairs of neighboring vertices. Each edge $(v_i, v_j) \in E$ has a corresponding weight $w(v_i, v_j)$, which is a non-negative measure of the dissimilarity between neighboring elements v_i and v_j. In the case of retinal image segmentation, the elements in V are pixels and weight of the edge (v_i, v_j) is the difference in feature vector between the pixels v_i and v_j $(w(v_i, v_j) = |f_{v_i} - f_{v_j}|)$.

In the graph-based method, a segmentation S is a partition of V into components such that each component (or region) $C \in S$ corresponds to a connected component in a graph $G' = (V, E')$, where $E' \in E$. We want the elements in a component to be similar and elements in different component to be dissimilar. This means that the edges between two vertices in the same component should b have relatively low weights, and edges between vertices in different components should have higher weights.

Felzenszwalb (2004) defines inter difference of a component $C \subseteq V$ to be the largest weight in the minimum spanning tree of the component, $MST(C, E)$. That is, $Int(C) = \max\limits_{e \in MST(C,E)} w(e)$. One intuition underlying this measure is that a given component C only remains connected when edges of weight at least $Int(C)$ are considered. Then, he defines the difference between two components $C_1, C_2 \subseteq V$ to be the minimum weight edge connecting the two components. That is, $Dif(C_1, C_2) = \min\limits_{v_i \in C_1, v_j \in C_2, (v_i, v_j) \in E} w(v_i, v_j)$. If there is no edge connecting C_1 and C_2, we let $dif(C_1, C_2) = \infty$.

The measure of difference only reflects the smallest edge weight between two components, but it works quite well in practice. Moreover, changing the definition in order to make it more robust to outliers, make the problem of finding a good segmentation NP-hard. Then, Felzerszwalb (2004) defines the pairwise comparison predicate as

$$D(C_1, C_2) = \begin{cases} \text{true} & \text{if } Dif(C_1, C_2) > MInt(C_1, C_2) \\ \text{false} & \text{otherwise} \end{cases},$$

where the minimum internal difference, $MInt$ is defined as, $MInt(C_1, C_2) = \min(Int(C_1) + \tau(C_1), Int(C_2) + \tau(C_2))$. The threshold function τ is defined as $\tau(C) = k / |C|$, where $|C|$ denotes the size of C, and k is some constants parameter. That is, for small components we require stronger evidence for a boundary. In practice k sets a scale of observation, in that a larger k causes a preference for larger components. Note, however, that k is not a minimum component size. Smaller components are allowed when there is a sufficiently large difference between neighboring components.

In our work, we exploit Felzenswalb's graph-based method (Felzenswalb et al., 2004) to partition the retinal images. Vessels of the retinal images belong to the detail information. To reserve the thin and small vessels in the segmentation result, k is not a larger value. So, there are multiple regions (vessel region or non-vessel regions) after graph-based segmentation.

2.4 Postprocessing

After clustering with Felzenszwalb's graph-based algorithm, there are multiple regions including vessel regions and non-vessel regions. We use the mean intensity feature of pixels in a region to estimate the region class.

K-mean method is used to label the region class based on the regions' inherent distance from each other. The algorithm assumes that the data features form a vector space and tries to find natural clustering in them. The points are clustered around centroids c_i, $i = 1$, 2 which are obtained by minimizing the objective

$$D = \sum_{i=1}^{2} \sum_{R_j \in C_i} \| f_{R_j} - c_i \|^2,$$

where there are *two* clusters C_i, $i = 1$, 2. f_{R_j} is the feature vector of region R_j, and c_i is the centroid of all the regions $R_j \in C_i$. The algorithm is composed of the following steps:

- Initial step: Initialize the centroids with K random vectors $c_i^{(1)}$, $i = 1$, 2, ..., K.
- Repeat the following steps until the clustering results of all the regions do not change anymore.
  - Assignment step: Assign each region to the closest centroid by $c_i^{(t)} = \{R_j : \| f_{R_j} - c_i^{(t)} \|^2 \leq \| f_{R_j} - c_m^{(t)} \|^2 \ \forall m, \ 1 \leq m \leq K\}$, where t is the iterative time.
  - Update step: Calculate the new centroids of the new clusters by $c_i^{(t+1)} = \dfrac{1}{|C_i^t|} \sum_{R_j \in C_i^{(t)}} f_{R_j}$.

Finally, in the visual inspection, small isolated regions misclassified as blood vessels are also observed. If the vessel region is connected with no more than 30 pixels, it will be reclassified as non-vessel. The segmentation result of our proposed method is shown in Figure 1 (e).

3 EXPERIMENTAL RESULTS

3.1 Database and similarity indices

The DRIVE database (Staal et al., 2004) is used in our experiments. This dataset is a public retinal image database, and is widely used by other researchers to test their blood vessel segmentation methods. Moreover, the DRIVE database provides two sets of manual segmentations made by two different observers for performance validation. In our experiments, performance is computed with the segmentation of the first observer as ground truth.

To quantify the overlap between the segmentation results and the ground truth for vessel pixels and non-vessel pixels, Accuracy (Acc) are adopted in our experiments. Acc is defined as $Acc = (TP + TN)/(TP + TN + FN + FP)$, where TP is the number of vessel pixels that are correctly classified as vessels (true positives), TN is the number of vessel pixels that are correctly classified as non-vessels (true negatives), FP is the number of pixels falsely classified as vessels (false positive), and FN is the number of pixels falsely classified as non-vessels (false negative).

3.2 Our method evaluation

For visual inspection, Figure 2 depicts the blood vessel segmentation results on different retinal images from DRIVE database. Figure 2 (a), (d) and (g) are original retinal images with different illumination conditions, and their segmentation results using our proposed method are shown in Figure 2 (b), (e) and (h) respectively. The manual segmentation results by the

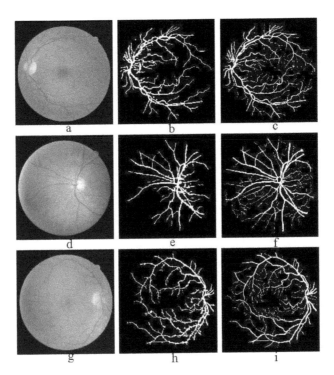

Figure 2. Examples of application of our segmentation method on three images with different illumination conditions. (a), (d), (g) Original RGB retinal images. (b), (e), (h) Segmentation results with our method. (c), (f), (i) The manual segmentation results by the first specialist.

Table 1. Performance results on DRIVE database images, according to Acc value.

Image	1	2	3	4	5
Acc	0.933840	0.937442	0.895387	0.955840	0.952073
Image	6	7	8	9	10
Acc	0.951261	0.934253	0.948812	0.956077	0.941138
Image	11	12	13	14	15
Acc	0.933566	0.954440	0.950155	0.930095	0.913721
Image	16	17	18	19	20
Acc	0.949621	0.951858	0.941231	0.954453	0.921477
Average	**0.940337**				

first specialist are presents in Figure 2 (c), (f) and (i) for visual comparison. It is evident that our method is robust to the low contrast and large variability in the retinal images.

In addition, we give a quantitative validation of our method on the DRIVE database with available gold standard images. Since the images dark background outside the Field-of-View (FOV) is provided, Accuracy (Acc) values are computed for each image considering FOV pixels only. The results are listed in Table 1, and the last row of the table shows average Acc value for 20 images in the database. As shown in Table 1, our method gets accurate segmentation results for retinal images on the DRIVE database.

3.3 Comparing with other methods

In order to compare our approach to other retinal vessel segmentation algorithms, the average Acc value is used as measures of method performance. We compare our method with

167

Table 2. Comparing the segmentation results of different algorithms with our method on DRIVE database in terms of average Acc value.

Method type	Method	DRIVE
Rule-based method	Martinez-Perez et al., 2007	0.9344
Model-based method	Jiang and Mojon, 2003	0.8911
Matched filter	Chaudhuri et al., 1989	0.8773
	Cinsdikici and Aydin, 2009	0.9293
Supervised method	Niemeijer et al., 2004	0.9417
Clustering method	Our proposed method	0.9403

the following published methods: Martinez-Parez et al. (2007), Jiang and Mojon (2003), Chaudhuri et al. (1989), Cinsdikici and Aydin (2009), and Niemeijer et al. (2004).

The comparison results are summarized in Table 2, which indicate our proposed method outperforms most of the other methods. Moreover, our method is a clustering method without the doctors' intervention and testing images, and gets the comparable segmentation results with the supervised method (Niemeijer et al., 2004).

4 CONCLUSIONS

This study proposes a blood vessel segmentation method based on graph-based algorithm. To overcome the problem of low contrast and large variability in retinal images, we enhance vessels by RGB to green conversion, background homogenization and morphological transformation. Then, we partition the retinal image into multiple regions with graph-based algorithm. Finally, the multiple regions are classified by K-mean algorithm.

Our method is validated on the DRIVE database with available gold standard images. In Section 3.2, it is evident that our method is robust to the low contrast and large variability in the retinal images, and gets accurate segmentation results, as shown in Table 1 and Figure 2. We compare our method with the state-of-art methods in Section 3.3, and experimental results in Table 2 indicate that out method outperforms most of the other methods.

In addition, the graph-based segmentation method is highly efficient, running in time nearly linear in the number of image pixels. So, our method can be used in the medical processing application. It should be noted that our method partitions vessels from retinal images without doctors' intervention. Therefore, the applicability of our method can be extended to diseased retinal images in the further research.

ACKNOWLEDGEMENTS

This project is supported in part by Shenzhen Science and Technology Plan Project (JCYJ20120615101059717), and Project of Shenzhen Institute of Information Technology (SYS201004).

REFERENCES

Al-Diri, B., Hunter, A. & Steel, D. 2009. An active contour model for segmenting and measuring retinal vessels. IEEE Transactions on Medical Imaging 28: 1488–1497.

Chaudhuri, S., Chatterjee, S., katz, N., Nelson, M. & Goldbaum, M. 1989. Detection of blood vessels in retinal images using two-dimensional matched filters. IEEE Transactions on Medical Imaging 8(3): 263–269.

Cinsdikici, M.G. & Aydin, D. 2009. Detection of blood vessels in ophthalmoscope images using MF/ant (matched filter/ant colony) algorithm. Comput. Methods Programs Biomed. 96: 85–95.

Felzenszwalb, P. & Huttenlocher, D.P. 2004. Effcient graph-based image segmentation. International Journal of Computer Vision 59(2): 167–181.

Jiang, X. & Mojon, D. 2003. Adaptive local thresholding by verification-based multithreshold probing with application to vessel detection in retinal images. IEEE Transactions on Pattern Analysis and Machine Intelligence 25(1): 131–137.

Kanski, J.J. 1989. Clinical Ophthalmology: A systematic approach. London: Butterworth-Heinemann.

Marinez-Perez, M.E., Hughes, A.D., Thom, S.A., Bharath, A.A. & Parker, K.H. 2007. Segmentation of blood vessels from red-free and fluorescein retinal images. Medical Imaging Anaysis 11: 47–61.

Niemeijer, M., Staal, J., Ginneken, B.V., Loog, M., Abramoff, M.D., Fitzpatrick, J. & Sonka, M. 2004. Comparative study of retinal vessel segmentation methods on a new publicly available database. In SPIE Med. Imag. 5370: 648–656.

Niemeijer, M., van Ginneken, B., Staal, J.J., Suttorp-Schulten, M.S.A. & Abramoff, M.D. 2005. Automatic detection of red lesions in digital color fundus photographs. IEEE Transactions on Medical Imaging 24(5): 584–592.

Odstrcilik, J., Kolar, R., Budai, A. & et al. 2013. Retinal vessel segmentation by improved matched filtering: evaluation on a new high-resolution fundus image database. IET image processing 7: 373–383.

Roychowdhury, S., Koozekanani, D.D. & Parhi, K.K. 2014. Blood vessel segmentation of fundus images by major vessel extraction and sub-image classification. IEEE journal of biomedical and health informatics 99, DOI: 10.1109/JBHI.2014.2335617.

Soares, J., Leandro, J., Cesar, R., Jelinek, H. & Cree, M. 2006. Retinal vessel segmentation using 2-D Gabor wavelet and supervised classification. IEEE Transactions on Medical imaging 25(9): 1214–1222.

Staal, J., Abramoff, M.D., Niemeijer, M., Viergever, M.A. & Ginneken, B. 2004. Ridge-based vessel segmentation in color images of the retina. IEEE Transactions on Medical Imaging 23: 501–509.

Vermeer, K.A., Vos, F.M., Lemij, H.G. & Vossepoel, A.M. 2004. A model based method for retinal blood vessel detection. Computers in Biology and Medicine 34: 209–219.

Zhang, B., Wu, X., You, J., Li, Q. & Karray, F. 2010. Detection of microaneurysms using multi-scale correlation coefficients. Pattern Recognition 43: 2237–2248.

Bioinformatics and Biomedical Engineering – Chou & Zhou (Eds)
© *2016 Taylor & Francis Group, London, ISBN 978-1-138-02784-8*

A fusion approach for dynamic skin detection

L. Chen & Y.H. Liu
College of Informational Science and Technology, Xiamen University, Xiamen, China

ABSTRACT: Image processing technology has driven the biomedical imaging to a profound change. Capable for a wide range of applications such as biomedical image processing, skin detection has developed rapidly with various approaches and methods of human skin segmentation. However, the principal obstacles that skin detection faces are still the different degrees of skin tone color, illumination conditions and skin color-like backgrounds. Given that, in this paper, we proposed a novel fusion strategy for dynamic skin detection, which is based on a smoothed 2-D histogram, Gaussian model and an online dynamic threshold calculated on face skin tone color, which reduces the training required, to a certain extent. In this approach, we adopted face detector to refine the skin model, as face is a prominent indicator of different characteristics of skin tone color, especially in images that include more than one face with different ethnicities. Qualitatively and quantitatively experimental results show that the proposed method is more robust and effective compared with state-of-the-art methods, because of its low computational costs and high accuracy.

Keywords: skin detection; biomedical imaging; skin model; image processing

1 INTRODUCTION

With the development of information technology recently, image processing has been playing a more and more significant role. What's more, image processing technology has driven the biomedical engineering to a profound change. Kinds of clinical applications of new biomedical imaging methods have brought the medical diagnosis and treatment technology great progress. Meanwhile, complementing the information got from various biomedical imaging techniques provides a strong scientific basis for clinical diagnosis and biomedical research. Among the wide range of image proccesions, skin detection, as a popular and useful technique, has received much attention because of its wide fields of application range from detecting and tracking human-body parts, face detection, face recognition, gesture analysis, image retrieval, biomedical image processing and visualization to kinds of human-computer interaction domains, which have been researched by Vadakkepat et al. (2008), Chan et al. (2007), Kubota & Nishida (2007), Linda & Manic (2010), Chan et al. (2006) and Pratl et al. (2007). Skin color detection is the foundation of skin detection, as a cue. However, there are various negative factors on standard skin color detection techniques, such as illumination conditions, skin color-like backgrounds and surfaces and ethnicity.

Typically, a skin detection process involves three primary steps. First, it finds the suitable color space for skin detection to represent the image pixels. And then, it uses the appropriate modeling technique to distribute the skin and non-skin pixels. Finally, it classifies the modeled distributions that rely on the right skin classification.

As the staple steps in skin detection, choosing the suitable one from variety of color spaces seems to be particularly important. Since digital images captured by a camera is normally stored and represented as RGB, it is the default color space for most available image formats. While representing matte surfaces with RGB values, ambient light is ignored. In order to provide robust parameters against varying illumination conditions, the RGB can be transformed into other color spaces by linear or non-linear methods. A detail survey of different color spaces such as RGB, normalized RGB, HSV (hue, saturation and value), HSI (hue, saturation and intensity), HSL (hue, saturation and lightness), TSL (tint, saturation and luminance), YCbCr, CIE-Lab and CIE-Luv has been researched by Kakumanu et al. (2007). With a profound effect on overall skin detection performance, color space transformations, namely RGB to HSV, HSI, normalized RGB, YCbCr and CIE-Lab, have been comprehensively studied by Gonzalez & Woods (2001) and Khan et al. (2012).

The pivotal essence of skin detection is to classify skin pixels and non-skin pixels using skin modeling techniques. However, a good skin classifier should be proved to overcome the negative factors. Explicitly, the general sorts of skin modeling are parametric, non-parametric and neural network techniques. From a large training dataset, Kakumanu et al. (2007) concluded one of the easiest and often used methods, which is to define skin-color thresholds for different color spaces in the images taken under illumination controlled conditions. With a very large training dataset, histogram technique is affected by the degree of overlap between the skin and non-skin classes. When considering about not only low False Positive Rate (FPR) but also high True Positive Rate (TPR), they suggested that Gaussian Mixture Models (GMM) may perform better than Single Gaussian Models (SGM). Although Multi Layer Perceptron (MLP) classifier performs similarly to Bayesian Network (BN) classifier, MLP requires very low storage, while BN gets the advantage of low-labeled data, which has been described in detail by Albiol et al. (2001), Terillon et al. (2000) and Fu et al. (2004).

Tan & Chan (2012) fused a smoothed 2-D histogram and Gaussian model into a novel framework to automatically detect human skin in images, which improved the accuracy in spite of various conditions of illumination, ethnicity and backgrounds. However, the performance of this method is dependent on eye detection as a subpar preprocessing method in the research of Yogarajah et al. (2010). Ibrahim et al. (2012) provided a dynamic skin detection method, in which relying on eye detector algorithms was replaced with acquiring a dynamic threshold from the face region. This method refined the performance by lowering the False Positive Rate (FPR), but not in the case of more than one face contained in one picture.

For the purpose of accurate and efficient skin detection in terms of various illumination conditions, skin color-like backgrounds and surfaces and ethnicities, we put forward a dynamic fusion approach. First, an online dynamic method based on face skin tone color, as in the research of Ibrahim et al. (2012), is adopted to compute the skin threshold. Second, we employ dynamic threshold got from the first step, 2-D histogram and Gaussian model to classify skin and non-skin distributions, separately. Finally, the products of three features are used to construct a fusion strategy framework to detect skin region in the whole image.

2 THE PROPOSED APPROACH

The developed framework for automatic skin detection is shown in Figure 1. First, we acquire the face in a designated image with a canal similar to that of Zhu et al. (2012). Second, an online dynamic approach based on face skin tone color is adopted to compute the skin threshold value. Third, we employ dynamic threshold got from the second step, 2-D histogram and Gaussian model to classify skin and non-skin distributions, separately. Finally, the products of three features are used to construct a fusion strategy framework to perform better in detecting the skin region in the whole image. In the paper, the RGB color space is transformed to HSV color space to obtain better performance.

2.1 Preprocessing

The steps include face detecting. For any given image, we first use self-adaptive luminance compensation and hybrid filtering technology to smooth. Then, median and morphological operations are adopted to segment skin color districts. As non-smooth textures of the skin color areas, eyes are detected relying on the matching relationship between luminance and human eyes color. The mouth is often situated on the perpendicular bisector between two

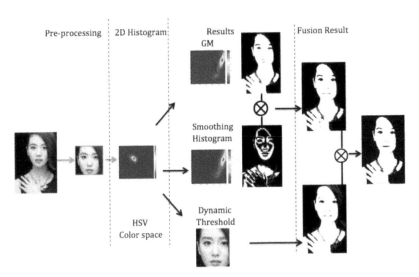

Figure 1.　Proposed framework: face detector, 2-D histogram, Gaussian model, dynamic threshold and fusion strategy.

eyes centers. Finally, the face is accurately detected from skin color areas through two eyes and mouth parts, as in the research of Zhu et al. (2012).

2.2　Color space

As a perceptual color space, HSV is used in this paper. The reason is twofold: first, Chaves-Gonzalez et al. (2010) conducted a brief study on ten commonly used color spaces in skin detection, in which HSV was the winner. And second, it is invariant to high intensity at white lights, ambient light and surface orientations relative to the light source for the transformation of RGB to HSV.

The transformation of the RGB model to the HSV model is as follows:

$$V = m \tag{1}$$
$$S = (m - n)/m \tag{2}$$

$$H = \begin{cases} 60 \times [(G - B)/(m - n)] & \text{if } m = R \\ 60 \times [2 + (B - R)/(m - n)] & \text{if } m = G \\ 60 \times [4 + (R - G)/(m - n)] & \text{if } m = B \end{cases} \tag{3}$$

where m and n, respectively, represent the maximum and minimum of R, G, and B values. While the H parameter indicates the color information, namely the position of spectral colors. This parameter is represented by the angular amount. So, if the H value is less than 0, it would be plus the 360.

2.3　Skin detection

2.3.1　The combination of smoothed 2-D histogram and Gaussian model

First, we use a 2D color histogram to represent the distribution of skin tones in color space. As the decisive factor of the distribution, the threshold value is adjusted to tradeoff between true and false positives. The smoothed 2-D histogram-based segmentation is given as (4):

$$D_H(q_s, q_n) = \begin{cases} 1 & \text{if } P(q_s, q_n) \geq 0.3 \\ 0 & \text{if } P(q_s, q_n) < 0.3 \end{cases} \tag{4}$$

where q_s and q_n, respectively, represent the ratio of the pixel counts of color c of the skin histogram and the non-skin histogram divided by the total counts of skin histogram and the total counts of non-skin histogram. And then, as the parametric model with little training data and storage requirements, the Gaussian mixture models used for skin classification on Probability Distribution Function (PDF) are defined as (5):

$$D_G(\Sigma_i, \lambda_i) = \begin{cases} 1 & if \ H_P(\Sigma_i, \lambda_i) \geq 0.05 \\ 0 & if \ H_P(\Sigma_i, \lambda_i) < 0.05 \end{cases} \tag{5}$$

where H_p is the value of PDF, namely $P(c)$, which is calculated as follows:

$$P(c) = \sum_{i=1}^{N} \omega_i \frac{1}{(2\pi)^{1/2} |\Sigma_i|^{1/2}} \exp\left(-\frac{1}{2}\lambda_i\right) \tag{6}$$

where λ_i is represented by Σ_i and μ_i; c is the color vector; and μ_i, Σ_i and ω_i, respectively, refer to the mean vector, the diagonal covariance matrix and the mixing weights, which satisfy the constraint $\sum_{i=1}^{N} \omega_i = 1$. Finally, we combine matching results of the independent features foregoing with the fusion rules defined as follows:

$$D_F(q_s, q_n, \Sigma_i, \lambda_i) = \Gamma\{D_H(q_s, q_n), D_G(\Sigma_i, \lambda_i)\} \tag{7}$$

2.3.2 The online dynamic threshold calculated on face skin tone color

This method gets the dynamic threshold from a trimmed image with static frontal face removed eyes and mouth parts. Then, the final threshold that is updated by increasing the pixels around the axes until 95% from face pixels are recognized as skin is used to identify skin pixels of the whole image, delimited as follows:

$$D_T(S_{min}, S_{max}, V_{min}, V_{max}) = \begin{cases} 1 & if \ S_{min} \leq S \leq S_{max} \\ 0 & if \ V_{min} \leq V \leq V_{max} \end{cases} \tag{8}$$

where $S_{min}, S_{max}, V_{min}$ and V_{max} are calculated by the final dynamic threshold in the HSV color space.

2.3.3 The fusion strategy

For the purpose of effective and robust skin detection, we propose a fusion strategy integrating the two above features to, respectively, classify skin and non-skin pixels into a combined single representation. The combined matching results can be obtained as (9):

$$D = \Gamma\{D_F(q_s, q_n, \Sigma_i, \lambda_i), D_T(S_{min}, S_{max}, V_{min}, V_{max})\} \tag{9}$$

where Γ is the selected fusion rule of image fusion technology, which represents the product $\otimes$.

3 EXPERIMENTS AND ANALYSIS

Considering various illumination conditions, skin color-like backgrounds and surfaces and ethnicities, we gathered from the web a group of images, each of which involves at least one frontal face. For the purpose of evaluation, we compared the results of the proposed method with the state-of-art methods of Tan & Chan (2012) and Ibrahim et al. (2012). Figure 2 shows the comparison of images of a single person, and Figure 3 compares the images of more than one person. From the comparison results of Figure 2 and Figure 3, we draw the conclusion that our proposed approach performs better than the other two, evidently. As we can see in the image sample at row 4 of Figure 2 and Figure 3, a lot of non-skin pixels are misidentified

Figure 2. Columns from left to right represent original images, the method of Tan & Chan (2012), the method of Ibrahim et al. (2012), and our proposed methods, respectively.

Figure 3. Columns from left to right represent original images, the method of Tan & Chan (2012), the method of Ibrahim et al. (2012), and our proposed method, respectively.

(a)　　　(b)　　　(c)　　　(d)　　　(e)

Figure 4. (a) Original Image. (b) 2-D histogram's result. (c) Gaussian model's result. (d) Dynamic threshold's result. (e) Fusion approach's result.

175

Table 1. Comparison between our proposed fusion and non-fusion approach in HSV color space using web images.

Classifier	Fusion	2D-Histogram	GMM	Dynamic threshold
Accuracy	**0.9057**	0.8854	0.858	0.8914
F-score	**0.6512**	0.6189	0.6145	0.629
True positive rate	**0.6521**	0.6695	0.8014	0.8377
False positive rate	**0.0679**	0.085	0.1156	0.1039

as skin in the performance of the method of Tan & Chan (2012), while many skin pixels as non-skin in the results of the approach of Ibrahim et al. (2012). It can be obviously observed from Figure 3 that certain image samples, which are full of rather complicated conclusions such as high illumination at row 5, skin color-like backgrounds at row 4 and surfaces at row 3 and constituted of humans from kinds of ethnicity at row 1, can be accurately detected by the proposed method with the least noise and the best robustness. In addition, the proposed approach affected little by image environment, resulting in effective performance because of little training stage.

The comparison results of adopting single feature—smoothed 2-D histograms, Gaussian mixture models, or dynamic threshold only, and the fusion approach of the three aforementioned features are shown in Figure 4. Evidently, one can notice from some particularly sample images the different performances of the four methods.

Further, a better quantitative analysis result is presented in Table 1 using web dataset. It is clear that the fusion strategy has the lowest False Positive Rate (FPR). Moreover, the fusion approach has higher accuracy and F-score compared with the single feature approach. For instance, the smoothed 2-D histograms method almost fails to detect the right skin region due to noise. The Gaussian mixture models confuse the skin color pixels with the skin color-like backgrounds such as clothes, trees, structures and hair, which has a very high False Positive Rate (FPR). The dynamic threshold approach based on face skin tone color loses some skin color pixels, detecting them as non-skin pixels incorrectly. However, it can be qualitatively visualized that our proposed method has achieved the best discrimination between skin color pixels and non-skin color pixels in terms of various illumination conditions, skin color-like backgrounds and surfaces and ethnicities.

4 CONCLUSIONS

The fusion strategy for dynamic skin detection is integrated by the smoothed 2-D histogram, Gaussian model and the dynamic threshold approach based on face skin tone color. This proposed method surmounts the state-of-art approach in terms of various illumination conditions, skin color-like backgrounds and surfaces and ethnicities. It lowers the False Positive Rate (FPR) with high accuracy. In addition, the proposed approach gets effective performance because of little training stage and low computing costs. Given that, it is applicable in a range of domains, especially in biomedical imaging field, as accurate and efficient skin detection provides a strong and robust technical support to medical image processing and analysis.

REFERENCES

Albiol, A (ed.). 2001. Optimum color spaces for skin detection. ICIP01.
Chan, C.S. (ed.). 2007. Recognition of human motion from qualitative normalized templates. *J. Intell. Robot* 48(1): 79–95.

Chan, C.S. (ed.). 2006. Human arm-motion classification using qualitative normalized templates. *Lecture Notes Artif. In- tell* 4251(1): 639–646.

Chaves-Gonzalez, J.M. (ed.). 2010. Detecting skin in face recognition system: A color spaces study. *Elsevier—Journal of Digital Signal Processing* 20:806–823.

Fu, Z. (ed.). 2004. Mixture clustering using multidimensional histograms for skin detection. ICPR04: 549–552.

Gonzalez, R.C. & Woods, R.E. 2001. Digital Image Processing. Upper Saddle River: Prentice Hall.

Ibrahim, N.B. (ed.). 2012. A dynamic skin detector based on face skin tone color. *8th International conference on Informatics and System*: 14–16.

Kubota, N. & Nishida, K. 2007. Perceptual control based on prediction for natural communication of a partner robot. *IEEE Trans. Ind. Electron* 54(2): 866–877.

Kakumanu, P. (ed.). 2007. A survey of skin-color modeling and detection methods. *Pattern Recognition* 40:1106–1122.

Khan, R. (ed.). 2012. Color based skin classification. *Pattern Recognition* 33: 157–163.

Linda, O. & Manic, M. 2010. Fuzzy force-feedback augmentation for manual control of multi-robot system. *IEEE Trans. Ind. Electron* 58(8): 3213–3220.

Pratl, G. (ed.). 2007. A new model for autonomous, networked control systems. *IEEE Trans. Ind. Informat* 3(1): 21–32.

Terillon, J.C. (ed.). 2000. Comparative performance of different skin chrominance models and chrominance spaces for the automatic detection of human faces in color images. AFGR00: 54–61.

Tan, W.R.& Chan, C.S. 2012. A fusion approach for efficient human skin detection. *IEEE Trans. Ind. Informat* 8(1).

Vadakkepat, P. (ed.). 2008. Multi- modal approach to human-face detection and tracking. *IEEE Trans. Ind. Electron* 55(3): 1385–1393.

Yogarajah, P. (ed.). 2010. A dynamic threshold approach for skin segmentation in color images. *In Proc. of IEEE Int. Conf. on Image Process*: 2225–2228.

Zhu, Y. (ed.). 2012. Face detection based on multi-feature fusion in YCbCr color space. *5th International conference on Image and Signal Processing*. CISP.

Bioinformatics and Biomedical Engineering – Chou & Zhou (Eds)
© 2016 Taylor & Francis Group, London, ISBN 978-1-138-02784-8

Local gradient thresholds computation in 3D images based on a vision model

J.Y. Li, P. Wang & L.S. Wang
Institute of Image Processing and Pattern Recognition in Department of Automation,
Shanghai Jiao Tong University, Shanghai, China

ABSTRACT: Gradient threshold computation is usually necessary for detecting bound-ary surfaces from 3D images, but it is a complex problem. Inspired by the adaptive con-trast detection mechanism of human vision, a method is proposed to compute local gradient thresholds in 3D images. By such local gradient thresholds, visible boundaries in many differ-ent 3D images can be adaptively detected and well reconstructed. Experimental results show the effectiveness of the proposed method.

1 INTRODUCTION

Modern 3D scanning techniques have generated lots of 3D images with internal structures. In many cases, different structures in a 3D image correspond to different intensities. Thus, their boundary surfaces locate where intensity sharply changes. Such boundary surfaces are called step-like boundary surfaces, just like step-like edges in 2D images. Their detection and reconstruction from 3D images is an important topic in 3D image analysis.

Each 3D image can be regarded as a discrete sampling of a three-dimensional function (denoted by $I(x,y,z)$) from a 3D regular grid. In the grid, eight adjacent grid points form a cube and all such cubes form a 3D sampling region of the 3D image. Step-like boundary surface of a structure in the 3D image can be regarded as a continuous implicit surface con-tained in the sampling region (Wang et al. 2007a, Wang et al. 2014), and is called the Step-Like Continuous Implicit Boundary Surface (SLCIBS). Let $f(x,y,z) = G(x,y,z,\sigma) * I(x,y,z)$ represent the smoothed image of $I(x,y,z)$ by the Gaussian function $G(x,y,z,\sigma)$ with the scale σ, $\nabla^2 f(x,y,z)$ and $\|\nabla f(x,y,z)\|$ represent the Laplacian function and the gradient magnitude function of $f(x,y,z)$, respectively. By the boundary surface detection theory (Bomans et al. 1990, Brejl & Sonka 2000, Marr & Hildreth 1980, Wang et al. 2007a), a SLCIBS (denoted by Ψ) in $I(x,y,z)$ is a zero-crossing surface patch with locally high gradient magnitudes in its local neighborhoods, i.e. satisfies:

$$\begin{cases} \nabla^2 f(x,y,z) = 0 \\ \|\nabla f(x,y,z)\| \geq T(U) \end{cases} \tag{1}$$

where, U represents different local small regions containing boundary points in the 3D image, and $T(U)$ is a local gradient threshold function used to separate boundary surface patches in different U from local backgrounds.

When users intend to detect and reconstruct Ψ from the 3D image, it is necessary to compute $T(U)$ in advance. If Ψ has uniformly high gradient magnitudes, then $T(U)$ can be replaced by a constant (i.e. global gradient threshold). For example, global gradient thresh-old was applied in 3D edge detection techniques (Bomans et al. 1990, Brejl & Sonka 2000, Cheng et al. 2010, Drew 2005), and in detection and reconstruction of SLCIBSes (Wang et al. 2007a, Wang et al. 2007b, Wang et al. 2003). However, when Ψ has varied gradient magni-tudes in different local regions of the 3D image, $T(U)$ cannot be replaced by any constant.

In practical applications, users usually do not know whether an interested SLCIBS Ψ has uniformly high gradient magnitudes. Hence, in such cases, local gradient thresholds rather than a global gradient threshold should be computed for Ψ.

As shown in 2D edge detection techniques (Henstock & Chelberg 1995, Rakesh et al. 2004), the automatic selection of an appropriate global gradient threshold remains a complex problem. Hence, it is more difficult to compute a local gradient threshold function $T(U)$ for a SLCIBS Ψ in a 3D image. This paper tries to develop a new and feasible method to compute automatically $T(U)$ in 3D images.

In (Wang et al. 2011), we computed $T(U)$ by an empirical method, but found that the method failed for SLCIBSes in many 3D images. It is known that human vision system can detect luminance contrast (or visual boundaries) adaptively in different conditions of luminance, and the adaptive contrast detection mechanism of human vision has been studied deeply in vision research area (Drew 2005, Thompson et al. 2011). So, we try to compute $T(U)$ adaptively in 3D images by drawing such visual mechanism. Along this line, researchers in (Hou et al. 2004) discussed the computation of local gradient thresholds in 2D images for detecting 2D edges. We recently proposed to compute local gradient thresholds in 3D images by combining context dependence with adaptive contrast detection as in the human vision system (Wang et al. 2014). The method in (Wang et al. 2014) can exclude uninterested surface patches and fragments connected with the interested SLCIBS while adaptively detecting the SLCIBS from 3D images. However, the method gradually expands a surface patch of the SLCIBS in a 3D image in a recursive calculation mode, and sometimes is a time-consuming process.

In order to explore and visualize the SLCIBS of interest in 3D images in real-time, it is necessary to compute $T(U)$ quickly. For this purpose, this paper will compute $T(U)$ based only on the adaptive contrast detection mechanism of human vision. Experimental results show that by such $T(U)$, SLCIBSes in many 3D images can still be well detected, while the computing speed is greatly improved compared to the method in (Wang et al. 2014).

2 METHOD

In the 3D image, the SLCIBS Ψ is contained in partial cubes, and these cubes are called edge-cubes of Ψ. Intuitively, Ψ can be detected and reconstructed from the 3D image by the following two steps:

- First, detect all of its edge-cubes from the 3D image;
- Second, compute boundary surface patches in each edge-cube.

Since each edge-cube has eight vertices, twelve edges and six faces, its detection differs from the detection of edge-voxels in 3D edge detection techniques, as shown in (Wang et al. 2007a, Wang et al. 2014). In a cube, an edge is called a zero-crossing edge if its two vertices p_1 and p_2 are a pair of zero-crossing points, i.e. satisfy $\nabla^2 f(p_1) \cdot \nabla^2 f(p_2) < 0$. A cube will be called a zero-crossing cube if it includes at least three zero-crossing edges. By (Wang et al. 2007a, Wang et al. 2003), a zero-crossing edge will be marked as the edge intersected by Ψ if its two vertices p_1 and p_2 have high gradient magnitudes in their local neighborhood U, i.e. satisfy $\|\nabla f(p_1)\| + \|\nabla f(p_2)\| \geq 2 \cdot T(U)$. Further, a cube will be marked as an edge-cube if it has at least three edges intersected by Ψ. This shows that the computation of $T(U)$ is a key for detecting edge-cubes from the 3D image.

2.1 Computation of local gradient thresholds

It is known that human vision system can detect automatically luminance contrast in any environment. Here, luminance contrast can be described by the weber contrast I_b, where I_b is average luminance of the background, I_o is luminance of an object in the background and $\Delta I = I_o - I_b$ is the luminance difference. Because of the adaptation of human eye, a small ΔI is negligible if I_b is high, while the same small difference is remarkable if I_b is low.

Particularly, Just Noticeable Luminance Difference (JNLD, denoted by $J(I_b)$) varies with different background luminance I_b, see Figure 1 and references (Drew 2005, Thompson et al. 2011). Figure 1 shows that when I_b is very low, $J(I_b)$ is changed with I_b but its value actually changes very small. In the middle range of brightness, $J(I_b)$ changes linearly with I_b and I_b is nearly a constant.

If we take gray intensity of the 3D image as the luminance, then we can detect all visible boundaries (i.e. the places where noticeable luminance difference exist) from the 3D image by $J(x)$. Here, x is the gray intensity of a local background. So, to some extents, $J(x)$ may be regarded as perceptional threshold function of visible boundaries in 3D images, and therefore, will be regarded as local gradient threshold function to detect SLCIBSes in 3D images in this paper.

According to (Hou et al. 2004, Wang et al. 2014), in 3D images with 256 levels of intensities, $J(x)$ may be described approximately as follows:

$$J(x)=\begin{cases} C, & 0 \le x \le d \\ \beta x, & d < x \le 210 \end{cases} \qquad (2)$$

where, the regions $[0,d]$, $[210,255]$ and $[d,210]$ corresponds to low, high and middle range of brightness, respectively. When $x \in [0,d]$, $J(x)$ is approximately seen as a constant $C = \beta \cdot d$. Here, d may be selected as a value between 25 and 80, and β may be selected as a constant value around 0.20. By our experience, $d = 36$ and $\beta = 0.25$ are usually acceptable for many SLCIBSes. A large β can facilitate to detect interested SLCIBSes from 3D images while excluding many uninteresting weak visible boundaries. In such cases, since $J(x)+x>255$ holds if $x \in (210,255]$ (i.e. visible boundaries cannot be perceived from backgrounds with high brightness), $J(x)$ does not need to compute when $x > 210$.

Since our purpose is to detect and reconstruct Ψ from the 3D image, we only need to estimate $T(U)$ in those local regions around Ψ. Edge-cubes of Ψ are also zero-crossing cubes. Hence, Ψ is contained within the zero-crossing cubes around Ψ, and we only need to estimate local gradient thresholds or $T(U)$ in those zero-crossing cubes around Ψ. In a zero-crossing cube Q, zero-crossing edges can be easily detected from it. Let $\delta(Q)$ denote the average value of gray intensities of vertices of all zero-crossing edges. $\delta(Q)$ will be regarded as the background luminance of the cube. By equation 2, a local gradient threshold in the cube Q is calculated as $J(\delta(Q))$. In this way, whenever a zero-crossing cube is marked, a local gradient threshold can be easily computed for it.

2.2 Detection and reconstruction of Ψ by local gradient thresholds

By observing 2D slices of the 3D image, users can select a 2D slice image and interactively draw a small rectangle which contains no other boundary curve except for a salient boundary curve of the interested Ψ. In this way, the interested Ψ is designated and an edge-cube of Ψ (i.e. a seed cube, denoted by Q) is determined in the 3D image (Cheng et al. 2010, Wang et al. 2011). The connectivity of Ψ provides the possibility for tracing the edge-cubes of Ψ from the 3D image based on the known seed cube Q, the surface connectivity and the computed $T(U)$.

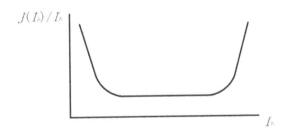

Figure 1.   Distribution of JNLD $J(I_b)$ along the luminance intensity I_b.

The seed cube Q is a zero-crossing cube. Thus, the local gradient threshold in Q can be computed as $J(\delta(Q))$, as shown in Section 2.1. In Q, there are at least three edges intersected by Ψ, and they can be detected from Q. Here, an edge with two vertices p_1 and p_2 will be marked as the edge intersected by Ψ if it is a zero-crossing edge and its two vertices satisfy

$$\|\nabla f(p_1)\| + \|\nabla f(p_2)\| \geq 2 \cdot J(\delta(Q)) \tag{3}$$

If an edge is intersected by Ψ, then two faces sharing the edge must be intersected by Ψ. Furthermore, by the continuity and connectivity properties of Ψ, if a face in Q is intersected by Ψ, then an adjacent cube that shares the face with Q must be intersected by Ψ as well (Wang et al. 2014). In this way, we can detect all new edge-cubes from six adjacent cubes (i.e. the six neighborhood) of Q. For each newly determined edge-cube, we can continue to compute its local gradient threshold, mark its edges intersected by Ψ, and detect new edge-cubes from its six neighborhood by the surface connectivity. By doing such operations recursively, all edge-cubes connecting with Q are detected from the 3D image. Ψ are just contained within them.

In each edge-cube, the contained boundary surface patches are some zero-crossing surface patches (i.e. isosurface patches satisfying $\nabla^2 f(x,y,z) = 0$). Thus, such surface patches can be computed from each edge cube by the Marching-cubes algorithm (Lorensen & Cline 1987, Wang et al. 2011). Consequently, in each edge-cube, such surface patches can be represented approximately by at least one and at most five triangles. In this way, a polygonal surface model of Ψ, which is formed by triangles in all edge-cubes, is reconstructed from the 3D image.

3 EXPERIMENTAL RESULTS AND DISCUSSIONS

The proposed method is applied to ten different 3D images, which belong to 3D CT, 3D MRI and 3D confocal microscope images, and their slices are shown in Figure 2, respectively. On each slice, a rectangle is drawn to designate a SLCIBS of interest in a 3D image and to determine the seed cube for the SLCIBS (denoted by BS1, BS2, ..., BS10, separately).

By the proposed method, we can compute $T(U)$ for each designed SLCIBS, detect and reconstruct its 3D surface model from 3D images by $T(U)$. The reconstructed surface models are shown in Figure 3, respectively. It can be seen that all these SLCIBSes are well detected and reconstructed. In the experiments, $\beta = 0.25$, $d = 36$ are used in detecting BS1–BS3 and BS5–BS10. However, they are adjusted as $\beta = 0.55$, $d = 45$ for well separating blood structures (i.e. BS4) from 3D MRA image.

BS1–BS10 have different ranges of gradient magnitudes and the changes of gradient magnitudes over these surfaces are different. Particularly, some have uniformly high gradient magnitudes and some have varied gradient magnitudes from low to high. In the upper part of Figure 4, distributions of gradient magnitudes over BS1–BS3, BS5–BS6 and BS8–BS9 are visualized. Meanwhile, local gradient thresholds used for detecting different boundary surface patches of BS1–BS3, BS5–BS6 and BS8–BS9 are visualized in the lower part of Figure 4, respectively. Figure 4 intuitively shows that while $T(U) \leq \|\nabla f(x,y,z)\|$, $T(U)$ are adaptively computed for different SLCIBSes. Figures 2–4 illustrate that based on the computed $T(U)$, SLCIBSes in many different 3D images, whether they have uniformly high gradient magnitudes or have gradually changed gradient magnitudes from low to high, can be well detected and reconstructed. This shows the effectiveness of the method for computing $T(U)$.

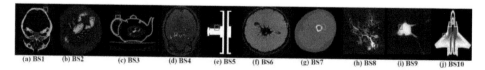

(a) BS1   (b) BS2   (c) BS3   (d) BS4   (e) BS5   (f) BS6   (g) BS7   (h) BS8   (i) BS9   (j) BS10

Figure 2.   Slices of different 3D images. Each rectangle designates a SLCIBS of interest.

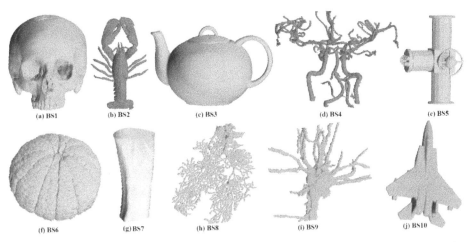

| (a) BS1 | (b) BS2 | (c) BS3 | (d) BS4 | (e) BS5 |
| (f) BS6 | (g) BS7 | (h) BS8 | (i) BS9 | (j) BS10 |

Figure 3.   3D surface models which are designated in Figure 2 and detected by the computed $T(U)$.

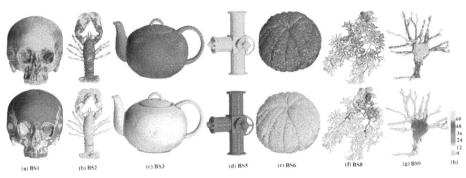

| (a) BS1 | (b) BS2 | (c) BS3 | (d) BS5 | (e) BS6 | (f) BS8 | (g) BS9 | (h) |

Figure 4.   Distributions of gradient magnitudes (the upper part) and local gradient thresholds (the lower part) over some boundary surfaces.

In equation (2), by adjusting β and d, the perceptional threshold function for weak or strong visible boundaries may be determined. Below, we discuss how the reconstructed SLCIBC is changed when β or d is changed. Figure 5 shows how the number of triangle patches of the reconstructed Ψ changes with the changing of one parameter when others remain almost unchanged. Figure 5 illustrates that the number of triangle patches changes slightly in BS2, BS6 when d is changed from 20 to 60 or β is changed from 0.175 to 0.325, and in BS8 when d is changed from 35 to 60 or β is changed from 0.25 to 0.325. Moreover, the visual effects of the reconstructed SLCIBSes have little difference when one of parameters changes in a relatively large range. Therefore, the setting of parameters (i.e. β and d) shows robustness in the proposed method.

By comparing Figure 3h in this paper with the Figure 5f in (Wang et al. 2014), we can see that the method in (Wang et al. 2014) is more effective than one in this paper in exclude uninteresting surface patches and fragments connected with the interested SLCIBS. Additionally, according to (Wang et al. 2014), certain zero-crossing surface patches contained in edge-cubes are possible to be incorrectly regarded as boundary surface patches. However, as shown in the following Table 1, the method in this paper has a great merit, namely, it can quickly compute $T(U)$, detect and reconstruct the interested SLCIBS in near real-time. Compared to the method in (Wang et al. 2014), the method in this paper greatly improves computing speed.

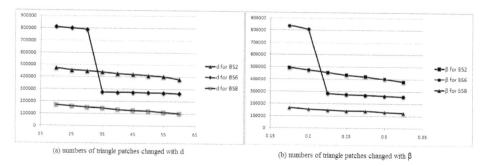

(a) numbers of triangle patches changed with d

(b) numbers of triangle patches changed with β

Figure 5.   Illustration of triangle patches numbers change with parameters. (a) d is changed from 20 to 60 and β remains 0.25. (b) β is changed from 0.175 to 0.325 and d remains 36.

Table 1.   Comparison of computational time in M1 and M2*.

SLCIBS	Time in M1 (ms)	Time in M2 (ms)
BS1	36926	1591
BS2	29157	1092
BS3	49608	1310
BS6	12980	717
BS8	2855	358

*M1 denotes method in (Wang et al. 2014) and M2 denotes method in this paper.

4   CONCLUSION

Local gradient thresholds are usually needed in the detection of interested boundary surfaces from 3D images, but their computation is a very complex problem. Inspired by the adaptive contrast detection mechanism of human vision, this paper proposes a feasible method to compute local gradient thresholds in 3D images. Experimental results show that the proposed method is effective for step-like boundary surfaces in many different 3D images.

ACKNOWLEDGEMENT

This work was supported in part by the NSFC of China (61375020), 973 program of China (2013CB329401) and Cross Research Fund of Biomedical Engineering of SJTU (YG2013ZD02, YG2012MS19).

REFERENCES

Bomans, M. Hohne, K.H. Tiede, U. & Riemer, M. 1990. 3-D Segmentation of MR images of the head for 3D diplay. *IEEE Trans. Medical Imaging* 9(2): 177–183.

Brejl, M. & Sonka, M. 2000. Directional 3D edge detection in anisotropic data: detector design and performance assessment. *Computer Vision & Image Understanding* 77(2): 84–110.

Cheng, L. Wang, L. & Ma, Y. 2010. A new strategy for boundary surface detection in 3D biomedical images. *IEEE Conference on Biomedical Engineering and Informatics (BMEI 2010)* Vol. 1: pp. 51–55.

Drew, S. 2005. Anatomy and physiology of the eye, contrast, contrast sensitivity, luminance perception and psychophysics. *Draft Report, Griffith University, Australia.*

Henstock, P. V. & Chelberg, D. M. 1995. Automatic gradient threshold determination for edge detection. *IEEE Trans. Image Process* 5(5): 784–787.

Hou, Z. Han, C. Zheng, L. & Xiao, L. 2004. Threshold selection tactics for an edge detection based on vision model. *Opto-Electronic Engineering* 31(2): 59–63 (in Chinese)

Lorensen, W.E. & Cline, H.E. 1987. Marching cubes: A high resolution 3D surface construction algorithm. *ACM siggraph computer graphics* 21(4): 163–169.

Marr, D. & Hildreth, E. 1980. Theory of edge detection. *Proceedings of the Royal Society of London. Series B. Biological Sciences* 207(1167): 187–217.

Rakesh, R.R. Chaudhuri, P. & Murthy, C. A. 2004. Thresholding in edge detection: a statistical approach. *IEEE Trans. Image Processing* 13(7): 927–936.

Thompson, W.B. Fleming, R.W. Creem-Regehr, S. H. & Stefanucci, J. K. 2011. *Visual perception from a computer graphics perspective*. New York: CRC Press.

Wang, L. Bai, J. He, P. Heng, P.A. & Yang, X. 2007. A computational framework for approximating boundary surfaces in 3D biomedical images. *IEEE Trans. Information Technology in Biomedicine* 11(6): 668–682.

Wang, L. Bai, J. Wong, T.T. & Heng, P.A. 2007. Isosurface computation for approximating boundary surfaces in 3D images. *Journal of Electronic Imaging* 16(1): 1–12.

Wang, L. Bai, J. & Ying, K. 2003. Adaptive approximation of the boundary surface of a neuron in confocal microscopy volumetric images. *Medical & Biological Engineering & Computing* 41(4): 601–607.

Wang, L. Wang, P. Cheng, L. Ma, Y. Wu, S. Wang, Y. & Xu, Z. 2014. Detection and reconstruction of an implicit boundary surface by adaptively expanding a small surface patch in a 3D image. *IEEE Trans. Visualization and Computer Graphics* 20(11): 1490–1506.

Wang, P. Cheng, L. Lin, B. & Wang, L. 2011. Boundary surface detection in 3D biomedical images based on local gradient thresholds. *IEEE Conference on Biomedical Engineering and Informatics (BMEI 2011)*: 175–179.

Bioinformatics and Biomedical Engineering – Chou & Zhou (Eds)
© *2016 Taylor & Francis Group, London, ISBN 978-1-138-02784-8*

Abnormality detection of specific brain structure in MR images based on multi-atlas and texture descriptor

B.Z. Chen, Y. Wang & L.S. Wang
*Department of Automation, Institute of Image Processing and Pattern Recognition,
Shanghai Jiao Tong University, Shanghai, China*

ABSTRACT: An abnormal brain structure might contain different types of lesions with different shapes, areas or textures, and the same normal brain structure in different persons might exhibit individual differences in shape or tissue texture. This makes it a challenging problem to judge whether or not a brain structure in a MR image is normal. In this paper, we present a framework for abnormality detection of the brain structure of interest (IBS) in MR images based on the multi-atlas and the texture descriptor. In the framework, a set of normal brain MR images are first collected, where different images are aligned and the atlases of different brain structures are marked in each image. Then, multi-atlas information is used to locate and segment the IBS in the test MR image. Subsequently, textural features of the IBS are computed in the test image and in each one of the collected normal images. Finally, by analyzing whether or not there is a significant difference between textural features of the test image and those of the collected images, a conclusion about the abnormality detection is drawn. This framework has been applied to MRI brain images with different abnormal subjects, and detection results are acceptable.

Keywords: abnormality detection; multi-atlas registration; textural feature; significant differences; normal atlas database

1 INTRODUCTION

In computer-aided diagnosis, an important task is to judge whether or not the brain structure of interest is normal based on a given MRI brain image. Since the same normal brain structure in different persons might exhibit individual differences in shape or tissue texture, and an abnormal brain structure might contain different types of lesions with different shapes, areas or textures, this becomes an extremely difficult task.

Over the past three decades, many different brain structure abnormality detection approaches have been proposed. These techniques can mainly be classified into three categories:

i. Utilizing the prior knowledge (lesion location, size, etc) of abnormal structures in the detection, such as Colliot's method (2006) that adopted cortical thickness, relative intensities and gradients of abnormal regions themselves to discriminate lesions and healthy structures. These techniques are limited to specific lesion types.
ii. Using prior information from both the abnormal and normal regions. For example, Anbeek and Vincken (2004) employed machine learning skills to train normal and lesion voxel samples annotated by radiologists and then learned a model or classifier for detecting abnormality. However, due to high variations among individuals and complexity of pathological changes, it is challenging to delineate as many types of abnormal cases as possible.
iii. Collecting a set of normal structure images for abnormality detection, where the test image is compared with that (or a specific one) in the set. Voxel-Based Morphometry (VBM) (Ashburner, 2000) employed a prior normal population to inspect neuroanatomical abnormalities. However, this method strongly relied on the choice of registration templates.

Aoki et al. (2012) applied a large database of normal chest radiographs to detect lung nodules by subtracting subjects, while this scheme needed the radiologist's subjective judgments. Erus et al. (2014) detected the patterns of abnormality by mapping a target image into the hypervolume of healthy ones. The method was not limited to a specific type of abnormality, but its detection accuracy was low and computation efficiency should be further improved.

It can be seen that techniques specified in (i) and (ii) are limited to a few specific lesion types. The techniques specified in (iii) are most probably to provide the possibility of detecting different types of abnormality. However, the detection result may be greatly affected by the collected set and the method used for comparing the test image and the set.

In this paper, we focus on the abnormality detection of the brain structure of interest in the MR image. So, we not only collect a set of normal brain MRI images and align them, but also mask the atlases of different brain structures in each image. By the set, the brain structure of interest (IBS) in the target MRI image can be located and segmented by the multi-atlas segmentation technique. Textural features are computed for the IBS in the target image and in each one of the collected normal images. By analyzing whether or not there is a significant difference between textural features of the target image and those of the collected images, a conclusion about the abnormality detection can be drawn. This method has been applied to MRI brain images with different abnormal subjects, and detection results are acceptable.

2 METHODS

After having collected a set of normal brain MR images, they are aligned and their atlases are marked. Then, the abnormality detection of the IBS in a target MRI image can be done through three steps as follows. First, we segment the IBS from the target image by aligning the normal set with the target image. Second, we calculate the textural features of the target brain patch and the corresponding regions of the normal brain atlases. Finally, we analyze significant differences in statistical textural characteristics. In the following sections, we elaborate these steps in detail.

2.1 Multi-atlas registration

Our aim is to identify as many abnormal tissues in the human brain as possible, and single atlas registration can inform us about the spatial information of different anatomy structures in the test image. If there are significance differences in texture descriptors between the target tissue patch and the corresponding atlas patch, we can confirm which type of tissues is abnormal. So, we first have to align the atlas images to the target image space. A brain skull and non-brain tissues occupy a large percentage of the whole brain, leading to large registration errors. Therefore, before registration, we should remove them to acquire a brain mask by the Brain Surface Extractor Algorithm (BSEA) (David et al, 2004). Because most of the structures in our brain belong to soft tissues, it is not able to align these brain MRIs by using a global registration step alone. In contrast, a local registration step works very well under the assumption that the registration MR images have similar sizes and sampling voxel dimensions. Therefore, in order to enhance the registration accuracy, the single atlas registration generally includes the global and local alignment steps.

However, sometimes, if there are large anatomical differences between a single atlas and a subject image, we cannot align them effectively, no matter what kind of single atlas registration algorithms is adopted. Therefore, multi-atlas registration (i.e. integrating multi-single-registration results) is a wise choice to align pairs of images; even if one of the atlases fails to be spatially normalized to the target MRI domain, it does not have a significant impact on the final alignment result.

2.1.1 Global and local registration steps

In order to reduce the influence of individual MR image anatomical differences and improve alignment efficiency, we use affine transform to align two images globally, and then the target and moving (atlas) MR scans have similar sampling sizes and voxel dimensions, which is helpful for local fine alignment. We regard affine transform parameters as local registration input.

Since the shape of the brain structures is irregular, nonlinear deformation transform can meet the need of the registration accuracy during the local registration step. Compared with other deformation registration algorithms, Maxwell's Demons (Thirion, 1998) registers more efficiently and has a higher alignment precision, especially when the target image has similar intensity distributions with the moving images (atlas image). So, we first normalize two MRIs to the same intensity space by linear mapping. Then, the moving image will have the similar intensity histogram with the subject by the histogram matching technique. It is worth nothing that the background of images affects the matching precision greatly, so we choose the mean gray value of the image to segment the background empirically.

After filtering the background and histogram matching, we adopt the original Demons algorithm to align two preprocessed MRI scans. However, we find that when there is large deformation among them or little gradient information from the subject image, the original Demons strategy may perform badly, so a modified Symmetric Demons (Rogelj, 2006) is used to improve its registration performance, simultaneously considering gradient information from the target and atlas MRI scans. In addition, the multi-resolution (i.e. register ranging from coarse to fine scales) skill in Symmetric Demons is used to enhance the registration speed and accuracy.

2.1.2 *Multi-atlas segmentation fusion*

After global and local registration, we can obtain each aligned atlas, label image, and their deformation fields (corresponding transformation relation matrix between the atlas and the subject image), as described in Section 2.1.1. As we want to determine which type of the brain anatomies is abnormal or normal, the first step is to segment the IBS. So, we employ the atlas label of the IBS as the segmentation threshold T, and then we acquire a series of the segmented atlas database results. Finally, the key step is to select an effective fusion strategy to integrate the IBS of each atlas. There are many excellent fusion algorithms, such as Weight Voting, COLLATE (Consensus Level, Labeler Accuracy and Truth Estimation) and STAPLE (Simultaneous Truth and Performance Level Estimation) (Warfield et al, 2004). Since STAPLE is a very fast and robust fusion strategy, we choose it to fuse the interested tissues of each atlas and obtain the final multi-atlas fusion result.

2.2 *Detection abnormality based on texture*

In this section, we present a robust detection dissimilarity algorithm based on significant differences in statistical textural characteristics by SPSS 20, under the assumption that the atlas images are spatially aligned to the test subject image very well. We use the registration information to map the atlas labels (i.e. each label represents a different anatomy tissue) to the target image, so we can determine what type of the structures in the subject MR image is abnormal, helping the doctors to do the early diagnosis and timely treatment.

2.2.1 *Textural feature selection*

Texture is widely used in MRI analysis. Since most of the visible abnormalities on the normal-appearing structure in the MR image have a different local textural distribution relative to normal anatomy, it is reasonable and effective to select textural features to discriminate the abnormal subject image. Existing methods for textural analysis mainly include the Gray Level Co-occurrence Matrices (GLCM), Gabor filtering, fractal dimension and discrete wavelet transform (Mahmoud-Ghoneim et al, 2003). It has been proved that GLCM analysis is useful and efficient to quantitatively measure the tissue difference among individual brain MR images (Mahmoud-Ghoneim et al, 2003). Three dimensional GLCMs are calculated by summing voxel pair frequencies within a 3D space. Since the range of gray levels in MR images tends to be large, if we only consider each different intensity as a gray level, the elements of GLCM are so many that it may lead to a high data dimensionality. In order to improve computing speed, we first compress the gray levels by dividing by 16, and then bin 1 in the compressed image represents gray levels from 0 to 15 in the original image. Moreover, in this paper, 5 GLCMs

189

are obtained on 5 different scales, so encapsulated GLCMs are still not convenient to be used to analyze attributes of the images intuitively. The usual way to do this is to compute some descriptors, representing the image textural information. Haralick et al. (1973) employed 14 textural features to describe GLCMs, but only 4 of these 14 descriptors are independent textural features (Ulaby et al, 1986). Based on this analysis, we utilize the textural variance to discriminate normal or abnormal tissues for its good classification ability in our experiments.

2.2.2 *Identifying significant differences of textural features*

Aligned atlases and the target image are, respectively, masked with the multi-atlas fusion result in Section 2.1.2, and we eventually acquire the segmentation result of the IBS in the test MR image and the corresponding structure of the normal atlases. Let S_{ROI} denote the IBS in the subject image, and A_i ($i = 1 \dots N$, N is the number of atlases) represent the corresponding structures in the atlases. We count GLCMs of S_{ROI} and A_i in the number of different scales (d), and then calculate their textural variances, respectively. Let $\{SF\}_j$ ($j = 1, \dots d$) denote the textural characteristics of the IBS in the test data and $\{AF_i\}_j$ ($i = 1 \dots N, j = 1, \dots d$) represent the features of the corresponding normal tissue sets. In order to reduce statistical error, we acquire final features of the interested target region SF and the normal textural feature set AF_i ($i = 1 \dots N$) by averaging $\{SF\}_j$ and $\{AF_i\}_j$, respectively.

We use different subject images to perform similar experiments as mentioned above, then we get SF_{num} and $\{AF_i\}_{num}$ ($num = 1, \dots M$, M is the number of test images). Since the subject tissue and the corresponding normal atlas regions are independent of each other, the independent sample t-test in SPSS is often used to analyze significant differences between SF and AF_i by comparing their averages. The significant level is set to 0.05. If the significant p-value is less than 0.05, we confirm that there is a statistically significant difference between the test interested tissue group and the corresponding normal atlas structure group in the brain. Therefore, the target brain structure is marked as abnormal. According to the atlas labels of the IBS, we can further determine which tissue of the brain is abnormal.

3 DATA AND EXPERIMENT

3.1 *Building a normal atlas database*

Our normal atlas database used in this paper is derived from The Internet Brain Segmentation Repository (IBSR). The database includes 18 T1-weighted MR images (atlases and the corresponding atlas label images). The MRIs have a size of $256 \times 256 \times 128$, and their voxel dimensions are from $0.8 \times 0.8 \times 1.5$ mm to $1.0 \times 1.0 \times 1.5$ mm. To make our normal atlas database more general, we choose 3 female atlases and 13 male ones, whose ages range from 7 to 71 years old ($N = 16$). The five ($M = 5$) test T1-Weighted Scans ($256 \times 256 \times 60$, $0.97 \times 0.97 \times 3.1$ mm) are also downloaded from the IBSR.

3.2 *Experiment and discussion*

Our experiment platform is based on Microsoft Visual Studio 2008 and ITK3.20.1. Besides, the IBM SPSS Statistics 20 is used to analyze our experimental data. Since white matter lesions (e.g. multiple sclerosis, Alzheimer's disease) are very common brain diseases, the validation experiments are applied on five abnormal subjects, whose abnormal voxels locate mainly in the regions of the left cerebral white matter marked by experts. We want to verify whether the significant difference analysis can effectively detect abnormalities in the left cerebral white matter. We align the normal atlas sets to every target MR image by our multi-atlas technique and select $T = 2$ (represents left cerebral white matter in the atlas label image) to acquire the segmentation fusion results of the left cerebral white matter (S_{ROI}) in the test MR image and the corresponding tissues of the normal atlases (A_i). We obtain SF and AF_i ($i = 1 \dots N$) by calculating textural variances of their corresponding regions. Then, the independent sample t-test is used to analyze the significant differences between SF and AF_i. Table 1 provides the 5 experimental p-values.

190

Table 1. The significant levels between SF_{num} ($num = 1, ..., 5$) and $\{AF_i\}_{num}$.

Test	Test 1	Test 2	Test 3	Test 4	Test 5
p-value*	0	0	0.0333	0	0

*p-value < 0.05 was considered to indicate a statistically significant difference.

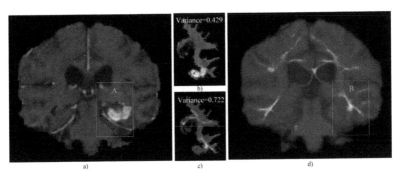

Figure 1. The abnormal and the corresponding normal specific brain segmented results by multi-atlas fusion. (a) Removal of the brain skull result of the target MRI and (d) one of the normal atlas databases, which has been aligned to the subject image space. (b) and (c) The segmented results of the IBS (A,B marked in green), respectively. The difference in texture between (b) and (c) indicates that the test specific brain tissue may be abnormal.

From Table 1, each experimental p-value is less than 0.05, that is to say, there are significant differences between the test brain structure and the corresponding normal atlas structures. The smaller the p-value, the greater the significant difference. For example, the p-value of Test 1 is zero, representing that SF_1 and $\{AF_i\}_1$ have surprisingly large variations. As all the five test images are abnormal in our experiments, the test results are in conformity with actual truth, which is proved that our proposed method is effective in detecting this type of abnormality.

In order to reduce the effect of noise on the overall statistical result, and enhance the credibility of significant difference analysis approaches, we regard SF_{num} (0.552, 0.429, 0.425, 0.273, 0.302) of the left cerebral white matter in 5 test images as the abnormal group and $((1/N)\Sigma AF_i)_{num}$ (0.930, 0.726, 0.998, 0.868, 1.125) of the corresponding tissue in the normal population as the control group, and use the Mann-Whitney U test to compare significant inter-group differences in addition to the significant level 0.05. The p-value obtained from the Mann-Whitney U test is 0.0079 ($p < 0.05$), which demonstrates that textural differences in the corresponding two groups are significant. Figure 1(a) shows removing brain skull result of the test target MRI. The green box area A, shown in Figure 1(a), approximately represents the IBS. Figure 1(d) shows one of the normal atlas databases, which has been aligned to the subject image space, and the green box B denotes the corresponding normal structure. Figure 1(b) and 1(c) shows the segmented results of the regions A and B, respectively. The corresponding textural variances of Figure 1(b) and 1(c) have a big difference, which illustrates that the subject may be abnormal in the region A relative to the normal region B.

Since different structures have similar intensity distributions and common abnormal regions in our brains exhibit rich textural information, then abnormalities in other locations of brains may be effectively detected by our algorithm such as recognition dissimilarities in the left cerebral white matter. We use SPSS to analyze significant levels between each target brain anatomy patch of interest and the corresponding regions of the normal brain MRI database. If there is a significant difference between SF and AF_i, we may conclude that the subject patch is abnormal, then according to the multi-atlas registration results, the atlas label can be propagated to the abnormal region, which may help doctors to make a decision that the abnormal area belongs to which part of the brain. Based on above analysis, our method is expected to detect all kinds of abnormalities with rich textural information in our brain just like Erus's algorithm (2014), but our feature dimensionality is much less than Erus's, which may improve computing efficiency. Our technique may identify more complex pathological changes than VBM, because the technique is based on the region, while VBM is based on voxels. In our algorithm, we use textural and multiple atlas information to distinguish abnormalities without the help of doctors, while Aoki (2012) has to rely on radiologists to do the similar job. In addition, our technique does not require any training in advance.

4 CONCLUSION

We discuss a novel detection "all kinds of abnormalities" method. We first establish a normal data set, and then normal MRIs are spatially aligned to the test subject image by the multi-atlas registration technique. We compare each target brain anatomy patch of interest with the corresponding regions of the normal brain MRI database. If we cannot find a similar one with the normal sets by analyzing the significance of texture descriptors between the two corresponding regions, we may conclude that the subject MRI is abnormal. According to prior knowledge provided by atlases, we can further determine the ownership structure of abnormal regions. The experimental results demonstrate that our algorithm is effective to discriminate abnormalities with rich textural information from T1-weighted brain MR images.

ACKNOWLEDGMENTS

This work was supported in part by the 973 program of China (2013CB329401) and the Cross Research Fund of Biomedical Engineering of SJTU (YG2013ZD02, YG2012MS19). The authors would like to thank the Center for Morphometric Analysis at Massachusetts General Hospital for making the MR brain data sets and their manual segmentations available.

REFERENCES

Anbeek, P. Vincken, K.L., Osch, M.J.P., et al. 2004. Probabilistic segmentation of white matter lesions in MR imaging. NeuroImage 21(3):1037–1044.
Aoki, T., Oda, N., Yamashita, Y., et al. 2012. Usefulness of Usefulness of computerized method for lung nodule detection on digital chest radiographs using similar subtraction images from different patients. European Journal of Radiology 81(5):1062–1067.
Colliot, O., Mansi, T., Bernasconi, N., et al. 2006. Segmentation of focal cortical dysplasia lesions on MR using level set evolution. NeuroImage 32(4):1621–1630.
David, E.R., David, W.S., Roger, P.W., et al. 2004. A meta-algorithm for brain extraction in MRI. NeuroImage 23(2):625–637.
Erus, G., Zacharaki, E.I., Davatzikos, C. 2014. Individualized statistical learning from medical image databases: Application to identification of brain lesions. Medical Image Analysis 18(3):542–554.
Haralick, R.M., Shanmugam, K., Dinstein, I. 1973. Textural Features for Image Classification. IEEE Transactions on Systems, Man and Cybernetics SMC-3(6):610–621.
Leemput, K.V., Maes, F., Vandermeulen, D., et al. 2001. Automated Segmentation of Multiple Sclerosis Lesions by Model Outlier Detection. IEEE Transactions on Medical Imaging 20(8):677–688.
Mahmoud-Ghoneim, D., Toussaint, G., Constans, J.M., et al. 2003. Three dimensional texture analysis in MRI: a preliminary evaluation in gliomas. Magnetic Resonance Imaging 21(9):983–987.
Rogelj, P., Kovacic, S. 2006. Symmetric Image Registration. Medical Image Analysis 10(3):484–493.
Thirion, J.P. 1998. Image matching as a diffusion process: an analogy with Maxwell's demons. Medical Image Analysis 2(3): 243–260.
Ulaby, F.T., Kouyate, F., Brisco, B., et al. Textural Information in SAR Images. 1986. IEEE Transactions on Geoscience and Remote Sensing GE-24(2):235–245.
Warfield, S.K. & Zou, K.H. & Wells, W.M. 2004. Simultaneous truth and performance level estimate (STAPLE): An algorithm for the validation of image segmentation. IEEE Transaction on Medical Imaging 23(7):903–921.

Bioinformatics and Biomedical Engineering – Chou & Zhou (Eds)
© 2016 Taylor & Francis Group, London, ISBN 978-1-138-02784-8

Retinal images change detection based on fusing multi-features differences

S. Yin, B.Z. Chen & L.S. Wang
Department of Automation, Institute of Image Processing and Pattern Recognition,
Shanghai Jiao Tong University, Shanghai, China

ABSTRACT: This paper proposes an approach to detect changes between two longitudinal retinal images of the same patient based on multi-attributes fusion. With the approach, two retinal images are firstly aligned spatially by registering them. Then illumination variations between these two images are reduced by employing the modified Single Scale Retinex method. Finally, changes between these two images are detected based on two attributes: intensity and texture. Experimental results show that the proposed approach is effective and robust in identifying small and weak dissimilar variations from retinal images, and has certain advantages compared with some state-of-the-art methods.

Keywords: longitudinal retinal images; change detection; illumination correction, multi-features

1 INTRODUCTION

Ophthalmologists associated with retinal surgery usually evaluate surgery results or clinical treatment strategies by observing change between longitudinal retinal images of the same patient. When there are lots of tiny abnormal regions or changed regions in retinal images, it is difficult to assess number and area of these small regions by visual observation. Currently, such minor and faint changes are usually inspected manually (Narasimha-Iyer et al. 2006), but it is very time-consuming and inaccurate. Therefore, it's necessary to develop an automatic or semi-automatic method to detect these changes in longitudinal retinal images.

In the past three decades, many change detection methods have been studied by researchers. Radke et al. (2005) present a systematic survey of the common processing steps and core decision rules in modern change detection algorithms. Li & Leung (2002) used a weighted combination of the Intensity Difference and a Gradient Difference (IDGD) measure to detect changes between two frames. This method is less sensitive to noise and illumination variations, but it behaves badly in detecting slight abnormalities. Gong et al. (2012) proposed an unsupervised learning change detection approach for Synthetic Aperture Radar (SAR). The approach can enhance the changed information and well detect big change regions, but it may be easily influenced by illumination changes and nonuniform local intensity. Narasimha-Iyer et al. (2006) detected and classified the changed area into multiple sets. By pre-detecting some irrelevant tissues like blood vessel, optic disk and fovea, they got some results with high accuracy. But their method is too complicated and their change mask still has too much noise.

In this paper we propose a novel approach to detect changes between two longitudinal retinal images. By the approach, the influence of the noise and inconsistent intensity of two images can be reduced, and accuracy and reliability in identifying minor and faint variation areas can be improved. In the approach, we mainly take three steps to reach our target. First, two retinal images are aligned spatially by registering them using Harris-PIIFD (Partial Intensity Invariant Feature Descriptor) registration method proposed in Chen et al. (2010). Then the illumination difference between two retinal images is corrected using the modified

SSR (Single Scale Retinex) algorithm and histogram specification. After that, we generate difference images based on intensity and contrast features and integrate them by Gaussian Models. Finally we get the binary "change mask" by setting a threshold empirically.

This paper is organized as follows: section 2 introduces our method in detail; section 3 is experimental results analysis and some discussion on our results; section 4 is some conclusions we get from experiments. Note that all images discussed in this paper are focused on gray images.

2 METHOD

In this paper, two retinal images are aligned spatially by registering them using Harris-PIIFD registration method. The method is fast and performs well in poor conditions, for instance, large lens distortion, different scales, arbitrary rotation and big non-overlapping areas. Below, we mainly introduce techniques for correcting illumination variations between two images and detecting change between two images.

2.1 Illumination adjustment

Illumination variations in images are usually caused by changes in the strength or position of light sources. In this paper, we try to reduce illumination variations between two retinal images by combining a modified Retinex technique with an intensity normalization technique. The former is used to reduce the influence of illumination in each image, and the later is used to make two images have the similar intensity histogram.

We find that most retina images have black background. And there is a gap between the gray level of background and that of the foreground, which makes us easy to separate the background and the foreground apart and leave the background unprocessed. Here we firstly use the K-means algorithm to find the boundary that separate the background and foreground apart.

2.1.1 Modified Single Scale Retinex algorithm

According to Retinex theory, intensities in an image can be described as $I(x,y) = R(x,y)*L(x,y)$. Where, $I \in [0,255]$ is the observed image; $R \in [0,1]$ is reflectance, which is invariable with respect to changed illumination; $L \in [0,255]$ is the illumination. And it is an ill-posed problem to solve R and L by using one observed image I (Fu et al. 2014). So, we use the convolution of $I(x,y)$ and a Gaussian filter $G(x,y)$ to estimate the $L(x,y)$ and then obtain $R(x,y)$ using division of $I(x,y)$ and $L(x,y)$. So, we have:

$$R(x,y) = \frac{I(x,y)}{I(x,y)*G(x,y)} \tag{1}$$

where $G(x,y)$ is the Gaussian kernel function, and the "*" means convolution.

It is notable that $R(x,y)$ is just a matrix consisting of many decimals. We use linear mapping to change the final range of $R(x,y)$ to [0,255]. We set $\bar{R}(x,y)$ denotes the image after linear mapping.

There are two parameters in (1) (i.e., SSR—Single Scale Retinex algorithm), the size of Gaussian kernel k (the size is $k*k$) and the standard deviation σ of the Gaussian function. k is the measurement of the area a single point can take effect. In this paper, we focus on small change regions. So, we set the k equal to 31. σ can influence $\bar{R}(x,y)$ significantly. A big σ can improve the contrast of $\bar{R}(x,y)$ while a small σ can makes the details more clear. We choose σ equals 10 to obtain a balance between the contrast and the details.

In order to reduce noise, the range of $\bar{R}(x,y)$ is further modified. It will be linearly mapped to [128, 255] from [0, 255]. Set $\max_1$, $\min_1$ denote the maximum and minimum of $R(x,y)$ (because the background is unprocessed, the $\max_1$ and $\min_1$ is from foreground). We set $\max_2 = \max_1$, $\min_2 = 2 \min_1 - \max_2$. Then we have $\max_2 - \min_2 = 2(\max_1 - \min_1)$. So, we can set the value of background area in $R(x,y)$ to min_2. Then after implementing linear mapping on $R(x,y)$, the value of foreground area in $\bar{R}(x,y)$ will change from [0,255] to [128,255]. Such processing can restrain some noise while keeping the "significant" differences. As shown in Figure 1, it's obvious that the noise in (g) is much less than that in (h). Also we can find that the abnormal areas marked on (a) is also evident in (g), but in (h), we could not see them clearly. So these results demonstrate that our improved SSR performs better than original SSR in detecting small changes.

2.1.2 Histogram specification

Although the modified Retinex algorithm is very helpful, it's sensitive to noise. Assume that there is a white point in the low intensity area at the point (x_1,y_1), which is a noise. We assume that $I(x_1,y_1) = 255$. Because the point is located in a low intensity area the Gaussian weighted average also won't be high. So, the $R(x_1,y_1)$ we get from equation (1) will be intolerably large. This value will determine the maximum value of the whole image, which will lead to contrast difference between two longitudinal images. So, we can use histogram specification to correct the contrast difference. Figure 2(c) and Figure 2(d) show results preprocessed by modified SSR, but their intensity levels are obviously different. After modified by using histogram specification, Figure 2(e) and Figure 2(f) have similar intensity histogram distributions.

2.2 Change detection based on the integration of intensity and texture

Intensity-based change detection technique can retain most changed pixels, but the change mask may be affected by noise and illumination variation easily. The texture-based change

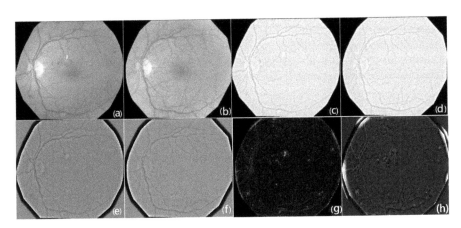

Figure 1. (a) and (b) are our origin images. The red regions marked on (a) are abnormal tissues. (c) and (d) are results we get after implementing our modified SSR method on (a) and (b). (e) and (f) are results after using SSR method. We keep k remains 31 and σ remains 10. (g) is difference image between (c) and (d); (h) is subtracted image using (c) and (d).

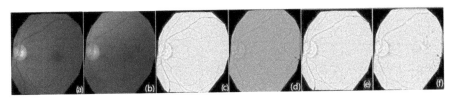

Figure 2. (a) and (b) are origin images, (c) and (d) are the result images of (a) and (b) processed by modified SSR method. Because of the poor quality of (b), the (d) is darker than (c). (e) is same as (c), while (f) is the result we get after using histogram specification on (c) and (d).

detection technique can restrain the influence of illumination changes and noise, but cannot well detect small changed regions. Therefore, it is reasonable to combine the intensity and texture for change detection.

In this paper, the following contrast is used as the texture descriptor (Unser 1986):

$$C(x,y) = \sum_j j^2 D(j) \qquad (2)$$

Here, $D(j)$ is the difference histograms of a small window centered at the point (x,y). It is an alternative to the usual co-occurrence matrices used (Nicho & Sarker, 2011). And it is less time-consuming and resource-consuming.

Then, each retinal image will be transformed into a new image whose pixel values are contrasts computed by (2). We calculate the difference between the two contrast images using direct subtraction, and let $diff_{con}(x,y)$ denotes the difference image of two contrast images. The range of $diff_{con}(x,y)$ is from −255 to +255. Then we calculate the histogram of $diff_{con}(x,y)$ and ignore the bin 0, because it is extremely large because of the black background. And we found that for many retinal images, the distributions of their such histograms usually can fit Gaussian distributions very well, especially when the indexes of bins in the histogram are small. This is illustrated in the lower part of Figure 3(f). This assumption can be used to reduce some false positive pixel points.

Let $diff_{int}(x,y)$ represents the intensity difference image between two retinal images, which is obtained via direct subtraction of two images. We find that the histogram of $diff_{int}(x,y)$ usually also fits a Gaussian distribution very well.

We assume that $diff_{con}(x,y) \sim N(\mu_1,\sigma_1^2)$, $diff_{int}(x,y) \sim N(\mu_2,\sigma_2^2)$. They are normalized and added together: $I_1(x,y) = diff_{con}(x,y) + diff_{int}(x,y)$. $diff_{con}(x,y)$ and $diff_{int}(x,y)$ have interdependency (e.g. the changed area in both images are evident) which is unknown to us. We have that $I_1(x,y)$ also fits Normal Distribution, $I_1 \sim N(0,\sigma^2)$, and we can use Law of Large Numbers to estimate its standard derivation σ. According to mathematical analysis, the percentage of points whose absolute values are less than 1.6σ is about 0.9. And most of the pixel points in the difference image are unchanged. So, we set:

$$I(x,y) = \begin{cases} 0, & \text{if } |I_1(x,y)| < 1.60\sigma \\ 1, & \text{if } |I_1(x,y)| \geq 1.60\sigma \end{cases} \qquad (3)$$

where $I(x,y)$ is a binary image. We use $I(x,y)$ as the change mask of the origin images.

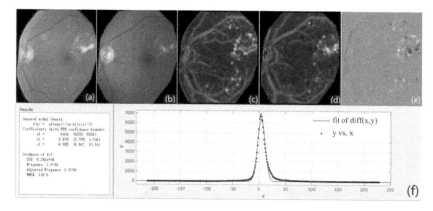

Figure 3. Contrast difference image fitting result, with Gaussian model. The bin 0 isn't plotted on the diagram. (a) and (b) are origin images, (c) and (d) are contrast images while (e) is the difference image. Dark region in (e) means that (d) is darker than (c) in that region while bright region means brighter. We fit the histogram of (e) using Matlab curve fitting toolbox with Gaussian model. The R-square of (f) is 0.9796.

196

3 EXPERIMENTS AND DISCUSSION

The proposed approach is compared with three approaches of Li's (2002), Gong's (2012) and the technique of Mr. Narasimha-Iyer (2006), as shown in Figure 4. Here, longitudinal retinal images used are from Mr. Narasimha-Iyer (2006). Figure 4(b) shows the some marked "significant changes" to be detected. As shown in Figure 4(c) 4(d) and 4(e), Li's (2002) and Gong's (2012) approaches fail to identify most of tiny changed areas, and the technique of Mr. Narasimha-Iyer (2006) may bring many noisy points which are difficult to distinguish interactively by doctors. However, our technique works better in detecting small and slight changed areas than the three techniques mentioned above, as shown in Figure 4(f). In Figure 4(f), most false positive pixel points locate near the edges and the blood vessels, and can be easily removed with the help of the ophthalmologists' experience.

The experimental results in Figure 4 show that our method can perform better than other techniques in identifying small and slight changed areas of retinal images. The Figure 5 below shows some other experiments results. We can see that most of true small and weak changes can be found in our change mask. There are also some false positive and false negative pixels in our change mask, but our proposed method can provide an interactive and quantitative

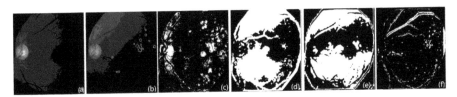

Figure 4. (a) and (b) are two longitudinal retinal images needed to be processed. (d), (e) are change detection results we get using Li's (2002) and Gong's (2012) methods respectively; (c) is change mask we get using Narasimha-Iyer's (2006) approach; (f) is change mask we get using our technique.

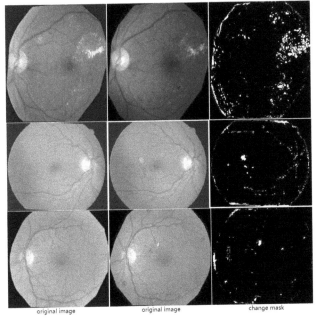

Figure 5. Three group experiments we took. The first two columns are origin images and the third column is the change mask using our method. We mark some significant changes on the origin images using red circles.

measurement means of retinal lesion evolution. These final results are accepted by the clinical ophthalmologists.

4 CONCLUSION

To monitor pathological varying tendency, we have developed a computer-aided lesion change detection and quantitative measurement changed areas system. Our method is more straightforward and works better than other state-of-the-art techniques. Experimental results show that it is effective in detecting changes (especially identifying small and weak variations) in longitudinal retinal images.

ACKNOWLEDGMENTS

This work was supported in part by Shanghai Jiao Tong University College Students' Innovative Practice (IPP10090) and the National Natural Science Foundation of China (31375020). The authors would like to thank Dr. Chen for providing their PIIFD source codes and Mr. Narasimha-Iyer et al. for making the longitudinal retinal images available.

REFERENCES

Chen, J., Tian, J., Lee, N., Zheng, et al. 2010. A partial intensity invariant feature descriptor for multi-modal retinal image registration. *IEEE Transactions on Biomedical Engineering* 57(7): 1707–1718.

Fu, X., Sun, Y, et al. 2014. A novel retinex based approach for image enhancement with illumination adjustment. In Florence, Italy (ed). *Acoustics, Speech and Signal Processing (ICASSP), 2014 IEEE International Conference on*: 1190–1194. Piscataway, NJ: IEEE.

Gong, M., Zhou, Z. & Ma, J. 2012. Change detection in synthetic aperture radar images based on image fusion and fuzzy clustering. *IEEE Transactions on Image Processing* 21(4): 2141–2151.

Li, L. & Leung, M.K.H. 2002. Integrating intensity and texture differences for robust change detection. *IEEE Transactions on Image Processing* 11(2): 105–112.

Narasimha-Iyer, H., Can, A., Roysam, A.B., et al. 2006. Robust detection and classification of longitudinal changes in color retinal fundus images for monitoring diabetic retinopathy. *IEEE Transactions on Biomedical Engineering* 53(6): 1–15.

Nicho, J.E. & Sarker, M.L.R. 2011. Improved biomass estimation using the texture parameters of two high-resolution optical sensors. *IEEE Transactions on Geoscience and Remote Sensing* 49(3): 930–948.

Radke, R.J., Andra, S., Al-Kofahi, O., et al. 2005. Image Change Detection Algorithms: A Systematic Survey. *IEEE Transactions on Image Processing* 14(3): 294–307.

Unser, M. 1986. Sum and Difference Histograms for Texture Classification. *IEEE Transactions on Pattern Analysis and Machine Intelligence* PAMI-8 (1): 118–125.

Bioinformatics and Biomedical Engineering – Chou & Zhou (Eds)
© 2016 Taylor & Francis Group, London, ISBN 978-1-138-02784-8

Effective method for extracting the characteristic value of photoplethysmography signals

Y.Z. Shang, S.S. Yong, D.X. Guo, R. Peng & X.A. Wang
The Key Laboratory of Integrated Microsystems, Peking University Shenzhen Graduate School, Shenzhen, Guangdong Province, China

ABSTRACT: This paper presents a method for extracting the characteristic values of PPG signals with greater efficiency, which can deal with multi-channel PPG signals at the same time. The removal of baseline wander and high-frequency noise is based on Discrete Wavelet Transform. Then, PPG signals are divided into segments, and the modulus maxima method is used to extract the characteristic value in each segment of PPG signals. The result demonstrates that the proposed method can simultaneously extract the characteristic values of four PPG signals with a time span of 3 minutes. In conclusion, the method can deal with several long-duration PPG signals at the same time, which can greatly improve the efficiency of research related to human health.

1 INTRODUCTION

PPG is an important signal that is related to the condition of human physiology (Vashist, 2012), and it is generated by heartbeat-induced blood flow along the arteries. PPG is a periodic signal whose period is almost the same as the heartbeat. Characteristic values of PPG waveform are closely associated with the characteristic parameters of the cardiovascular system (Zhaopeng Fan, 2009), which contain extremely rich information on cardiovascular physiology and pathology system. PPG is obtained non-invasively by kinds of sensors, such as strain gauge pressure sensor, photoelectric sensor, electrical impedance sensor and piezo-electric crystal pressure sensor.

In recent years, use of photoplethysmography to measure blood glucose non-invasively and monitor blood pressure without the airbag has become the hot research topic (Tura et al., 2007). Research mainly consists of obtaining PPG, extracting characteristic values of PPG and processing data. During health-related research, many experiments need to be carried out repeatedly to explore the relationship between PPG signals and human health. Since those experiments will get a large number of PPG signal data, accurately and quickly extracting the characteristic values of PPG is the key for conducting research. The existing basic idea of extracting the characteristic value of PPG is to divide PPG signals into cycles and then to locate the PPG feature points in each cycle. The main method of dividing PPG into cycles is based on setting the threshold, which has a risk of judging the cycles of signal incorrectly. The longer the PPG signals are, the larger the risk of error is. Because the existing method cannot divide PPG signals into cycles accurately, wrong characteristic values are likely to be extracted at any time during characteristic value extraction; it severely limits the efficiency of using computer to processing PPG signals.

For the current problems, this paper presents a method for multi-channel PPG signals that cuts signals into segments and then exacts the characteristic values in each segment.

2 ALGORITHM

The method presented in this paper consists of two parts: signal pre-processing and extracting characteristic values. Baseline wander and high-frequency noise of four PPG signals are removed during signal pre-processing, and then the method extracts the characteristic values of four PPG signals piecewise by nesting the inner and outer loops during the second part (Cuiwei et al., 1995). Eventually, the whole algorithm accomplishes the extraction of the characteristic values of four PPG signals. The flow chart of the algorithm is shown in Figure 1.

2.1 Signal pre-processing

In the PPG process, the movements of the body and breathing can cause a low frequency baseline drift (Xu et al., 2007), random noise and environmental interference, leading to the emergence of high-frequency noise. The low-frequency baseline drift, high-frequency noise and PPG signal are clearly different in frequency (Ding et al., 2012, REN Jie, 2010). Based on multi-resolution wavelet decomposition, the baseline drift, high-frequency noise and PPG signals are decomposed into different layers (Pal and Mitra, 2010).

Then, on the decomposition layer related to the baseline drift, wavelet coefficients are cleared. On the decomposition layer of high-frequency noise, wavelet coefficients are processed by using the soft-threshold method. Finally, the PPG signal removed the baseline drift and high-frequency noise is reconstructed.

Extracting the characteristic values of the PPG signal piecewise is the main idea of the method presented in this paper, which is shown in Figure 2. The number of segments is determined as follows: Length/5120 = cycle_num. Here, Length is the length of PPG signals and 5120 is the length of a segment.

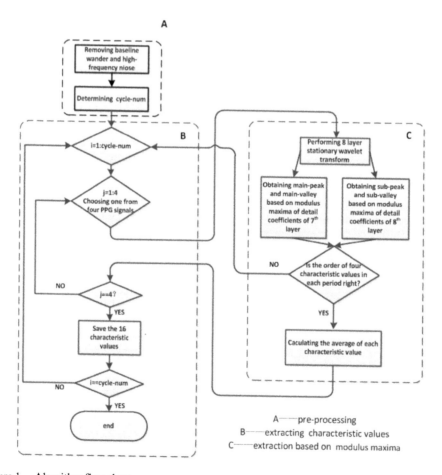

Figure 1.   Algorithm flow chart.

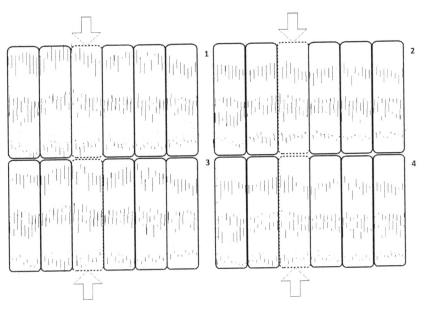

Figure 2.   Dividing four PPG signals into segments.

2.2   Extracting characteristic values

In this section, the method mainly consists of two nested loops. Feature extraction algorithm core is performed multiple cycles under the control of two nested loops. Ultimately, the result of characteristic value extraction is output, which contains main_peak, sub_peak, main_valley and sub_valley.

In the outer loop, the loop variable i determines which segment of four PPG signals should be selected. In the inner loop, the loop variable j determines which one of the four PPG signals should be selected. Under the control of i and j, one segment of one PPG signal is selected. The characteristic value extraction of four PPG signals is accomplished at the end of the two nested loops. The algorithm core is located in the center of the two nested loops.

The core of the algorithm based on the modulus maxima performs extracting the characteristic values of a segment of the PPG signal with a length of 5120. It consists of the following steps:

- Performing the 8-layer stationary wavelet transform on the PPG signal, obtaining the detail coefficients of layer 7 and layer 8;
- Calculating the modulus maxima series of the detail coefficients of layer 7, determining a reasonable threshold for the modulus maxima series; dividing the modulus maxima series into cycles according to the obtained threshold; locating the main_peak and main_valley in each signal cycle based on the modulus maxima; if the algorithm cannot determine a reasonable threshold, it concludes that the waveform of this segment of the PPG signal is abnormal. The algorithm quits extracting the characteristic values in this segment of the four PPG signals by jumping out of the inner loop;
- Performing similar operations on the 8th layer of detail coefficients to locate the sub_peak and sub_valley in each signal period.

If the difference among four kinds of characteristic values (main_peak, sub_peak, main_valley and sub_valley) is greater than 1 or if the order of the four characteristic values is not the same as main_peak->sub_peak->main_valley->sub_valley, then the algorithm will conclude that there must be something wrong with the waveform of the current segment of the PPG signal. The algorithm quits extracting the characteristic values in this segment of the four PPG signals by jumping out of the inner loop.

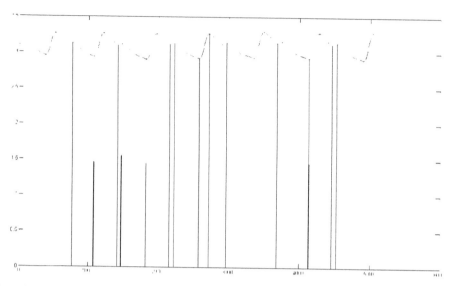

Figure 3. The result of characteristic value extraction.

Having obtained four characteristic values in each period of the current segment of the PPG signal, the algorithm calculates the average of each characteristic value among all the periods of current signal segment, which will be considered as the result of characteristic value extraction in the current signal segment.

3 RESULTS

The result of characteristic value extraction is shown in Figure 3. From the figure, we can see that the method proposed in this paper can accurately locate the characteristic values of each PPG signal period. Given that the performance of the algorithm on each segment of four PPG signals is identical, we show only one segment of the PPG signal.

4 CONCLUSION

The method presented in this paper can extract the characteristic values of four PPG signals, which are of long duration. In the proposed method, the algorithm can recognize the abnormal segments of PPG signals and quit performing the extraction in those segments, which will avoid introducing errors into the extracting result and enable the method to deal with long-duration PPG signals. In this paper, the proposed method is performed on four PPG signals; in fact, it is still suitable when the count of PPG signals is changed. There are several kinds of pulse wave signals whose waveforms are similar, so the method presented in this paper is also suitable for other kinds of pulse wave signal.

REFERENCES

Cuiwei, L., Chongxun, Z. & Changfeng, T. 1995. Detection of ECG characteristic points using wavelet transforms. *Biomedical Engineering, IEEE Transactions on,* 42, 21–28.
Ding, W., Qin, S., Miao, L., Xi, N. & Li, H. 2012. Study on the removal method of electrogastrogram baseline wander based on wavelet transformation. *Sheng wu yi xue gong cheng xue za zhi = Journal of biomedical engineering = Shengwu yixue gongchengxue zazhi,* 29, 1189-92, 1196.

Pal, S. & Mitra, M. 2010. Detection of ECG characteristic points using Multiresolution Wavelet Analysis based Selective Coefficient Method. *Measurement,* 43, 255–261.

Ren Jie, Y.L.-X. 2010. An Methods for Rectifying ECG Baseline Wander Based on Wavelet Coefficient. *Chinese Medical Equipment Journal,* 31, 24–26.

Tura, A., Maran, A. & Pacini, G. 2007. Non-invasive glucose monitoring: Assessment of technologies and devices according to quantitative criteria. *Diabetes Research and Clinical Practice,* 77, 16–40.

Vashist, S.K. 2012. Non-invasive glucose monitoring technology in diabetes management: a review. *Anal Chim Acta,* 750, 16–27.

Xu, L., Zhang, D., Wang, K., Li, N. & Wang, X. 2007. Baseline wander correction in pulse waveforms using wavelet-based cascaded adaptive filter. *Comput Biol Med,* 37, 716–31.

Zhaopeng Fan, G.Z., Liao 2009. clinical analysis for cardiovascular disease by calculating stiffness index, cardiac output from pulse wave, 478–481.

Biomedical materials and products

Bioinformatics and Biomedical Engineering – Chou & Zhou (Eds)
© 2016 Taylor & Francis Group, London, ISBN 978-1-138-02784-8

TEOS hydrolysis method synthesize nano silica and its biological toxicity research

L.J. Hu, R. Li & B. Liu
School of Stomatology, Lanzhou University, Lanzhou, Gansu, China

X. Han, X. Zheng & B.B. Tong
Gansu College of Traditional Chinese Medicine, Lanzhou, Gansu, China

ABSTRACT: The nano silica spheres with different particle sizes were synthesized by the Tetraethyl Orthosilicate (TEOS) hydrolysis method. The X-Ray Diffraction (XRD) patterns revealed that the sample was of a single phase and its strongest peak was significantly broadened. From the images of Transmission Electron Microscopy (TEM), we can see that the samples' surfaces have a good dispersion and uniform spherical morphology. The proliferating activity of Hela cell was assessed by the cell toxicity test (MTT) assay. Cell morphology and Annexin Hoechst 33342 staining were employed to determine apoptosis in Hela cells. The results show that when the Hela cell was treated with different concentrations of 200 nm nano silica, lower concentrations of nano silica showed better cytocompatibility. The 24 h cell-stained image shows that under the concentrations of 100 $\mu g \cdot ml^{-1}$ of nano silica, the cell structure was normal and the nucleus was complete.

Keywords: TEOS hydrolysis method; nano silica; biocompatibility

1   INTRODUCTION

Since the 21st century, nanotechnology has been gradually developing as one of the main driving forces of development in the world. Nano-silicon with a small particle size, large specific surface area and good property of dispersion characteristics, as well as superior stability, reinforcing, thixotropy and excellent optical and mechanical properties has a very broad application prospects in photonic crystals, catalyst carrier, precision ceramics, rubber, plastic, paint, chromatographic filler, polymer composite materials and many other fields. In aspects such as medical field, nano-silicon has also received much attention. There are many preparation methods for nanometer silicon dioxide at present, including the mechanical crushing method, vapor deposition method, sol-gel method, precipitation method and microemulsion method (Jiang L.C. 2013, Li G.B. 2011, Guo Y. 2005, He Y.F. 2008, Wang Z. 1998, Shi L.Y. 1998 & Jia H. 2001). In recent years, the research of nano silicon dioxide used for the continuous dosing system was a major concern. In 2008, Jia Husheng et al. used nano silicon dioxide as a carrier for synthesizing zinc-silver composites. The antibacterial performance showed that the antibacterial properties of nano silicon dioxide that carried zinc-silver is better than the nano silicon dioxide that carried silver; both of these dental composite materials can be used as the base material with an antibacterial effect (Jia H.S. et al. 2008). In 2009, Chun-mei Lin et al. modified the silica surface with the silane coupling agent KH-570, and studied the adsorption and slow-release performance in abamectin. The adsorption rate of the modified silica in abamectin increased from 13.98% to 31.36%, with a favorable slow release effect. The controlled release time of the modified silica can last for 80 h in dissolution medium of abamectin. Therefore, the nano silicon dioxide modified by the silane coupling agent can be used as a hydrophobic drug delivery system (Lin C.M. et al. 2009). In addition, in 2011, Linlin Li et al. found that nano silicon dioxide can be used as a drug carrier, and built a new type of anticancer drug delivery system, which fixed the mesenchymal stem cells effectively, made the release of adriamycin wider and lengthened the retention time, thus significantly improving the mortality rate of tumor cells and increased the anti-cancer efficiency (Li L.L. et al. 2011). The research on nano silicon dioxide in compound prescription has been reported frequently, but the research on nano-silica single biological cell toxicity has been reported rarely. Therefore, we have performed a series of *in vitro* cell toxicity experiments on nano-silica and attempted to lay the foundation for the toxicology research of nano-silica drug-loading composite systems.

2 EXPERIMENTS

2.1 *The preparation of nano silica spheres*

Nano silica spheres were synthesized by the tetraethyl orthosilicate hydrolysis method. The starting materials were TEOS (92.9%–93.6%), CH_2CHOH (>99.0%) and NH_4OH (>99.0%). All the reagents used were of analytical grade. TEOS–CH_2CHOH (1:4) was mixed with solution A and stirred at 500 rev/min for 4 h at a constant temperature. A magnetic stirrer was used to get emulsion. Distilled water (7:1) was slowly added to solution A in order to get solution B. NH_4OH was used to adjust the pH of solution B to get solution C, then placed in a drying oven, at 80°C for 48 h to get solid A. Finally, a muffle furnace was used at 580°C and sintered for 4 h to get the product.

The Rigaku D/Max-2400 X-ray diffractometer was employed to check the phase of the phosphor powder using CuKαradiation at room temperature. The wavelength was 1.54056×10^{-4} um. The Scanning Electron Microscopy (SEM) images of the nanoparticles were obtained by using SEM (JEM-2100F Electron Microscope/JEOL Co. 200 kV). The pH value of solution B was measured and adjusted using the Sartorius pH meter. The solution C was dried by the electric blast oven. Solid A was heated in the muffle furnace.

2.2 *MTT cell experiment*

The logarithmic phase Hela was selected, and 0.25% trypsin was used for digestion. The cells were made into single cell suspension and inoculated on 96-well plates (corning Inc, NY, USA) with a density of 3×10^3 cells per hole and divided into control groups and experimental groups. The experimental groups were added to the complete medium containing nano silica at the concentrations of 100 µg·ml^{-1}, 80 µg·ml^{-1},60 µg·ml^{-1}, 40 µg·ml^{-1} and 20 µg·ml^{-1}. The blank control group was only added to the complete medium. The 96-well plates were placed in a 5% CO_2 water-saturated culture incubator at 37°C. After incubation for 24 h, 48 h, and 72 h, 20 µL (5 mg·ml^{-1}) of the MTT were added to the 96-well plates and placed in the incubator for 4 h reaction. The 96-well plates were removed quickly and the liquid was discarded. A solution of 150 µL dimethyl sulfoxide was added and oscillated for 10 min. The absorbance values were measured at a wavelength of 570 nm in the ELISA analyzer (Bio-Rad, Hercules, CA, USA). MTT assay procedure was repeated three times. Hela cell activity was measured by the OD value in the laboratory. The Hela growth inhibition rate was calculated by using formula (1), and all the experiments were repeated three times.

$$[1 - A490 \text{ (test)}/A490 \text{ (control)}] \times 100\% \qquad (1)$$

2.3 *Hoechst 33342 cell staining to observe the influence of nano-silica to Hela apoptosis*

The Hela cells 1×10^3 per well were seeded onto a 96-well cell culture plate, and the cells were divided into control and experimental groups. After the cells were attached, the complete medium was added to the blank control group. The experimental groups were added to the complete medium containing SiO$_2$ nanoparticles with the concentrations of 100 µg ml^{-1}, 80 µg ml^{-1}, 60 µg ml^{-1}, 40 µg ml^{-1}, and 20 µg ml^{-1}. The cells were cultured for 24 h. After incubation, the culture medium was removed and each well was washed three times with PBS buffer for 5 minutes. The cells were observed under an inverted phase contrast microscope (push around—DP72).

2.4 *Statistical analysis*

The SPSS 20.0 statistical software was used for statistical analysis. Pairwise comparisons were made using the t test. P < 0.05 was considered statistically significant.

3 RESULTS AND DISCUSSION

3.1 *The sample analysis of the XRD and surface photography*

Figure 1 shows the XRD pattern of the samples. As can be seen from the figure, all samples are of a single phase, and the strongest peak position are the same with the standard JCPDF no. 75-0923. The width of the diffraction peak changed with the increase in pH constantly.

From Figure 2, we can see that the particle size of the sample was around 200 nm. We can also see that the sample has good dispersibility, smaller particle size and a better morphology. This is because we reduced the concentrations of the reaction system, adjusted the pH and performed ultrasonic dispersion, which changed the morphology and particle size of the silica balls.

3.2 *The analysis of the nano-silica sample for Hela cell toxicity*

Figure 3 shows that the Hela cell survival rate of different concentrations of SiO$_2$ group was higher than 92%. After incubation with nano silica spheres for 24 h, 48 h and 72 h, the survival rate of Hela increased over time. There was a significant difference in comparison between the 100 µg·ml^{-1} group and the 80 µg·ml^{-1} group (P < 0.05). In addition, the 60 µg·ml^{-1}, 40 µg·ml^{-1} and 20 µg·ml^{-1} groups were not statistically different (P > 0.05). This showed that the spheres had little irritation on the growth of Hela. The influence was weaker at the concentration of 20 µg·ml^{-1} compared with the control group, indicating the nano silica had good cell compatibility.

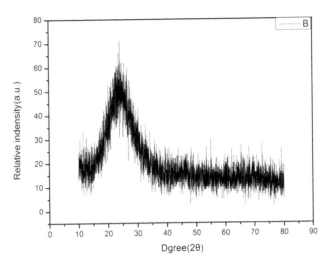

Figure 1. XRD of the nano silica sample.

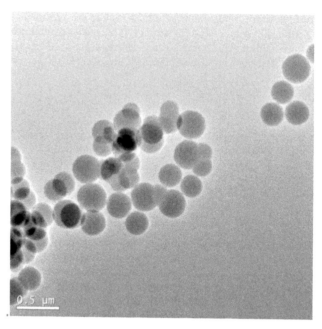

Figure 2.　The best sample's TEM image.

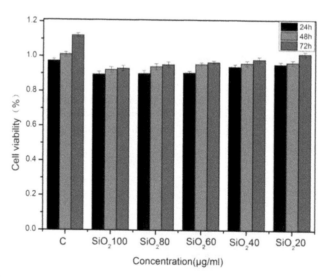

Figure 3.　MTT assay in different concentrations of 200 nm SiO$_2$ combined with Hela cells.

　　From Figure 4, we can see that the Hela cells were affected by the different concentrations of the nano silica ball for 24 h. The image was observed by an inverted microscope. It was observed that Hela cells showed a normal morphological structure, with a long spindle shape and no large areas of abnormal death. It was possible to observe a 10%–15% of the bubble formation in apoptotic bodies and cell detachment, at a nano-silica concentration of 100 μg·ml^{-1}. At the concentration of 20 μg·ml^{-1} of nano-silica, 95% of Hela cells showed a normal cell structure and no significant difference compared with the control group.

　　Figure 5 shows the stained images of Hela cells at different concentrations of SiO$_2$ after 24 h. Hela cells were dyed with Hoechst 33342. From the image, we can see that Hela cells

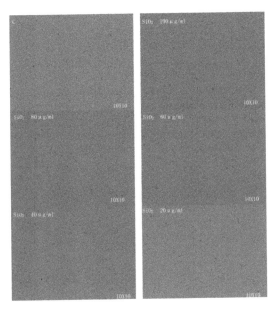

Figure 4. The image showing that the Hela cells were affected by the different concentrations of nano silica for 24 h.

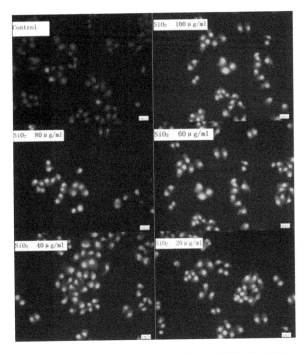

Figure 5. Hela cells' stained images at different concentrations of SiO$_2$ after 24 h.

treated with different concentrations of the SiO$_2$ ball after 24 h had a concentrated nucleus and occasionally had small, round apoptotic bodies at the concentration of 100 μg/ml. The Hela cell structure was normal, the nucleus was complete, and the cells dispersed normally at concentrations less than 80 μg/ml. It showed that SiO$_2$ at low concentrations had no apparent toxicity on Hela cells.

4 CONCLUSION

A series of samples were synthesized by the TEOS hydrolysis method, showing that all samples were single phases and particle sizes were approximately 200 nm. The surface morphology was spherical. Factors affecting the particle size of the sample included the magnetic stirring rate and the pH of the reaction solution. However, the large-sized samples have a reunion phenomenon, which was due to the fact that the reactant concentration of the system was too high and the samples were not treated by ultrasonic dispersion after being generated. We confirmed the inhibitory effect of nano silica on the proliferation of Hela cells was weak through the *in vitro* experiment. The Hela cells were cultured for 48 h with nano silica at the concentrations of 100 µg/ml, 80 µg/ml, 60 µg/ml, 40 µg/ml and 20 µg/ml. The survival rate of each concentration group was more than 92%. The Hela structure observed by using an inverted phase contrast microscope showed a normal structure and no abnormal death. MTT cytotoxicity assay confirmed that the nano silica has a smaller influence on the growth of Hela, with good biological compatibility. Thus, nano silica has a potential application in controlled release as a drug carrier.

REFERENCES

Guo Y. et al. 2005. Sol-gel Preparation of Nano silica[J]. Tianjin Chemical 39:336–341.

He Y.F. et al. 2008. Research progress of preparation of nano-silica powders[J]. Materials development and application.p.187–201.

Jia H. et al. 2001. The super gravity preparation of nano-silica[J]. Journal of materials research 11:17–21.

Jia H.S. et al. 2008. The structures and antibacterial properties of nano-silica supported silver/zinc-silver materials[J]. Dental Materials 24:244–249.

Jiang L.C. et al. 2013. Research progress in preparation and application of nano-silica[J]. 41:124–127.

Li G.B. et al. 2011. Technical progress of SiO2 nanoparticles' preparation[J]. New Chemical Materials 39: 177–182.

Li L.L. et al. 2011. Silica nanorattle-doxo—rubicin—anchored mesenchymal stem cells for tumor-tropic therapy[J]. ACS Nano 9:7462–7470.

Lin C.M. et al. 2009. Nano-silica surface modification and its adsorption of abamectin and slow-release performance[J]. Journal of Ecological Environment 17:197–200.

Shi L.Y. et al. 1998. The structure of microemulsion and its application in the preparation of ultrafine particles[J]. Functional materials.

Wang Z. et al. 1998. High purity superfine SiO2 powder preparation technology research[J]. Journal of Railway 12:56–62.

Bioinformatics and Biomedical Engineering – Chou & Zhou (Eds)
© 2016 Taylor & Francis Group, London, ISBN 978-1-138-02784-8

Solvothermal method preparation of nano $Y_2O_2S:Eu^{3+}$ and research of luminescence property and cytotoxicity

C. Zhou
The First Clinical Medical College of Lanzhou University, Lanzhou, Gansu, China

W.J. Yan
Anesthesiology Department of the People's Hospital of Gansu Province, Gansu, China

J. Jin & X.X. Li
Qingyang City Hospital of Traditional Chinese Medicine, Gansu, China

Y. Li
Basic Medical College of Lanzhou University, Lanzhou, Gansu, China

P. Xie
Cardiology Department of the People's Hospital of Gansu Province, Gansu, China

ABSTRACT: A series of $Y_2O_2S:Eu^{3+}$ samples were prepared by the solvothermal method using ethylenediamine as the solvent at 220°C for 24 h. All samples were white powders. X-Ray Diffraction (XRD) proved that all of them were of a single phase. The Laser granularity cryoscopy (rise 2008) was used to measure the size of the samples, and most of the samples sizes ranged from 150 nm to 200 nm. The Scanning Electron Microscopy (SEM) image showed that the morphologies of $Y_2O_2S:Eu^{3+}$ were spherical, but these morphologies had an aggregation phenomenon. The cell toxicity test (MTT) showed that the cytotoxicity of the sample is relatively low (growth inhibition rate is low) when the concentration of the $Y_2O_2S:Eu^{3+}_{0.01}$ samples ranged from 10 to 50 µg/ml. This study explored the preparation of Y_2O_2S and its biocompatibility, and thus aimed to provide a scientific basis for Y_2O_2S as a drug carrier.

Keywords: solvothermal; nano particle; $Y_2O_2S:Eu^{3+}_{0.01}$; biological toxicity

1 INTRODUCTION

Sulfur oxide phosphor activated by europium (Jau-Ho Jean & Szu-Ming Yang. 2000) is widely used in red light-emitting materials (Bao H et al. 2004). It has the following characteristics: color purity, color distortion, brightness of the current saturation and stability (Gaponik N et al. 2002).

$Y_2O_2S:Eu^{3+}$ as a material with a powerful light-emitting property is commonly used in the industry (Wang P.T et al. 2008) and gradually applied in medicine (Han P.D et al. 2012). Recently, Zheng Long-jiang, Gao Xiao-yang, Liu Hai-long, Li Bing, Xu Chen-xi et al have researched the heating effect of the Er^{3+}/Yb^{3+} doped Y_2O_3 nanometer powder by 980 nm Laser Diode Pumping (Zheng L.J et al. 2013). The phosphor samples excited by a 980 nm diode laser have a surface temperature rise phenomenon. This phenomenon plays an important role in the power saturation phenomena that occurred in the analysis of rare-earth ions for the up-conversion process, and shows broad application prospects in high-temperature sensing materials and medical biological cells that burn holes. In 2013, the article by Yuen Shun Wu, Dan Zeng Hui, Sangyong Yuan has reported the Asia ultrafine (about 500 nm) up-conversion phosphors $Y_2O_2S:Er^{3+}$, Yb^{3+} prepared by the high-temperature solid-phase method and by adding additives such as Bi_2O_3. This made the material to play an important role in medicine (Yuen S.W et al. 2013). Human blood is transparent to NIR, so a good biomarker material should have an emission peak near 780 nm–900 nm (NIR), and $Y_2O_2S:Eu^{3+}$ has the potential. As a potential biomarker material, $Y_2O_2S:Eu^{3+}$ has a better anti-decay property compared with organic dyes, such as Rhodamine B.

The preparation methods include the high-temperature solid-phase method and co-precipitation method (Zhang H et al. 2003). Sample particle sizes ranged between 3 and 5 μm (Sondi I 2004 & Cui C et al. 2012). In recent years, the combustion method has been applied and improved constantly, which can obtain the preferred particle size (Franciszek B et al. 2003). Y_2O_2S prepared by the solid state method often has a larger size than nanoparticles, and is water-soluble. Therefore, we used the solvothermal method to synthesize Y_2O_2S (Deng S.Q et al. 2003) and $Y_2O_2S:Eu^{3+}$ nano particles, and then obtained the final powders, which have the hydrophilic characteristic after modification with thioglycolic acid (Lü D et al. 2013).

2 MATERIALS AND METHODS

2.1 The preparation of $Y_2O_2S:Eu^{3+}$

$Y_2O_2S:Eu^{3+}$ was prepared as follows. (1) Dissolved Y_2O_3 (99.999%) and Eu_2O_3 (99.999%) (different proportions) in concentrated nitric acid and heated to remove excess nitric acid until a transparent thin film yttrium nitrate was formed. (2) Made the gel to fully form by adjusting the pH using ammonia. (3) Filtered and washed with deionized water several times, finally dried at 100°C in reaction precursors. (4) Added the samples with excess 10% $CS(NH_2)_2$ to a 50 mL stainless steel reactor, filled with 30 ml ethylenediamine and the appropriate amount of surface active agents, covered the thermostat tightly and placed in the furnace at 220°C for 24 h. (5) Cooled at room temperature naturally, then washed the product by deionized water and ethanol several times, and finally dried at 100°C (Shao M.F et al).

The Rigaku D/Max-2400 X-ray diffractometer was employed to check the phase of the phosphor powder using CuKα radiation at room temperature. The wavelength was 1.54056×10^{-4} um. The Scanning Electron Microscopy (SEM) images of the nanoparticles were obtained by using SEM (JEM-2100F Electron Microscope/JEOL Co. 200 kV). The measurement and pH value adjustment of solution B were processed by using the Sartorius pH meter adjustment. Solution C was dried in an electric blast oven. The solid A was heated by muffle furnace.

2.2 MTT cell experiment

Logarithmic phase Hela cells were selected and digested by 0.25% trypsin, and made into single cell suspension. The cells were inoculated on the 96-well plates (corning Inc, NY, USA) with a density of 3×10^3 cells per hole and divided into control groups and experimental groups. The experimental groups were added to the complete medium containing $Y_2O_2S:Eu^{3+}$ at concentrations of 50 μg·ml^{-1}, 40 μg·ml^{-1}, 30 μg·ml^{-1}, 20 μg·ml^{-1} and 10 μg·ml^{-1}. The control groups were added to the complete medium. The 96-well plates were placed at 37°C and 5% CO_2 in a water-saturated culture incubator. After incubating for 24 h and 48 h, 20 μL (5 mg · ml^{-1}) MTT was added to the 96-well plates and placed in the incubator for 4 h. The 96-well plates were removed quickly and the liquid was discarded. A solution of 150 μL dimethyl sulfoxide was added and oscillated for 10 min. The absorbance values were measured at the wavelength of 570 nm in the ELISA analyzer (Bio-Rad, Hercules, CA, USA). MTT assay procedure was repeated three times. Hela cell activity was measured by the OD value in the laboratory. The growth inhibition rate of Hela was calculated by formula (1), and all the experiments were repeated three times.

$$[1 - A490 \text{ (test)}/A490 \text{ (control)}] \times 100\% \qquad (1)$$

2.3 Hoechst 33342 cell staining to observe the influence of nano-silica to Hela apoptosis

The Hela cells at 1×10^3 per well were seeded onto the 96-well cell culture plate, and the cells were divided into control and experimental groups. After the cells were attached, the complete medium was added to the blank control groups, and the experimental groups were added to the complete medium containing $Y_2O_2S:Eu^{3+}$ at the concentrations of 50 µg·ml⁻¹, 40 µg·ml⁻¹, 30 µg·ml⁻¹, 20 µg·ml⁻¹ and 10 µg·ml⁻¹, cultured for 24 h. After incubation, the culture medium was removed and washed three times with PBS buffer for 5 minutes. The cells were observed under the inverted phase contrast microscope (push around—DP72).

2.4 Statistical analysis

The SPSS 20.0 statistical software was used for statistical analysis. T test was used for pairwise comparisons. $P < 0.05$ was considered statistically significant.

3 RESULTS AND DISCUSSION

Figure 1 shows the XRD patterns of typical samples of solvothermal precipitation. Compared with the standard sample, sulfur yttria (PDF24–1424), the series samples are of a single phase. The XRD strongest peak showed the product as Y_2O_2S, and no impurity peaks in the XRD data.

The Laser granularity cryoscopy was used to investigate the size distribution of the samples. From Figure 2, we can see that the sample sizes were located at 150 nm–200 nm. By effectively changing the temperature and the magnetic stirring speed during the preparation of the experiment, we obtained a small particle diameter sample, and the sample particle size was controlled within a certain range.

From Figure 3, we can see that the resulting samples were spherical at approximately 200 nanometers. Yet, the SEM image showed that the sample still has an aggregation phenomenon. This phenomenon may be caused by the excessive concentration of the reaction system, or by an inappropriate choice of the surface active agent. However, the polymerization sample of the figure does not appear to be whole, which may due to the slow stirring.

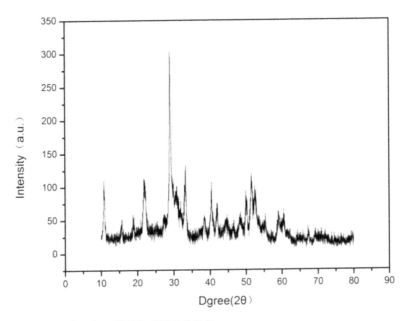

Figure 1.   X-Ray Diffraction (XRD) of $Y_2O_2S:Eu^{3+}$.

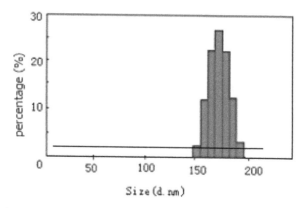

Figure 2.   The optimum sample size distribution of Y_2O_2S.

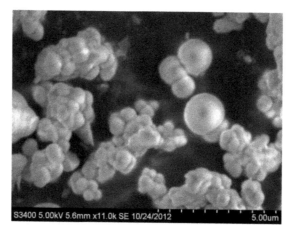

S3400 5.00kV 5.6mm x11.0k SE 10/24/2012   5.00um

Figure 3.   SEM photograph of the $Y_2O_2S:Eu^{3+}$ sample.

From Figure 4, we can see that when the samples of $Y_2O_2S:Eu^{3+}_{0.01}$ were monitored at 254 nm, the position of the strongest intensity emission peak was located at 626 nm. It corresponded to the transition $^5D_0 \rightarrow {}^7F_2$ level of Eu^{3+} ions. The emission spectrum of $Y_2O_2S:Eu^{3+}_{0.01}$ is shown in Figure 4.

From Figure 5, we can see that the survival rate of Hela cells at different concentrations of the $Y_2O_2S:Eu^{3+}_{0.01}$ group was higher than 89% after 24 h, with 48 h period extending to 48 h. There was a significant difference in comparison between the 50 $\mu g \cdot ml^{-1}$ and 40 $\mu g \cdot ml^{-1}$ groups ($P < 0.05$). In addition, the 30 $\mu g \cdot ml^{-1}$, 20 $\mu g \cdot ml^{-1}$ and 10 $\mu g \cdot ml^{-1}$ groups had no statistical difference ($P > 0.05$). This showed that nano $Y_2O_2S:Eu^{3+}_{0.01}$ had no strong stimulating effect on the growth of Hela. At the concentration of 10 $\mu g \cdot ml^{-1}$, compared with the control group, the influence was more weak, indicating that the material had good cell compatibility.

Figure 6 shows the stained images of Hela cells at different concentrations of $Y_2O_2S:Eu^{3+}_{0.01}$ after 24 h. The Hela cells were dyed by Hoechst 33342. From the images, we can see that the Hela cells were treated at different concentrations of $Y_2O_2S:Eu^{3+}_{0.01}$ after 24 h. The concentrated nucleus and small, round apoptotic bodies can be seen occasionally at a concentration of 50 $\mu g/ml$. The structure of Hela was normal, the nucleus was complete, and the cells dispersed normally when the concentrations were less than 30 $\mu g/ml$. It followed that $Y_2O_2S:Eu^{3+}_{0.01}$ at low concentrations had no apparent toxicity on Hela cells.

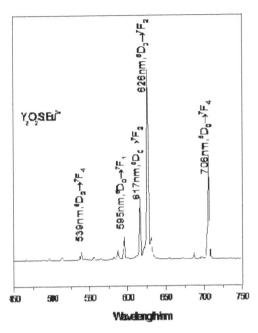

Figure 4. Emission spectrum of $Y_2O_2S:Eu^{3+}_{0.01}$ (Ex = 254 nm).

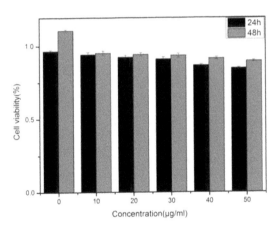

Figure 5. Different concentrations of the effect of 200 nm $Y_2O_2S:Eu^{3+}_{0.01}$ on Hela cells' MTT results.

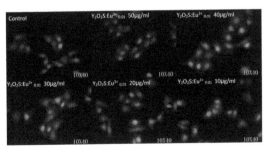

Figure 6. Hoechst33342-stained images at different concentrations of $Y_2O_2S:Eu^{3+}_{0.01}$ after 24 h by MTT (10×40).

4 CONCLUSION

A series of $Y_2O_2S:Eu^{3+}$ samples were synthesized by the solvothermal method. All samples were of single phases and particle sizes ranged approximately from 150 nm to 200 nm. Besides, the SEM of the sample and the particle size distribution showed that the surface morphology of the sample was roughly spherical and had some aggregation. We confirmed through the experiment that the inhibitory effect of $Y_2O_2S:Eu^{3+}$ on the *in vitro* proliferation of Hela cells was weak. The survival rate of Hela cells cultured with $Y_2O_2S:Eu^{3+}$ at the concentrations of 50 $\mu g \cdot ml^{-1}$, 40 $\mu g \cdot ml^{-1}$, 30 $\mu g \cdot ml^{-1}$, 20 $\mu g \cdot ml^{-1}$ and 10 $\mu g \cdot ml^{-1}$ was more than 84% for 48 h. MTT cytotoxicity assay at 24 h/48 h confirmed that $Y_2O_2S:Eu^{3+}$ had a smaller influence on the growth of Hela with good biological compatibility, and can be used as a potential biomarker.

REFERENCES

Bao H. et al. 2004. Chem. Mater 16: 3853–3859.

Cui C. et al. 2012. Synthesis and Luminescence Properties of Red Long Afterglow Phosphor Y2O2S:Eu3+, Mg2+, Ti4+ by Sol-Gel Process. Journal of the Chinese Ceramic Society 01: 40–49.

Deng S.Q. et al. 2003. Nanoscale La(OH)3 Hollow Spheres and Fine-Tuning of Its Outer Diameter and Cavity Size China. Journal of Materials Science & Technology [J]. Journal of Materials Science & Technology 107: 1886–1891.

Franciszek B. et al. 2003. X-ray fluorescence solution semi-microanalysis of the luminophoretype materials using scattered radiation and attenuation coefficients. Spectrochimica Acta 58: 1917–1925.

Gaponik N. et al. 2002. Phys. Chem. B 106: 7177–7185.

Han P.D. et al. 2012. Preparation and Properties of Upconversion Luminescent Materials of Y2O2S:Yb, Ho, [J]. Journal of the Chinese Ceramic Society 24: 145–153.

Jau-Ho Jean & Szu-Ming Yang. 2000. Y2O2S:Eu Red Phosphor Powders Coated with Silica [J]. J. Am. Ceram. Soc 839: 1928–1934.

Lü D. et al. 2013. State Key Laboratory of Cotton Biology, Key Laboratory of Plant Stress Biology, Henan University, Kaifeng 475004, China. ATHK1 acts downstream of hydrogen peroxide to mediate ABA signaling through regulation of calcium channel activity in Arabidopsis guard cells, [J]. Chinese Science Bulletin 58: 336–341.

Shao M.F. et al. 2011. Preparation of Fe3O4@SiO2@Layered Double Hydroxide Core–Shell Microspheres for Magnetic Separation of Protein [J]. American Chemical Society 134: 1071–1077.

Sondi I. et al. 2004. Journal of Colloid and Interface Science 275: 503–507.

Yuen S.W. et al. 2013. Asian superfine in Y2O2S: of Er3 +, Yb3 + phosphor prepared upconversion luminescence characteristics [J]. Journal of the Chinese Ceramic Society 50: 505–513.

Wang P.T. et al. 2008. Guard-cell signalling for hydrogen peroxide and abscisic acid 178: 703–718.

Zhang H. et al. 2003. Gene for gene alignment between the Brassica and Arabidopsis genomes by direct transcriptome mapping [J]. Phys. Chem.B. 107: 168–180.

Zheng L.J. et al. 2013. The Heating Effect of the Er3+/Yb3+ Doped Y2O3 Nanometer Powder by 980 nm Laser Diode Pumping, [J]. Spectroscopy and Spectral Analysis 01: 0151-04.

Enzyme assisted extraction of polysaccharides from *Flammulina velutipes* and its radicals scavenging activity

G.T. Chen, Y.R. Dong, G.H. Qi, Z.P. Yang, H.X. Wang & S.L. Wang
Department of Food Quality and Safety, China Pharmaceutical University, Nanjing, China

ABSTRACT: An efficient enzyme-assisted extraction procedure was developed and optimized for the extraction of polysaccharides from *Flammulina Velutipes* (FVPs). Based on the single-factor experiments, Response Surface Methodology (RSM) was used to estimate and optimize the experimental variables. The results showed that the optimal extraction conditions for the polysaccharides were extraction temperature 55.0 °C, extraction pH 5.24, and extraction time 3.4 h. Under these conditions, the experimental yield of polysaccharides was $8.60 \pm 0.22\%$, which was close with the predicted yield value (8.66%). In addition, the radicals scavenging activity of the FVPs were investigated *in vitro*. The results showed that FVPs possessed significant inhibitory effects on 1,1-Diphenyl-2-Picrylhydrazyl (DPPH) radical, hydroxyl radical and superoxide radical. These results suggested that enzyme-assisted extraction technology is a great extraction technology on the polysaccharides from *F. velutipes*, and FVPs could be a suitable natural antioxidant and may be the functional foods for humans.

1 INTRODUCTION

Flammulina velutipes is one of the most popular edible mushrooms, which has being under a large-scale artificial cultivation and increasingly consumed in China, Japan, and other Asian countries owing to its high nutritional values and attractive taste. Therefore, much attention has been paid on its active constituents and pharmacological properties. Recently, various active substances isolated from *F. velutipes* have been recorded in the literature, and polysaccharide was one of the important substances. Up to date, several polysaccharides have been isolated from *F. velutipes* fruit bodies or mycelium, and their pharmacological efficacies were investigated. The extraction methods of polysaccharides commonly including traditional hot-water extraction, Soxhlet extraction, and methods assisted by ultrasonic wave and microwave to improve the extraction efficacy (Sun, Liu, & Kennedy, 2010; Zhong et al., 2012); however, it is usually associated with longer extraction time, higher temperature, or high investment costs and energy requirements. Recently, enzyme-assisted extraction, which is considered as a mild, efficient and environmentally friendly method, has been employed to extract different compounds from plants, and has been proved to be effective in improving the yield of the target component (Zhang et al., 2011) compared to conventional extraction method. Thus, enzyme assisted extraction may be an effective and advisable technique for the extraction of polysaccharides. To the best of our knowledge, there is hardly any report that enzyme assisted extraction has been applied to extract polysaccharides from *F. velutipes*. In this study, the enzyme-assisted extraction parameters (ratio of liquid to solid, extraction temperature and time) of polysaccharides from *F. velutipes* was firstly investigated and optimized using a three-level, three-variable Box–Behnken Design (BBD). In addition, the scavenging effects on DPPH radical, hydroxyl radical and superoxide radical were used to evaluate the antioxidant activities of FVPs.

2 MATERIALS AND METHODS

2.1 Materials and reagents

The fruiting bodies of F. Velutipes were purchased from local supermarket (Nanjing, China). Samples were dried at 60 °C for 12 h and grinded by a FW-177 herbal medicine smashing machines (Tianjin Taisite Instrument Co. LTD, Tianjin, China) and were sieved through a 100 mesh sieve, the powder was stored at 4 °C until used.

Papain (500 U/mg), cellulase (15 U/mg) were obtained from Sinopharm Chemical Reagent Co., Ltd. (Shanghai, China). 1,1-Diphenyl-2-Picrylhydrazyl (DPPH) was purchased from Sigma Chemical Co. (St. Louis, MO, USA). All other chemicals used in this investigation were analytical grade and purchased from Shanghai Chemical Co. (Shanghai, China).

2.2 Preparation of polysaccharides

2.2.1 Enzyme extraction and determination of polysaccharides

The powder of F. Velutipes was refluxed in 80% ethanol for 6 h. Then the cooled extract was discarded and the residue was washed with 95% ethanol, anhydrous ethanol, acetone and diethyl ether respectively to remove impurities and small lipophilic molecules. The residue was dried at room temperature (25 ± 2 °C) for 24 h. Ten grams of the pretreated dried powder was immersed in 200 ml citric acid–sodium hydroxide–chlorhydric acid buffer in a 1000 ml beaker and the sample was then extracted with complex enzyme (the ratio of papain: cellulase was 1:1) at a given pH for a certain hours, while the temperature of the water bath was kept steady at a given temperature during the entire extraction process.

The suspension was centrifuged (4500 rpm, 10 min) and the insoluble residue was treated again for 2 times as mentioned above. The supernatant was incorporated and concentrated to one-fifth of the initial volume using a rotary vacuum evaporator at 65 °C. The supernatant was precipitated by the addition of anhydrous ethanol to a final concentration of 80% (v/v) and then incubated at 4 °C for 24 h. The precipitates as crude extract were collected by centrifugation. After being washed three times with anhydrous ethanol, the precipitate was air-dried at 50 °C until its weight was constant. The content of the polysaccharides was measured by the phenol-sulfuric acid method (Dubois et al., 1956) with D-glucose as a standard at 490 nm. The yield of polysaccharides was calculated by the following equation:

$$\text{Polysaccharide yield } (\%) = \frac{\text{polysaccharides content of extraction (g)}}{\text{weight of } F. \, velutipes \text{ powder (g)}} \times 100 \qquad (1)$$

2.2.2 Experimental design and statistical analysis

Single-factor-test was employed to determine the preliminary range of the extraction variables including complex enzyme amount, ratio of water to raw material, enzyme action temperature, pH and extraction time. Then, a three-level-three-factor Box–Behnken factorial Design (BBD) was used in this study. Enzyme action temperature (X_1), pH (X_2), and extraction time (X_3) were chosen for independent variables to be optimized for the extraction of FVPs. The coded and uncoded (actual) levels of the independent variables are presented in Table 1. Seventeen experiments were carried out in BBD (Table 2).

A quadratic polynomial model was fitted to correlate the response variable (yield of polysaccharide) to the independent variables. The general form of quadratic polynomial equation is as follows:

$$Y = \beta_0 + \sum_{i=3}^{3} \beta_i X_i + \sum_{i=3}^{3} \beta_{ii} X_i^2 + \sum_{i<j=1}^{3} \beta_{ij} X_i X_j \qquad (2)$$

where Y is the response variable, and β_0, β_i, β_{ii}, and β_{ij} are the regression coefficients for intercept, linearity, square, and interaction, respectively, while X_i and X_j are the independent variables.

Table 1.   Box-Behnken design of the levels of factors.

	Factor levels		
Independent variables	−1	0	+1
X_1: Temperature (°C)	50	55	60
X_2: pH	4	5	6
X_3: Time (h)	2	3	4

Table 2.   Response surface central composite design and response values for the yield of FVPs.

Run	X_1-Temperature (°C)	X_2-pH	X_3-Time (h)	Yield of FVPs (%)	
				Actual values	Predicted values
1	50	4.5	3	4.83	5.05
2	60	4.5	3	4.43	4.69
3	50	5.5	3	7.63	7.73
4	60	5.5	3	7.05	6.83
5	50	5	2	6.12	6.16
6	60	5	2	4.18	4.18
7	50	5	4	6.35	6.35
8	60	5	4	7.47	7.43
9	55	4.5	2	4.22	3.95
10	55	5.5	2	6.55	6.77
11	55	4.5	4	6.47	6.25
12	55	5.4	4	7.63	7.89
13	55	5	3	8.21	8.23
14	55	5	3	8.30	8.23
15	55	5	3	8.21	8.23
16	55	5	3	8.23	8.23
17	55	5	3	8.20	8.23

Data were expressed as means of three replicated determinations. Design Expert (Trial Version 7.0.3) was employed for experimental design, analysis of variance (ANOVA), and model building. SPSS 12.0 software was used for statistical calculations and correlation analysis. Values of $P < 0.05$ were considered to be statistically significant.

2.3   Radicals scavenging activity assay

2.3.1   Measurement of scavenging effect on DPPH radical
The scavenging effects of the FVPs on DPPH radical were measured by the method of Suda (2000).

2.3.2   Measurement of hydroxyl radical scavenging activity
The hydroxyl radical scavenging activities of FVPs were investigated by the method of Wang et al. (2008).

2.3.3   Measurement of superoxide radical scavenging activity
The superoxide radical scavenging activities were performed by the method of Liu et al. (1997).

2.3.4   Statistical analysis
All data were presented as mean ± SD of triplicate experiments (n = 3). Differences between groups were assessed using a t-test. P values <0.05 were considered statistically significant.

3 RESULTS AND DISCUSSION

3.1 Effect of ratio of water to raw material on extraction yield of FVPs

The yield of FVPs extracted by different ratio of water to raw material from 10 ml/g to 50 ml/g was shown in Figure 1(A). The enzyme amount, extraction time, pH and temperature were fixed at 2%, 3 h, 5.5, and 50 °C, respectively. It could be founded that the extraction yields of the polysaccharides increased from 6.80% to 7.23% with the ratio increasing from 10 ml/g to 20 ml/g. Then, it tended to decreased as the ratio of water to raw material increased. Therefore, extraction ratio 20 ml/g was favorable for polysaccharides production.

3.2 Effect of enzyme amount on extraction yield of FVPs

Extraction process was carried out using the enzyme amount 0.5, 1.0, 1.5, 2.0 and 2.5%, when other extraction parameters were as following: ratio of water to raw material 20 ml/g,

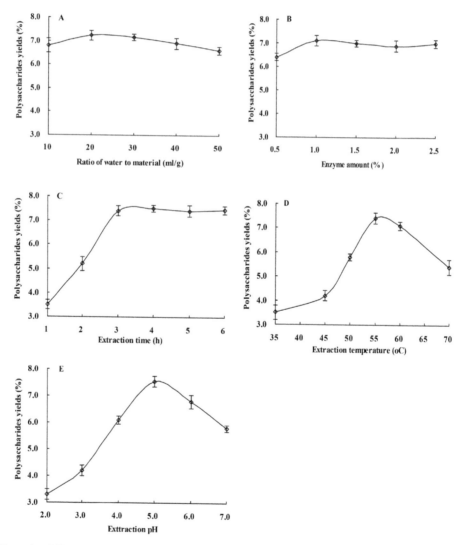

Figure 1. Effects of ratio of water to material (A), enzyme amount (B), extraction time (C), extraction temperature (D), and extraction pH (E) on the extraction yield of FVPs. The error bars indicate the standard deviation of 3 replicates.

extraction time 3 h, temperature 50 °C, and pH 5.5. It was found in Figure 1(B) that with increasing ultrasonic power from 0.5% to 1.0%, the traction yield of FVPs slowly increased from 6.4% to 7.1%. When the enzyme amount above 1%, the traction yield of FVPs no longer changed. This indicated that the complex enzyme amount of 1% was sufficient to obtain good polysaccharides yield. Thus, 1% was considered to be optimal complex enzyme amount in this experiment.

3.3 Effect of extraction time on extraction yield of FVPs

The effects of extraction time on FVPs yield are shown in Figure 1(C). The time was changed from 1 h to 6 h, while other extraction variables were set as follows: ratio of water to raw material 20 ml/g, enzyme amount 1%, temperature 50 °C, and pH 5.5. It could be found that the extraction yield increased as the extraction time ascended from 1 to 3 h, the maximum yield of polysaccharides (7.4%) was observed when the extraction time was 3 h, then the extraction yield of polysaccharides maintained a dynamic equilibrium with the increasing of the time. Therefore, extraction time of 3 h was adopted in the present work.

3.4 Effect of extraction temperature on extraction yield of FVPs

Extraction process was carried out by using different temperatures of 35, 45, 50, 55, 60 and 70 °C while the other extraction conditions were: complex enzyme amount 1%, extraction pH 5.0, and extraction time 3 h. As shown in Figure 1D the extraction yield of polysaccharides increased when temperature increased from 35 to 55 °C. The maximum yield (7.40%) of polysaccharides was observed when extraction temperature was 55 °C.

3.5 Effect of extraction pH on extraction yield of FVPs

The effect of pH on the extraction yield of polysaccharides is shown in Figure 1E. Extraction operation was carried out at different pH conditions while other extraction variables were set as follow: complex enzyme amount 1%, and extraction time 3 h, temperature 55 °C. The result showed that the extraction yield of polysaccharides continued to increase with the increase of pH value (2.0–5.0) and reached the peak value (7.55%) at pH value 5.0.

3.6 Optimization of extraction conditions of FVPs

3.6.1 Fitting the model and evaluation of the model predictability

Based on the results of single-factor-test, extraction temperature, pH, and time were adopted to research their effects on the yield of FVPs using RSM. The whole design consisted of seventeen experimental points and responses under different treatment conditions, including predicted values and actual values were shown in Table 2. The mathematical model representing the yield of polysaccharides as a function of the independent variables within the region under investigation was expressed as follows:

$$Y = 8.23 - 0.22X_1 + 1.11X_2 + 0.86X_3 - 0.045X_1X_2 + 0.77X_1X_3 - 0.29X_2X_3$$
$$- 1.22X_1X_1 - 1.03\ X_2X_2 - 0.98X_3X_3 \quad (3)$$

where Y is the yield of polysaccharides, and X_1, X_2, and X_3 represent extraction temperature, pH and time, respectively. Predicted response values for the yield of polysaccharides could be obtained using this quadratic polynomial equation in terms of independent variables values.

The statistical significance of the regression model was checked by F-test and p-value, and the analysis of variance (ANOVA) for the response surface quadratic model is shown in Table 3. The corresponding variables would be more significant if the F-value becomes greater and the P-value turns to smaller. So the F-value ($F = 57.30$) and P-value ($P < 0.0001$) implied the model was highly significant. The quadratic regression model showed the value of the determination coefficient (R^2) was 0.987, which indicated that 98.7% of the variations

223

Table 3. Analysis of variance for the fitted quadratic polynomial model of extraction of polysaccharides.

Source	SS	DF	MS	F-value	P-value Prob>F
X_1	0.40	1	0.40	5.91	0.0454
X_2	9.92	1	9.92	144.74	<0.0001
X_3	5.87	1	5.87	85.55	<0.0001
$X_1 X_2$	8.10×10^{-3}	1	8.10×10^{-3}	0.12	0.7411
$X_1 X_3$	2.34	1	2.34	34.14	0.0006
$X_2 X_3$	0.34	1	0.34	4.99	0.0606
X_1^2	6.23	1	6.23	90.85	<0.0001
X_2^2	4.46	1	4.46	65.00	<0.0001
X_3^2	4.07	1	4.07	59.43	0.0001
Model	35.36	9	3.93	57.30	<0.0001
Residual	0.48	7	0.069		
Pure error	6.60×10^{-3}	4	1.65×10^{-3}		
Cor total	35.84	16			
$R^2 = 0.987$		$R^2_{Adj} = 0.969$	CV = 3.90%		Adeq Precision = 21.29

could be illustrated by the fitted model. For a well statistical model, the adjusted determination coefficient (R^2_{Adj}) should be close to R^2. As shown in Table 3, R^2_{Adj} was 0.969, which implied that only 3.1% of the total variations were not explained by the model. A relatively low value of CV (the coefficient of variation) indicated a better reliability of the experiments values. In this study, CV = 3.90% showed the accuracy and the general availability of the polynomial model were adequate. It indicated a good degree of correlation between the actual values and the predicted values of FVPs yield. Adeq precision measures the signal to noise ratio, and a ratio greater than 4 is desirable (Zhu et al., 2010). An adequate ratio (Adeq precision = 21.29) of this fitted model indicated that it can be used to navigate the design space.

The significance of each coefficient was checked by F-test and P-value (Table 3). The p-value was used as a tool to check the significance of each coefficient, which in turn may indicate the pattern of the interactions between the variables. The smaller the value of P, the more significant is the corresponding coefficient. It can be seen from Table 3 that the linear coefficients (X_1, X_2, X_3), a quadratic term coefficient (X_1^2, X_2^2, X_3^2) and cross product coefficients ($X_1 X_3$) were significant, with very small P-values ($P < 0.05$). The other term coefficients were not significant ($P > 0.05$), which suggested that all the three independent variables significantly influenced the yield of FVSP.

3.6.2 The optimal conditions and validation of the model

By prediction of computing program, the optimal conditions for the highest yield of FVPs were as follows: extraction temperature 55.07 °C, pH 5.24, and extraction time 3.37 h. A predicted value of 8.66% was obtained for yield of polysaccharides under the optimal conditions. In order to ensure the predicted result was not bias the practical value, the optimal conditions were modified as follows: temperature 55.0 °C, pH 5.2, and time 3.4 h. The modified conditions were used to validate the suitability of the fitted model equation for accurately predicting the responses values. The results showed that the actual values of polysaccharides yield were 8.60 ± 0.22% under the modified conditions, which were in agreement with the predicted values significantly ($P > 0.05$). The results of analysis confirmed that the response model was adequate for reflecting the expected optimization, and the model of Eq. (3) was satisfactory and accurate.

Furthermore, FVPs was extracted with a traditional hot water extraction method (ratio of water to material of 20 ml/g, extracted in 90 °C water bath for 3 h) in the same time, and a yield of 6.37% was obtained, which was significantly less than that obtained by the enzyme extraction method. The results suggested that enzyme assisted extraction technology is a great extraction technology on the polysaccharides from F. velutipes.

3.7 Radical scavenging activities of FVPs

3.7.1 DPPH radical scavenging activity
It was found in Figure 2 (A) that FVPs exhibited notable DPPH radical-scavenging activity, and the DPPH radical scavenging effects were increased with increasing concentrations. At concentrations of 0.02–1.0 mg/ml, the scavenging abilities of FVPs on DPPH radicals were in the range of 1.24–83.0%. The results indicated that FVPs had a noticeable effect on scavenging DPPH free radicals, especially at high concentrations.

3.7.2 Hydroxyl radical scavenging activities
The results of hydroxyl radical scavenging activities of FVPs and ascorbic acid were given in Figure 2B. Results showed that ascorbic acid has a high level of hydroxyl radical scavenging effect. For the FVPs samples, the effects of scavenging hydroxyl radicals were in a concentration-dependent manner. The highest scavenging effect (86.3%) was obtained with a concentration of 1.0 mg/ml, but this effect was still lower than that of ascorbic acid (90.6%) with a concentration of 0.2 mg/ml. Despite all that, it also showed that the FVPs can be the effective scavenger of hydroxyl radical.

3.7.3 Superoxide radical scavenging activity
The scavenging effects of different concentration of FVPs and ascorbic acid on superoxide radical were tested and significantly exhibited in a concentration-dependent manner (Fig. 2C). For the FVPs samples, the peak value (89.5%) was observed at the concentration of 0.6 mg/ml, which was close to that of ascorbic acid (90.6%) at the concentration of 0.2 mg/ml. It indicates that FVPs is also an excellent superoxide radical scavenger.

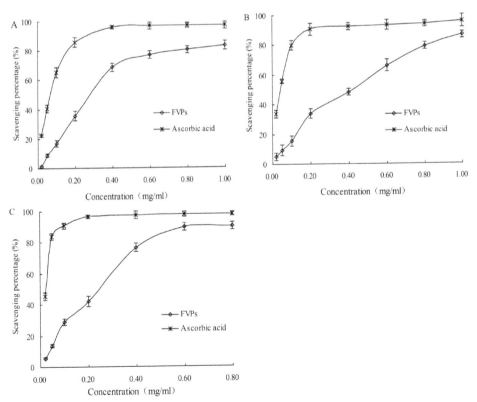

Figure 2. Scavenging effects of different concentration of FVPs and ascorbic acid on DPPH radical (A), hydroxyl radical (B), and superoxide radical (C). The error bars indicate the standard deviation of 3 replicates.

4 CONCLUSIONS

Based on the single-factor experiments, RSM was used to estimate and optimize the experimental variables: extraction temperature (°C), extraction pH, and extraction time (h). All the independent variables, quadratic of all the independent variables had highly significant effects on the response values ($P < 0.05$). The optimal extraction conditions for the polysaccharides were as follows: extraction temperature of 55.0 °C, extraction pH 5.24, and extraction time 3.4 h. Under these conditions, the experimental yield of polysaccharides was 8.60 ± 0.22%, which was close with the predicted yield value (8.66%).

The results from different *in vitro* assay systems demonstrated that the crude *F. Velutipes* polysaccharides has effective antioxidant activities. Hence, the *F. Velutipes* polysaccharides can be used as an easily accessible source of natural antioxidants, as a food supplement, or in the pharmaceutical and medical industries.

ACKNOWLEDGEMENTS

The authors gratefully acknowledged the financial supports by the National Natural Science Foundation of China (No. 31101255) and the Fundamental Research Funds for the Central Universities (JKQ2011006).

REFERENCES

Dubois, M., Gilles, K.A., Hamilton, J.K., Rebers, P.A. & Smith, F. 1956. Colorimetric method for determination of sugars and related substances. *Analytical Chemistry* 28: 350–356.

Jian-hua, Xie., Ming-yue, Shen., Ming-yong, Xie., Shao-ping, Nie., Yi, Chen., Chang, Li., Dan-fei, Huang. & Yuan-xing, Wang. 2012. Ultrasonic-assisted extraction, antimicrobial and antioxidant activities of Cyclocarya paliurus (Batal.) Iljinskaja polysaccharides. *Carbohydrate Polymers* 89: 177–184.

Suda, I. 2000. Antioxidative activity. Shinohara K, Suzuki T & Kaminogaw S(eds), *The methods of food functions analysis*: 218–220. Japan: Korin.

Sun, Y.X., Liu, J.C. & Kennedy, J.F. 2010. Application of response surface methodology for optimization of polysaccharides production parameters from the roots of Codonopsis pilosula by a central composite design. *Carbohydrate Polymers* 80: 949–953.

Wang, X., He, L. & Liu, B. 2008. Study on extraction, purification and scavenging activity to hydroxyl radicals of polysaccharides from leaves of Ilex kudincha C.J. Tseng. *Food Science* 29(6): 37–40.

Zhang, J., Jia, S., Liu, Y., Wu, S. & Ran J. 2011. Optimization of enzyme-assisted extraction of the Lycium barbarum polysaccharides using response surface methodology. *Carbohydrate Polymers* 86: 1089–1092.

Zhong, K., Lin, W., Wang, Q. & Zhou, S. 2012. Extraction and radicals scavenging activity of polysaccharides with microwave extraction from mung bean hulls. *International Journal of Biological Macromolecules* 51: 612–617.

Zhu, T., Heo, H.J. & Row, K.H. 2010. Optimization of crude polysaccharides extraction from Hizikia fusiformis using response surface methodology. *Carbohydrate Polymers* 82: 106–110.

Bioinformatics and Biomedical Engineering – Chou & Zhou (Eds)
© 2016 Taylor & Francis Group, London, ISBN 978-1-138-02784-8

Preparation of CdTe quantum dots in aqueous phase

Y. Hao

Department of Biomedical Engineering, China Jiliang University, Hangzhou, Zhejiang, China

ABSTRACT: CdTe Quantum Dots (QDs) were synthesized in aqueous phase when Mercaptopropionic Acid (MPA) acts as capping reagent, and $CdCl_2$ and NaHTe act as predecessor. The effects of reaction time, reaction temperature and pH on the property of QDs were studied. The samples were characterized by Ultraviolet-Visible (UV-Vis), Transmission Electron Microscopy (TEM) and fluorescence spectrum, respectively. The results of experiment show that CdTe QDs are spherical with good dispersibility, particle diameter is 5–8 nm, and yield is up to 27.3%.

1 INTRODUCTION

In the last decade, quantum dots have emerged as potential components suitable to be integrated in devices specially designed for sensing applications such as chemical sensing, biosensing or photo-detection. Their use in detecting and identifying of biological molecules has been demonstrated to be very promising due to the novel optical properties and functionalities derived from their high surface-to-volume ratio, controlled surface interactions, efficient nanoscale transduction mechanisms, and quantum confinement effects (Wang, B.B. et al. 2014); (Wang, Q.S. et al. 2011); (Wang, Q.S. et al. 2012); (Fu, X. et al. 2009); (Xue, X.H. et al. 2009); (Cao, Y.C. et al. 2006). Quantum dots offer attractive features that superior to classical fluorescent labels, including broad excitation spectra, narrow symmetric emission spectrum, large Stokes shift and high fluorescence quantum yield. Water-solubility and biocompatibility are critical factors when quantum dots are applied to biomedical field which depend on reaction mechanism of quantum dots. How to control the size, surface, dispersibility and microstructure in the process of growth is one of the most important research of quantum dots.

CdTe quantum dots were synthesized in aqueous phase with mercaptopropionic acid as stabilizer. The synthesis conditions were systematically optimized, which included reaction time, reaction temperature and pH. The physical, chemical and optical properties of the prepared sample were investigated by ultraviolet-visible, transmission electron microscopy and fluorescence spectrum.

2 EXPERIMENT

2.1 *Materials and instruments*

$NaBH_4$ (purity, 96%), $CdCl_2$ (analytical reagent) and Te (analytical reagent) were obtained from Sinopharm Chemical Reagent Co., Ltd., China. MPA (analytical reagent) was obtained from Sigma-Aldrich Co. LLC., USA. The other reagents used were analytical reagents and not purified before use. Deionized water was used in all experiments.

Testing instruments: ultraviolet spectrophotometer, CARY 100, was purchased from Varian, USA. Transmission Electron Microscope (TEM), JEM-1230EX, was purchased from JEOL, Japan. Spectrofluorophotometer, RF5301PC, was purchased from Shimadzu Corporation, Japan.

2.2 Preparation of NaHTe solution

$NaBH_4$ and Te with mass ratio 1:2 were added in conical flask, then deionized water was added. Conical flask was filled with N_2 gas to eliminate O_2 gas. Conical flask was sealed by rubber plug with syringe needle, which can discharge the gas produced in reaction. After 2 hours by magnetic stirring, NaHTe solution was obtained. The reaction function is shown as:

$$4NaBH_4 + 2Te + 7H_2O = 2NaHTe + Na\ B_4O_7\downarrow + 14H_2\uparrow$$

2.3 Preparation of CdTe quantum dots

$CdCl_2$ (200 mmol L^{-1}) and MPA (200 mmol L^{-1}) with mole ratio 1:1 were added in conical flask, then deionized water was added. NaOH (1 mol L^{-1}) was used to adjust pH to 12. Conical flask was sealed by rubber plug with syringe needle, which can discharge the gas produced in reaction. NaHTe solution was added with volume ratio 1:5. Conical flask was immersed in a constant temperature bath with 100 C for 1.5 hours, and kept away from light. Then cooled to room temperature, and CdTe quantum dots solution was obtained. The reaction function is shown as:

$$Cd^{2+} + MPA = Cd.MPA^{2+}$$
$$Cd.MPA^{2+} + Te^{2-} = CdTe.MPA$$

2.4 Purification of CdTe quantum dots

CdTe quantum dots solution and acetone were added in centrifuge tube. After centrifuge (<3000 rpm), supernatant was removed and precipitate of quantum dots was kept. Repeating 10 times, the purified CdTe quantum dots were obtained.

3 DISCUSSION

3.1 Effects of reaction condition

The core growth of CdTe quantum dots is affected by many factors, such as reaction time, temperature, pH, etc. The size of CdTe quantum dots, wavelength of peak emission and width of peak emission are also affected. Through the experiments, it was found that when reaction temperature is lower than 90°C, CdTe quantum dots could not be prepared; higher than 120°C, fluorescence intensity of CdTe quantum dots decreases evidently and defect quantum dots increase. When reaction pH is from 10 to 12, wavelength of peak emission emerges red shift, and fluorescence intensity of CdTe quantum dots increases first and then decreases. When reaction pH is 10.5, fluorescence intensity of CdTe quantum dots is maximum. When reaction time is from 90 min to 300 min, wavelength of peak emission emerges red shift from 537 nm to 623 nm and half width of peak emission is from 48 nm to 65 nm. By quantum size effect, peak of fluorescence emission spectrum is shift. Stokes shift is 90 nm. It indicates that CdTe quantum dots have good uniformity and dispersibility.

According to the optimization of reaction condition, reaction temperature 100°C and pH 10.5 were selected. When reaction time is 90 min, 140 min, 190 min, 240 min and 300 min, quantum dots with different color were obtained. Under the ultra-violet lamp with 395 nm wavelength, quantum dots are green, yellow, orange, red and purple, respectively.

3.2 Characterization of CdTe quantum dots

CdTe quantum dots were observed by TEM, as shown in Figure 1. Quantum dots are spherical with good dispersibility, and particle diameter is 5–8 nm.

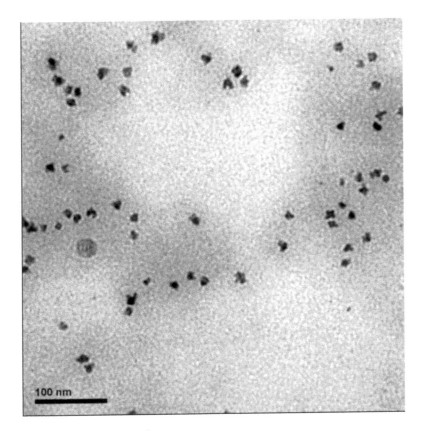

Figure 1.   TEM of CdTe quantum dots.

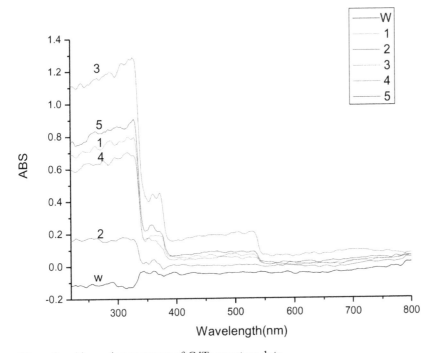

Figure 2.   Absorption spectrum of CdTe quantum dots.

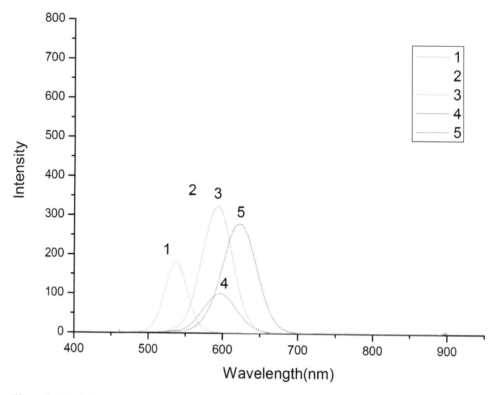

Figure 3. Emission spectrum and fluorescence intensity of CdTe quantum dots.

Ultraviolet spectrophotometer was used to measure absorption spectrum of CdTe quantum dots to confirm excitation wavelength. Scanning wavelength is 220–800 nm. Absorption spectrum graph is shown in Figure 2. The lines 1–5 are quantum dots with different color which are mentioned above. The line W is pure water as background.

Spectrofluorophotometer was used to measure emission spectrum and fluorescence intensity of CdTe quantum dots. According to absorption spectrum, excitation wavelength with 450 nm was selected. Emission spectrum and fluorescence intensity graph is shown in Figure 3. The lines 1–5 are quantum dots with different color which are mentioned above.

3.3 Fluorescent quantum yield

Rhodamine B was selected as standard substance in fluorescent quantum yield. Absorption and emission spectrum of rhodamine B solution (0.625 μg mL^{-1}) and CdTe solution (0.074 μmol mL^{-1}) were measured by ultraviolet spectrophotometer and spectrofluorophotometer. Excitation wavelength is 450 nm. And fluorescent quantum yield is 27.3%.

4 CONCLUSIONS

CdTe quantum dots are synthesized in aqueous phase with MPA as capping reagent, and CdCl$_2$ and NaHTe as predecessor. Reaction time, reaction temperature and pH have effect on the property of quantum dots. Under the condition of reaction temperature 100°C, pH 10.5, and reaction time 90 min, 140 min, 190 min, 240 min and 300 min, quantum dots with different color were prepared. The samples were characterized by ultraviolet-visible, transmission electron microscopy and fluorescence spectrum, respectively. The results of experiment show that CdTe quantum dots are spherical with good dispersibility, particle diameter is 5–8 nm, and yield is up to 27.3%.

ACKNOWLEDGMENTS

Research supported by Zhejiang Provincial Natural Science Foundation of China (No. LY12C10002).

REFERENCES

Cao, Y.C. Huang, Z.L. Liu, T.C. etc. 2006. Preparation of silica encapsulated quantum dot encoded beads for multiplex assay and its properties. Analytical Biochemistry 351(2):193–200.

Fu, X. Huang, K.L. Liu, S.Q. 2009. A robust and fast bacteria counting method using CdSe/ZnS core/shell quantum dots as labels. Journal of Microbiological Methods 79(3):367–370.

Wang, B.B. Huang, X. Ma, M.H. etc. 2014. A simple quantum dot-based fluoroimmunoassay method for selective capturing and rapid detection of Salmonella Enteritidis on eggs. Food Control 35(1): 26–32.

Wang, Q.S. Ye, F.Y. Fang, T.T. etc. 2011. Bovine serum albumin-directed synthesis of biocompatible CdSe quantum dots and bacteria labeling. Journal of Colloid and Interface Science 355(1):9–14.

Wang, Q.S. Li, S. Liu, P. etc. 2012. Bio-templated CdSe quantum dots green synthesis in the functional protein, lysozyme, and biological activity investigation. Materials Chemistry and Physics 137(2): 580–585.

Xue, X.H. Pan, J. Xie, H.M. etc. 2009. Fluorescence detection of total count of Escherichia coli and Staphylococcus aureus on water-soluble CdSe quantum dots coupled with bacteria. Talanta 77(5): 1808–1813.

Bioinformatics and Biomedical Engineering – Chou & Zhou (Eds)
© *2016 Taylor & Francis Group, London, ISBN 978-1-138-02784-8*

Optimization of the preparation of nalmefene hydrochloride injection

M.H. Duan, Q. Hao, Y.X. Sun, Z. Zhang, Z. Wang & Y. Yang
Jilin University, Changchun, Jilin, China

L.N. Wang & Y.H. Jiang
Changchun University of Chinese Medicine, Changchun, Jilin, China

C.L. Zhou
Training Corps one of the Aeronautical University, Changchun, Jilin, China

J. Pei
Jilin University, Changchun, Jilin, China

ABSTRACT: Nalmefene hydrochloride injection is not stable, and the formation of dimers may occur quickly, limiting the use of nalmefene hydrochloride in the clinic. Therefore, in this study, we aimed to optimize the preparation of nalmefene hydrochloride injection in order to increase the stability of the drug.

1 INTRODUCTION

Nalmefene hydrochloride is a specific neuroprotective agent and a first-line drug for the emergency management of respiratory depression (Gani D.U. et al. 2006). However, due to the poor stability of this drug (Murthy S.S. et al. 1996), the contents of two known impurities (naltrexone hydrochloride and bisnalmefene hydrochloride) regularly exceed quality standards. Therefore, nalmefene hydrochloride is no longer used clinically in the United States, where it was originally developed.

In this study, we designed a process for optimal production of nalmefene hydrochloride injection to ensure storage stability of the product.

2 MATERIALS AND METHODS

2.1 *Materials*

Nalmefene hydrochloride was purchased from Haikou (Hainan, China). Sodium chloride, activated carbon, and membranes were purchased from Changchun (Jilin, China).

2.2 *Formulation design*

According to commercial prescriptions, nalmefene hydrochloride was formulated as shown in Table 1.

2.3 *Choice of membrane*

First, we analyzed the appropriate membrane. For this, the nalmefene hydrochloride injection was prepared, and the content was detected by analysis of the main peak area. The sample

Table 1. Prescription of nalmefene hydrochloride injection.

Ingredient	Amount	Action	Implementation of standards
Nalmefene hydrochloride $(C_{21}H_{25}NO_3)$	0.1 mg	Main drug	Corporate internal control standards
Sodium chloride	9.0 mg	Accessory	Chinese Pharmacopoeia
Water for injection	1.0 mL	Solvent	Chinese Pharmacopoeia

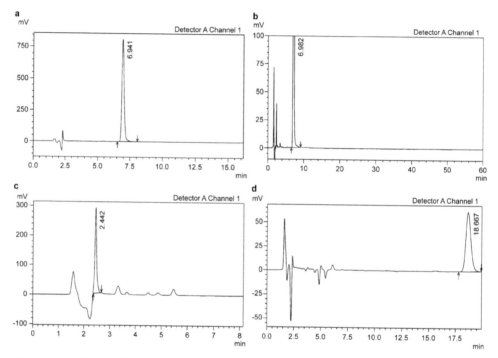

Figure 1. a: Content of nalmefene hydrochloride; b: Related substances of nalmefene hydrochloride; c: Naltrexone hydrochloride; d: Bisnalmefene hydrochloride.

was then filtered using a 0.45-μm membrane (Quality standard) and the content was analyzed again. Next, the sample was filtered using a 0.22-μm PES filter, and the content was detected. Absorption of the drug by the filter was determined by comparing the main peak areas of the unfiltered and filtered samples.

2.4 Choice of activated carbon content

For analysis of the appropriate activated carbon content, nalmefene hydrochloride injection was prepared as a 1.0 mg/mL solution. Different concentrations of activated carbon were added (0.05%, 0.075%, 0.1%, or 0.2% [v/w]), and the samples were incubated with stirring for 15 min. Samples were then filtered using the 0.45-μm membrane and 0.22-μm PES, and content was analyzed as described in the previous section, comparing the main peak areas of the original solution and the solution after addition of activated carbon (Quality standard). A content chromatogram is shown in Figure 1a.

2.5 Analysis of sample stability after exposure to air over time

We analyzed the effects of exposure to air over time using (Quality standard) the optimized conditions above, for up to 12 h. Peak detection and sample preparation were as

234

described above. Chromatograms for related substances, naltrexone, and dimers are shown in Figures 1b–d, respectively.

3 RESULTS

3.1 Choice of the appropriate membrane

Nalmefene hydrochloride injection was prepared under sterile conditions. Therefore, it was necessary to filter the solution to remove bacteria. According to the content determination method for nalmefene hydrochloride injection quality criteria (Yuan X. et al. 2010), the optimal membrane was chosen based on the information in Table 2.

As shown in Table 2, no adsorption of nalmefene hydrochloride was observed with the 0.45-μm membrane and 0.22-μm PES. Therefore, we chose the 0.45-μm membrane and 0.22-μm PES as the membrane for preparation of nalmefene hydrochloride.

3.2 Choice of activated carbon content

Activated carbon content was analyzed as shown in Table 3.

As shown above, when the content of activated carbon was 0.05% (v/w), the adsorption rate of nalmefene hydrochloride was low. Therefore, we chose an activated carbon of 0.05% (v/w).

3.3 Results of content analysis over time

Content of the nalmefene hydrochloride injection solution was assessed in air, as shown in Table 4.

As shown in the table, the reduction in AUC concentration was less than 2%, even at up to 12 h. Therefore, these data showed that nalmefene hydrochloride injection was stable for at least a 12-h preparation time.

3.4 Analysis of the contents of related substances over time

According to the related substances determination method in nalmefene hydrochloride injection quality criteria (Hou W. et al. 2012), we analyzed the presence of related substances, as shown in Table 5.

Table 2. Analysis of the appropriate membrane.

Membrane	AUC	Conc. (%)	Adsorption rate (%)
Reference	9934702	99.9	
Original solution	9925677	99.9	0.1
0.45 μm	9822471	98.87	1.13
0.45 μm + 0.22 μm (PES)	9828653	99.02	0.98

Table 3. Results of activated carbon content analysis.

	AUC	Adsorption rate (%)
Reference	10096785	
0.05%	7879160	21.96
0.075%	7699438	23.74
0.1%	6934646	31.32
0.2%	4556027	54.88

Table 4. Results of content analysis in the sample solution in air during preparation.

Time (h)	AUC	Conc. (%)	Change in the AUC ratio (%)
Reference	10043257		
0	10039881		
2	10037118	99.97	0.03
4	10056891	100.17	0.17
6	10010555	99.71	0.29
8	10089370	100.49	0.49
10	10048994	100.09	0.09
12	10026209	99.86	0.14

Table 5. Detection of related substances during preparation of the sample solution in air.

Time (h)	Naltrexone content (%)	Dimer content (%)	Content of the largest single impurity (%)	Total impurity content (%)
0	0.099	0.018	0.051	0.249
2	0.091	0.015	0.045	0.228
4	0.092	0.022	0.046	0.232
6	0.092	0.017	0.047	0.233
8	0.093	0.023	0.044	0.238
10	0.096	0.022	0.045	0.238
12	0.095	0.021	0.044	0.250

As shown in the above table, nalmefene hydrochloride injection was stable in air over a 12-h preparation time.

4 DISCUSSION

In this study, we examined factors affecting the stability of nalmefene hydrochloride injection. From this analysis, we developed an optimized protocol for production of this drug.

The stability of nalmefene hydrochloride injection is determined primarily by the dimer content (Pharmacopoeia 2010). Therefore, in the optimization process, the most important factor is reduction of dimer content. Importantly, temperature, but not light, affects the production of dimers. Therefore, it is also necessary to determine the optimal preparation temperature for the solution to prevent the formation of dimers. In this study, we were able to develop a protocol in which the dimer content remained low, even after 6 h at room temperature under lighted conditions.

Bacterial endotoxin content is used as an indicator of safety. Low bacterial endotoxin content is needed to ensure the safety of the solution. During production, activated carbon is used to adsorb bacterial endotoxin. However, it is also necessary to ensure the lowest possible level of activated carbon for optimal efficacy of the drug because activated carbon will adsorb the drug as well. Therefore, during sample preparation, it is necessary to increase the sample concentration to counteract the necessity for activated carbon (Li P. et al. 2007).

To study the filter material, the PES membrane is better. Use activated carbon to remove bacterial endotoxin, the concentration of 0.5% (V/W) is better. The product can be stable for at least 12 hours when the sample placed in the air.

In summary, we achieved improved stability of nalmefene hydrochloride injection without production of harmful metabolites using an optimized protocol. This method is suitable for producing nalmefene hydrochloride that is not sensitive to temperature or light. Further studies are needed to determine the mechanism of dimer formation.

REFERENCES

Gani D.U. & Yongliang G.A.O. & Shufang N.I.E. 2006. The permeation of nalmefene hydrochloride across different regions of ovine nasal mucosa. Chem Pharm Bull 54:1722–1724.

Hou W. 2012. Determination of the content of naltrexone hydrochloride and double nalmefene hydrochloride in nalmefene hydrochloride injection by HPLC. World Health Digest Medical Periodical 9:128–129.

Li P. & Chen X. & Dai X. & Wen A. & Zhang Y. & Zhong D. 2007. Application of a sensitive liquid chromatographic/tandem mass spectrometric method to pharmacokinetic study of nalmefene in humans. J Chromatogr B Analyt Technol Biomed Life Sci 852:479–484.

Murthy S.S. & Brittain H.G. 1996. Stability of revex, nalmefene hydrochloride injection, in injectable solutions. J Pharm Biomed Anal 15:221–226.

Nalmefene Hydrochloride injection YBH14222008 SFDA Quality standard.

Nalmefene Hydrochloride injection YBH07072010 SFDA Quality standard.

Nalmefene Hydrochloride injection YBH11242008 SFDA Quality standard.

The People's Republic of China Pharmacopoeia (2010 Edition) National Pharmacopoeia Committee.

Yuan X. & Pan W. & Wu L. 2010. Determination of related substances in nalmefene hydrochloride injection by HPLC. Chin J Pharma 41:208–210.

Bioinformatics and Biomedical Engineering – Chou & Zhou (Eds)
© 2016 Taylor & Francis Group, London, ISBN 978-1-138-02784-8

Development and validation of a method for detecting contaminants in nalmefene hydrochloride injections

M.H. Duan, Q. Hao, Y.X. Sun, Z. Zhang, Z. Wang & Y. Yang
Jilin University, Changchun, Jilin, China

Y.H. Jiang & L.N. Wang
Changchun University of Chinese Medicine, Changchun, Jilin, China

C.L. Zhou
Training Corps one of the Aeronautical University, Changchun, Jilin, China

J. Pei
Jilin University, Changchun, Jilin, China

ABSTRACT: Nalmefene hydrochloride injection is a first-line drug in emergency management to relieve respiratory depression, but its poor stability limits its clinical application. We aimed to develop and thoroughly validate a detection method for impurities in nalmefene hydrochloride injection and provide a foundation for studies of product stability.

1 INTRODUCTION

Nalmefene hydrochloride is a specific neuroprotective agent and a first-line drug for emergency management to relieve respiratory depression. Due to poor stability in the preparation and storage of this product, the contents of two known impurities (naltrexone hydrochloride and bisnalmefene hydrochloride) regularly exceeded quality standards; for this reason, nalmefene hydrochloride is no longer available in the United States, where it was originally developed. Studies have addressed the clinical application of nalmefene hydrochloride injection, but none have explored ways to improve product stability. The aim of this study was to develop an accurate method for the detection of contaminating substances and to accurately determine the content of two known impurities. This method could be used to guide the preparation and processing of nalmefene hydrochloride and accurately determine the presence of degradation intermediates. The method could also be used to precisely determine product shelf life and optimal storage conditions.

2 EXPERIMENTAL

2.1 Materials

Nalmefene hydrochloride injection was provided by Changchun Sanshun Pharmaceutical Co. (batch numbers 1203061, 1203071, and 1203081). All sample indexes for these batches were consistent with relevant regulations.

2.2 Durability test

According to guidelines (Gani Du et al. 2006) specific variables (Table 1) were used (Quality standard) to test for interferents (Quality standard) in the new and original (Quality standard)

Table 1. Durability variables.

Condition	Original mobile phase	Variation A	Variation B
Composition	Acetonitrile:phosphate buffer* = 20:80	Acetonitrile:phosphate buffer* = 18:82	Acetonitrile:phosphate buffer* = 22:78
Flow rate	1.0 mL/min	1.2 mL/min	0.8 mL/min
pH value	4.2	4.0	4.4
Chromatography column		Same brand, different batch	Same brand, different batch
Buffer solution	To 2 mL triethylamine, 7.8 g sodium dihydrogen phosphate was added in a final volume of 1 L, and adjusted to pH 4.2 ± 0.02 with 85% phosphoric acid	To 2 mL triethylamine, 8.0 g (or 7.6 g) sodium dihydrogen phosphate was added in a final volume of 1 L, and adjusted to pH 4.2 ± 0.02 with 85% phosphoric acid	To 1.8 mL (or 2.2 mL) triethylamine, 7.8 g sodium dihydrogen phosphate was added in a final volume of 1 mL, and adjusted to pH 4.2 ± 0.02 with 85% phosphoric acid

*Phosphate buffer was prepared by mixing 2 mL trimethylamine with 7.8 g sodium dihydrogen phosphate in a final volume of 1 L and adjusted to pH 4.2 ± 0.02 with 85% phosphoric acid.

Table 2. Validation of the durability test.

Experimental conditions	Nalmefene (%)	Bisnalmefene hydrochloride (%)	Most abundant single impurity (%)	Total impurities (%)	Δ main peak and impurity peak(s)	
Original conditions	0.091	0.008	0.057	0.278	3.424	9.576
Mobile phase ratio						
22:78	0.097	0.012	0.027	0.252	3.521	10.356
18:82	0.101	0.016	0.064	0.287	2.257	1.278
Flow rate						
0.8 mL/min	0.092	0.008	0.063	0.298	1.949	–
1.2 mL/min	0.088	0.013	0.052	0.249	2.031	9.045
Mobile phase pH						
4.0	0.092	0.012	0.055	0.267	1.968	6.066
4.4	0.087	0.012	0.057	0.264	1.963	8.889
Sodium dihydrogen phosphate						
7.6 g	0.093	0.011	0.055	0.262	3.484	3.412
8.0 g	0.093	0.009	0.057	0.267	3.496	10.328
Trimethylamine						
1.8 mL	0.087	0.017	0.052	0.297	3.635	10.372
2.2 mL	0.092	0.010	0.054	0.314	1.759	8.334
Column						
Elite	0.085	0.0012	0.053	0.235	5.167	7.264
Phenomenex	0.097	0.014	0.056	0.256	4.031	2.404

(Table 2). The results showed that the changes in each condition were consistent with predefined performance characteristics, indicating that with small variations, the system passed the suitability test and the detection method for related substances was effective.

Blanks were generated by testing 100 μl of each of the excipient and mobile phase. Naltrexone hydrochloride and bisnalmefene hydrochloride are two known impurities (Murthy SS et al. 1996) in nalmefene. Aliquots (100 μl) of each of these known impurities were injected into the HPLC system (Li P et al. 2007), and chromatograms were recorded. Results are shown in Figure 1a–d.

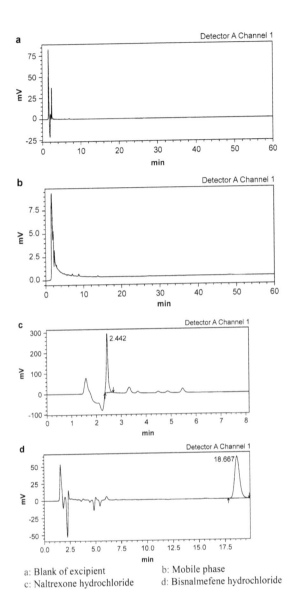

a: Blank of excipient
c: Naltrexone hydrochloride

b: Mobile phase
d: Bisnalmefene hydrochloride

Figure 1. Results

Table 3.   Solution stability.

Standing time (h)	Naltrexone hydrochloride (%)	Bisnalmefene hydrochloride (%)	The most abundant impurity (%)	Total impurities (%)
0	0.087	0.011	0.054	0.303
2	0.091	0.009	0.113	0.414
4	0.095	0.009	0.110	0.545
6	0.073	0.008	0.073	0.429
8	0.096	0.008	0.137	0.448
10	0.084	0.010	0.113	0.444
12	0.082	0.011	0.172	0.519
Result	No significant change	No significant change	No significant change	No significant change

2.3 Solution stability

To test solution stability (Quality standard), nalmefene hydrochloride (Yuan X et al. 2010) (batch number: 1203061) content was measured after the product was allowed to stand at room temperature for 0 h, 2 h, 4 h, 6 h, 8 h, 10 h, and 12 h (Table 3). Measurements of naltrexone hydrochloride solution (1.0 µg/mL) were taken after it was allowed to stand at room temperature for 0 h, 2 h, 4 h, 6 h, 8 h, 10 h, and 12 h (Table 4). Measurements of bisnalmefene hydrochloride solution (1.0 µg/mL) were taken after it was allowed to stand at room temperature for 0 h, 2 h, 4 h, 6 h, 8 h, 10 h, and 12 h (Table 5).

2.4 Precision

According to the proposed method (Hou W et al. 2012), six replicates of the product (batch number 1203061) were measured (Pharmacopoeia. 2010) by six different analysts. The results are shown in Table 6. Sampling precision was determined by injecting the same sample six times and calculating the RSD. The test sample was nalmefene hydrochloride solution (0.1 mg/mL; batch number: 1203061) (Table 7).

Table 4.   Naltrexone hydrochloride solution stability.

Standing time (h)	Peak area	RSD (%)
0	169535	1.07
2	172541	
4	169666	
6	173057	
8	172116	
10	171046	
12	168086	

Table 5.   Bisnalmefene hydrochloride stability test.

Standing time (h)	Peak area	RSD (%)
0	65530	0.79
2	64602	
4	64778	
6	66131	
8	65173	
10	65622	
12	65215	

Table 6.   Reproducibility.

No.	Naltrexone content (%)	Bisnalmefene hydrochloride content (%)	The single most abundant impurity (%)	Total impurities (%)
1	0.085	0.014	0.056	0.200
2	0.099	0.052	0.053	0.244
3	0.096	0.056	0.054	0.266
4	0.093	0.046	0.054	0.233
5	0.101	0.042	0.055	0.274
6	0.100	0.018	0.055	0.265
Max. Δ	0.016	0.042	0.003	0.074

Table 7. Sampling precision.

No.	Naltrexone content (%)	Cisnalmefene hydrochloride (%)	The single most abundant impurity (%)	Total impurities (%)
1	0.095	0.018	0.060	0.263
2	0.103	0.016	0.051	0.310
3	0.096	0.014	0.018	0.308
4	0.095	0.012	0.088	0.324
5	0.094	0.014	0.074	0.320
6	0.097	0.014	0.082	0.319
Max. Δ	0.009	0.006	0.037	0.016

3 INVESTIGATION AND RESULTS

3.1 *Durability test*

The results showed that the changes in each condition were within performance specifications, indicating that with small variations, the system passed the suitability test and the detection method for related substances was effective.

3.2 *Solution stability*

The results showed no significant changes in the content of naltrexone hydrochloride, bisnalmefene hydrochloride, the most abundant impurity, or total impurities. In addition, no new impurities appeared.

No significant changes in peak area were found in the test sample, and the RSD of peak areas within 12 h was 1.07%. Naltrexone hydrochloride solution may be allowed to stand at room temperature for 12 h.

No significant changes in the peak area were found in the test sample, and the RSD of peak areas within 12 h was 1.07%. Bisnalmefene hydrochloride solution may be allowed to stand at room temperature for 12 h.

3.3 *Precision*

Six different analysts performed the analysis and their results showed a maximum difference of less than 0.1%, thus demonstrating good reproducibility. Six-fold injection of the same sample yielded results in which the maximum difference was less than 0.1%. Thus, the method provided good sampling precision, feasible execution, and can be used to detect contaminating substances in nalmefene hydrochloride.

4 DISCUSSION

To validate the method for impurity detection, we must analyze the factors that lead to the generation of impurities; the results of such study will guide production and product storage. Impurity detection methods are specific to known impurities; the detection of unknown impurities requires further investigation. The quality of nalmefene hydrochloride has been investigated, but previous studies have not corrected for known impurities. Such studies cannot provide accurate determinations of impurity content and their results must be evaluated in this context. In the future, we plan to manufacture nalmefene hydrochloride injection, investigate the influence of each process step on product stability, optimize the process parameters, and control product impurity levels during production. We also plan to explore ways to ensure product stability during storage.

We validated a new contaminant detection method suitable for analysis of nalmefene hydrochloride injection. This validation was more thorough than that of prior similar studies of nalmefene hydrochloride injection, because it included a determination of the correction factors and durability testing. With two known impurities as controls, this method accurately determined the content of both known impurities with good reproducibility and stability. In addition, the method is accurate and feasible for laboratory implementation.

To validate the related substance includes: durability, solution stability and precision. Durability results show that the method applicability. Solution stability according to the results; Product into 12 hours as a solution mode, the product is still stable. Precision of testing results showed that transformation and conditions such as time, accurate and reliable detection results.

We validated a new contaminant detection method suitable for analysis of nalmefene hydrochloride injection. This validation was more thorough than that of prior similar studies of nalmefene hydrochloride injection, because it included a determination of the correction factors and durability testing. With two known impurities as controls, this method accurately determined the content of both known impurities with good reproducibility and stability. In addition, the method is accurate and feasible for laboratory implementation.

REFERENCES

Gani Du & Yongliang Gao & Shufang Nie. 2006. The permeation of nalmefene hydrochloride across different regions of ovine nasal mucosa. Chem Pharm Bull 54:1722–1724.

Hou W. 2012. Determination of the content of naltrexone hydrochloride and double nalmefene hydrochloride in nalmefene hydrochloride injection by HPLC. World Health Digest Medical Periodical 9:128–129.

Li P. & Chen X. & Dai X. & Wen A. & Zhang Y. & Zhong D. 2007. Application of a sensitive liquid chromatographic/tandem mass spectrometric method to pharmacokinetic study of nalmefene in humans. J Chromatogr B Analyt Technol Biomed Life Sci 852:479–484.

Murthy SS. & Brittain HG. 1996. Stability of revex, nalmefene hydrochloride injection, in injectable solutions. J Pharm Biomed Anal 15:221–226.

Nalmefene Hydrochloride injection YBH14222008 SFDA Quality standard.

Nalmefene Hydrochloride injection YBH07072010 SFDA Quality standard.

Nalmefene Hydrochloride injection YBH11242008 SFDA Quality standard.

The People's Republic of China Pharmacopoeia (2010 Edition) National Pharmacopoeia Committee.

Yuan X. & Pan W. & Wu L. 2010. Determination of related substances in nalmefene hydrochloride injection by HPLC. Chin J Pharma 41:208–210.

Bioinformatics and Biomedical Engineering – Chou & Zhou (Eds)
© 2016 Taylor & Francis Group, London, ISBN 978-1-138-02784-8

A new technology for the preparation of Deproteinized Extract of Calf Blood

H.Y. Li, L.N. Chen, G.X. Yuan, G.Y. Xu, Y. Sheng, P.G. Du & L.P. An
School of Pharmacy, Beihua University, Jilin, China

T.C. Li
Food and Drug Inspection Institution of Changchun City, Changchun, China

ABSTRACT: In this study, the technology process of DECB preparation was optimized, which was expected to improve the quality of the product, protect its biological activities and shorten its production cycle for the achievement of an industrial production of DECB. Under a sterile condition, the blood from calves at ages of 2–6-months was sterilized and filtrated by pressing calves; The filtrate was adjusted to different solutions at different pH values, which were ultra-filtrated to remove the proteins, adjusted their pH to the neutral, and then concentrated, sterilized and packaged; the breathing activity (QO_2) of it was measured with a Warburg micro breathing pressure meter, and the Stimulation Index (SI) was calculated; According to the pyrogenic reaction test described in Pharmacopoeia of Peoples republic of China, the pyrogenic reaction of samples was observed. The results showed that when the filtrate was adjusted to one at an acidic pH of 3 or an alkaline pH of 9, the breathing activity was strongest and the pyrogenic reaction was lowest. The results indicated that DECB prepared by the improved technology in this study should have a higher biological activity with less pyrogenic reaction, and the preparation method of DECB should be simple, easily operated and feasible.

Keywords: Deproteinized Extract of Calf Blood (DECB); preparation; breathing activity (QO_2); Stimulation Index (SI); pyrogenic reaction

1 INTRODUCTION

Deproteinated Extract of Calf Blood (DECB) is a small molecule with biological activities from calves at ages of 2–6 months by the deproteinization, and its primary components are amino acids, small peptides, nucleic acids and other compounds containing nitrogen. DECB is the precursor substance for the preparation of Deproteinated Calf Blood Extractives Injection and Deproteinated Calf Blood Extractives Injection was successfully developed by Jaeger KH in the 1960s (Obermaier Kusser B, 1989), followed by its production throughout the world. For decades, it has attracted a widespread attention as a drug used for the improvement of cellular energy metabolism due to its outstanding efficacy at home and abroad. The widely used Deproteinated Calf Blood Extractives for a long time Injection at home and abroad is Actovegin Injection made in Austria (Siegfried Hoyer, 1989) and Solcoseryl Injection made by Solcoseryl Pharmaceutical Factory in Swiss (Ding Yajun, 2002), which can significantly improve the activity of the reticuloendothelial system and some enzymes, accelerate the oxidative phosphorylation and ATP synthesis of cells, and make the anaerobic metabolism become the aerobic metabolism of glucose to promote an increased production of energy substances, so that they can prolong the survival of cells, and promote the metabolism of tissues and cells, functional recovery and tissue repair (Wang Xin, 2008). The drugs have showed a positive therapeutic effect on cerebral nerve injuries, stroke and cerebral infarction, as well as cornea repair, and improvement of memory, so that they have been widely used in clinic

(Pan Xiaofeng, 2006). However, organic solvents are used as the solvents (mostly ethanol and acetone) in the current production process of DECB to remove the macromolecular proteins, and the extracts obtained by this method are low in the activity and yield, and in addition show some adverse reactions. DECB made by domestic different manufacturers are quite different in their biological activities and pyrogens. In this study, two related issues were focused on, and the preparation process of DECB was optimized to improve its biological activity and quality standards, reduce its side effects, lower its costs, alleviate the pollution caused by the process, technically solve the low yield and long production cycle and avoid the higher loss of activity to achieve an industrial production of it (Guo Dongyu, 2007).

2   INSTRUMENTS AND REAGENTS

2.1   *Instruments*

Autoclave tank and vertical electric pressure steam sterilizer (LDZX-40 type); ceramic membranes, ultrafiltration membranes, reverse osmosis membrane, pH meter (mine magnetic PHS-3C) and Warburg micro breathing pressure meter (SKW-3) (Shanghai Shenan Medical Apparatus Factory); high performance liquid chromatograph, gel chromatographic column (TSK-G2000SWxl) weighing bottle, surgical instruments, dryer and homogenizer (Shanghai University).

2.2   *Reagents*

Blood of 2–6-month-old calf, hydrochloric acid, sodium hydroxide, sea sand, phosphorous pentoxide drying agent, 20% sulfosalicylic acid solution, sodium taurocholate, Evans blue, sodium chloride, disodium hydrogen phosphate and potassium dihydrogen phosphate.

2.3   *Experimental animals*

Clean-grade male guinea pigs weighing 250 ± 10 g were used in this experiment. After fasted for 24 hours, guinea pigs were sacrificed by striking their necks, then their livers were quickly removed and 5.5 g of the liver tissue were dissolved in 49 ml of Soerensen buffer for the homogenization.

3   EXPERIMENTAL METHODS

3.1   *Sample preparation*

Under sterile conditions, the blood from 2–6-month-old calves was continuously stirred with a surface-sterilized glass rod to make the fibrin attach to the sterile glass rod for removing the fibrin; the liquid was heated for the sterilization and then pressed. The supernatant obtained by pressing was filtrated through an inorganic membrane (7 μm ceramic membrane); the filtrate was adjusted into different solutions at pH 2.0, 3.0, 4.0, 5.0 and 6.0 with hydrochloric acid, respectively, the resulting solutions were adjusted into different solutions at pH 8.0, 8.5, 9.0, 9.5 and 10.0 with NaOH, respectively, and the adjusted solutions were filter with an ultrafiltration membrane (20,000 daltons); the filtrates were collected and adjusted their pH to neutral, and then were concentrated with a reverse osmosis membrane; the proteins in the concentrates were removed with 5000 molecular weight ultrafiltration, which were sterilized to be prepare the samples.

3.2   *Determination of DECB biological activities*

Based on the micro pressure measurement method, the capacity of DECB samples in promoting the breathing activity of guinea pig liver homogenates were determined. The DECB produced by Austria Nycomed Company (Actovegin) was used as the positive control, and changes in the pressure in the calibration tubes, blank tubes, sample tubes and positive tubes were measured. The breathing activities (QO_2) and Stimulation Indexes (SI) were calculated based on the related formulas.

The sample and reagents were put into a reaction flask that had been pre-washed and pre-dried, the reaction flask was connected with a corresponding piezometric tube, which was immersed in a bath water at a constant temperature of 37 ± 1 °C, and was shaken for 10 min with an oscillator to make the temperature inside and outside the reaction flask similar. The liquid level at the right side of piezometric tube was adjusted to 150 mm, then the reading at the left side (A) was recorded; the three-dimensional piston was closed for the reaction for 30 min, then the liquid level at the right side of the column was adjusted to 150 mm, and finally the reading at the left side was recorded (B). The three-dimensional piston was opened, the process described above was repeated for the next reaction for 30 min, and the liquid level readings at the left side initially (C) and after the 30 min (D) were recorded.

A set of temperature and pressure gauge for the calibration must be attached during the experiment. 2.5 ml of Soerensen buffer were add into the reaction flask, so that the experiment could be performed under the same experimental conditions. The change in the pressure (ΔC) was recorded to eliminate the effect of temperature and atmospheric pressure during the experimental process.

2 ml liver homogenate was put into a weighing bottle containing sea sand and dried to a constant weight at 110 °C, and the twice weight differences ΔW (mg) were counted.

$$\text{Calculation: } QO_2 \text{ (ul}O_2/mg \cdot h) = (AB\ CD) \times K1 - \Delta C \times K2]/(G \cdot T)$$
$$SI = \text{test sample } QO_2/\text{blank } QO_2$$

3.3 Pyrogen reaction test

The pyrogen reaction test was carried out according to the pyrogen test stipulated by Pharmacopoeia of People's Republic of China (Edition 2010). A rabbit was weighed and its normal body temperature was measured. A certain amount of the prepared DECB was given to the rabbit in ear vein injection. 30 min later, the body temperature of rabbit was measured again, and changes in the body temperature of rabbit was recorded.

4 RESULTS

4.1 Effects of PH values on the activity of DECB

The DECB sample filtrates were adjusted at different pH values and activities of DECB sample filtrates at different pH values were detected. The results showed that the active SI was 4.8 when the acidic pH value was at 3 or the alkaline pH was at 9, which was strongest (Table 1).

Table 1. Effects of pH values on the activity of DECB (Stimulation Index SI).

Acidic pH Alkaline pH	2.0	3.0	4.0	5.0	6.0
8.0	3.5	3.7	2.9	2.5	3.1
8.5	3.2	3.9	3.1	2.9	3.2
9.0	4.1	4.8	4.0	3.4	3.6
9.5	3.9	4.3	4.2	2.9	2.7
10.0	3.9	3.4	4.3	4.0	3.9

Table 2. Effects of DECB at different pH values on the pyrogenic reaction (°C).

Acidic pH Alkaline pH	2.0	3.0	4.0	5.0	6.0
8.0	0.31	0.28	0.31	0.22	0.25
8.5	0.23	0.20	0.39	0.35	0.28
9.0	0.22	0.18	0.18	0.26	0.31
9.5	0.25	0.23	0.19	0.25	0.32
10.0	0.29	0.31	0.23	0.29	0.33

4.2 Effects of DECB at different pH values on the pyrogenic reaction

The DECB sample filtrates were adjusted at different pH values and the effect of DECB sample filtrates on rabbit's pyrogen reaction at different pH values was detected. It was found that the changes in the rabbit's body temperature were minimum and the pyrogenic reaction results were optimal when the filtrate was adjusted to the acidic pH value at 3 or 4, or the alkaline pH at 9 (Table 2).

5 DISCUSSION

Deproteinated Calf Blood Extractives Injection has been widely used in clinical practice, but its preparation process is not so perfect. Currently, DECB is prepared primarily by the concentration and the ultrafiltration or dialysis of calf blood, but the current process is relatively rough, biological activities of the prepared DECB are low, its SI is about 2.5, and the pyrogenic reaction induced by it is commonly found. Pyrogenic reaction is a common infusion reaction in clinic, which may cause chills, shivering, fever, sweating, nausea, vomiting and other symptoms, even threaten lives when it is severe. In this study, on the basis of the traditional preparation process, the preparation technology of DECB was improved, the traditional chemical extraction method was replaced by a physical extraction to remove the proteins in the calf blood for the protection of biological activities of DECB. Moreover, multiple acidic pH and alkaline pH points were selected for the orthogonal experiment to seek a maximum yield. The results showed that when the acidic PH value was adjusted to 3 and the alkaline pH value of DECB solution was adjusted to 9, biological activities of DECB were strongest, its SI was 4.8, the change in the pyrogenic reaction was minimum, 0.18 °C, indicating that a strict limitation in pH values of DECB filtrate may optimize the preparation process to prepare a more active DECB with less pyrogenic reaction. And it is reported that the biological activities of Deproteinated Calf Blood Extractives Injections made through different technology processes by different manufacturers are different (Zhao Zongge, 2008).

Differences in the materials and the preparation process cause the differnces in the peptide content and the breathing activity of products from different manufacturers. Due to the higher content of active polypeptides in product D and E (Yu Z.S., 2006), their pharmacological effects are stronger. Some reports believed that polypeptides in Deproteinated Calf Blood Extractives Injection should be the active ingredients to exert its therapeutic effect (Yu Ru, 2000; Lv Yuan, 2006). The preparation process of DECB in this study should be simple and easily operated; the activity of DECB prepared in this study could be higher than that of products by the five manufacturers, with less pyrogenic reaction. However, the definitive active substances to play a therapeutic role in Deproteinated Calf Blood Extractives Injection are still unclear currently, which requires further studies.

ACKNOWLEDGMENT

We gratefully acknowledge the support from the special fund project of pharmaceutical industry development in Jilin province (YYZX20120), "The Twelfth Five Year Plan" science and technology research project of Education Department in Jilin province (2014-184) and the research project of health and family planning commission in Jilin province (20142078). Correspondence should be addressed to Liping An, School of Pharmacy, Beihua University, Jilin, Jilin, China, 132013, E-mail: 630114113@qq.com, Tel: +86 18604499109, and Peige Du, School of Pharmacy, Beihua University, Jilin, Jilin, China, 132013, E-mail: dupeige2001@126.com, Tel: +86 18604498601.

REFERENCES

Ding Yajun, Wang Ying, Wang Hongxin. 2002. Basic research and clinical evaluation of Adegold. *Drugs use evaluation and analysis of Chinese hospital* 2 (1):15–18.

Guo Dongyu, Zhao Hongmei, Lu Huan. 2007. A Deproteinized Hemoderivative of Calf Blood Injection and its preparing method. *Journal of management in Chinese medicine* 15 (2):110–112.

Lv Yuan, Quan Juxiang. 2006. Deproteinized calf serum injection's protective effect on hypoxia and ischemia of brain cell and clinical application. *Journal of Chinese clinical pharmacy* 22 (2):141.

Obermaier Kusser B, MBhlbacher C, Mushack J. et al. 1989. Further evidence for a two-step model of glucose-transport regulation. *Biochem J* 261:6991.

Pan Xiaofeng, Lin Congli, ye Xiao. 2006. New clinical application of Actovegin. *Journal of pharmaceutical practice* 24 (5):275–277.

Siegfried Hoyer, Karen. 1989. Betz1Elimination of the delayed postischemic energy deficit in cerebral cortex and hippocampus of aged rats with a dried, deproteinized blood extract (Actovegin). *Arch Gerontol Geriatr* 9:1811.

Wang Xin, Ju Yang, Luo Qin et al. 2008. The anti hypoxia effect on mice of total glucosides of Acanthopanax giraldii Harms and its mechanism. *Journal of Lanzhou University (Medicine Edition)* 34 (4):41–43.

Yu Ru, Xiao Jianying, Liu Yongzhang et al. 2000. The biological activity study of PCMA. *Journal of Chinese new drugs* 9 (2):961.

Yu Z.S., Johnston K.P., Williams R.O. 2006. Spray freezing into liquid versus spray-freeze drying: Influence of atomization on protein aggregation and biological activity. *Eur J Pharm* 27 (1):9–181.

Zhao Zongge, Wang Zunwen, Xu Kangsen et al. 2008. Quality analysis of Deproteinized extract of calf blood, of fetal calf serum injection and its antihypoxia effect. *Journal of pharmaceutical analysis* 28 (10):1637–1640.

Bioinformatics and Biomedical Engineering – Chou & Zhou (Eds)
© *2016 Taylor & Francis Group, London, ISBN 978-1-138-02784-8*

Preparation of mouse amelotin antibody by synthetic peptides

J.J. Zhang & Y. Sun
School of Stomotalogy, Weifang Medical University, Weifang, Shandong Province, China

ABSTRACT: Amelotin is a new kind of enamel protein gene discovered recently, and it is associated with enamel mineralization process. In order to further study the function of amelotin, we prepare the amelotin polypeptide antibody by synthetic specific peptides coupled with the carrier protein KLH. The amelotin antibody titer reached l:1,000,000, which was higher than that reported previously. The immunohistochemical analysis revealed that amelotin was detected in enamel full-thickness of 3 d and 7 d mice, which was similar to our previous findings. The anti-amelotin peptide antibody obtained through immunizing rabbits was of high titer and specificity, which was helpful for further research of amelotin.

1 INTRODUCTION

Amelotin is a kind of specific expression gene in the human ameloblast, which is highly conservative in species such as mice, rats and pigs, with a molecular weight of 21,000–38,000 (Trueb et al., 2007). The mouse amelotin full-length cDNA sequence is 1,022 bp, encoding 213 amino acids (Moffatt et al., 2006), rich in leucine, proline and threonine residues, and highly homologous to human amelotin (Iwasaki et al., 2005). Amelotin family members in the signal peptide sequence at 5 'end consist of 16 amino acids (Lacruz et al., 2012), so they may be a kind of secreted protein. Research has found that amelotin is involved in the development and maturation of the enamel (Somogyi-Ganss et al., 2012), and the change in its structure and function is closely related to the development of enamel diseases (Gasse, Silvent, & Sire, 2012). In order to further explore the effects of amelotin on enamel development and maturation process, we developed the amelotin polypeptide antibody. Based on the amino acid sequence of the mouse amelotin cDNA code, we applied correlation protein analysis software to design peptide fragments, and immunized rabbits after coupling with the carrier protein KLH to prepare the amelotin polypeptide antibody. Therefore, this provided a useful tool to study the molecular function and proteomics of amelotin.

2 MATERIALS AND METHODS

2.1 *Materials*

Complete Freund's Adjuvant (CFA), Incomplete Freund's Adjuvant (IFA) and Acrylamide were purchased from American Sigma company. Bovine Serum Albumin (BSA), Poly lysine, HRP-labeled goat anti rabbit IgG, and ABC Kit and DAB Kit for immunohistochemical detection were purchased from Beijing Zhong Shan Jinqiao Biological Technology. The RIPA lysate kit and the BCA protein assay kit were bought from Shanghai Biyuntian company. The PVDF membrane was obtained from Roche company. The ECL Plus chemiluminescence kit was purchased from Santa Cruz company. Kunming mice aged 3 d and 7 d and 2, 2 New Zealand rabbits (3 kg each) were provided by the Experimental Animal Center of Weifang Medical University.

2.2 Methods

2.2.1 Analysis of the amelotin antigen

We obtained the amino acid sequence of the amelotin protein from the GenBank protein database. Then, we used the online protein analysis tool to acquire a comprehensive forecast about hydrophilicity, surface probability, flexibility, antigenicity and two levels of structure parameters, in order to determine the required synthesis of polypeptide sequence.

2.2.2 Mouse anti-amelotin polypeptide antibody preparation

Amelotin peptide was synthesized by Wuhan Mitaka Biological Co. Ltd. The synthesis of amelotin polypeptide chains was crosslinked with Keyhole Limpet Hemocyanin (KLH) by the glutaraldehyde connection method to format amelotin-KLH conjugates, which was a high-molecular-weight compound used as a semi-antigen to immunize the animals. Briefly, 2 mg (1,000 µL) of peptide KLH conjugates and the same volume of the CFA hybrid were mixed and fully emulsified. A total of 4 injection points were selected to inject on the back of New Zealand rabbits, each point was injected at a volume of 250 µL. The first strengthened immunization, which was a half dose of the peptide antigen emulsified with the incomplete Freund's adjuvant injected on the back, was carried out after 4 weeks. The immunity was strengthened once every two weeks. The antiserum titer was determined by sampling the rabbit's ear border vein. After achieving the desired titer, amelotin-KLH antigen without the adjuvant was injected at a volume of 250 µL per point to strengthen the immunization once. After 3 d, blood samples were collected from the rabbit carotid artery, serum was separated, sterilized for shipment, and stored in the laboratory at −80 °C for further use.

2.2.3 Purification of amelotin polypeptide antibody

Peptide antibodies were purified by using affinity chromatography. A sample of 1 mg amelotin polypeptide was coupled with Cyanogen Bromide (CNBr) Sepharose 4B after activation to prepare peptide affinity chromatography gel. A solution of 20 mL rabbit serum against amelotin and the 1.6 mL polypeptide coupled gel were thoroughly mixed for 6 hours at 4 °C and then joined the column. The balanced gel was washed with PBS, and the glycine buffer was eluted. Protein concentrations were measured by the BCA kit to quantify the protein, and SDS–PAGE Coomassie Blue staining was used to determine the purification effect.

2.2.4 Determination of antibody titer by indirect ELISA

A 96-microtiter plate was coated with 1 mg/L of KLH polypeptide conjugates (100 µL/hole), at 4 °C overnight. Then, 5 g/L of BSA (160 µL/hole) were added and closed for 2 h at 37 °C. The diluted purified polypeptide antibody, rabbit antiserum or non-immune serum (100 µL/hole) was added and incubated for 1 h at 37 °C, added HRP Goat anti rabbit IgG (1:3000 dilution), colored for 15 min by TMB color liquid in the dark at room temperature, terminated color in 2 mol/L sulfuric acid, and determined the OD value of the samples at 450 nm.

2.2.5 Identification of amelotin polypeptide antibody specificity by Western blot

Total protein was isolated from ameloblast by using the RIPA lysate kit, the mouse amelotin fusion protein in *Escherichia coli* (stored in our laboratory), containing active amelotin mouse ameloblastin lysate and KLH were suspended in the 2 × SDS sample buffer, heated for 7 min at 95 °C, and cooled on ice for 10 min. Proteins were separated by SDS–PAGE, and bound to the PVDF membrane. The samples were incubated for 2 h at room temperature after addition of purified polypeptide antibody at 1:1,000 dilution. After addition of 1:5,000 HRP goat anti rabbit IgG, the samples were incubated for 1 h at room temperature. The proteins were detected by using the ECL Plus chemiluminescence kit for 5 min, and exposed to X-ray light film in the chamber.

2.2.6 Application of amelotin polypeptide antibody

The mandibular organization of 3 d and 7 d Kunming mice was placed in 400 mL/L formaldehyde, fixed for 16 h at 4 °C, demineralized for 6 d at 4 °C by 100 g/L EDTA solution, and made paraffin sections. Tissue sections were dewaxed in water, and then the sections were put in sodium citrate (pH 6) at 95 °C for 40 min in order to restore the antigen. Endogenous peroxidase activity was removed by incubating the sections with 50 mL/L of H_2O_2/methanol for 15 min, 1:50 normal goat serum was added for 30 min at 37 °C, purified rabbit anti mouse amelotin polypeptide antibody was added at 1:50 dilution, incubated overnight at 4 °C, detected immunohistochemically by using the ABC kit, colored by the DAB kit, mounted on neutral gum, and observed under a microscope.

3 RESULTS

3.1 *Analysis of the amino acid sequence of amelotin*

We used protein online tools to analyze the amino acid sequence of the amelotin protein, in order to determine the synthesized polypeptide sequence. The 14 amino acids at position 197–210 (ATHTTEGTTIDPPN) of amelotin were determined according to the prediction results of comprehensive research on hydrophilicity, flexibility, two-level structure, antigenicity and surface properties of the protein.

3.2 *Determination of amelotin antiserum titer and purification of polypeptide antibody*

The amelotin peptide KLH conjugates were used as an antigen to immunize New Zealand white rabbits to obtain rabbit anti mouse antiserum. The purified amelotin antiserum was displayed by the SDS-PAGE band single. The concentrations of the polypeptide antibody acquired by immunizing two rabbits were 300 g/mL and 450 mg/mL, which were determined through the BCA protein concentration determination kit. The high concentration of the antibody was selected for the next experiment. The indirect ELISA method showed that the titer of antiserum against amelotin could reach 1:100,000, and the purified amelotin polypeptide antibody titer was 1:1,000,000 (Fig. 1).

3.3 *The identification of amelotin polypeptide antibody specificity*

The Western blot showed the interaction between the polypeptide antibody and the *Escherichia coli* amelotin fusion expression protein (1–2 bands): the appearance of a clear zone, matched with the fusion expression protein molecular weight; 3–5 bands showed the interaction with

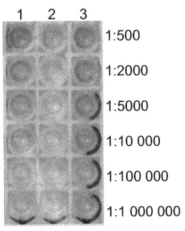

1: Rabbit anti mouse amelotin antiserum; 2: Negative control; 3: Purified amelotin polypeptide antibody.

Figure 1.   The results of indirect ELISA for the detection of antibody titer obtained from rabbit anti mouse amelotin antisera and purified antibody.

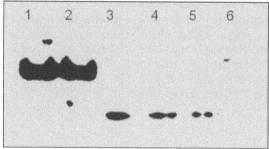

1–2: Interaction of polypeptide antibody and *Escherichia coli* amelotin fusion protein; 3–5: Interaction of polypeptide antibody and ameloblast protein lysate; 6:KLH.

Figure 2. Analysis of antisera specificity by Western blot.

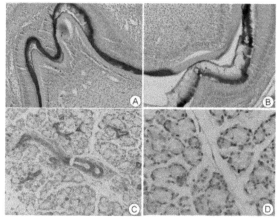

A, B: Strong positive signal of amelotin was observed in the 3 d and 7 d mouse molar enamel full layer; C: Amelotin expressed in the 7 d mouse submandibular gland duct epithelial cell cytoplasm; D: In the 3 d mouse submandibular gland was not found amelotin positive signal.

Figure 3. 3 d and 7 d mouse mandibular tissue immunohistochemical staining results.

active amelotin ameloblast protein lysate, consistent with the activity of amelotin; no band appeared for the interaction with KLH (Fig. 2). This showed that the prepared antibody can not only identify the semi-antigen epitope protein, but also combine with the natural amelotin protein.

3.4 *Detection of amelotin expression in mouse mandibular tissue by the polypeptide antibody*

The purified amelotin polypeptide antibody was examined in 3 d and 7 d mouse mandibular tissues by immunohistochemical staining. The results showed strong positive signals in the molar enamel full-thickness of 3 d and 7 d mice (Fig. 3 A, B). In addition, amelotin expression was also observed in the 7 d mouse submandibular gland duct epithelial cell cytoplasm, but the acinar cells showed a negative response. However, the 3 d mouse submandibular glands showed no amelotin expression (Fig. 3 C, D).

4 DISCUSSION

Amelotin is a new kind of enamel protein gene discovered recently. It is located in the q13.3 region on human chromosome 4 and closely linked with enamelin and ameloblastin

(Sawada et al., 2014; Crivelini et al., 2012). There is no obvious amelotin expression in the secretory stage of the enamel. The expression of amelotin gradually increases with enamel maturation in ameloblasts, and this protein may be associated with the enamel mineralization process (Gao et al., 2010; Gadhia et al., 2012). In order to study the function of amelotin, we prepared the amelotin polypeptide antibody.

Using the ELISA and Western blot experiments, we found that the prepared amelotin polypeptide KLH had good immunogenicity, antiserum with high titer, could reach 1:1,000,000, and could meet the applications of the subsequent experiment. Immunohistochemical experiment results by using prepared polypeptide antibody showed high expression of amelotin in the molar enamel of 3 d and 7 d mice. In addition, the expression of amelotin was observed in the 7 d mouse submandibular gland duct epithelial cell cytoplasm, while there was no expression in the 3 d mouse. As a specific protein participating in the developmental maturation of the enamel, expressed in glands, and the expression level showing temporal and spatial specificity, it remains to be known whether amelotin is also associated with the anomaly of gland differentiation and development. The above experiments proved that we successfully designed and synthesized amelotin polypeptide fragment, the prepared polypeptide antibody was of high titer, good specificity, and could meet the requirements. The successful preparation of amelotin synthesis peptides and antibodies will be a good basis to study the molecular function and proteomics of amelotin.

5 CONCLUSIONS

The prepared polypeptide amelotin antibody is of high titer and good specificity. Amelotin is highly expressed in the molar enamel of 3 d and 7 d mice. It can also be observed in the 7 d mouse submandibular gland duct epithelial cell cytoplasm.

ACKNOWLEDGMENT

This research was financially supported by the Natural science foundation of China (no. 81441107), the Natural science foundation of Shandong province (no. ZR2012HQ036), the Science and technology project of Shandong Province (no. J12LL51), and the Campus project Weifang Medical University (no. K1302015).

REFERENCES

Crivelini, M.M. et al. 2012. Expression of odontogenic ameloblast-associated protein, amelotin, ameloblastin, and amelogenin in odontogenic tumors: immunohistochemical analysis and pathogenetic considerations. *J Oral Pathol Med* 41(3):272–280.
Gasse, B., Silvent, J., & Sire, J.Y. 2012. Evolutionary analysis suggests that AMTN is enamel-specific and a candidate for AI. *J Dent Res* 91(11):1085–1089.
Gao, Y. et al. 2010. Distribution of amelotin in mouse tooth development. *Anat Rec (Hoboken)* 293(1):135–140.
Gadhia, K. et al. 2012. Amelogenesis imperfecta: an introduction. *Br Dent J* 212(8):377–379.
Iwasaki, K. et al. 2005. Amelotin—a Novel Secreted, Ameloblast-specific Protein. *J Dent Res* 84(12):1127–1132.
Lacruz, R.S. et al. 2012. Targeted overexpression of amelotin disrupts the microstructure of dental enamel. *PLoS One* 7(4):e35200.
Moffatt, P. et al. 2006. Cloning of rat amelotin and localization of the protein to the basal lamina of maturation stageameloblasts and junctional epithelium. *Biochem J* 399(1):37–46.
Somogyi-Ganss, E. et al. 2012. Comparative temporospatial expression profiling of murine amelotin protein during amelogenesis. *Cells Tissues Organs* 195(6):535–549.
Sawada, T. et al. 2014. Expression and localization of laminin 5, laminin 10, type IV collagen, and amelotin in adult murine gingiva. *J Mol Histol* 45(3):293–302.
Trueb, B. et al. 2007. Expression of phosphoproteins and amelotin in teeth. *Int J Mol Med* 19(1):49–54.

Bioinformatics and Biomedical Engineering – Chou & Zhou (Eds)
© *2016 Taylor & Francis Group, London, ISBN 978-1-138-02784-8*

Optimization of protein depletion technology for polysaccharides from *Angelica* and *Astragalus* by orthogonal test design

X.Y. Pu, X.L. Ma, L. Liu, J. Ren, X.Y. Li & H.B. Li
College of Life Science and Engineering, Lanzhou University of Technology, Lanzhou, Gansu, China

ABSTRACT: In this study, we developed and optimized an efficient enzyme-assisted protein depletion procedure for polysaccharides from *Angelica* and *Astragalus*. The Percentage of the Removed Protein (PRP) served as an evaluation index to compare the deproteinization effects of the Sevag method, TCA method and enzyme-assisted method. Processing parameters for enzyme-assisted deproteinization conditions were optimized using four factors at three levels of orthogonal design based on single-factor experiments. The results showed that the optimal conditions were as follows: papain concentration of 0.3%, deproteinization time of 120 min, enzyme action temperature of 55°C and pH of 5.0. Under these conditions, the PRP was 75.4 ± 0.10%, which is basically in accord with the model predicted value of 75.0%.

1 INTRODUCTION

The roots of *Angelica* and *Astragalus* are a herb, widely distributed in the northwest part of the People's Republic of China, and used as a sedative or a tonic agent (Ru, 1983). Polysaccharides are the main chemical composition, which have biological activities such as anti-tumor, anti-inflammatory, anti-oxidative and immunity function (Yang et al., 2003; Cai et al., 2007). Pu reported that a combination of polysaccharides from *Angelica* and *Astragalus in vitro* and *in vivo* exhibits a range of antioxidant and anti-aging activities (Pu & Li et al., 2011a; 2011b; 2012). Polysaccharides extracted from wild growing plants that contain a large amount of protein will exhibit changes in their biological activity and lower pharmacological effects (Ou, Huang, & Cao, 2012). Therefore, deproteinization is a very important step in the extraction of polysaccharides. The common deproteinization methods are the Sevag method, the TCA method and the enzyme-assisted method (Liu, Zhu, 2008). In the present study, these three methods were compared to remove protein from polysaccharides. However, reports on the deproteinization of *Angelica* and *Astragalus* (AAP) are still lacking. We evaluate the Percentages of the Removed Protein (PRP) and optimize the best deproteinization method. To our knowledge, this is the first report on enzyme-assisted deproteinization, which is applied to extract polysaccharides from *Angelica* and *Astragalus* (AAP).

2 EXPERIMENTAL

2.1 Materials and chemicals

The roots of *Angelica* and *Astragalus* were purchased from Minxian Shunfa Medicinal Material Company (Minxian City, China). Papain (6000 U/mg) was obtained from Nanning Pangbo Biotechnology Co. Ethanol, trichloroacetic acid, chloroform and n-butanol were obtained from Tianjin Reagent Co. (Tianjin, China). All other reagents were of analytical grade.

2.2 Extraction procedure

To obtain a fine powder, 360 g roots of *Angelica* and *Astragalus* (1:5, w/w) were ground in a blender. The sample was extracted with 80% ethanol for 2 h at 70–80°C to remove lipophilic

Table 1. Factors and levels of orthogonal experimental design.

Level	(A) Concentration (%)	(B) Temperature (°C)	(C) Time (min)	(D) pH
1	0.2	50	60	5.0
2	0.3	55	90	6.0
3	0.4	60	120	7.0

molecules. The degreased powders were dried and extracted twice with distilled water (4000 ml) for 2h at 95°C. The aqueous extract was centrifuged at 4500 rpm for 20 min to remove the pellet. The supernatant was concentrated to 20 ml by a rotary evaporator (RE-52CS, Yarong Biochemical Equipment Co., Shanghai, China) at 60°C under vacuum. Then, it was precipitated in 80% (v/v) ethanol by adding dehydrated ethanol and kept overnight at 4°C. The precipitate was collected at 4500 rpm for 10 min, washed successively with anhydrous ethanol and acetone, respectively, to obtain the crude AAP. Finally, the protein was estimated in the AAP using the method of Bradford, and BSA was used as a standard reagent (Bradford, 1976).

2.3 Sevag method and TCA method

The crude AAP was dissolved in distilled water. The proteins in the solute were removed by Sevag reagent (chloroform:n-butanol 3:1,4:1,5:1,6:1,7:1). After removal of the Sevag reagent, the supernatant was used to measure the protein content. Simultaneously, the proteins in the solute were removed by TCA (3 mol/L trichloroacetic acid 1, 2, 3, 4 and 5 mL) and then kept overnight at 4°C to compare the efficiency of deproteinization. Then, the supernatant was centrifuged at 4000 rpm to determine the protein content.

2.4 Orthogonal test design to optimize deproteinization technology

On the basis of the single-factor test, an orthogonal $L_9(3)^4$ test design was used for optimization the deproteinization parameters. In this study, the effect of four single factors, including pH value (5.0, 6.0 and 7.0), papain concentration (0.2%, 0.3% and 0.4%), enzyme action temperature (50°C, 55°C and 60°C) and deproteinization time (60 min, 90 min and 120 min), on the deproteinization of AAP were investigated (see Table 1).

3 RESULTS AND DISCUSSION

3.1 Effect of different proportions of Sevag reagent on the PRP of AAP

In this work, the effects of different proportions (chloroform:n-butanol) on the deproteinization of AAP were studied, and the results are shown in Figure 1A. As shown in Figure 1, the PRP continued to increase and reach the peak value (67.9%) when chloroform:n-butanol was 5:1. The PRP of AAP started to decrease after chloroform:n-butanol exceeded the proportion of 5:1.

3.2 Effect of different consumption of TCA on the PRP of AAP

In our work, the effects of different consumption of TCA on the deproteinization of AAP were investigated, and the results are shown in Figure 1B. As shown in Figure 1B, when 3 mol/L TCA was 2 ml, the PRP reached the peak value (54.5%), so the best consumption of trichloroacetic acid was 2 ml.

3.3 Effect of different enzyme concentrations on the PRP of AAP

The effect of different enzyme concentrations on the PRP of AAP is shown in Figure 1C. First, papain concentration was set at 0.1%, 0.2%, 0.3%, 0.4% and 0.5%, while other

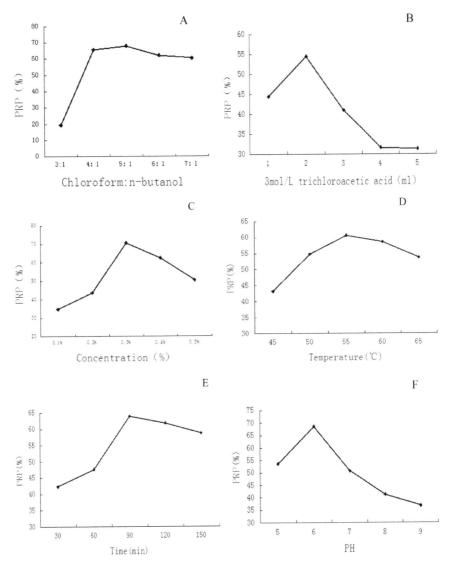

Figure 1. Effect of different results on the PRP of AAP based on single-factor experiments. (A) The Sevag method; (B) the TCA method; (C) the enzyme concentration; (D) the enzyme temperature; (E) the enzyme time; and (F) the enzyme pH.

deproteinization parameters were as follows: deproteinization time of 120 min, enzyme action temperature of 55°C and pH of 5.0. With increasing papain concentrations from 0.1% to 0.3%, the PRP of AAP increased from low to high until reaching a maximum concentration at 0.3%, and then dropped as the action proceeded. This indicated that the enzyme concentration of 0.3% was sufficient to obtain a good deproteinization effect. Thus, 0.3% was considered to be the optimal enzyme concentration in this experiment.

3.4   Effect of different enzyme temperatures on the PRP of AAP

In this work, the effects of enzyme temperature on the deproteinization of AAP were investigated, and the results are shown in Figure 1D. As shown in Figure 1D, the PRP continued to increase with the deproteinization temperature and reached the peak value (60.6%) when the deproteinization temperature was 55°C. The PRP of polysaccharides from *Angelica* and *Astragalus* started to decrease after the deproteinization temperature exceeded 55°C. Thus, 55°C was considered to be the optimal enzyme temperature in this experiment.

3.5 Effect of different enzyme times on the PRP of AAP

In our work, the effects of enzyme time on the deproteinization of AAP were investigated, and the results are shown in Figure 1E. First, a deproteinization time was set at 30 min, 60 min, 90 min, 120 min and 150 min, while other deproteinization parameters were as follows: 0.3% papain concentration, enzyme action temperature of 55°C and pH of 5.0. With increasing enzyme times of raw material from 30 min to 90 min, the PRP of AAP first increased and then dropped. The maximum value was 63.9% when the deproteinization time was 90 min. Thus, 90 min was considered to be the optimal enzyme time in this experiment.

3.6 Effect of different enzyme pH values on the PRP of AAP

The effect of different enzyme pH values on the PRP of AAP is shown in Figure 1F. First, deproteinization pH was set at 5.0, 6.0, 7.0, 8.0 and 9.0, while other deproteinization parameters were as follows: 0.3% papain concentration, deproteinization time of 120 min and enzyme action temperature of 55°C. With increasing papain concentrations from 5.0 to 7.0, the PRP of AAP increased from low to high until reaching a maximum value at 6.0, and then dropped. Thus, 6.0 was considered to be the optimal enzyme pH in this experiment.

3.7 Optimization experimental design for the deproteinization parameters of AAP

In fact, papain concentration, deproteinization time, enzyme action temperature and pH were generally considered to be the most important factors. In our study, an optimization of suitable conditions in the deproteinization of polysaccharides was carried out. All selected factors were examined using an orthogonal $L_9(3)^4$ test design through single-factor experiments. We analyzed the total evaluation index by the statistical method. In Table 2, the results of the experiments indicated that the maximum PRP of the crude extract was 75.4%. However, we cannot select the best deproteinization conditions based on only these outcomes in Table 2, and warranted a further orthogonal analysis. Thus, we calculated the K and R values and listed in Table 2. From Table 2, we can find that the degree of influence on the PRP of AAP decreased in the order: D > C > B > A according to the R values. The deproteinization pH

Table 2. Analysis of orthogonal experimental design $L_9(3)^4$ results.

NO.	(A) Concentration (%)	(B) Temperature (°C)	(C) Time (min)	(D) pH	PRP (%)
1	1	1	1	1	73.1
2	1	2	2	2	70.6
3	1	3	3	3	68.9
4	2	1	2	3	68.9
5	2	2	3	1	75.4
6	2	3	1	2	69.1
7	3	1	3	2	72.2
8	3	2	1	3	63.3
9	3	3	2	1	74.3
K1	70.86	71.40	68.50	74.26	
K2	70.76	69.40	71.63	70.63	
K3	70.30	71.13	71.80	67.03	
R	0.567	2.000	3.300	7.234	

R refers to the result of extreme analysis.

was found to be the most important determinant of the PRP. In general, the maximum PRP of AAP was obtained when the optimal conditions were as follows: 0.3% papain concentration, deproteinization time of 120 min, enzyme action temperature of 55°C and pH of 5.0, respectively.

4   CONCLUSION

Our preliminary data demonstrated that enzyme-assisted deproteinization was a green and efficient technique, which could be used to improve the Percentage of Removed Protein (PRP) of AAP by comparing the deproteinization effects of the Sevag method and TCA method. Next, the papain concentration, deproteinization time, enzyme action temperature and pH were chosen for optimization based on single-factor experiments. The results showed that the further optimal conditions for the enzyme-assisted deproteinization of AAP were as follows: 0.3% papain concentration, deproteinization time of 120 min, enzyme action temperature of 55°C and pH of 5.0 by the orthogonal test design. Under the optimal conditions, the PRP was $75.4 \pm 0.10\%$, which agreed closely with the predicted value. In conclusion, these results indicated that the enzyme-assisted method was the best way to remove the protein, and that the proposed method was superior to the Sevag method or the TCA method.

ACKNOWLEDGMENTS

This research was supported by research grants from the National Natural Science Foundation of China (NO 81260070) and the Project of Science and Technology of Lanzhou (NO 2011-1-71).

REFERENCES

Bradford M.M. 1976. A rapid and sensitive method for the quantitation of microgram quantities of protein utilizing the principle of protein–dye binding. Anal Biochem 72: 248–254.
Cai L. & Zhu J. 2007. Research Status and Development of Astragalus Polysaccharide, Chinese Herbal Medicine (34): 896–900.
Liu Y.J. & Zhu H.Y. 2008. Optimization of De-protein Process for Polysaccharide of Hippophae rhamnoides. Journal Of Henan Agricultural Sciences: 84–87.
Ou W., Huang R.B. & Cao X. 2012. Study on Deproteinization from Polysaccharide Extract in Shepherd's-purse. Technology& Development of Chemical Industry (41): 7–10.
Pu X.Y., Li Y. & Wang P. 2011a. Study on Anti-oxidation of Guiqi Polysaccharides in Vitro. China Food Industry (193): 64–66.
Pu X.Y., Li Y. & Zhou L.Y. 2011b. Deferring senile effect of polysaccharides from angelica and astragalus on aging mice. Human Health and Biomedical Engineering (HHBE): 289–292.
Pu X.Y., Li Y. & Zhang W.J. 2012. Study on Anti-aging Effect of Guiqi Polysaccharides. Natural Product Research and Development (11), 1630–1633.
Ru D. 1983. The national assembly of Chinese herbal medicine. Beijing People Press: 151.
Yang T.H. & Lu B.H. 2003. Immunoregulation effect of Angelica polysaccharide isolated from Angelica sinensis. Chinese Pharmacological Bulletin (19): 448–451.

Biomedical devices and systems

Bioinformatics and Biomedical Engineering – Chou & Zhou (Eds)
© 2016 Taylor & Francis Group, London, ISBN 978-1-138-02784-8

Detection of C-Reactive Protein based on a Dynabeads-labeled sandwich immunoassay by using a GMI biosensor

Z. Yang, C. Lei & Y. Zhou
Key Laboratory for Thin Film and Microfabrication of the Ministry of Education,
Department of Micro/Nano Electronics, School of Electronic Information and Electrical Engineering,
Shanghai Jiao Tong University, Shanghai, China

ABSTRACT: A Dynabead-labeled sandwich immunoassay had been developed using a Giant Magnetoimpedance (GMI) biosensor for the detection of C-Reactive Protein (CRP). The biosensor involving the sensing elements of Cr/Cu/NiFe/Cu/NiFe was fabricated by Micro Electro-Mechanical System (MEMS) technology. The immune reaction of biomarkers CRP was accomplished on a separated Au film substrate surface with a self-assembled layer. The fundamental principle for detection of CRP based on GMI biosensor was that Dynabeads were employed as magnetic labels of CRP, and CRP can be monitored by detecting the fringe field of Dynabeads using magnetic sensing elements. We observed that the GMI ratio were significantly enhanced due to the presence of CRP combined with Dynabeads. The rise decreased as the CRP concentration increased. A lower detectable concentration of 10 ng/ml was achieved in present work. The GMI-based biosensor provides a new method to rapid and sensitive detection CRP, which has a large potential for bio-application.

1 INTRODUCTION

C-Reactive Protein (CRP) is an acute-phase serum protein, which can react with pneumococcal C polysaccharide body and form compounds. It was discovered firstly in 1930 by Tillett and Francis at the Rockefeller University (Tillett et al. 1930). Nowadays, the CRP is routinely checked in blood counts and other medical diagnostics because of its relevance as a significant biomarker for infections and inflammatory processes in human blood serum (Mygind et al. 2011; Pai et al. 2008). CRP has also evolved from being a postulated biomarker that can possibly predict cardiovascular events and mortality to a proven direct participant in the pathogenesis of atherosclerotic Cardiovascular Disease (CVD) (Nagai et al. 2011). Detection and quantification of CRP in an easy, cheap, and fast way are important in the diagnosis of cardiovascular diseases. Immunomagnetic detection or magnetically labeled immunoassay has recently become a focus of interest for researchers because of its high sensitivity, versatile diagnostic methods, convenient processes and high accuracy.

The Giant Magneto-Impedance (GMI) effect is a large change in the ac impedance of a ferromagnetic conductor at high frequency (usually >0.1 MHz) subject to a dc magnetic field. The GMI-based magnetic sensors have several advantages, such as smaller size, quick response, high sensitivity, high stability and lower cost. Recently, the GMI sensors have been introduced into the field of biosensing as a biosensor prototype (Kurlyandskaya et al. 2003; Chen et al. 2011; Wang et al. 2014) in order to develop a new generation of bioanalytical system.

In this paper, a Giant Magneto-Impedance (GMI)-based biosensor was developed for detection of CRP with different concentration (10–100 ng/ml) labeled by Dynabeads. It is significant to establish a new method for detection of biomarker of cardiovascular disease, which provides the basis of early warning and diagnosis system for the research on the new major diseases. Besides, this method paves the way for future development on a cardiac panel electrochemical point-of-care diagnostic device.

2 MATERIALS AND METHODS

2.1 Chemical and reagents

Human CRP antigen, mouse CRP monoclonal antibody and biotinylated mouse CRP polyclonal antibody were purchased from Linc-Bio Science Co. Ltd. (Shanghai, China). 11-mercaptoundecanoic acid (11-MUA) was purchased from Aladdin Chemistry Co. Ltd (USA). 1-ethyl-3-[3-dimet-hylaminopropyl] carbodiimide (EDC) hydrochloride was purchased from Aladdin Chemistry Co. Ltd (USA). N-Hydroxysuccinimide (NHS) was purchased from Medpep (Shanghai Medpep Co. Ltd.). Phosphate buffer tablets (PBS PH 7.4) were purchased from Medicago AB (Uppsala, Sweden). Albumin from Bovine Serum (BSA) was purchased from Via-gene pro bio Technologies Co. Ltd. (Shanghai, China). Deionized water was used to prepare phosphate buffered solution.

2.2 Design and fabrication of GMI sensor

The GMI sensor was designed for meander shape as seen in Figure 1. The length of the straight line segment in the tortuous structure was 5 mm. The total turn number was 10. The widths of NiFe and Cu film were 140 and 100 μm respectively. The space between neighbor segments was 60 μm. The thicknesses of NiFe and Cu film were about 3 and 2 μm, respectively. The area of electrode was 2 mm × 3 mm. The total area of sensing elements was 5 mm × 3.94 mm.

The manufacturing process had been reported elsewhere (Wang et al. 2014a). It can be depicted briefly as follows: a) the 100 nm thick Cr/Cu seed layer was deposited on a glass substrate by radio frequency magnetron sputtering. b) Photo etching. c) The bottom NiFe layer was coated by electrodeposition. d) The Cu layer was accomplish with similar process e) the top NiFe layer was got. f) The seed layer was removed by reactive ion etching.

2.3 Preparation of CRP samples under test

In this section, gold substrates were prepared by deposition of gold film (300 nm) on a glass wafer by sputtering. After that, a photoresist layer was spun on the Au layer and patterned to several rectangles with dimensions of 5 × 3 mm, then the uncovered part of the Au layer was removed by reactive ion etching in the mixed solution (KI: I_2: H_2O − 4 g: 2 g: 10 ml). The wafer was sliced into several chips each of which had one Au film on it. And then, the Au films were bathed in 1 mol/l NaOH solution and 1 mol/l HCl in turn for 10 min, the gold substrates were rinsed with deionized water and alcohol, at last it was dried using a stream of nitrogen gas.

The detail of the fabrication of the biomarker samples was similar to previous report (Lei et al. 2013). The process can be depicted briefly as follows: 1. the gold film was modified with

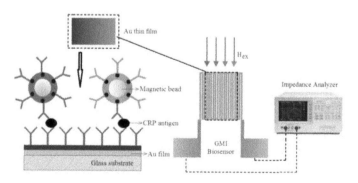

Figure 1. Graphical illustration of the test setup.

30 ml 11-mercaptoundecanoic acid at the room temperature for 3 h. 2. The gold film was activated with EDC and NHS for 30 min. 3. CRP monoclonal antibody was immobilized on the gold film. 4. The gold film was sealed with 100 uL BSA solution at 4 C for 2 h. 5. CRP antigens were combined with CRP monoclonal antibodies. 6. Biotinylated CRP polyclonal antibody was immobilized on the gold film. 7. 20 μl 10 μg/ml Dynabeads were conjugated to CRP polyclonal antibody by streptavidin-biotin system. The process of specifically capturing and labeling of CRP was showed in Figure 1.

2.4 *Detection of CRP based GMI measuring method*

The detection of CRP was based on a chemical conjugation of the magnetic beads and CRP. The basic principle on which the CRP detection based was that Dynabeads were employed as magnetic labels of CRP, and CRP can be monitored by detecting the fringe field of the Dynabeads using magnetic sensing elements. Different concentration CRP samples (10–100 ng/ml) were placed on the surface of GMI sensor respectively. The GMI responses were measured by an impedance analyzer (E4991 A). An external magnetic field (H_{ex}) of 0–120 Oe was applied along the longitudinal direction of the sample in order to induce strong changes in the skin depth. The relative change in impedance (GMI ratio) was defined as:

$$\text{GMI (\%)} = 100\% \times \frac{Z(H) - Z(H_0)}{Z(H_0)} \tag{1}$$

where $Z(H)$ and $Z(H_0)$ are the magnetoimpedance with and without magnetic field respectively. The test setup was shown in Figure 1.

3 RESULTS AND DISCUSSION

Scanning Electron Microscopy (SEM) was used to observe the CRP-conjugated Dynabeads (10 ng/ml) as seen in Figure 2, and we can find that Dynabeads on the surface of Au film were nearly saturated. Other concentrations of CRP-conjugated Dynabeads were also observed and not showed here. High concentration Dynabeads had cause the high-density clusters of Dynabeads. The SEM observations confirmed that Dynabeads-labeled CRP was immobilized on the Au film, and the numbers of Dynabeads increased with increasing CRP concentration.

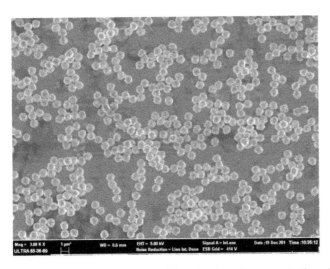

Figure 2.   SEM characterizations CRP-conjugated Dynabeads at the concentration of 10 ng/ml.

The GMI ratio of the fabricated sensor was studied, and the maximum GMI ratio was 195.2%. From a practical point of view, the sensitivity was estimated and the maximum was 21.7%/Oe at 2.4 MHz. Figure 3 showed the GMI ratios of biosensors without and with different concentrations CRP antigens obtained at frequency f = 2.4 MHz. Evidently, the GMI ratio increased first and then decreased as H_{ex} increased. This can be explained in terms of magnetization rotation model (Panina et al. 1994): the rotational magnetic permeability related to the GMI effect was first increased and then decreased with increasing Hex, the maximum permeability was achieved as $H_{ex} = H_k$. H_k was the anisotropy field.

The GMI ratios had risen in varying degrees due to the presence of CRP samples with different concentrations near to GMI sensor. It was worthwhile to note that the rise decreased with increasing the concentration of CRP. Because high concentration Dynabeads had cause the high-density clusters of Dynabeads, which led to a reduction in fringe field. The rises were 16.53%, 12.3%, 11.22%, 9.62% and 6.5% when the 10 ng/ml, 20 ng/ml, 40 ng/ml, 80 ng/ml and 100 ng/ml CRP were used respectively as seen in Figure 4. The GMI-based biosensor in this work possesses lower minimum detectable concentration of 10 ng/ml. This can be attributed to the unique advantage of the GMI sensor, namely, the frequency-sensitive inductance contributes to magnetoimpedance. Especially the mutual inductance of flexural sandwich structure was ultrasensitive at high frequencies.

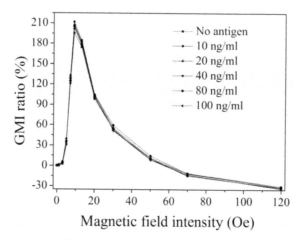

Figure 3.   Field dependence of GMI responses at f = 2.4 MHz under the different concentration of CRP.

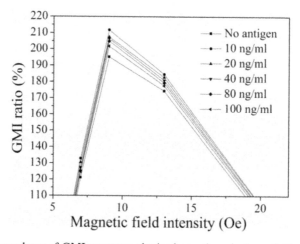

Figure 4.   Field dependence of GMI responses obtained near the anisotropy field.

268

In our early report (Wang et al. 2014b), the GMI ratio was improved owing to the presence of superparamagnetic beads on the surface of the sensor, and it was found that high field sensitivity in detection of magnetic beads can be obtained near H_k, the present result was in agreement with it. Many similar researches were also reported (Kurlyandskaya et al. 2003; Devkota et al. 2013) previously, and related theories were put forward to explain the phenomenon. Kurlyandskaya suggested that the presence of the Dynabeads may change the superposition of the constant applied field and the alternating field, thereby changing the magnetic charge distribution near the surface of the sensing element, the GMI effect was thus enhanced.

4 CONCLUSIONS

In this work, a GMI-based biosensor was developed for detection of CRP. The multilayered Cr/Cu/NiFe/Cu/NiFe GMI sensor was fabricated by MEMS technology. The immune reaction of biomarkers CRP was accomplished on a separated Au film substrate surface with a self-assembled layer. The results showed that the presence of CRP antigens on the biosensors improved the GMI effect owing to the induced magnetic dipole of superparamagnetic Dynabeads, and the GMI ratios showed distinctive changes at high frequency. Detection of CRP antigens using the GMI-based biosensor was fully realized, and a low detection limit (10 ng/ml) was achieved at f = 2.4 MHz. It is significant to establish a new method for detection of biomarker of cardiovascular disease, which provides the basis of early warning and diagnosis system for the research on the new major diseases.

REFERENCES

Chen, L. & Bao, C.C. 2011. A prototype of giant magnetoimpedance-based biosensing system for targeted detection of gastric cancer cells. Biosens. Bioelectron. 26: 3246–3253.

Devkota, J. & Ruiz, A. 2013. Magneto-impedance biosensor with enhanced sensitivity for highly sensitive detection of nanomag-D beads. IEEE Trans. Magn. 49: 4060.

Kurlyandskaya, G.V. & Sanchez M.L. 2003. Giant-magnetoimpedance-based sensitive element as a model for biosensors. Appl. Phys. Lett. 82: 3053–3055.

Lei, J. & Lei, C. 2013. Detection of targeted carcinoembryonic antigens using a micro-fluxgate-based bio sensor. Appl. Phys. Lett, 103: 203705.

Mygind, N.D. & Harutyunyan M.J. 2011. The influence of statin treatment on the inflammatory biomarkers YKL-40 and Hs CRP in patients with stable coronary artery disease Inflammation Research. 60: 281–287.

Nagai, T. & Anzai, T. 2011. C-Reactive protein Overexpression Exacerbates Pressure Overload–Induced Cardiac Remodeling through Enhanced Inflammatory Response. Hypertension. 57: 208–215.

Pai, J.K. & Mukamal, K.J. 2008. C-Reactive Protein (CRP) Gene Polymorphisms, CRP Levels, and Risk of Incident Coronary Heart Disease in Two Nested Case-Control Studies. PlosOne3.

Panina, L.V. & Mohri, K. 1994. Magneto-impedance effect in amorphous wires. Appl. Phys. Lett. 65: 1189.

Tillett, W.S. & Francis, T. 1930. Serological reactions in pneumonia with a non-protein somatic fraction of pneumococcus. J Exp Med. 52: 561–571.

Wang, T. & Yang, Z. 2014a. An integrated giant magnetoimpedance biosensor for detection of biomarker. Biosens. Bioelectron. 58: 338–344.

Wang, T. & Yang, Z. 2014b. A giant magnetoimpedance sensor for sensitive detection of streptavidin-coupled Dynabeads. Physica status solidi (a). 211(6): 1389–1394.

Bioinformatics and Biomedical Engineering – Chou & Zhou (Eds)
© *2016 Taylor & Francis Group, London, ISBN 978-1-138-02784-8*

Combining eye gaze detection with vision-based object recognition for a portable Human-Machine Interface

C.X. Yuan, J. Jiang, J.S. Tang, Z.T. Zhou & D.W. Hu
*College of Mechatronics and Automation, National University of Defense Technology,
Changsha, Hunan, China*

ABSTRACT: Eye gaze is an ideal input channel to construct convenient and natural
Human-Machine Interfaces (HMI). However, the Midas Touch problem must be taken into
consideration under the gaze-based HMI systems. In this paper, We incorporated the vision-
based object recognition method into the traditional gaze-based HMI system to avoid the
error commands caused by unintentional gazes. A double selection was designed to produce different commands. Dwell-time-based approach could select a possible input gaze and
extract the ROI (the region of interest). Object recognition based on shape matching could
avoid errors caused by noninput gazes and recognize the gazed object. A mobile eye tracker
with a scene camera was used to detect the eye gaze as well as the objects needed to control.
The system proposed in this paper was applied to control a real air conditioning. Four functions of switch, mode, heating and cooling could be controlled by gazing four different icons.
The results showed that the proposed eye control system could provide more control degree
and reliable accuracy with the incorporation of the vision-based object recognition method.

Keywords: HMI; eye gaze detection; object recognition; the Midas Touch problem

1 INTRODUCTION

Some time ago, the Ice Bucket Challenge swept the globe. The Ice Bucket Challenge is an activity to promote awareness of the disease Amyotrophic Lateral Sclerosis (ALS) and encourage
donations to research. ALS is the most common motor neuron disease. For people who are with
ALS, eye movement is the only way to communicate with the world in late stages. Eye control
is also the most convenient and effective technology to help handicapped and vocally disabled
people to regain the ability of communication and control (Hutchinson, 1989).
 The Human-Machine Interface (HMI) based on eye gaze is a convenient and natural way in
which humans interact with machines effectively. Gaze means the stay on the observed target,
which usually lasts at least stay 100 ms ~ 200 ms or more. Gaze is one form of eye movement
(Collewijn, 1975). Eye gaze contains the current status of tasks and human internal status and
other information, so eye gaze is an ideal candidate input to make HMI easy and natural to
happen. Gaze detection makes extracting useful information of human-computer interaction
as possible, for purpose of achieving a natural, intuitive and effective interaction. Therefore,
the gaze tracking technology and its applications in HMI has a special price (Carlos, 2005).
 In past research, eye gaze controlled interfaces have been applied to HMI. Sibert and Jacob
established a gaze controlled interface and showed that the eye-gaze input system made an
even faster interaction process than using a mouse (Sibert & Jacob, 2000). At the 2011 CeBIT
show, Tobii together with Lenovo unveiled a concept notebook which can be operated by eye
movement.
 However, the gaze-based system has to take the unintentional fixations and sporadic dwell-
ings on objects into consideration, which typically occur in the process of visual searching or
when people are involved in demanding mental activity (Jacob, 1991). The fact is so-called the

Midas Touch problem. In gaze-based systems, unintentional and sporadic gazes would also be regarded as input gazes to cause error commands. Dwell-time-based approach was regarded as the best technique to detect gazes. The approach means that a gaze occurs after users fixate somewhere for a predefined duration. A long predefined dwell time duration is chosen to make fewer errors caused by noninput gazes. But it leads to users' discomfort and weakens the advantage of gaze-based systems. Using eye gaze channel in cooperation with other channels (voice, keyboard and BCI) has been tried to solve the mentioned problem by adding one more selection channel. Zander et al. combined eye gaze input with a BCI for touchless HCI (Zander et al., 2011). The hybrid HMI could deal with different stimulus complexities.

Vision-based object recognition can also be an additional input channel used in gaze-based HMI. It is a fundamental task in computer vision applications. A special target can be recognized based on the gray, color or shape information. Mature applications include gesture, text and fingerprint recognition. The specific recognition method used in this paper is based on shape matching. The shape context is a feature descriptor of describing shapes that allows for measuring shape similarity. A system uses shape contexts for shape matching by two approaches: (1) feature-based, which uses extracted features such as moments or junctions, and (2) brightness-based, which utilizes pixel brightness directly.

In this paper, we implemented the portable eye-gaze-controlled system with the incorporation of the vision-based object recognition method that provided a portable human-machine interaction technique. Eye gaze could select a possible input gaze. Object recognition could exclude noninput gazes. This hybrid HMI is a promising way in which the disabled interact with machines.

2 METHOD

2.1 Participants

Five volunteer participants (5 males) were recruited from the local university campus. Participants ranged from 24 to 29 years (mean = 25). None had prior experience with eye tracking. All participants had normal vision.

2.2 Apparatus

2.2.1 Eye tracker

We used ASL Mobile Eye XG as the eye tracker. XG is an excellent product. Its biggest advantage is the portability. The accuracy of XG is 0.5–1 degree. It supports unfettered eye, head and hand movement in different lighting conditions. (see the top of Fig. 1a).

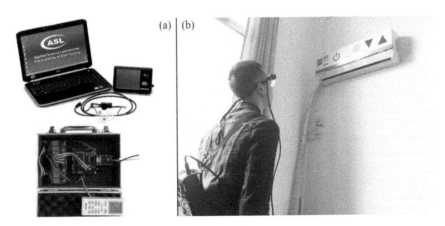

Figure 1. (a) Apparatus schematics. The top shows the ASL Mobile Eye XG. The bottom shows the appliance control module. (b) System in use.

2.2.2 Appliance control module

We designed a microcontroller based switching circuit of relay group. The circuit contains 8 relay branches. Through the serial port, we achieve strobe control of any one or more channels under the upper computer software command. By connecting suitably between the board of appliance remote control and relay group, the appliance is controlled by the computer. (see the bottom of Fig. 1a).

2.3 Task

In the experiment, an air conditioning was the actual controlled object. Specific model was Gree KFR-32GW/K(32556)G1-N2. On the surface of the air conditioning we posted four flags which represent switch, mode, heating and cooling. The subjects were asked to gaze icons to trigger off the air conditioning executing the corresponding function. In the experiments, subjects tried to gaze a central location of a icon. Of course, certain gaze direction offset was allowed. (see Fig. 1b).

2.4 Algorithm

2.4.1 Calculate gaze coordinates

The eye tracking technique used is "Pupil to CR" Tracking. This method uses the relationship between two eye features, the black pupil and mirror reflections from the front surface of the cornea (Corneal Reflections, or CRs), to compute eye direction within a scene.

The mobile eye PC can receive real eye direction coordinates with respect to the scene image in scene image pixels. The eye data is updated at 30 HZ. Because the main consideration is speed, we chose 200 ms as the dwell time duration. Meanwhile, ensure the accuracy to some extent. Dwell-time-based approach is adopted to calculate gaze coordinates.

Taking 200 ms as the time interval, the procedure to determine gaze coordinates is as follows:

1. Collect six eye direction coordinates in latest 200 ms.
2. Decide whether these coordinates located in a square with side length of T. We collected some gaze samples for analyze and assigned 8.62 to the value of T.
3. If in the square, get the rounded mean values of six coordinates as gaze coordinates. If not, repeat the previous steps.

2.4.2 How to combine eye gaze detection with vision-based object recognition

After eye gaze coordinates have been calculated, collect the scene image data. The scene image is from the same frame as the latest eye direction coordinates of the six. Taking gaze coordinates as the center point, we extract ROI from the scene image. The ROI is a square area and segmented for subsequent recognition.

2.4.3 Object recognition

Morphological operations must to be done for preprocessing the image of ROI. These operations include Otsu binarization, open operation and close operation. Then we use the shape matching method based on Hu moments. Hu moments are proved to be invariant to the image translation, rotation and scaling (Hu, 1962). Seven Hu invariants are calculated. The method uses the Hu invariants as follows (A denotes object 1, B denotes object 2):

$$I(A,B) = \sum_{i=1...7} \left| \frac{1}{k_i^A} - \frac{1}{k_i^B} \right| \tag{1}$$

where:

$$k_i^A = \text{sign}(h_i^A) \cdot \log h_i^A \tag{2}$$
$$k_i^B = \text{sign}(h_i^B) \cdot \log h_i^B \tag{3}$$

and h_i^A, h_i^B are the Hu moment of A and B.

The value of I noted as the match score is used to measure the similarity between two shapes. The square image was contrast to four icon shape (switch, mode, heating and cooling) template to get four match scores noted as I_1, I_2, I_3 and I_4 respectively. The smaller the value of I is, the higher the similarity between two shapes becomes. If I_1 is the minimum of four scores and smaller than threshold T_1, the switching function is executed. If I_2 is the minimum of four scores and smaller than threshold T_2, the mode function is executed. If I_3 or I_4 is the minimum of four scores and smaller than threshold T_3, heating or cooling icon is chosen. Hu moments are proved to be invariant to the image rotation, so the two need to be judged further. See Equation 4 below:

$$flag = \text{sign}\left(\sum_{i=1}^{N-1} \text{sign}\left(W_{i+1} - W_i\right) \right) \qquad (4)$$

where W_i is the number of white pixels per row in closed image. N is the number of rows. If *flag* is greater than zero, the heating function is executed. If *flag* is less than zero, the cooling function is executed.

3 RESULT AND DISCUSSION

3.1 *Implementation of a task*

To make readers clear about the work mechanism of the system, intermediate results are shown in this chapter.

1. Detect the eye direction coordinates in scene image pixels.
   Figure 2 shows how to use "Pupil-CR" method to detect eye direction coordinates. The last image is recorded by the scene camera. The others are recorded by the eye camera. The Corneal Reflection spots (CRs) are derived from the near infra-red lights. The eye images are processed to calculate the eye direction position relative to the corresponding field of scene data.
2. Extract the ROI
   Figure 3 shows how to extract the ROI. The data is from a task to make the air conditioning execute cooling function. In the Figure 3a, input is six eye direction coordinates in latest 200 ms. The output is gaze coordinates. Taking gaze coordinates as the center point, we extract the ROI from the scene image. The red square is the ROI corresponding to the gaze.

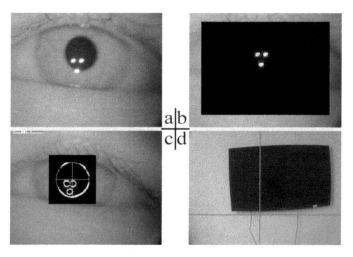

Figure 2. (a) Positioning the eye image. (b) Adjusting the corneal reflection spots detection. (c) Adjusting the pupil detection. (d) Positioning and calibrating the scene image.

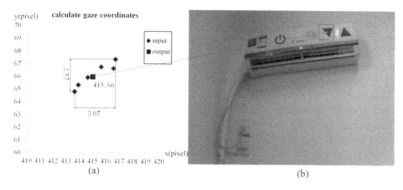

(a)　　　　　　　　　　　　　　　　(b)

Figure 3.　(a) Schematic showing how to calculate gaze coordinates. (b) The corresponding scene image.

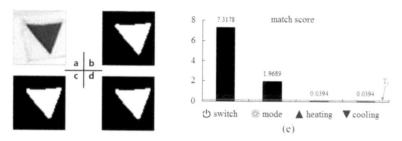

(e)

Figure 4.　(a) ROI. (b) Binary image. (c) Opened image. (d) Closed image. (e) Match score schematic.

Table 1.　The results of the experiment.

	Accuracy				
	Switch	Mode	Heating	Cooling	Mean
Sub. 1	80.0%	73.3%	100%	93.3%	86.7%
Sub. 2	86.7%	66.7%	93.3%	100%	86.7%
Sub. 3	86.7%	73.3%	93.3%	93.3%	86.7%
Sub. 4	80.0%	80.0%	93.3%	86.7%	85.0%
Sub. 5	86.7%	86.7%	100%	100%	93.3%
Mean	84.0%	76.0%	96.0%	94.7%	

3. Icon recognition

Figures 4a–4d show the morphological operation procedure to process the image of ROI. Figure 4e shows the calculated match score. Notably. the smaller the value of I, the higher the similarity between two shapes is. I_3 and I_4 are smaller than I_1, I_2 and T_3. So, heating or cooling icon is chosen. Further judgment was needed to distinguish $flag = sign(-20) < 0$. So, cooling function is executed by the control module.

3.2　*Performance of the system*

Each subject was asked to perform five runs. Between runs, participants took a short break. A run contained three blocks. In one block, a cued participant performed four different tasks to make the air conditioning execute switch, mode, heating and cooling functions. The order of four tasks was randomized in one block. So subjects performed each task 15 times. A task was regard as a success if the assigned function was executed in 10 seconds. Conversely, a fail appeared if the assigned function was not executed or executed wrongly. The results of the experiment are shown in Table 1.

3.3 *Discussion*

The deviation of gaze coordinates is from two main aspects which are the scene calibration procedure and gaze procedure. The scene calibration procedure teaches how eye movements relate to eye direction within the scene. The calibration is semi-manual procedure which inevitably leads to deviation. Using more calibration points can improve tracking accuracy. The other cause is head movements and nystagmus. We assume that scene image is not change in gaze procedure, which means head must be kept still. Nystagmus is an involuntary eye movement. Luckily, it affects little.

The results showed that the differences across subjects were narrow. Sub.5 did the best. Other subjects performed nearly. The standard deviation of the last column accuracies was marked as σ_{fun}. The standard deviation of the last row accuracies was marked as σ_{fun}. $\sigma_{sub} = 0.029$, $\sigma_{fun} = 0.082$. So the differences across functions were significant. The heating and cooling functions were executed best, followed by the switch function. The precision of mode were terrible. The mode icon shape is complex, which is not conductive to use shape matching method. We can use more effective recognition methods or just replace a simple icon to improve accuracy. In addition, it is easy to see that the system could provide more control degree by providing more icons to recognize.

4 CONCLUSION

In order to implement a portable and natural HMI, we combined eye gaze detection with vision-based object recognition. For gaze-based systems, to differentiate an input gaze and a noninput gaze is difficult. In previous dwell-time-based systems, a long dwell time duration is chosen to make fewer errors caused by noninput gazes. But a too long duration leads to users' discomfort and weakens the advantage of gaze-based systems. Adding vision-based object recognition to dwell-time-based interaction could solve the described problem by building a double selection. A short dwell time duration is chosen to select a possible input gaze and extract the ROI. Object recognition based on shape matching could exclude noninput gazes and recognize the icon.

The results showed the system could provide more control degree and reliable accuracy. In addition, previous eye-control systems restricted users to work in front of the computer screen. This system is excellent in flexibility and practicality. Users could move freely within the environment, which accorded with people's habits of interaction and brought a great increase to people's comfort and satisfactory. But at the same time, in order to ensure the real-time control, we used the algorithm of high efficiency, which led to the relatively low accuracy of switch and mode icon recognizing. The next step of our research is to find an algorithm with both high executing efficiency and high recognizing efficiency.

REFERENCES

Carlos, H. et al. 2005. Eye gaze tracking techniques for interactive applications. *Computer Vision.*
Collewijn, H. & Vander Mark F. 1975. Precise recording of human eye movements. *Vision Research.*
Hu, M.K. 1962. Visual pattern recognition by moment invariants. *IRE Transactions on Information Theory.*
Hutchinson T.E. et al. 1989. Human-computer interaction using eye-gaze input. *IEEE Transactions on Systems, Man, and Cybernetics* 19(6): 1527–1533.
Jacob, R.J.K. 1990. What you look at is what you get: eye movement-based interaction techniques. *CHI'90 Proceedings. ACM.*
Jacob, R.J.K. 1991. The use of eye movements in human computer interaction techniques: what you look at is what you get. *ACM Transactions on Information Systems* 9: 152–169.
Jacob, R.J.K. & Keith S.K. 2003. Eye tracking in human computer interaction and usability research: Ready to deliver the promises. *Computer Vision and Image Understanding.*
Sibert, L.E. & Jacob, R.J.K. 2000. Evaluation of eye gaze interaction. *Proceedings of CHI2000:* 281–288.
Yarbus, A.L. et al. 1967. Eye movements during perception of complex objects. *Eye movements and vision.* 171–196.
Zander, T.O. et al. 2011. Combining eye gaze input with a brain-computer interface for touchless human-computer interaction. *Journal of human-computer interaction* 27(1): 38–51.

Bioinformatics and Biomedical Engineering – Chou & Zhou (Eds)
© *2016 Taylor & Francis Group, London, ISBN 978-1-138-02784-8*

The design of a rehabilitation training system with EMG feedback for stroke patients

C.X. Yu, J.Y. Guo, Z.G. Yu, H.J. Niu & Y.B. Fan
School of Biological Science and Medical Engineering, Beihang University, Beijing, China

Z. Wang, W.R. Zhao & H.H. Zhao
Rehabilitation Hospital of the National Research Center on Technical Aids, Beijing, China

ABSTRACT: Rehabilitation training systems have been widely used to help patients recover muscle function in recent years. However, most of the systems are not so suitable for stroke patients. Combined with the neuromuscular function characteristics of stroke patients, this study introduces a rehabilitation training system with surface electromyography (sEMG) feedback based on the ARM embedded system and LabVIEW. This system could not only perform real-time multi-channel surface electromyography (sEMG) signal acquisition, processing and multi-monitor, but also compute muscle fatigue level of patients' trained parts and other related parameters during the training process. The verification results showed that the whole system was stable and had good interactivity. More importantly, the open system provided a convenient method to update design according to clinical feedback.

Keywords: embedded system; LabVIEW; stroke; sEMG feedback; characteristic parameters

1 INTRODUCTION

Stroke caused by the poor circulation of cerebrovascular blood has become a leading cause of disability. In China, more than 1.5 million people suffer from stroke every year (Dinevan et al., 2011, Johnston et al., 2009, Mohr, 1997). Rehabilitation can effectively reduce stroke patients' physical disabilities, recover function, as well as improve their ability to perform daily activities. Therefore, increasing attention has been paid to research in order to develop efficient rehabilitation training methods and training systems.

Rehabilitation training systems have been widely used in clinical settings. Liu et al. designed an electromyography biofeedback apparatus through EMG signal real-time acquisition, with which doctors could encourage patients to complete the training tasks by setting thresholds (Liu et al., 2009). As the development of embedded technology, Zhuang et al. designed a new apparatus for rehabilitation therapy. The graphical interface with biofeedback function could make patients actively participate in rehabilitation training (Zhuang et al., 2014). In recent years, Zhang et al. designed an EMG feedback diagnosis system, which could not only acquire multi-channel sEMG signals, but also use four kinds of electromyography parameters to assess the training effect (Zhang, 2014). However, present training systems have some defects, such as sole function, poor interactivity and insufficient effective feedback parameters. Furthermore, doctors cannot master the recovery status of the patients' muscle in real time or quantitatively judge the muscle fatigue level.

Besides, most of these systems are not suitable for stoke patients. There are obvious differences between neuromuscular function decline caused by stroke and other factors, as well as healthy people. Qi et al. statistically analyzed integrated electromyography (iEMG) and co-contraction ratio of the biceps and the triceps around the uninjured side and the affected side in the process of Maximum Isometric Voluntary Contraction (MIVC). Their results showed that the iEMG of the biceps around the uninjured side was greater than that around the affected side during elbow flexion, while co-contraction of triceps in the affected side was larger than that in the uninjured side during elbow extension (Qi et al., 2006). The sEMG signal amplitudes of stroke patients were mainly under 200 μV. Wan studied the upper limb muscle fatigue of stroke patients at different stages. Their results indicated that the Mean Power Frequency (MPF), which could reflect a detailed change in fatigue in the process of muscle activity, was more sensitive than the Median Frequency (MF). At the lower recovery stage, muscle fatigue occurred more easily and MPF declined faster (Wan, 2013). Studies about the sEMG signal feature of the lateral gastrocnemius in sitting and standing positions by Li et al. showed that there were obvious differences in the Root Mean Square (RMS) between stroke patients' uninjured side and affected side, as well as the affected side and corresponding side of healthy people (Li et al., 2007). These results provide an important guide for rehabilitation training of stroke patients.

This study introduces a rehabilitation training system with electromyography feedback for stroke patients. Based on the ARM embedded system and LabVIEW, the system could not only achieve real-time multi-channel sEMG signal acquisition, processing and multi-monitor, but also compute related feature parameters, including iEMG, RMS, MPF and co-contraction ratio. As a result, this system provides a better method for patients to monitor their status of rehabilitation. Moreover, it can help doctors make more specific training plans, in order to improve the effect of rehabilitation training.

2 SYSTEM DESIGN

The rehabilitation training system with sEMG feedback, as shown in Figure 1, consists of two parts: hardware and software. Magnification, filter and signal acquisition are included in the hardware, while the software is mainly used for signal processing and analysis. Figure 2 shows the system physical map.

2.1 The hardware design

2.1.1 Amplifier and filter circuits
In this paper, three-stage amplification circuits were applied. The instrumentation amplifier INA128 was used to form a pre-amplifier and its gain was set to 26 times. The filter circuit was combined with the second stage. In third stage, the potentiometer was used for continuously adjustable gain. The gain can be realized approximately 400–4000 times, so that it could adapt to different intensities of the electromyography signal.

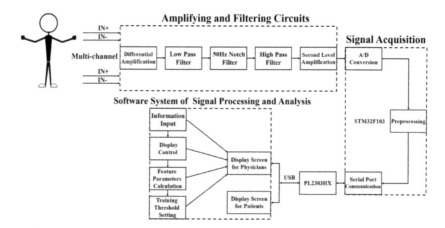

Figure 1.   Overall structure of the system.

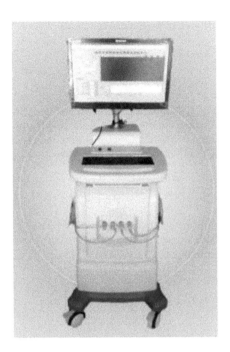

Figure 2.   The system physical map.

The focused energy range of sEMG was about 20–500 Hz, and a second-order band-pass active filter was utilized to restrain noise. Meanwhile, a 50 Hz notch filter was connected to remove power-line interference in the circuit. In addition, to satisfy the dynamic range of A/D conversion voltage, the analog output voltage of the system was controlled approximately 0–3.6 V by the level-up circuit.

2.1.2 Signal acquisition

In this paper, STM32F103 Cortex-M was used for multi-channel sEMGs' A/D conversion, pre-processing and data communication. Its frequency could reach to 72 MHz and was equipped with serial and USB2.0 ports. A multi-channel of the ADC subsystem with a 12-bit sampling precision was integrated on the chip. Its maximum speed of conversion was 1 MHz. It could fully satisfy the design demands.

The system interrupted the acquisition of sEMG signals every 400 μs, so the sampling frequency was 2500 Hz. After A/D conversion, a 5-point average filter was used to restrain thermal noise and periodic interference existing in the electronic components. The lower computer accepted and confirmed the startup command sent by PC. Then, the packets combined by 12 bytes per frame of sEMG data were transferred to PC. The PC extracted the data of each channel according to the communication protocol. At the end of training, the PC automatically sent terminating command to control the lower computer to stop the acquisition. The error rate of data transmission was estimated to be 8.3×10^{-4}.

2.2 The software design

2.2.1 Feature extraction

The core of the software system is the real-time analysis, which provides valuable parameters to doctors for muscle function evaluation. On the basis of current research on sEMG feature extraction and the neuromuscular function characteristics of stroke patients, four feature parameters are selected as feedback of training effectiveness: iEMG, RMS, MPF and co-contraction ratio. Particularly, iEMG, RMS and co-contraction ratio are time domain features, while MPF is a kind of frequency domain feature.

279

iEMG reflects the intensity change of sEMG signals. The iEMG x_{iemg} can be written as follows:

$$x_{iemg} = \frac{1}{N} \sum_{i=0}^{N-1} |x(i)|$$ (1)

where $x(i)$ is the amplitude of sEMG signals and N is the sampling number.

The co-contraction ratio is selected mainly for indicating the motion condition of the agonist and the antagonist, which could help doctors make more pertinent training plans. The Co-Contraction Ratio (CCR) can be computed as follows:

$$CCR = \frac{ATI}{AI + ATI}$$ (2)

where ATI is the iEMG of the antagonist and AI is the iEMG of the agonist.

The variation of the sEMG signal amplitude can also be found with RMS, which reflects the virtual value of neuromuscular discharge. The RMS can be expressed as follows:

$$RMS = \left(\frac{\int_t^{t+T} |EMG^2(t)|dt}{T} \right)^{\frac{1}{2}} = \left(\frac{\sum_{i=1}^{N} X_i}{N} \right)^{\frac{1}{2}}$$ (3)

where $EMG(t)$ is a time function of electromyography; X_i is the sample value of $EMG(t)$; N is the sampling number; and T is the length of the time window.

The fatigue level of muscle is another significant factor. MPF is a sensitive feedback parameter to catch the changes during the training process. This parameter can be calculated by the following equation:

$$MPF = \frac{\int_0^\infty fP(f)df}{\int_0^\infty P(f)df}$$ (4)

where $P(f)$ refers to the power spectral density function of sEMG signals.

2.2.2 Main software interface

The user interface of the software system is shown in Figure 3. It included five parts: information input, display control, calculation of feature parameters, threshold setting, and query of data.

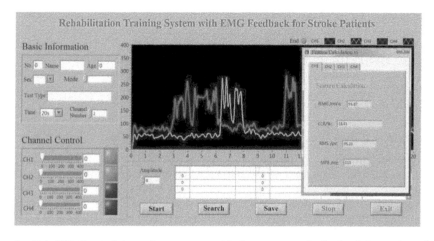

Figure 3. Rehabilitation training software system with electromyography feedback for stroke patients.

Table 1. System test data.

No.	1*	2**	3***	4***	5***	6***	7***
Amplitude (μV)	94.1	552.5	62.8	134.5	87.6	101.3	113.2

*For signal source.
**For healthy people.
***For stoke patients.

Doctors could record the information of patients, set the display mode, test type, acquisition time and channel number according to the schedule of training. Moreover, three kinds of display modes could be chosen to meet the patients of different ages, especially for the elderly. In addition, doctors could set the color and width of each channel waveform. The bipolar feature of raw sEMG signals leads doctors to make inaccurate judgment. Given that, the low-pass filtering and smoothing were added in this software design, which can provide a better way to effectively extract the signal envelope. Patients could be encouraged by threshold indicator lights to complete the training program. After signal acquisition, data of sEMG were saved automatically and feature parameters were presented, respectively, and the status of patients' muscle could be conveyed to doctors. Basic information and training results including sEMG data and value of feature parameters could be saved and searched conveniently.

3 SYSTEM VERIFICATION TEST

In order to testify the validity and stability of the system, 5 stroke subjects were chosen from the Rehabilitation Hospital of the National Research Center on Technical Aids. Signals of sEMG were acquired from their affected muscles of the upper limb involved in limb motion. Three kinds of limb motion were recognized: arm lift, elbow flexion and elbow extension. According to the experimental data, most of the sEMG signals of the patients' affected muscles were weak. The value of each feature parameter, except for the co-contraction ratio, was small; however, sEMG signals of some patients showed a tetanic feature. The system test data are given in Table 1.

A 100 μV signal source, referred by No. 1, was used as an input signal to test the system. The amplitude acquired by the system could reach to 94.1 μV. We could get the 5.9% error from the result. 552.5 μV reflected the maximal EMG of the biceps of healthy people with elbow flexion. The maximum values of sEMG signals of stroke patients with the same part and limb motion were all below 200 μV.

After the testing, the results showed that this system was stable and had good interactivity. The objective feedback parameters coincided with the subjective evaluation and requirements of the clinical treatment could be fully satisfied. Besides, this system effectively protected against electromagnetic radiation disturbance and immunity. In addition, the input noise of the system was less than 1 μV, the differential-mode input impedance was greater than 5 MΩ, and the common-mode rejection ratio was more than 100dB. All indicators had satisfied the relevant criterion.

4 CONCLUSION

From the experimental results, the effectiveness and stability of the system were fully verified according to the subjective evaluation by doctors and patients and the objective tests by the related departments. In the future, this system will have more user-friendly designs based on clinical feedback. Moreover, the existing evaluation parameters will also be improved combined with features of signal changes in stroke patients in the training process. This method will further enhance the effectiveness of rehabilitation.

ACKNOWLEDGMENT

This work was supported by the National Key Technology R&D Program in the 12th Five Year Plan of China (2012BAI33B03) and the Program for New Century Excellent Talents in University (NCET-11-0772).

REFERENCES

Dinevan, A., Aung, Y.M. & Al-Jumaily, A. Human computer interactive system for fast recovery based stroke rehabilitation. Hybrid Intelligent Systems (HIS), 2011 11th International Conference on, 2011. IEEE, 647–652.

Johnston, S.C., Mendis, S. & Mathers, C.D. 2009. Global variation in stroke burden and mortality: estimates from monitoring, surveillance, and modelling. *The Lancet Neurology,* 8, 345–354.

Li, Z., Xie, B. & Luo, C. 2007. Features of Surface Electromyographic Signal of Tibial Anterior Muscle and Gastrocnemius Muscle in the Stroke Patients when Sitting and Standing. *Chinese Journal of Rehabilitation Theory and Practice,* 12, 020.

Liu, Q., Tian, X., Li, F., Ge, G., Tang, H., Xu, J. & Wen, H. 2009. Development of the stroke rehabilitation apparatus based on EMG-biofeedback. *Journal of Biomedical Engineering,* 26, 417–420.

Mohr, J. 1997. Some Clinical Aspects of Acute Stroke Excellence in Clinical Stroke Award Lecture. *Stroke,* 28, 1835–1839.

Qi, R., Yan, J., Fang, M., Zhu, Y. & Zhang, H. 2006. Features of surface myoelectric signals taken from the triceps brachii and biceps brachii of stroke patients. *Chinese Journal of Physical Medicine and Rehabilitation,* 28, 399–401.

Wan, Z. 2013. *Surface electromyography decomposition and muscle fatigue research of upper extremity in stroke patients.* Changchun: Changchun institute of optics, fine mechanics and physics, Chinese academy of sciences.

Zhang, Y. 2014. *sEMG feedback diagnosis and treatment system.* Zhejiang University.

Zhuang, P., Tian, X. & Zhu, L. 2014. Design of an embedded stroke rehabilitation apparatus system based on Linux computer engineering. *Journal of Biomedical Engineering,* 31, 288–292.

Bioinformatics and Biomedical Engineering – Chou & Zhou (Eds)
© 2016 Taylor & Francis Group, London, ISBN 978-1-138-02784-8

Image Guided Surgery system for Optic Nerve Decompression Operation

Y.J. Wu
Department of Automation, Institute of Image Processing and Pattern Recognition, Shanghai Jiao Tong University, Shanghai, China

C.W. Xiao
Department of Ophthalmology, Shanghai Ninth People's Hospital Affiliated to Shanghai Jiaotong University, Shanghai, China

C.L. Fang & L.S. Wang
Department of Automation, Institute of Image Processing and Pattern Recognition, Shanghai Jiao Tong University, Shanghai, China

ABSTRACT: Endoscopic optic nerve decompression is a high-risk operation due to the fact that the scalpel may be very close to the optic nerve and internal carotid artery, and possibly injure them. In order to reduce the risk and enhance the safety of the operation, we apply an Image Guided Surgery (IGS) system into the operation. By using this system, surgeon can know how close the scalpel is from the optic nerve and the internal carotid artery during the surgery. This paper introduces how 3D models of the optic nerve and the internal carotid artery are segmented and reconstructed from CT images, and how such segmentation results are used and inputted in an independent IGS system that has no appropriate segmentation approach of the optic nerve and the internal carotid artery. Clinical applications have shown that such IGS system is practical and helpful in operation.

1 INTRODUCTION

Traumatic Optic Neuropathy (TON) is an uncommon and severe ophthalmic disease, which often results in a partial or complete loss of visual function (Chen 2006, Lübben 2001). Clinical experiences have shown that endoscopic optic nerve decompression is an effective treatment for TON (Pletcher 2007). The surgery involves operating both the eye and the sinus. Due to the fact that the scalpel may be very close to the optic nerve and the internal carotid artery during the operation, the surgery has a high potential risk of carotid artery injury and bleeding (Levin 2003). Therefore, a full understanding and analysis of the spatial distribution of the optic canal, the internal carotid artery and the scalpel is crucial to a successful operation. Particularly, surgeons need to know how close the scalpel is to the optic nerve and the internal carotid artery during the surgery. A feasible solution to such problem is to apply an appropriate IGS system into the operation.

In recent decades, with the development of 3D imaging and visualization technology, advances in IGS are making it possible for surgeons to perform difficult operations (Tietjen 2005). In an IGS system, detailed patient-specific anatomic models are usually reconstructed properly. Surgical instruments can be tracked by registering in reconstructed models with the actual patient position, and doctors can get real-time feedback about the relationship between the position of the instrument and the surrounding anatomic structures (Grimson 1999, Wagner 1996). In Optic Nerve Decompression Operation (ONDO), the IGS can be used to track the scalpel in real time. When the distance between the scalpel and the internal carotid artery or the optic nerve is too close, the surgeon will be alerted about the risk of carotid artery injury. In this way, the safety and efficacy are greatly enhanced.

In the IGS system applied in ONDO, the segmentation of the optic canal and the internal carotid artery plays an important role. However, such segmentation task is very complex, and many IGS systems actually do not contain an appropriate approach for segmenting the optic canal and the internal carotid artery from CT images. As a result, in the IGS system, a CT image without segmentation is directly shown in the screen with their three slices along the axial, sagittal and coronal planes. The optic canal and the internal carotid artery in each slice are recognized by surgeons based on their observation. Similarly, the distance between the scalpel and the internal carotid artery or the optic nerve is also estimated in this way. This is usually difficult for the surgeons without enough imaging experience. Additionally, in such cases, surgeons will not be alerted about the risk of carotid artery or optic nerve injury if the optic canal and the internal carotid artery are not segmented from CT images. Therefore, it is necessary to segment the optic canal and the internal carotid artery from CT images and input such segmentation result into IGS systems.

In this paper, we use an interactive method to segment and reconstruct 3D models of the optic canal and the internal carotid artery from CT images. However, many IGS systems do not provide an interface to input such segmentation results. Instead, they only provide an interface to read the CT DICOM images of patients. So, we further write the segmentation results of the optic nerve and the internal carotid artery into the original DICOM images of the patient through voxelization (Dong 2004, Huang 1998). By using such DICOM images, the segmentation result can be easily inputted into the IGS system. Then, the IGS system can be well applied in ONDO. Clinical applications have shown that such IGS system is practical and helpful in the operation.

2 METHODS

We first introduce the method for segmenting and reconstructing the optic canal and the internal carotid artery from CT images. Subsequently, we introduce how the segmentation result can be easily inputted into an independent IGS system that only provides an interface to read the CT DICOM images of patients.

2.1 *Segmentation and reconstruction of the internal carotid artery and the optic canal*

The internal carotid artery and the optic canal can be regarded as tube-like structures in 3D CT images. The internal carotid artery can be interactively segmented or reconstructed from a 3D CT image by the following steps:

First, we roughly determine the center line of the internal carotid artery. For this purpose, a 3D surface model of the head skull is reconstructed from the CT image by the Marching Cubes algorithm and is visualized on the screen, as shown in Figure 1 (a). Based on the observation, the virtual sagittal plane is adjusted interactively, so that it cuts through the

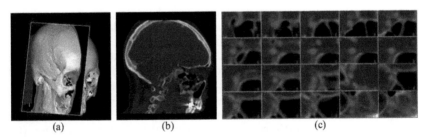

(a)   (b)   (c)

Figure 1. (a) Reconstructed 3D head model: the iso-surface intensity is 2000 Hounsfield unit in the evaluated bone. (b) Virtual sagittal cutting plane of the internal carotid artery. (c) Resampled image series of the internal carotid artery. All figures are from the same patient.

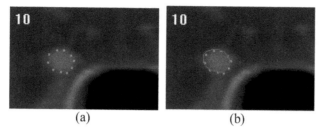
(a)   (b)

Figure 2.   (a) Marked points of a contour. (b) Cardinal spline fit of the marked contour points.

internal carotid artery and generates a 2D slice, as shown in Figure 1 (b). In the 2D slice, we interactively mark some points along the center line of the internal carotid artery (shown in 2D slice) and fitting them by a B-spline curve, as shown in Figure 1 (b). Then, a curved 3D center line of the internal carotid artery is obtained.

Second, we regenerate a series of 2D slices of the internal carotid artery by resampling the original CT image, as shown in Figure 1 (c). These 2D slices are perpendicular to the fitted center line.

Third, on each 2D slice, we interactively mark some boundary points of the carotid artery and fit them using a closed cardinal spline, as shown in Figure 2 (a) (b), respectively. Then, a series of contours of the internal carotid artery are obtained on different 2D slices.

Finally, a 3D triangular surface model of the internal carotid artery is generated from a series of contours by using the improved shortest diagonal algorithm. In the process, the vertices of triangular facets are chosen adaptively, so that each edge of a triangular facet is as close as possible in length. The generated models are smoother than those reported in earlier work [Wu 2014]. The 3D model of the optic canal can be reconstructed from the CT image by using the same method as one of the internal carotid artery.

The contours of the internal carotid artery are marked by doctors and fitted using a closed cardinal spline, as shown in Figure 2 (a) (b), respectively.

2.2   *Inputting the segmentation results into an independent IGS system*

We first voxelize 3D surface models of the internal carotid artery and the optic canal. Subsequently, these discrete voxels of the models will be written into appropriate positions of original CT DICOM images. In this way, the new DICOM images will contain segmentation results of the internal carotid artery and the optic canal.

In order to voxelize 3D surface models of the internal carotid artery and the optic canal, we need to compute intersection curves between 2D slices of the original CT DICOM images and the 3D surface models. There are three intersection cases between each triangular facet and a 2D slice: a point, a line segment and the whole facet. In the first and the third case, we directly get the set of interaction points between the 3D models and the 2D slice. In the second case, we first use Bresenham's line algorithm to discretely draw the intersected line segment on the original DICOM slices, and calculate 2D connectivity constrained by intersected 2D segments for each of the 2D slices after all facets are intersected.

On each 2D slice, the voxels that are enclosed by the intersection curves in 2D regions will have artificially endowed CT values far larger than those of bone structures. In this way, segmentation results of the internal carotid artery and the optic canal are written into the original CT DICOM image.

3   RESULTS AND DISCUSSION

We evaluate our method and the IGS system by a real clinical surgery of optic nerve decompression. The internal carotid artery and the optic canal are interactively segmented from a

285

CT image by a surgeon with rich imaging experience, and the segmentation results are written into the original CT DICOM image. The revised CT DICOM image is inputted into an independent IGS system for real-time navigation guidance. Clinical surgeons ensure that the accuracy of the distance between the scalpel and the segmented optic nerve and the internal carotid artery is acceptable, and such IGS system is helpful and efficient in guiding their surgery.

The segmentation results of the internal carotid artery and the optic canal are shown in Figure 3 (a) (b), respectively. Figure 3 (b) shows the reconstructed 3D model of the internal carotid artery and Figure 3 (a) shows the reconstructed 3D model of the optic canal. Figure 3 (c) compares the 3D surface model of the internal carotid artery and its voxelization model.

Figure 4 (a) (b) compares the original CT DICOM slice with the revised slice wherein the segmentation result of the internal carotid artery is written. Figure 4 (c) (d) shows the comparison between the original CT DICOM slice and the revised one wherein the segmentation result of the optic canal is written.

Figure 5 (a) shows the visualization result of the reconstructed 3D models of the optic canal and the internal carotid artery. Figure 5 (b) shows a real scene in an optic nerve decompression operation. Figure 5 (c) shows the original CT DICOM image (left) and the revised one with the segmented optic canal (right), which are generated from an independent IGS system in its real-time navigation in an optic nerve decompression operation.

The segmentation and reconstruction method in this paper has improved in its accuracy as well as interactivity. The vertices of the reconstructed 3D models are resampled adaptively, thus the results generated by the shortest diagonal algorithm are smoother than those reported in earlier work [Wu 2014]. Moreover, the reconstructed results in earlier work [Wu 2014] are 3D surface models, which cannot be directly used in independent clinical IGS systems, while in this paper, models are voxelized and the results are standard DICOM files that can be easily inputted into clinical devices. Our system has realized both pre-operational analysis and in-operational guidance for the operation. Furthermore, whenever the IGS system

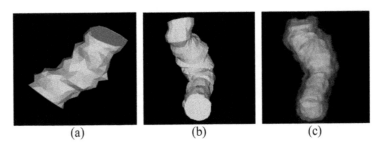

| (a) | (b) | (c) |

Figure 3. (a) Reconstructed 3D model of optic canal. (b) Reconstructed 3D model of the internal carotid artery. (c) Voxelized 3D model (yellow, 30% translucent) of the internal carotid artery and its original reconstructed 3D model (cyan, 50% translucent).

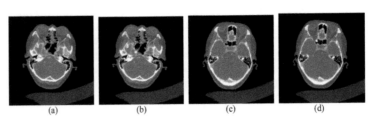

| (a) | (b) | (c) | (d) |

Figure 4. (a) The original CT DICOM slice. (b) The rewritten CT DICOM of (a) with the voxelized 3D model of the internal carotid artery rewritten area is marked by a red circle. (c) The original CT DICOM slice. (d) The rewritten CT DICOM of (c) with the voxelized 3D model of the optic canal rewritten area marked by a red circle.

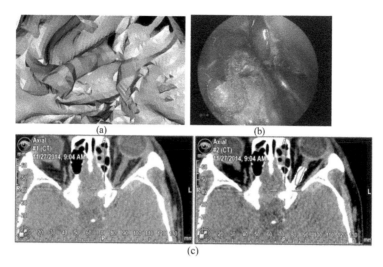

(a)   (b)

(c)

Figure 5. (a) A 3D scene with the optic canal and internal carotid artery in the head observation model. (b) Real scene in an operation. (c) Original CT DICOM image (left), rewritten CT DICOM image of the optic canal (right) in the CT bone image window. The scalpel is marked by a green cross.

provides an interface to output the spatial position information of the scalpel, we can even intuitively visualize spatial positions of the scalpel, the internal carotid artery and the optic nerve in the same 3D space. By observing the visualization result, surgeons can easily and intuitively know how close the scalpel is to the optic nerve and the internal carotid artery during the surgery.

4 CONCLUSION

Optic nerve decompression operation is an effective treatment for TON with a high potential risk of carotid artery injury and bleeding. In this paper, we apply an IGS system into the operation. 3D models of the optic nerve and the internal carotid artery are interactively segmented and reconstructed from CT images, and the segmentation results are inputted and used in an independent IGS system that has no appropriate segmentation approach of the optic nerve and the internal carotid artery. Clinical experiments have shown the feasibility and validity of our approach for the operation.

ACKNOWLEDGMENTS

This work was supported in part by the Cross Research Fund of Biomedical Engineering of SJTU (YG2012MS19, YG2013ZD02) and the NSFC of China (61375020), 973 program of China (2013CB329401).

REFERENCES

Chen C., Selva D., Floreani S., et al. 2006. Endoscopic optic nerve decompression for traumatic optic neuropathy: an alternative. *Otolaryngology—Head and Neck Surgery* 135(1): 155–157.
Dong Z., Chen W., Bao H., et al. 2004. Real-time voxelization for complex polygonal models. *Computer Graphics and Applications, PG 2004. Proceedings. 12th Pacific Conference on. IEEE*: 43–50.
Grimson W.E.L., Kikinis R., Jolesz F.A., et al. 1999. Image-guided surgery. *Scientific American* 280(6): 54–61.

Huang J., Yagel R., Filippov V., et al. 1998. An accurate method for voxelizing polygon meshes. *Volume Visualization. IEEE Symposium on. IEEE:* 119–126.

Levin L.A., Baker R.S. 2003. Management of traumatic optic neuropathy. Journal of Neuro-Ophthalmology, 23(1): 72–75.

Lübben B., Stoll W., Grenzebach U. 2001. Optic nerve decompression in the comatose and conscious patients after trauma. *The Laryngoscope* 111(2): 320–328.

Pletcher S.D., Metson R. 2006. Endoscopic optic nerve decompression for nontraumatic optic neuropathy. *Archives of Otolaryngology–Head & Neck Surgery* 133(8): 780–783.

Tietjen C., Isenberg T., Preim B. 2005. Combining Silhouettes, Surface, and Volume Rendering for Surgery Education and Planning. *EuroVis:* 303–310.

Wu L., Xiao C., Fang C., et al. 2014. 3D Reconstruction of the Optic Canal and Internal Carotid Artery. *International Conference on Bioinformatics and Biomedical Engineering:* 159–165.

Wagner A., Ploder O., Enislidis G., et al. 1996. Image-guided surgery. *International journal of oral and maxillofacial surgery* 25(2): 147–151.

Bioinformatics and Biomedical Engineering – Chou & Zhou (Eds)
© 2016 Taylor & Francis Group, London, ISBN 978-1-138-02784-8

Denitrification under aerobic condition by a reactor packed with corncobs as carbon source and bio-carrier

L. Shao & Y. Ling
College of Fisheries and Life Science, Water Environment and Engineering Research Center,
Shanghai Cooperative Innovation Center for Aquatic Animal Genetics and Breeding,
Shanghai Ocean University, Shanghai, China

ABSTRACT: In this study, a synthetic river environment reactor packed with corncobs was used to purify wastewater contaminated with nitrate by biological denitrification, where corncob was used as carbon source and bio-carrier. The operation efficiency and the influence of DO on the system were tested. The experimental results showed that well nitrogen removal efficiency was found even in higher DO condition. During more than 30-days process under aerobic condition (DO > 3 mg/L), the nitrate removal efficiency was still higher than 60%. It means biological denitrification with corncob as carbon source and substrate was less affected by DO. Structural biofilm was found on corncob surface by electron microscope scanning, which could well prevent diffusion of DO into the interior region and minimized the negative effect of oxygen. By preserving stability against DO inhibition, carbon source/substrate integration denitrification process will be a reasonable way for denitrification in low C/N ratio and high DO environment.

1 INTRODUCTION

With the development of agriculture and industry, more rivers were polluted by nitrogen pollutants. As nitrogen pollutants are responsible for promoting eutrophication and can have harmful effect on human health and aquatic life, the removal of nitrogen compounds from wastewater is of increasing importance. Since the denitrification process actually removed the nitrogenous from wastewaters, more researches are focus on biological denitrification in recent years (Rocca et al. 2007; Shao et al. 2009; Xu et al. 2009). A common phenomenon in the southern polluted rivers of China is low C/N ratio and aerobic concentration. In our previous work, a new solid carbon source was developed in which corncobs were used as the carbon source and the only physical support for microorganisms (Xu et al. 2009). And well nitrate removal rate was found. It was always reported in studies that DO have a negative effect on biological denitrification and denitrification always occur in anoxic environment (Richardson, 1992). DO is reported to be a limiting factor in the biological denitrification process because oxygen functions as the electron acceptor for microorganisms over nitrate, and aerobic conditions repress enzymes involved in denitrification (Zumft, 1997). However, DO in the river is always high. Therefore, how to remove nitrate from river under aerobic condition is technical bottleneck of the study. Fortunately, there are also some literatures reported that denitrification doesn't stop even if under high DO condition (Chang et al. 2006). A similar conclusion was also found in our previous works. The results showed that no significant difference was observed when DO change from 1.5 to 4 mg/L (Xu et al. 2009). In this research, a novel denitrification technology for low C/N wastewater was studied in a synthetic river environment reactor system, where corncobs were used as the carbon source and the only physical support for microorganisms. And in the present research, DO concentration was studied as a crucial environmental factor on the performances of the denitrification systems. In order to explain the diffuse of oxygen in a better way, electron microscope scanning was also used.

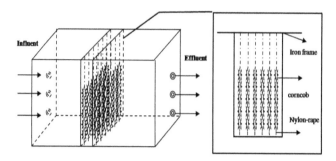

Figure 1. Experimental scheme.

2 MATERIALS AND METHOD

2.1 Pretreatment of corncobs

Corncobs were gathered from a local village in Shanghai. The same material was used in all experiments. The corncobs were washed with tap water before air drying (30°C). The material was kept in a moisture-free container.

2.2 Seed sludge

Fresh activated sludge was taken from a local municipal wastewater treatment plant, Shanghai. The sludge was cultured for 3 d in a liquid medium (KNO$_3$ 2 g/L; K$_2$PO$_4$ 0.5 g/L; MgSO$_4$·7H$_2$O 0.2 g/L; C$_4$H$_4$ KNaO$_6$·4H$_2$O 20 g/L) and then used as inoculum seeding in the denitrification reactor.

2.3 Experimental apparatus

The reaction tank which synthetic urban river was a cuboids' PVC reactor of 2000 mm length, 400 mm width and 1000 mm height (Fig. 1). The effective volume of the reactor after operation was 42 L. Two iron frames which fixed with corncobs (together 2.2 kg) were placed in the middle of the reactor as showed in Figure 1. Then medium and inoculum were dosed into the reactor. The inoculated reactor was cycle operated for 3 days before continuous flow operation, in order to leave enough time for microorganism's attachment and culturing. The flow created a horizontal-flow through the reactor with a rate regulated by peristaltic pumps.

2.4 Synthetic wastewater

Synthetic wastewater was prepared daily by tap water supplemented with KNO$_3$ as the nitrogen source and K$_2$PO$_4$ as the phosphorus source. In order to establish different dissolved oxygen condition in the reactor, the media was swept by nitrogen gas or air.

2.5 Analytical methods

Samples were collected daily from the inlet and outlet of the reactor. NO$_3^-$, NO$_2^-$, TN, COD, pH, DO, flow rate and temperature were measured daily. The pH, temperature and DO were measured with a standard electrode (HACH). NO$_3^-$, NO$_2^-$, TN, COD were measured using standard methods (Chinese NEPA Standard Methods, 2002).

3 RESULTS AND DISCUSSION

3.1 Start of the reactor

Plant operating parameters were shown as following: water temperature, 25 ± 1°C; inlet nitrate concentration, 12.5–13.5 mg/L; flow rate, 250–288 L/d; DO 1.5–2 mg/L. The operation condition and treated water quality were shown in Figures 2–4. As can be seen from Figure 2, high nitrate removal efficiency more than 90% was found at the first day of operation. However, TN removal efficiency was lower than 85% until operated for 7 days. Therefore, it was considered that one week of continuous system operation was needed to establish a steady-state condition in this experiment. Anyway, more time was saved when compared with other reports (Jin et al. 2004). The quick start-up of the reactor can be explained as

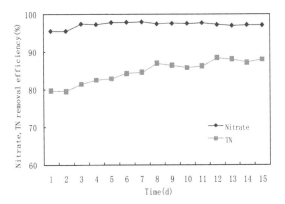

Figure 2.　Nitrate and TN removal efficiency according to time (d).

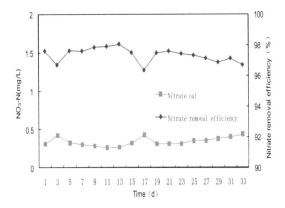

Figure 3.　Nitrate removal efficiency and change of nitrate concentration in the effluent after steady-state condition of the reactor was established.

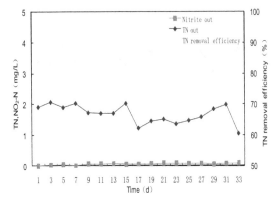

Figure 4.　TN removal efficiency and change of TN and nitrite in the effluent.

following: (1) The inoculated reactor was allowed to stand for 3 days before operation, and enough time was left for biofilm culturing; Moreover, the soluble fraction of carbon presenting in the fresh corncobs accelerated microbial growth. (2) A larger bacterial population always results in a faster startup of a reactor. Corncob which was used in the reactor has large external specific surface area, which makes bacteria easier to attach to the surface. Therefore, only 7 days of operation was needed to startup the reactor.

After steady-state condition of the reactor was established, NO_3^-, NO_2^- and TN were measured daily. The results were show in Figures 3–4. As can be seen from Figure 3, a correspondingly steady state of nitrate removal efficiency of higher than 96% was observed. The nitrate concentration in the effluent water was lower than 0.5 mg/L. As can be seen from Figure 4, a correspondingly steady state of TN removal efficiency of higher than 83% was found and the TN in the effluent was almost below 2 mg/L. No nitrite was accumulated and nitrite in the effluent was always lower than 1 mg/L.

3.2 Operation under aerobic condition

To establish aerobic condition (DO > 3 mg/L) in the reactor, the media was swept by oxygen gas. During the study, nitrate condition in the influent was about 6.8 ± 0.5 mg/L. As can be seen from Figure 5, nitrate removal efficiency was always higher than 60% when DO condition was higher than 3 mg/L. The great nitrate removal efficiency got in this study under aerobic condition can be explained as following: Due to the fact that oxygen has to be transported into the biofilm by diffusion, oxygen's negative effect could be minimized by structural biofilm adhere on the corncobs. As can be seen from Figure 6, microstructure of corncobs

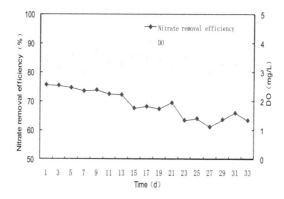

Figure 5. Nitrate removal efficiency under aerobic condition.

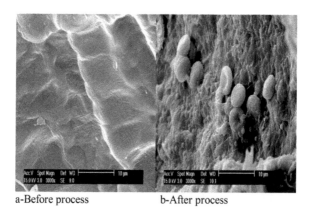

a-Before process   b-After process

Figure 6. Electron microscopes scanning of corncobs surface.

292

surface was studied by Electron microscope scanning. At the beginning of the process, only microstructure of corncobs surface was found and no biofilm was existed on the surface. After days of process, structural biofilm was found on the surface of corncobs and diffusion rate of DO was minimized. Oxygen was almost completely used before it got to the inner zone. Therefore, when DO in the water was higher than 3 mg/L there were still many micro-environment in the inner zone kept at anoxic environment. Inner denitrifying bacteria were protected and denitrification not stopped at those special anoxic microenvironments. Besides, it was considered that the denitrification rates will be limited because of an inadequate carbon supply even when microenvironments exist close to the substrate (Chang et al. 2006). However, in this research, corncobs were used as carbon source and sole biofilm substrate. COD concentration in the inner zone was far higher than outside and more electron donors can be provided. At the same time, decomposition of carbon source may also consume oxygen and hinder denitrification under aerobic conditions. That is to say high COD is helpful to reduce DO concentration in the environment. Therefore, great nitrate removal efficiency was observed under high DO condition in carbon source/substrate integration denitrification process. Gómeza et al. (2002) used a unidirectional submereged filter system to purify nitrate contaminated groundwater. The results showed that nitrogen removal was almost constant under condition in which DO concentration was below 4.5 mg/l. Other research (Chang et al. 2006) also found similar results as this research. It means that the effect of DO on denitrification should take into bioreactor characteristics, operation condition, carbon source kinds and substrates. Nitrate can be removed under high DO once the right way was chose.

On the other hand, compared with low DO condition (2 < mg/L), a reduction about 37% for nitrate removal efficiency was found. It means that the denitrification process in the system was partly inhibited in the presence of high concentration of DO. As can be seen from Figure 5, slightly decrease of nitrate removal efficiency was found as time went by. And it can be explained as following: with the passage of time, more dissolved oxygen was diffused into inner biofilm and the space of anoxic microenvironments was reduced. Due to the concentration of oxygen in the inner increased, part denitrifying bacteria choose oxygen, a more energetically favourable electron acceptor, as an electron acceptor. Therefore, nitrate removal efficiency of the system was reduced.

3.3 Effect of carbon source dosing quality/DO on denitrification

Chemical interactions between DO and denitrification will be influenced by changes in wastewater composition. For example, COD concentration in the wastewater will also influence the effect of DO on denitrification. There are indications that in a certain range, increased COD concentration will mitigate the effect of DO on denitrification (Jack, 1985). The increasing COD levels would help to create anaerobic condition and possibly provided conditions favourable

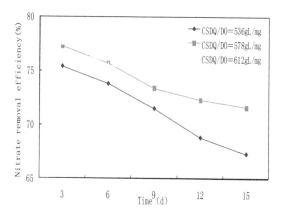

Figure 7. Effect of carbon source dosing quality/DO on nitrate removal rate.

for denitrification. In sum, the degree of negative effect on denitrification by DO related to carbon source used. Therefore, Experiments with different Carbon Source Dosing Quality/DO (CSDQ/DO) were carried out. As can be seen from Figure 7, the effect of DO on denitrification decreased with the carbon source dosing quality increased. During the experiment, When CSDQ/DO was 612 gL/mg, a reduction only about 6.4% was found. And nitrate removal efficiency of the reactor was always higher than 70%. However, When CSDQ/DO decreased to 536 gL/mg, a reduction about 10.7% was found. The result shows that the effect of DO on denitrification can be minimized by carbon source. The result can be explained as following: with the increase of carbon source, more carbon source and substrate were supplied to denitrifiers. On the other, oxygen in the water may consumed by carbon source at the same time. Plósz et al. (2003) reported that when the substrate concentration in the bioreactor is high enough, the DO concentration will kept at low level. And the effect of oxygen on denitrification is minimized. Guo et al. (2008) also reported that oxygen consumed rapidly when more carbon source was supplied.

4 CONCLUSIONS

In this research, it is interesting to note that the inhibition of denitrification by oxygen may be prevented for both carbon source and biofilm barrier. In the presence of higher than 3 mg/L O_2, only about 37% reduction of the denitrification rate was found when compared with anaerobic condition. Carbon source/substrate integration denitrification system used in this research is less affected by DO concentration. And offers an alternative way to purity nitrate contaminated river in south China.

ACKNOWLEDGEMENTS

This work was supported by Natural Science Foundation of Shanghai (12ZR1444900) and Construction of key courses of Shanghai Ocean University (A1-0209-15-0302-7).

REFERENCES

Chang, Y.J., Ho, C.M., Chang C.C., Tseng S.K., 2006. Denitrification under High Dissolved Oxygen by a Membrane-attached Biofilm Reactor. Journal of the Chinese Institute of Engineers. 29(4), 741–745.

Gómez, M.A., Hontoria, E., González-López, J., 2002. Effect of dissolved oxygen concentration on nitrate removal from groundwater using a denitrifying submerged filter. Journal of Hazardous Materials B90, 267–278.

Guo, H.Y., Zhou, J.T., Zhang, X.H., 2008. Influence of influent C/N ratio and aeration rate on the nitrogen removal in an airlift combined biofilm reactor. The 2nd International Conference on Bioinformatics and Biomedical Engineering (iCBBE 2008), May 16–18, Shanghai.

Jack T. T., 1985. The influence of oxygen concentrations on denitrification in soil, Applied Microbiology Biotechnology. 23, 152–155.

Jin, Z.F., Chen, Y.X., Ogura, N., 2004. Denitrification of Groundwater Using Cotton as Energy Source. Journal of Agro-Environment Science. 23(3), 512–515.

Plósz, B.G., Jobbágy, A., Grady C.P. Jr., 2003. Factors influencing deterioration of denitrification by oxygen entering an anoxic reactor through the surface. Water Research. 37(4),853–863.

Richardson, D.J., Ferguson, S.J., 1992. The influence of carbon substrate on the activity of the periplasmic nitrate reductase in aerobically grown Thiosphaera pantotropha. Arch. Microbiol. 157, 535–537.

Rocca, C.D., Belgiorno, V., Meric, S., 2007. Overview of in-situ applicable nitrate removal processes. Desalination. 204, 46–62.

Shao, L., Xu, Z.X., Jin, W., 2009. Rice Husk as Carbon Source and Biofilm Carrier for Water Denireification. Polish Journal of Environmental Studies. 18(4), 693–699.

Xu Z.X., Shao, L., Yin, H.L., 2009. Biological Denitrification Using Corncob as Carbon Source and Biofilm Carrier. Water Environment Research. 81(3), 242–247.

Zumft W.G., 1997. Cell biology and molecular basis of denitrification. Microbiology and molecular biology reviews. 61(4), 533–616.

Bioinformatics and Biomedical Engineering – Chou & Zhou (Eds)
© 2016 Taylor & Francis Group, London, ISBN 978-1-138-02784-8

Wireless platform for real-time Electrocardiography (ECG) recording and analysis

M. Al-Qahtani, M. Alwahiby, M. Abdelhamid & E.H. Mirza
Department of Biomedical Technology, College of Applied Medical Sciences, King Saud University, Riyadh, KSA

X. Yang
Biomedical Engineering Research Group, Cardiff School of Engineering, Cardiff University, Cardiff, UK

ABSTRACT: Cardiovascular diseases have high mortality rates. Electrocardiography (ECG) provides best diagnosis to date for any disorder. A conventional ECG device has a monitor and printer attached to the device that is vendor specific, and may not be used with any other platform or device. Conventionally ECG traces are printed over a paper which is then analysed by the clinician. All this process takes ample time which may risk patient's quality of life, as in some occasions the clinician might not be present on location. Demands for data to be analysed remotely have increased recently as more than one expert opinions can be taken for a particular case, which eventually reduces any human errors. We report a comprehensive platform for ECG measurement and interpretation. This involves a dedicated ECG hardware device that is capable of working on 9.0V battery. Furthermore, a separate software based on LabView with ECG trace display, acquisition and interpretation can be communicated wirelessly over any hand-held device through a Wi-Fi module.

1 INTRODUCTION

Cardiovascular disease is among the diseases with highest mortality rates. According to World Health Organization (WHO), it is estimated that each year 17 million people die due to cardiovascular diseases, and this figure is expected to increase gradually (Chen and Zhang, 2014).

In the world of clinical cardiac medicine, the Electrocardiography (ECG) is among the most widely used diagnostic tool. Various types of ECG exists, such as: 3-lead, 5-lead and 12-lead in general (Pipberger et al., 1961, Solanki et al., 2013, Maron et al., 2014). Usually an ECG device comprises of a display to monitor the ECG signal and a printer to print the ECG trace from an individual (Hsieh and Hsu, 2012). However, the monitor and the printer attached to the device are vendor specific and cannot be used by other systems or a platform to monitor or print the ECG trace. An ECG device usually comprises of electrodes, a multiplexer, pre-amplifiers, filters and Analogue to Digital (A/D) convertors (Fig. 1). Results of ECG trace are conventionally taken either on a paper or over the display. However, these days there is an increasing demand for data to be accessed more remotely via internet or wirelessly on a mobile device. Our objective was to create a platform for ECG and heart rate measurement that provides ease of access for patients and doctors. Here we report a complete

Figure 1. Block diagram of a conventional ECG system demonstrating the process of flow.

platform that enables the display of real-time ECG trace and heart rate utilising only a 9 volts (V) battery on a mobile device, laptop or desktop computer wirelessly.

2 MATERIALS AND METHODS

2.1 Design of ECG platform

A design of platform for ECG detection and analysis is provided in Figure 2. A surge arrestor is used to protect ECG circuitry from excess voltage as a result of defibrillation. An EMI

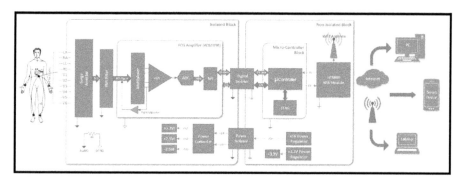

Figure 2. Block diagram of the complete platform, demonstrating the connections from the ECG leads to final retrieval of ECG trace via wireless transmission on portable devices.

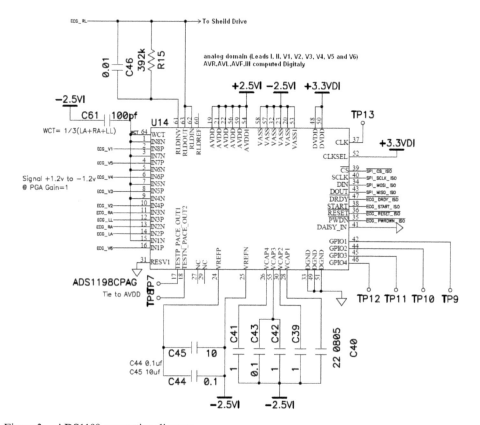

Figure 3. ADS1198 connection diagram.

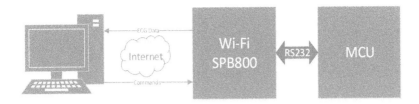

Figure 4.   Basic Wi-Fi connectivity block diagram.

filter is then provided to protect interference of ECG signal with electromagnetic waves from other sources. ADS1198 is used to develop a complete ECG system. An RC filter at the input acts as an Electromagnetic Interference (EMI) filter on all of the channels.

2.2   *Schematics of ECG platform*

Connections used on ADS1198 to obtain ECG trace and heart beat are given in Figure 3. Only 8 ECG lead traces can be calculated through ADS1198 so, the other remaining 4 leads were calculated via a small program designed on LabVIEW.

2.3   *Wireless connectivity*

A Wireless module SPB800 was used to transfer data wirelessly. The module has a high performance chip antenna as the primary RF interface. Digitised ECG signals are sent via module by reading them from microcontroller unit. A basic process diagram for Wi-Fi connectivity is displayed in Figure 4.

3   RESULTS AND DISCUSSION

Our results demonstrate a significant difference that can be seen through the software.

Four different conditions from the ECG simulator were tested; (A) Normal ECG trace, (B) Pair Premature Ventricular Complex (PVC) ECG, (C) Trigeminy and (D) Ventricular Fibrilation. ECG trace results are displayed in Figure 5 (A–D). All the traces clearly show a change in frequency and amplitude of the signals when compared to normal ECG rhythm. Figure 5B reveals an abnormal trace that is of Pair PVC which displays an irregular heartbeat. PVC is generally termed as skipping of a heart beat meaning it decreases the number of beats per minute (Yazawa and Katsuyama, 2009). Furthermore, Table 1 shows an increased RR mean while a decreased heart beat per minute for Pair PVC when compared to normal ECG. Figure 5C shows ECG trace of Trigeminy. Trigeminy is appearance of PVC every third beat, this phenomenon is visible in Figure 5C. The heart rate for trigeminy is 53.2 ± 23.1 which is having a great amount of variation and trigemini have an RR interval of 1.255 ± 0.359, this shows that due to trigeminy there is a substantial gap between the beats.

The Power Spectral Density (PSD) result in Figure 6C illustrates an increased PSD for trigeminy when compared to PSD of normal ECG. Furthermore, ECG trace of ventricular fibrillation displays an abnormal activity. This was confirmed by RR interval and heartbeat as observed in Table 1. These results are similar to that reported by Beck C et al. (Beck et al., 1947). PSD results in Figure 6D demonstrate a highly increased PSD for increase in frequency.

Our results demonstrate that, the device fabricated for ECG measurement is helpful in not just taking the ECG trace but it will also help the physicians to pin-point the finding present in a certain individual. The fabricated device is a real time, quick and easy to install device with features that may help clinician in better diagnosis even if they are not present in close vicinity of the patient, through a Wi-Fi device like laptop or any other hand-held device.

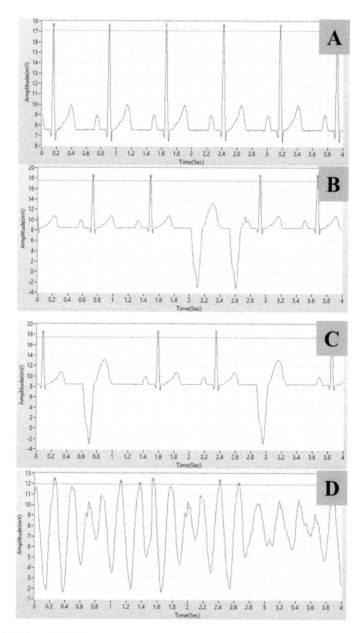

Figure 5. (A) Normal ECG trace, (B) Pair PVC (Premature Ventricular Complex) ECG trace, (C) Trigeminy ECG trace and (D) Ventricular Fibrilation ECG trace taken from ECG simulator.

Table 1. Representation of normal and pathological Heart Rate Variability (HRV) parameters.

Type of ECG	RR mean ± SD	HR mean ± SD
Normal	0.753 ± 0.036	79.7 ± 0.2
Pair PVC (premature ventricular complex)	0.977 ± 0.346	67.2 ± 21.8
Trigeminy	1.255 ± 0.359	53.2 ± 23.1
Ventricular fibrilation	0.479 ± 0.335	200 ± 120

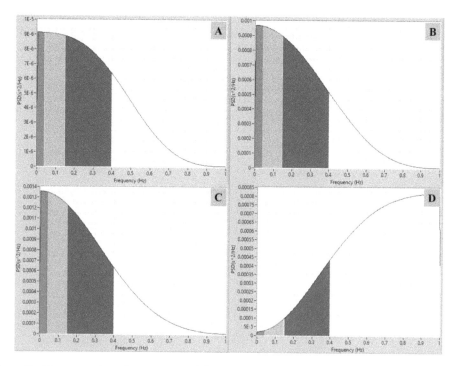

Figure 6. Representation of power spectral density against frequency of (A) Normal ECG trace, (B) Pair PVC (Premature Ventricular Complex) ECG trace, (C) Trigeminy ECG trace and (D) Ventricular Fibrilation ECG trace taken from ECG simulator.

4 CONCLUSION

A portable ECG device with real time data acquisition was fabricated for better diagnosis of individuals with critical heart conditions. This device will be helpful for the clinicians and the patients as well for better diagnostic of pathological condition.

ACKNOWLEDGEMENT

The authors extend their appreciation to the College of Applied Medical Sciences Research Center and the Deanship of Scientific Research at King Saud University for funding this research.

REFERENCES

Beck, C.S., Pritchard, W.H. & Feil, H.S. 1947. VEntricular fibrillation of long duration abolished by electric shock. *Journal of the American Medical Association,* 135, 985–986.
Chen, W. & Zhang, Y. Design and implementation of remote multiple physiological parameters monitoring system. Software Engineering and Service Science (ICSESS), 2014 5th IEEE International Conference on, 27–29 June 2014 2014. 740–743.
Hsieh, J.-C. & Hsu, M.-W. 2012. A cloud computing based 12-lead ECG telemedicine service. *BMC Medical Informatics and Decision Making,* 12, 77.
Maron, B.J., Friedman, R.A., Kligfield, P., Levine, B.D., Viskin, S., Chaitman, B.R., Okin, P.M., Saul, J.P., Salberg, L. & Van Hare, G.F. 2014. Assessment of the 12-Lead ECG as a Screening Test for Detection of Cardiovascular Disease in Healthy General Populations of Young People (12–25 Years of Age) A Scientific Statement From the American Heart Association and the American College of Cardiology. *Circulation,* 130, 1303–1334.

Pipberger, H.V., Bialek, S.M., Perloff, J.K. & Schnaper, H.W. 1961. Correlation of clinical information in the standard 12-lead ECG and in a corrected orthogonal 3-lead ECG. *Am Heart J*, 61, 34–43.

Solanki, S.L., Kishore, K., Goyal, V.K. & Yadav, R. 2013. Electrocautery induced artifactual ST segment depression in leads II, III and aVF on intra operative 5-lead electrocardiogram. *Journal of clinical monitoring and computing*, 1–2.

Yazawa, T. & Katsuyama, T. 2009. Premature ventricular contractions, a typical extra-systole arrhythmia, lowers the scaling exponent: DFA as a beneficial biomedical computation. *Proceedings of the 2nd International Conference on Interaction Sciences: Information Technology, Culture and Human*. Seoul, Korea: ACM.

Bioinformatics and Biomedical Engineering – Chou & Zhou (Eds)
© *2016 Taylor & Francis Group, London, ISBN 978-1-138-02784-8*

Design of the Invasive Blood Pressure simulator

Q.C. Liu & B. Xiao

Guangdong Food And Drug Vocational College, Guangzhou, China

ABSTRACT: The accuracy test of Invasive Blood Pressure (IBP) monitoring devices is of importance as the Blood Pressure (BP) is a crucial factor for doctors to judge the physiological condition of patients. In this work, a novel IBP simulator is designed based on ARM platform according to the standards in pharmaceutical industry of YY0783-2010. The rotor of a stepper motor reciprocally hits an elastic film and leads to the periodic change of pressure in the liquid path, which is simultaneously detected by a pressure sensor and transferred to a periodic signal. Experimental results show that the signal can be accurately controlled and simulates the dynamic changes of BP, which meets the requirements of the latest standards in pharmaceutical industry in terms of both accuracy and fast response.

1 INTRODUCTION

IBP monitoring is a commonly used method in first aid, cardiovascular surgery, Intensive Care Unit (ICU) and anesthesia (Hu Xiangqin et al. 2008). This technique mainly involves direct measurement of arterial pressure by the use of a cannula needle inserting to an artery. It benefits in high accuracy and intuition, and independent of external factors such as artificial compression, decompression, and the width and tightness of cuff. The detected BP is one of the important factors to determine patient's physiological condition. Therefore, the accuracy test of IBP monitoring devices is highly standardized by the regulatory authorities in China.

China's pharmaceutical industry standard YY0783-2010 has a clearly defined accuracy for IBP monitoring equipment and standardized test methods (YY0783-2010.2010). A number of instruments including liquid path system, power supply, signal generator, power amplifier driver, oscilloscopes and pressure meter are required to facilitate the test, which is difficult and tedious to operate. With this as the background, we present a novel IBP simulator for the accuracy test based on ARM platform. With the required functions integrated in a single device, the test procedure is significant simplified. We demonstrate a variety of mimic BP signal generated, and prove that the performance of the device meets the requirements of regulatory standards.

2 REGULATORY STANDARD

The regulatory standard (YY0783-2010) for the accuracy of IBP monitoring equipment mainly include three aspects as followed:

1. Accuracy of static pressure: apply pressure in the form of a percentage of the maximum full-scale pressure, such as 0, 10, 20, 50, 80, 100, 80, 50, 20, 10, 0, −10, −15 and 0%. The output pressure should not exceed ± 4% of the reading, or ± 0.5 kPa (± 4 mmHg) comparing with the reference value, whichever is larger.
2. Accuracy of systolic and diastolic pressure: adjust the static pressure to 90 mmHg, and then apply a sinusoidal pressure signal of 1 Hz with the peak pressure of 120 mmHg, the bottom peak value of 60 mmHg, and the peak-to-peak value of 60 mmHg.

The frequency accuracy of 1 Hz signal should be at least 1%. The magnitude accuracy of systolic and diastolic signal should not exceed 0.5 kPa (± 4 mmHg).

3. Frequency Response: The frequency response range of devices and sensors should be at least from DC to 10 Hz. The attenuation of test equipment for 10 Hz sinusoidal output signals should not exceed 3 dB with respect to the signal of 1 Hz.

3 SYSTEM DESIGN

3.1 System frame

The overall design of the system is shown in Figure 1. The embedded ARM controls the drive circuit of both static and dynamic motor, and the static bias pressure is adjusted through the connected valve by the static motor. The dynamic motor with pressure regulation piston reciprocally pushes the elastic membrane, results in periodic changing of the hydraulic pressure of liquid chamber, mimicking the dynamic blood pressure.

The pressure in the fluidic chamber is subsequently measured by the pressure sensor, and transferred to the embedded ARM by 16-bit AD converter. The measured value is then compared with the preset pressure to adjust the magnitude of the reciprocal movement, i.e., the pressure curve, and mimic the variation of real BP.

The control software is developed using embedded systems, which can be used to set the operating parameters, record the data measured by the pressure sensor, and converted it to pressure curve. Thus, it can provide various pressure parameters to users as a user-friend interface system.

3.2 The generation of sinusoidal pressure signal

According to the regulatory standards, the dynamic pressure varies in the form of sinusoidal wave, with the peak-to-peak value of 60 mmHg, and the peak error not exceeding 4 mmHg. The simulation of dynamic pressure is realized by the vibration of the elastic membrane induced by the rotor of periodical rotating motor. The periodical rotation is controlled by a algorithm with linear, exponential acceleration and deceleration characteristics and traditional algorithm with S-curve acceleration and deceleration characteristics. Combining the actual needs of a desktop robot, the five-stage algorithm with S-curve acceleration and deceleration characteristics is the most appropriate.

As shown in Figure 2, the time slots of T1, T2, T3, T4, T5 represent the phases of increasing acceleration, decreasing acceleration, constant speed, increasing deceleration and decreasing deceleration respectively. The phases of increasing acceleration T1 and decreasing acceleration T2 are the increasing-sensitive periods. The phases of increasing deceleration T4 and

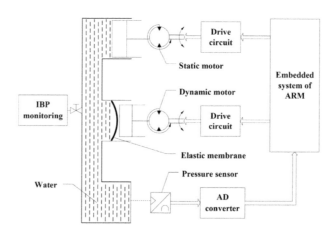

Figure 1.   The illustration schematic of the system.

decreasing deceleration T5 are the decreasing-sensitive periods. In order to achieve the acceleration derivative equal to zero at the start and end jerks at decreasing deceleration stage, T1 should be equal to T2. Herein, j represents acceleration derivative, to achieve a smooth operation process, the absolute value of j is set to a constant. As shown in Figure 2, the velocity at the start point and end point for the five-stage algorithm is zero, fulfilling the system requirements. Furthermore, the acceleration derivative is continuous for the algorithm, which facilitate the system flexibility (Ma yongchao et al. 2014).

In the control system of the stepper motor, one pulse signal leads to one-step moving or rotating. Assuming the pulse number is N, at a certain time T, its frequency f with respect to t can be expressed as:

$$f(t) = f_m - f_m \exp(-t/\tau) \tag{1}$$

In the above equation, f_m is the maximum frequency of the stepper motor, τ is the time constant to determine the increasing speed. The uniform speed for a certain system and the time to achieve the maximum speed can be experimentally determined.

Our system utilizes an embedded ARM timing interrupt method to control the velocity of the motor. The speed tuning is realized by changing the load value of timer. Considering the speed increasing process as discrete segments, and set the acceleration time as: T = T1 + T2, i.e., increasing acceleration stage and decreasing acceleration stage. If T is equally divided to 40 time slots, with both T1 and T2 divided to 20 time slots respectively, so that the time interval for neighboring slots is Δt = T/40. The frequency for each gear can be calculated by Eq. (1) and the number of steps for each shift can be figured out as well.

3.3   Core control chip

Our system utilizes STM32F 103 processor as the controller. The processor is a 32 bit standard RISC processor with ARMv7-M architecture based on Cortex-M3 core, featuring in high coding efficiency and fully utilizing the performance of ARM core in the storage space of 8-bit and 16-bit systems. This series of microprocessor has a working frequency of 72 MHz, a built-in flash memory up to 128 K, a 20 K-bit SRAM, and numbers of I/O ports, thus have great developing potential in many fields such as motor driving, real-time control, PC game peripherals

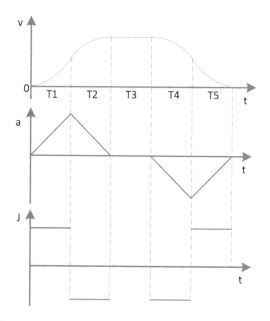

Figure 2.   Five-stage S-curve.

and air conditioning (Oxenham AJ et al. 2003). The embedded processing has advantages in terms of low-cost, high performance, rapid response, and abundant on-chip resource, guaranteeing its capability on both functional control and data transmission for our system.

3.4 Design of motor drive circuit

The motors for simulating both the static and dynamic pressure signal are driven by the motor driver chip TA8435H. It is a sinusoidal micro step bipolar stepping motor driver, featuring in simple circuit and high reliability (Wei Houjie. 2011).

The chopping frequency of the chip is determined by the capacitance of the external capacitor connected to Pin 4. In this work, 0.01 μF is used. The electrical potential is assumed to be high for REF IN pin, as the driving current needed for the motor is 0.1 A, Rnf (R17, R18) = 0.8 Ω. Stepper motor interface requires the use of fast recovery diodes (D2–D5), to discharge the winding current, as shown in Figure 3.

In order to improve the reliability of the hardware and effectively suppress interference, an electrical potential isolation circuit consisting of TLP521-4 TLP521-2 optocouplers is added between STM32F103 and TA8435H chips, to isolating the control signal from IO port for the STM32F103 processor. As shown in Figure 3, the isolated conversion of electrical potential is realized through the optocoupler TLP 52 connected between PWM output pin and IO port of the STM32F103 processor.

3.5 Pressure detection circuit

Real-time pressure detection in the liquid chamber is required in this system. Based on the pressure curve, the operating parameters of the stepper motor is corrected, while the pressure value is used to depict the pressure variation waveform with the relevant parameters displayed simultaneously. In this system, high-precision pressure sensor is adopted to concert the pressure value to electrical voltage between 0 and 5 V, which is then converted to digital data by AD7607. AD7607 is a fully integrated, multi-channel data acquisition chip, enabling 16-bit conversion without missing code even with high-noise power supply (Lu Yapu et al. 2014).

3.6 User interfacial module

The system adopts an 8-inch TFT 800*600 color LCD touch screen as a display terminal. This screen is free of the problems such as fonts and image display, and directly uses a RS232 serial port to communicate with the host STM32F103 processor. The GUI graphics is directly programmed to the built-in flash, and the host ARM sends HMI command offered by Devon for display control. Users can input data or control command through touch screen, which is sent to STM32F103 processor by RS232 for further process.

3.7 Design of the system software

To implement the graphic display functionality for user interaction, our system adopts an upgradable embedded Graphical User Interface (GUI) platform. Besides, a TFT-LCD touch

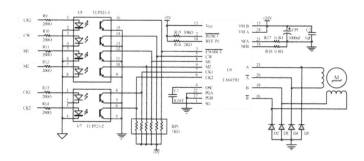

Figure 3.   Motor drive circuit.

304

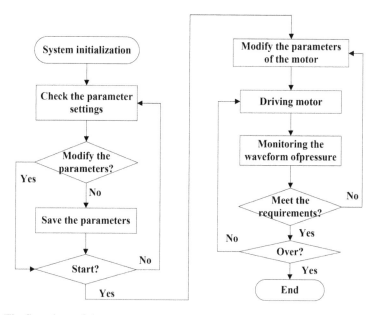

Figure 4.   The flow chart of the master program.

Table 1.   Static pressure simulation results.

Reference/ mmHg	Test times	Average value	Standard deviation
300	10	300.23	0.46
240	10	240.20	0.39
150	10	150.18	0.33
60	10	60.14	0.17
30	10	29.98	0.17
0	10	0.02	0.26
−30	10	−29.96	0.20
−60	10	−59.91	0.15

Table 2.   Simulation results of dynamic pressure at 1 Hz.

	Reference	Test times	Average value	Standard deviation
Systolic pressure/mmHg	120	10	119.86	0.56
Diastolic pressure/mmHg	60	10	59.91	0.47
Frequency/Hz	1	10	1.03	0.11

screen with flash memories used, enabling store all the images of graphical interface. Relevant parameters and commands communicate with the master chip STM32F103 through the serial port, no additional embedded operating system. Our design allows the prototype compact, efficient and shorten developing cycle.

Firmware mainly consists of master program, motor drive, touch screen module, display module, AD conversion module, and RS232 communication. Entire program uses modular design method. The initial configuration and display monitoring is implemented by the master function, while the rest modules adopt interrupt-driven approach to improve the systematic efficiency. The flow chart of the master program is shown in Figure 4.

Table 3. Simulation results of dynamic pressure at 10 Hz.

	Reference	Test times	Average value	Standard deviation
Systolic pressure/mmHg	120	10	114.06	1.62
Diastolic pressure/mmHg	60	10	63.47	0.73
Frequency/Hz	10	10	9.65	0.36

4 EXPERIMENTAL RESULTS

The prototype is tested with the preset static pressure between −60 mmHg and 300 mmHg. For the dynamic pressure, the frequency is set to be 1 Hz, and the systolic and diastolic pressure is set to be 120 mmHg and 60 mmHg respectively. A pressure transducer TS110 and a Tektronix oscilloscope DPO5054B are used to monitor the simulation results, and evaluate the repeatability and measurement accuracy. The tranducer has a measurement range of −600 mmHg ~ +600 mmHg, with the detection accuracy of 0.15% and the output voltage up to 5 V.

The generated static pressure is adjusted to 300 mmHg, 240 mmHg, 150 mmHg, 60 mmHg, 30 mmHg, 0 mmHg, −30 mmHg, −60 mmHg for subsequent tests. As shown in Table 1, the repeatability and pressure error is acceptable for our prototype.

The test result of dynamic pressure is shown in Table 2. The error of the dynamic pressure at 1 Hz is less than 1%, the attenuation of the frequency response doesn't exceed 0.5 dB for sinusoidal pressure signal of 10 Hz with respect to 1 Hz.

5 CONCLUSION

In this work, we designed an IBP simulator based on the latest standards for IBP monitoring equipment. The use of stepper motors for controlled pushing of the elastic membrane results in the pressure change in the liquid channel. The pressure change is real-time monitored and displayed based on an ARM platform. Experimental result of the prototype demonstrates that the simulator can produce high accurate static and dynamic pressure, which totally fulfills the requirements of the industry standards, thus showing its feasibility. Our approach represents a convenient alternative for the test of IBP monitoring equipment according to the standards.

ACKNOWLEDGMENTS

The author thanks the support of the Medical Research Foundation of Guangdong, grant B2014071.

REFERENCES

Hu Xiangqin, WangChunmei. Research progress on invasive monitoring of blood pressure for patients [J]. *Chinese Nursing Research*, 2008, 22(193–195).

Lu Yapu, Wang Haiyan Zhang Qingpengetl. Application and design of AD7606 in the logging acquisition module [J]. *Electronic Measurement Technology*, 2014, 37(1):105–108.

Ma yongchao. Study of a Five–segments–curve Acceleration and Deceleration Algorithm [J]. *Industrial control computer*, 2014, 27(12):60–62.

Oxenham AJ, Shera CA. Estimates of human cochlear tuning at low levels using forward and simultaneous masking. *J Assoc Res Otolaryngol*. 2003;4(4):541–554.

WeiHoujie. Design for Automatic Control System of Sampling [J]. *China Medical Device Information*, 2011, 12:19–22.

YY0783-2010, Medical electrical equipment—Part 2–34: Particular requirements for the safety, including essential performance, of invasive blood pressure monitoring equipments [S]. *Standards Press of China*, 2010:11–12.

Bioinformatics and Biomedical Engineering – Chou & Zhou (Eds)
© 2016 Taylor & Francis Group, London, ISBN 978-1-138-02784-8

Needle guide device development for CT system-based biopsy

H.K. Yun, G.R. Park, K.C. Choi, T.S. Shin & M.K. Kim
Department of Biomedical Engineering, Seoul Asan Medical Center, Seoul, Korea

S.K. Joo
College of Medicine, University of Ulsan, Ulsan, Korea

ABSTRACT: If the CT system is utilized in the biopsy of a patients' lesion area, the entry angle of the biopsy needle is critical for precise and safe biopsy needle penetration to a desired tissue area. CT imaging is a way to utilize the CT system in biopsy, in which the needle entry angle is planned and a physician follows the planned angle while an assistant watches an angle meter with the unaided eyes to provide assistance for biopsy precision. As such observation with unaided eyes is prone to large errors, the CT imaging and correction procedures should be repeated several times to improve the precision of the needle entry angle, elevating the patients' pain and exposure dose. This study researched methods necessary for a precise and efficient CT system-based biopsy and presented relevant examples.

1 INTRODUCTION

For the purpose of this study, biopsy physicians, biopsy-assisting radiologists and medical device-managing biomedical engineers met together to discuss the most cost-effective method and primarily how to reduce the patients' pain and exposure dose.

1.1 *CT system-based biopsy*

Needle biopsy is performed in order to acquire a tissue necessary for pathological diagnosis. It can be done by utilizing various imaging devices for diagnosis. Of them, the CT system-based method is useful to detect small lesions by offering 3D localization of a lesion area for users to check the needle and tissue locations. This method is illustrated in Figure 1. As for areas with overlapping tissues, in particular, the CT system method provides a more excellent detection performance than the ultrasound-guided or X-ray fluoroscopic-guided procedures. As the method shows the relationship of the tissue of interest with its surrounding tissues, physicians can easily set a bypass route around the lesion according to operation procedures in more diversified positions of the patient.

1.2 *Necessity of research*

The practitioner checks the needle entry point and the calculated angle in the CT image, and performs biopsy by following the guide wires and the guided angle while the assistant watches the angle meter with unaided eyes.

Such a conventional method determines a needle entry angle by relying on the assistant's eye measurement, embedding a strong likelihood of angle errors. This method is shown in Figure 2. For this reason, the entry angle should be modified several times, so the CT image-taking frequency rises along with examination time delay, patients' exposure dose and pain levels.

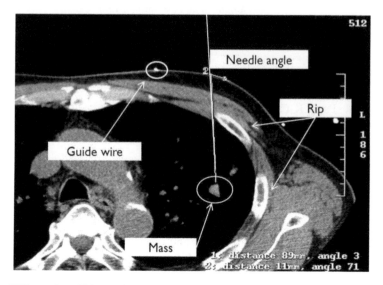

Figure 1. CT image-based biopsy plan. Guide wires attached to a patient's skin for CT biopsy is easily identifiable in CT images. Based on the CT image, the tissue of the biopsy area, needle entry point and angle are planned for biopsy.

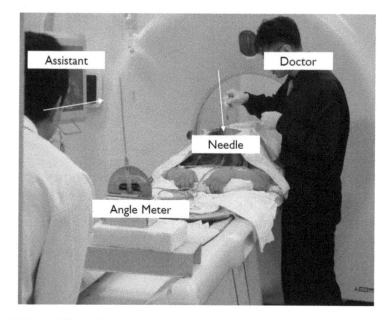

Figure 2. Existing CT-based biopsy.

2 LASER GUIDE METHOD

In order to improve the needle entry angular precision, the line laser was used in this study to guide the needle (Haesang J et al. 2012). It was found that using one laser unit did not provide an easily discernable laser line for practitioners. So, three line laser units were utilized herein and installed by spacing 2.5 mm each on their left and right sides, as shown in Figures 3 and 4.

The laser module-fixing base was made of acryl for easy mobility, whose height is 30 cm and width 80 cm that is equal to the patient's table width of the CT system. The laser

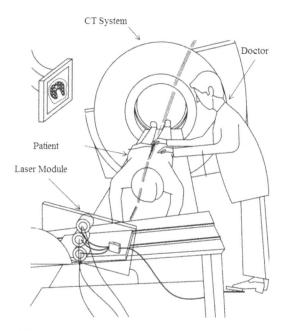

Figure 3. Image of applying the model to the actual biopsy.

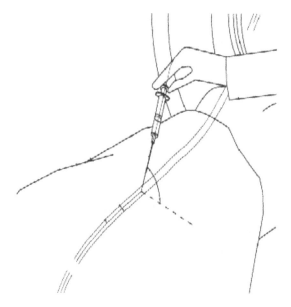

Figure 4. Enlarged image of the actual biopsy.

module-fixing areas were adjusted to different angles. To move the laser to the left and right sides, rails were installed on the acrylic base (Albert P M et al. 2002). At the left end of the rail, an angle meter was placed to set the angles. 6V batteries and a switch were installed to supply power to the laser unit, as shown in Figure 5.

A preliminary model was tested, and it was found that the laser applied to the curved areas was less visible. So, the acrylic support height and table length were calculated to make the sliding rail tilt by 15 degrees, where the laser module was mounted, as shown in Figure 6.

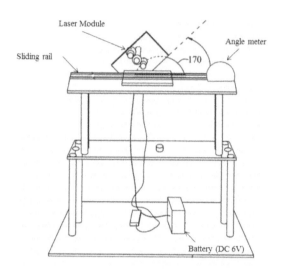

Figure 5.   Preliminary model.

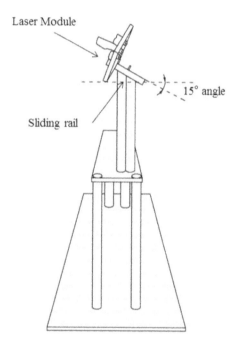

Figure 6.   2nd model.

3   RESULTS

The model was utilized clinically for approximately 100 times, and benefits were found in diverse aspects.

3.1   *Procedure time*

The procedure time was compared between the laser device developed herein and the conventional eye measurement-based method.

Table 1. Procedure time.

| Examination | Procedure time (min) | | Outcome |
	Existing method	Laser device-based method	
PCNA	30~40	20~30	29% reduction
RFA	30~50	30~40	12% reduction

Table 2. CT scanning frequency.

| Examination | No. of CT scanning frequency (times) | | Outcome |
	Existing method	Laser device-based method	
Lung bx	8~10	6~8	23% reduction
Bone bx	5~8	3~6	31% reduction
RFA	8~10	6~8	23% reduction

Table 3. Radiation exposure dose.

| Examination | Dose length product (mGy) | | Outcome |
	Existing method	Laser device-based method	
Lung bx	30~60	20~40	33% reduction
Bone bx	200~250	150~200	33% reduction
RFA	500~600	400~500	18% reduction

The former shortened the procedure time by 29% in the PCNA (Percutaneous Needle Aspiration) procedure on average and by 12% in the RFA (Radio Frequency Ablation) on average.

3.2 CT scanning frequency

The Needle Guide Device method reduced the CT scanning frequency by up to 31%, thanks to the more precise targeting of patient lesion areas compared with the eye measurement method.

3.3 Radiation exposure dose

DLP (Dose Length Product) is defined as the total measured value of the dose and is calculated by multiplying the CTDI volume value by the scan length. It refers to the dose to which a patient is actually exposed. With the laser device in place, as high as 33% of radiation exposure has been found to decrease compared with the conventional method. This is a considerable change and important for patient safety.

3.4 Other benefits

By guiding with the laser beam in a precisely calculated angle, the device helps practitioners operate faster and safely with confidence. Once an angle is set, it is maintained throughout the procedure, and the device could help reduce the number of personnel staying in the aseptic operation space, further lowering the infection risk of the patient.

4 CONCLUSION

The devices developed herein are at the stage of idea realization, so they need to be improved in several aspects before applying to clinical cases. First, it is a cost issue. The total device frame was made of low-cost acryl, posing a potential safety and size issues. It may not be the optimal design to put on the CT table. Further improvements in material and shape are necessary.

Second, the angle meter attached for angle check could also cause errors due to a leveling issue. Concerning the angle adjustment of the sliding mobile laser module, addition of an automatic operation device using a motor is necessary, so that the angle adjustment could be done in an outside control room instead of doing it inside the CT room manually. The device can be continuously improved for a safer and reliable operation based on the constant end-user opinion collection.

REFERENCES

Albert, P.M. 2002. *Electronic Principles*, Seoul: McGraw-Hill Korea.
Haesang, J. 2012. *Optics and Lasers*, Seoul: Iljinsa.

Bioinformatics and Biomedical Engineering – Chou & Zhou (Eds)
© 2016 Taylor & Francis Group, London, ISBN 978-1-138-02784-8

A novel target selection approach by incorporating image segmentation into P300-based Brain-Computer Interfaces

P. Du, Y. Yu, E.W. Yin, J. Jiang, Y.D. Liu & D.W. Hu
College of Mechatronics and Automation, National University of Defense Technology, Changsha, Hunan, China

ABSTRACT: A P300-based Brain-Computer Interface (BCI) can achieve a target selection task by detecting only the human brain activities. In conventional P300-based BCIs, the row-column mode was widely used to modulate the stimulations of the targets. However, when extending the P300-based BCIs to practical applications, the regular stimulation mode is insufficient to reflect the complex target information in actual environments. To address this problem, we propose a novel target selection approach by incorporating the image segmentation method into the P300-based BCIs. In this approach, the image of the environment was captured by a camera, and partitioned using the Entropy Rate Super-pixel Segmentation (ERS) algorithm. Then, a random flash stimulation was embedded on each segment of the image to evoke the P300 signal. A two-step mechanism was used in our BCI system, where a group containing the target was selected first and then the target was selected from this group. To verify the performance of our approach, a target selection experiment was performed in different real environments. The average online accuracy in the experiment for five subjects was found to be 83.4% using our proposed approach. The results showed that the feasibility and practicality of the P300-based BCIs for target selection was improved by incorporating the image segmentation method.

Keywords: Brain-Computer Interface (BCI); P300; image segmentation; Entropy Rate Super-pixel Segmentation (ERS); target selection

1 INTRODUCTION

A Brain-Computer Interface (BCI) is a type of system that can straightly acquire signals from the human brain and translate them into digital commands (Wolpaw, 2002). The BCI establishes an alternative communication channel between the human brain and the computer, which can help disabled patients to promote their quality of life, and make them more independent or with less costs in social work (Scherer et al, 2004). EEG is a non-invasive signal collection method, which is safe, convenient and acceptable (Brunner, 2009). In recent years, different types of brain activity have been reflected in EEG signals and used in the BCI such as the P300 component event including potentials (ERPs), mu and beta rhythms, slow cortical potentials and visual evoked potentials (Pfurtscheller et al, 1993; Kostov & Polak, 2000; Polich, 1999; Farwell & Donchin, 1988). One type of the BCI system speller is based on the P300 event-related potential: a positive peak about 300 ms after the identification of a stimulus evoked in the brain can be observed in EEG signals.

Since its proposal by Farwell and Donchin in 1988 (Farwell & Donchin, 1988), P300 paradigms have been widely investigated in different ways. Conventional paradigms are based on the Row-Column mode (RC), in which all the targets are distributed in a regular matrix, such as a 6×6 matrix displayed on a screen. These targets are flashed in rows and columns in random order and the flashing of the rows and columns evokes the P300 feature. Then, the character is selected by the intersection of the target row and column. Various transformations of the RC mode have also been developed, e.g. Region-Based paradigm (RB), Single Character paradigm (SC) and chessboard paradigm (Fazel-Rezai, 2009; Guan et al, 2004; Townsend, 2010), to improve the performance. To date, the P300-based BCIs have been widely used in various areas, such as character speller systems. However, extending this type of the BCI system to target selection in real environments is still a challenge (Sellers, 2010; Cecotti, 2011). In the real environment, the desired target is more complex, which always varies in shape, location and size. The regular RC mode stimulus cover in the actual target is insufficient to reflect the complex target information. Moreover, it is difficult for the traditional P300 stimulation modes to locate the targets with a random distribution in actual conditions.

In this paper, a novel target selection approach by incorporating image segmentation into the P300-based BCI paradigm is proposed. Using the image segmentation method, the targets can be extracted from the environment with detailed shape, size and location information. The Entropy Rate Super-pixel Segmentation (ERS) algorithm was used to divide the picture into multiple homogeneous regions, which include the desired targets (Ming-Yu et al, 2014; Chanchan et al, 2014). After partitioning by ERS, an adaptive stimulation is embedded on the image segments to evoke the P300 feature. To improve the accuracy, a two-step mechanism was used in our BCI system. To verify the performance of our approach, some experiments were performed to show the feasibility and practicality of the two-step P300-based BCIs.

This paper is organized as follows: image segmentation by ERS, the paradigm design, experimental procedures and datasets are described in Section 2; the experimental results are presented in Section 3; the discussion is presented in Section 4; the conclusions and ideas about future work are presented in Section 5.

2 METHOLOGY AND MATERIALS

2.1 *Image segmentation by ERS*

In our approach, the picture obtained from the camera (Fig. 1a) can be partitioned into a desired number (in this paper, the number is 36). To fulfill the requirement, a super-pixel segmentation method was chosen. It has been proved that the Entropy Rate Super-pixel Segmentation (ERS) proposed by Ming-Yu Liu can divide the picture into multiple homogeneous regions, which can achieve our requirement (Ming-Yu et al, 2014; Chanchan et al, 2014). As shown in Figure 1b, the picture (Fig. 1a) was partitioned by using the ERS algorithm and then we obtained thirty-six compact, homogeneous and balanced clusters (regions). The two dogs and the desk were the desired targets and the other regions are partitioned to help evoke the P300 feature. Then, each region is embedded in the stimulus in the same shape as itself.

Figure 1.   The proposed approach processing: (a) the background picture, (b) the picture was divided into 36 regions (by ERS), (c) all the 36 regions were placed into six groups and different colors represented different groups, (d) the group containing the target dog is selected from six groups, (e) six different regions in the group are selected in step 1, and (f) the target dog is selected in step 2.

2.2 Stimulation paradigm

After the image segmentation, a random flash stimulation was embedded on each segment of the image to evoke the P300 feature. The operation processing of our paradigm run was as follows: read a picture from the camera (Fig. 1a); do image segmentation (Fig. 1b); divide the regions into six groups (Fig. 1c); select the group containing the target in step 1 (Fig. 1d); and select the target in step 2 (Fig. 1f). The target recognition is done with a two-step mechanism. In the first step, the picture after image segmentation is placed into six groups based on the pseudo-random principle, as shown in Figure 1c. In the second step, the target was selected from the group chosen in step 1, which contained six regions and the target was among them. For example, Figure 1d shows one group (purple color) selected in the first step. In the second step, the target is selected in the same way as in the first step: while the subject concentrated on the dog with a purple color, the dog was selected (Fig. 1d).

2.3 Experimental design

2.3.1 Subjects and data
Five healthy subjects who had no history of psychological or neurological disorders (5 males, age 23–29 years, mean age 25.6 years) voluntarily participated in our experiments. All subjects signed the written informed consent form, and all the experimental procedures conformed to the guidelines laid down in the Declaration of Helsinki. The EEG signals were recorded utilizing a BrainAmp DC Amplifier (Brain Products Gmbh, Germany). Applying the 64-channel international 10/20 system, five-channel active electrodes were placed at Fz, Cz, P3, P4, Oz and referenced to PO8 and grounded to Fpz. During the experiment, the impedance of each electrode was maintained below 10 kΩ before recording. The EEG signals were sampled at 250 Hz and band-pass filtered at 0.1–50 Hz.

2.3.2 Experimental procedure
During the experiment, subjects sat in a comfortable chair situated approximately 80 cm away from the monitor. The design and the purpose of the system were explained to each subject before the experiment. Additionally, each subject spent some time to be familiar with the experiment before the initiation. Before the online test, a training session was performed to estimate the parameters of the classifier. Then, this classifier was loaded and used to finish three runs in the online selection. In each run, subjects were required to complete six trials. For each region or target, the stimulation flashed 5 times in one trial, and the

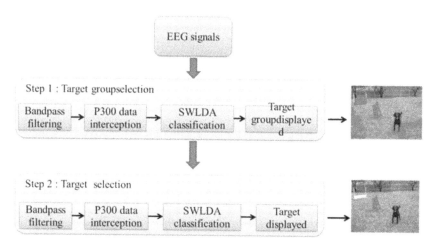

Figure 2. The data analysis framework of our target selection approach. The P300 signal processing was working as shown in the diagram. The target selection process system is also illustrated.

total number of flash times in one trial was $6 \times 5 \times 2 = 60$ times, occurring in a total of 60×240 ms $+ 2$ s $= 14.6$ s. The time between two trials was 2 s.

2.4 EEG signal processing

The EEG data were filtered, as mentioned above, with a band-pass filter at 0.1–50 Hz before the extraction of the P300 feature. Then, the 800 ms segments of data with 200 samples, obtained from the onset of each stimulus, were extracted for P300 feature analysis. Stepwise Linear Discrimination Analysis (SWLDA) was chosen to predict the suitable variables, which discriminates the target and non-target flashes (Fig. 2). It has been shown that the approach performs better in P300 signal processing compared with other classification approaches (Krusienski et al, 2006). The SWLDA was selected to train a classifier to calculate the scores of the P300 response.

3 EXPERIMENTAL RESULTS

3.1 Selection accuracy

Table 1 presents the online performance for the five subjects. The error in Table 1 recorded the wrong selection times made by the subjects against the suggested target in step 1 and step 2. The error times help us to calculate the selective accuracy for our paradigm. Because the decision is made by two steps, the accuracy is not straightforward compared with conventional paradigms. We used a method to calculate the accuracy in both step 1 and step 2. It is worth noting that if the wrong region was selected at the first step, an error was considered for target detection. The specific calculation method is given by

$$SA_1 = e_1/N \tag{1}$$

$$SA_2 = e_2/(N - e_1) \tag{2}$$

where SA_1 denotes the accuracy in step 1; SA_2 denotes the accuracy in step 2; e_1 and e_2 denote the error number in step 1 and step 2, respectively; and N denotes the total number of selection times in the experiment of each subject. Then, we have

$$\text{Total accuracy} = SA_1 \times SA_2 (\%) \tag{3}$$

From Table 1, it can be seen that the average accuracy was 88.1% in step 1 and 94.0% in step 2, respectively. The total target selection average accuracy for all subjects in the experiment was 83.4%. All the subjects performed better in step 2 than in step 1. In Table 1, it can also be seen that different subjects performed with a high discrimination of accuracy. YCX had the worst performance with 76.7% total accuracy, and DP had the best performance with 90% total accuracy.

Table 1. The accuracies for the online performances of the five subjects.

Subject	Task number	Step 1 error	Step 2 error	Step 1 accuracy	Step 2 accuracy	Total accuracy
DP	30	2	1	93.3%	96.4%	90.0%
YCX	30	5	1	83.3%	96.0%	80.0%
YY	30	3	1	90.0%	96.3%	86.7%
WJJ	30	3	2	90.0%	92.6%	83.4%
ZNN	30	4	3	86.7%	88.5%	76.7%
AVG	30	3.4	1.6	88.1%	94.0%	83.4%

 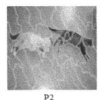  

P1   P2   P3   P4

Figure 3.   Four background pictures divided by ERS.

Table 2.   The total accuracy (%) in different background pictures.

	P1	P2	P3	P4	AVG
DP	86.7	90.0	90.0	84.6	87.8
WJJ	76.7	86.7	83.4	76.7	80.9
YCX	75.3	86.7	80.0	73.3	78.8
AVG	79.6	87.8	84.5	78.2	82.5

3.2   Performance in different real environments

Figure 3 shows four different scenarios containing people, animals and a complex environment divided by ERS. By segmentation, we sketch the contours of the desired targets (as shown in Figure 3P1–P4). Furthermore, the details of the targets are partitioned, in order to benefit a more complicated operation. For example, if the face of the female, shown in Figure 3P1, is the desired target, then the subject only needs to concentrate on the face in our experiment.

In this section, three subjects were asked to perform the same experiment under the different background pictures (see Fig. 3). Table 2 shows the selection accuracy of the three subjects in the four background pictures. As shown in Table 2, the overall accuracy of the experiment was 82.5%. The discrimination among the subjects was significant across all the pictures, with 87.8% higher accuracy and 78.8% lower accuracy. The subjects performed better in P2 and P3 with 87.8% accuracy and 84.5% accuracy, respectively.

4   DISCUSSION

To extend the application of P300 in the real environment, in this paper, we propose a novel target selection approach by incorporating the image segmentation method into the P300-based BCIs. The average total accuracy was 83.4% (see Table 1), proving that the approach is feasible and practical. Generally, the target in the real environment is complicated and the outline of the target is always irregular or peculiar-looking. In our paper, the proposed approach can overcome the problem. From, Figure 3, it has been shown that no matter what shape the target is, we can separate it from the picture or get a region containing the target through the image segmentation approach. Furthermore, the results shown in Figure 1d, f suggested that the stimulation, which was similar to the segment region, can be embedded on each group or target to evoke P300 signals successfully.

A two-step mechanism was used in our BCI system, and has been proved that it can achieve a higher accuracy than the single target paradigm (Pan, 2013). In our study, the flash time in one trail was 14.6 s. If the target is selected in only one step similar to SC, the flash times will be $36 \times 5 = 180$ times, occurring in a total of 180×240 ms $= 43.2$ s. The efficiency of our two-step mechanism is significantly higher than the one-step mechanism only. Furthermore, our approach is proposed for the application in the real environment, which require a higher efficiency. Thus, a two-step mechanism is used in our research.

Additionally, in Table 2, it was shown that the performance of all subjects in step 2 was better than that in step 1. Previous studies have reported the adjacent problems and showed

317

that the flashes of the non-target near to the target may attract the user's attention and produce P300. The findings that step 2 performed better than step 1 can be interpreted that the adjacent area of contact in step 1 was larger than that in step 2. As a result, the subjects in step 1 scattered more attention in the non-target area near to the target than those in step 2.

Partitioned by ERS, the target in the graph can be sketched and retained its most original information (see Fig. 3). The discrimination among the subjects is significant in Table 2, which indicates that the accuracy is closely related to the complexity of the background image. The three subjects obtained an average selection accuracy of more than 75% (Table 2) in four different background pictures, which suggested that the segmentation region-based paradigm can help to select the target in different conditions.

5 CONCLUSION

In this paper, we proposed a novel target selection approach by incorporating image segmentation into P300-based BCIs. The target was selected in two steps, which were based on the regions divided by ERS. Additionally, the experimental results with the five subjects suggested that the approach can be executed effectively and the accuracy can be achieved in the application of P300. In future work, to improve the performance of our system, we will perform additional studies that will adopt advanced image segmentation methods and more reasonable grouping methods to reduce the adjacent problem. Furthermore, we will conduct more research to increase its accuracy and reduce human error to extend the application of P300.

REFERENCES

Brunner, P. & Ritaccio, A.L. & Emrich, J.F. & Bischof, H. & Schalk G. 2009. A Brain-Computer Interface Using Event-Related Potentials (Erps) and Electrocorticographic Signals (Ecog) in Humans. *Epilepsia* 50: 389–389.
Cecotti, H. 2011. Spelling with non-invasive brain-computer interfaces-current and future trends. *Journal Physiol* 105: 106–114.
Chanchan, Qin. & Guoping, Zhang. & Yicong, Zhou. et al. 2014. Integration of the Saliency-based Seed Extraction and Random Walks for Image Segmentation. *Neurocomputing* 129: 378–391.
Farwell, L.A. & Donchin, E. 1988. Talking off the top of your head: Toward a mental prosthesis utilizing event-related brain potentials. Electroenceph. *Clin. Neurophysiol* 70(6): 510–523.
Kostov, A. & Polak, M. 2000. Parallel man-machine training in development of EEG-based cursor control. *IEEE Trans. Rehab. Eng* 8(2): 203–205.
Krusienski, D.J. & Sellers, E.W. & Cabestaing, F. et al. 2006. A comparison of classification techniques for the P300 speller *J. Neural Eng* 3: 299–305.
Ming-Yu, Liu. & Tuzel, O. & Ramalingam, S. & Chellappa, R. 2014. Entropy-Rate Clustering: Cluster Analysis via Maximizing a Submodular Function Subject to a Matroid Constraint. *IEEE Trans Pattern Anal Mach Intell* 36(1): 99–112.
Pfurtscheller, G. & Flotzinger, D. & Kalcher, J. 1993. Brain computer interface: A new communication device for handicapped persons. *J. Microcomput. Applicat* 16(3): 293–299.
Polich, J. 1999. P300 in clinical applications in Electroencephalography: Basic Principles, Clinical Applications and Related Fields (ed.). *Urban and Schwartzenberger:* 1073–1091.
Scherer, R. et al. 2004. An asynchronously controlled EEG-based virtual keyboard: Improvement of the spelling rate. *IEEE Trans. Bio-Med. Eng* 51(6): 979–985.
Sellers, E.W. & Vaughan, T.M. & Wolpaw, J.R. 2010. A brain-computer interface for long-term independent home use. *Amyotrophic Lateral Sclerosis* 11: 449–455.
Townsend, G. & Lapallo, B. et al. 2010. A novel P300-based brain-computer interface stimulus presentation paradigm moving beyond rows and columns. *Clin. Neurophysiol* 121: 1109–1120.
Wolpaw, J.R. & Loeb G.E. et al. 2006. BCI Meeting 2005: workshop on signals and recording methods. *IEEE Trans Neural Syst Rehabil Eng* 14: 138–141.
Wolpaw, J.R. et al. 2002. Brain-computer interfaces for communication and control. *Clin. Neurophysiol* 113(6): 767–791.

Bioinformatics and Biomedical Engineering – Chou & Zhou (Eds)
© *2016 Taylor & Francis Group, London, ISBN 978-1-138-02784-8*

Rapid identification of *Panax ginseng* and *Panax quinquefolius* using SNP-based probes

R.L. Wang, D.J. Gu, C.R. Hou & Q.J. Liu
State Key Laboratory of Bioelectronics, Southeast University, Nanjing, Jiangsu, China

ABSTRACT: An effective, easy and rapid method of distinguishing *Panax* species is urgently needed, for traditional methods are time-consuming and inaccurate. Here, universal primer for multiplex PCR, and probes containing the 2 SNPs and an additional mismatch were designed in ITS2 region to distinguish *P. ginseng* and *P. quinquefolius*. In this system, multiplex amplification was integrated with hybridization, making the single test costing only 2–4 h. The detection sensitivity of this system reached 2 pg per reaction. Results showed the system can be applied to effectively identification of *P. ginseng* and *P. quinquefolius*.

1 INTRODUCTION

SNP genotyping is the measurement of genetic variations of Single Nucleotide Polymorphisms (SNPs) between members of a species (Curk et al., 2015, Silva-Junior et al., 2015, Drywa et al., 2014). SNPs provide a genetic fingerprint for use in identity testing (Caceres and Gonzalez, 2015, Chagne, 2015). Recently, SNP microarrays were widely used in identifying allele genotyping for high throughput (Dalton-Morgan et al., 2014, Lapegue et al., 2014, Chagne, 2015, Silva-Junior et al., 2015).

P. ginseng, also known as Chinese ginseng, owning strong tonic efficacy. *P. quinquefolius*, named American ginseng, is similar in appearance to Chinese ginseng, and taken as sweet and a little bitter in flavor, however, cold in nature (Chen et al., 2008). These two medicines are very similar in morphological features, making it very difficult to identify by traditional methods alone.

Lots of methods have been developed to identify *Panax* species for such as DNA barcoding identification method, composed of PCR-based and other molecular identification methods (Choi et al., 2011, Kwon et al., 2009, Shaw and But, 1995, Shim et al., 2005). Researchers found that Internal Transcribed Spacer (ITS) exhibited the highest sequence divergence. Chen and co-workers (Chen et al., 2010) firstly validated the ITS2 region as a novel DNA barcode for identifying medicinal plant species.

Based on the published GenBank ITS2 sequences of *P. ginseng* and *P. quinquefolius*, two SNPs in ITS2 were found to exist stably. A combination of C and T at positions 32 and 43 (Chen et al., 2013), respectively, indicates *P. ginseng,* while a combination of T and C indicates *P. quinquefolius*.

Here, we used a pair of universal primers, through multiplex PCR and designed special probes containing the 2 existed SNPs and an additional mismatch to distinguish *P. ginseng* and *P. quinquefolius*. This system is composed of a PCR for multiplex amplification and hybridization with microarrays. The system was 100% consistent with expectations. This system could be applied to effectively distinguish traditional Chinese medicinal materials.

2 MATERIALS AND EXPERIMENTAL PROCEDURES

2.1 PCR material

All the Original PCR products of *Panax* species were genially offered by Jianping Han from Institute of Botany (The Chinese Academy of Sciences). The concentration of the DNA samples was evaluated by a NanoDrop 1000 spectrophotometer (NanoDrop Technologies, LLC, Wilmington, DE).

2.2 Chemicals and reagents

All the chemicals were from Sigma-Aldrich (Missouri, America). Taq DNA polymerase and dNTPs were purchased from TaKaRa (Dalian, China), DNA spotting buffer was purchased from Capitalbio Corporation (Beijing, China), Hybridization solution includes 3 × SSC (saline sodium citrate), 30% formamide, and 0.2% (w/v) SDS.

2.3 Primers and probes design

Each reverse primer was modified by adding a universal reverse primer (5′-CCACT ACGCCTCCGCTTTCCTCTGA-3′). The sequence of positive position probe was 5′-NH_2-poly (T) $_{20}$-cy3-3′ (Fig. 1a), and positive quality probe was 5′-NH_2-poly (T) $_{12}$-GAAAAATAAACTGTAAATCATATTC-3′ (Fig. 1b). Apart from the probes of 2 SNPs locations (Fig. 1c), another mismatch was designed the between the 2 SNP location at the site of guanine (Fig. 1d). All the primers and probes were synthesized from Invitrogen Co., Ltd (Shanghai, China).

2.4 Multiplex PCR reaction

The detection PCR was performed in a thermal cycler (Eastwin Biotech, China) with total volume of 25 µl, including the following reagents: 2 µl Template, 7.5 µl TaKaRa PCR-Taq master mix (Dalian, China), 5 µl primer mix and 10.5 µl of dd H_2O. The multiplex PCR program was set as following: 94 °C for 5 min; 8 cycles of 94 °C for 30 s, 51 °C for 30 s, 72 °C for 5 min; 20 cycles of 94 °C for 30 s, 62 °C for 30 s, 72 °C for 30 s, 10 cycles of 94 °C for 30 s, 65 °C for 30 s, 72 °C for 30 s; and 72 °C for 5 min; 4 °C for forever (Han et al., 2006).

2.5 Detection chip

In this study, the microarray chip was produced on the microscope slides. After washing, modified by organic chemical reagent, probes and spotting buffer (Capitalbio, Beijing, China) mixed up with a final concentration of 80 µM and printed on the slides by a Smart Array 48 Microarray Spotter. Figure 2 shows patterns of the array. Probes 1, 2, 3 and 4 were for *P. ginseng* and 5, 6, 7 and 8 for *P. quinquefolius*. Probe 1 and 5 contain 2 SNPs (Fig. 1c) and others contain 3 SNPs (Fig. 1d).

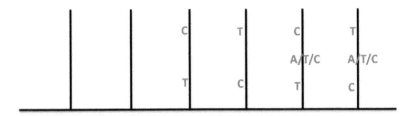

Figure 1. Schematic illustrations of probes designed in this study.
a) Position probes. b) Quality control probes. c) probes contains 2 SNPs. d) Probes contains 3 SNPs.

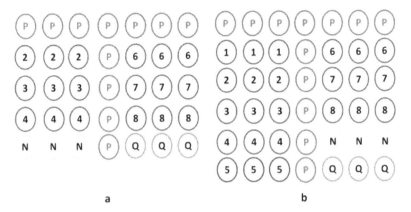

a   b

Figure 2.   The 2 patterns of the array used in this study.
a) probes contains 3 SNPs. b) probes contain 2 SNPs and 3 SNPs. P-positive position probe, N-negative
control probe, Q-positive quality control probe.

Microarray chip printed with probes then incubated at 37 °C for an hour. For the hybridization experiment, PCR products were diluted to 10 times added with hybrid quality with a
final concentration of 10 nM, incubated at 37 °C for half an hour. Following the hybridization, the microarray images were recorded by LuxScan 10 K microarray Scanner (Capitalbio, Beijing, China). The ratio of signal to noise (SNR) was used to judge the hybridization
results, and SNR of the positive spot was set as >1.5.

3   RESULTS

3.1   Detection system

The detection fragments were amplified by multiplex PCR. The probes of aminated oligo-
nucleotide were printed on the slides. The space and size of spot was set as 400 μm. PCR
products hybridized with probes printed on the slides at 37 °C and the slides were scanned by
the LuxScan 10 K microarray Scanner. And the SNR of all probes was calculated with the
background was set as 1.

3.2   Template gradient

Suitable template for PCR will benefit the hybridization results of the products. Template of
the PCR Products was 10 ng/μl and diluted to 10^{-2} ng/μl, 10^{-3} ng/μl, and 10^{-4} ng/μl. After the
multiplex-PCR, the PCR products were detected by agarose gel of 2% (Fig. 3). The result
shows template concentration of 10^{-3} ng/μl was better and chosen for the next step.

3.3   Hybridization results of all probes

To detect the distinguishing capacity of all SNP-based probes. PCR products of *P. ginseng*
were hybridized with all probes (pattern of Fig. 2b). Figure 4 shows the results and gives the
SNR of all probes. The perfectly matched probe P1 give the highest SNR of all the probes.
Probes of *P. ginseng* contains 1 SNP give higher SNR (at least 22) of all probes. Probes of
P. quinquefolius give the smallest SNR (the highest is about 5). There was 0 mismatch at the
P1 probe sites, 1 mismatch at P2, P3 and P4, 2 mismatches at P5, 3 mismatches at P6, P7
and P8. The SNR differs apparently when compared with sites of 0 and 3 mismatches, often
occurred in sites of 1 and 3 mismatches. The adding SNP in the probes helps the distinguishing process and differing in the SNR of the probes.

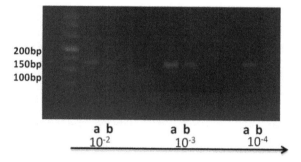

Figure 3. PCR product of template gradients (from 10^{-2} ng/μl to 10^{-4} ng/μl). a) *P. ginseng*. b) *P. quinquefolius*.

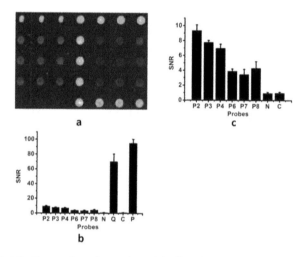

Figure 4. The hybridization results using probes with all probes.
a) Hybridization results. b) SNR of probes with background set as 1. c) SNR without position probe and quality positive probe.

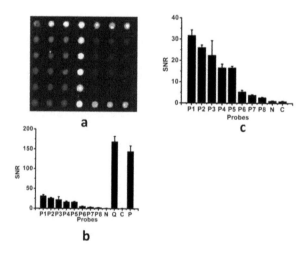

Figure 5. The hybridization results using probes with 3 SNPs.
a) Schematic illustrations of hybridization results. b) SNR of probes. c) SNR without position probe and quality positive probe.

3.4 Hybridization results of probes containing 3 SNPs

To detect the distinguishing capacity of the probes of 3 SNPs. PCR products of *P. ginseng* was hybridized with all probes (pattern of Fig. 2a). Figure 5 shows the results and gives the SNR of all probes. Probes of *P. ginseng* contains 1 SNP give higher SNR of all probe when compared with the Probes of *P. quinquefolius*, almost 2 times of that of *P. quinquefolius*.

4 DISCUSSIONS

SNP genotyping microarrays are promising tools with property of high-throughput in parallel and become research hotspots in lots of fields (Silva-Junior et al., 2015, Oh et al., 2015, Obidiegwu et al., 2015). Measuring SNPs in *Panax ginseng* and *Panax quinquefolius* applied for fast identification and quality control of the *Panax* species.

In this study, Multiplex PCR and SNP-based probes were designed. The results were accurate, the sensitivity of the detection system was 1pg per test, and the detection need only 2–4 hours, saving sample and time. Compared to other methodologies (Chen et al., 2013, Choi et al., 2011, Kwon et al., 2009), except for high-throughput, accuracy, sensitivity, and simplification, the system has other advantages, the low cost is attractive, and the price can be controlled within 1 dollar per test in mass commercialize production.

ACKNOWLEDGMENTS

Jianping Han from Institute of Botany (The Chinese Academy of Sciences) for origin PCR materials. This work was supported by National Basic Research Program of China (2011CB707600), the National Natural Science Foundation of China (61071050, 61372031), Tsinghua National Laboratory for Information Science and Technology (TNList), Cross-discipline Foundation and "the Fundamental Research Funds for the Central Universities", and the Foundation for the author of National excellent doctoral dissertation.

REFERENCES

Caceres, A. & Gonzalez, J.R. 2015. Following the footprints of polymorphic inversions on SNP data: from detection to association tests. *Nucleic Acids Res,* 43, e53.

Chagne, D. 2015. Application of the high-resolution melting technique for gene mapping and SNP detection in plants. *Methods Mol Biol,* 1245, 151–9.

Chen, C.F., Chiou, W.F. & Zhang, J.T. 2008. Comparison of the pharmacological effects of Panax ginseng and Panax quinquefolium. *Acta Pharmacol Sin,* 29, 1103–1108.

Chen, S.L., Yao, H., Han, J.P., Liu, C., Song, J.Y., Shi, L.C., Zhu, Y.J., Ma, X.Y., Gao, T., Pang, X.H., Luo, K., Li, Y., Li, X.W., Jia, X.C., Lin, Y.L. & Leon, C. 2010. Validation of the ITS2 Region as a Novel DNA Barcode for Identifying Medicinal Plant Species. *PLoS One,* 5.

Chen, X., Liao, B., Song, J., Pang, X., Han, J. & Chen, S. 2013. A fast SNP identification and analysis of intraspecific variation in the medicinal Panax species based on DNA barcoding. *Gene,* 530, 39–43.

Choi, H.I., Kim, N.H., Kim, J.H., Choi, B.S., Ahn, I.O., Lee, J.S. & Yang, T.J. 2011. Development of Reproducible EST-derived SSR Markers and Assessment of Genetic Diversity in Panax ginseng Cultivars and Related Species. *Journal of Ginseng Research,* 35, 399–412.

Curk, F., Ancillo, G., Ollitrault, F., Perrier, X., Jacquemoud-Collet, J.P., Garcia-Lor, A., Navarro, L. & Ollitrault, P. 2015. Nuclear species-diagnostic SNP markers mined from 454 amplicon sequencing reveal admixture genomic structure of modern citrus varieties. *PLoS One,* 10, e0125628.

Dalton-Morgan, J., Hayward, A., Alamery, S., Tollenaere, R., Mason, A.S., Campbell, E., Patel, D., Lorenc, M.T., Yi, B., Long, Y., Meng, J., Raman, R., Raman, H., Lawley, C., Edwards, D. & Batley, J. 2014. A high-throughput SNP array in the amphidiploid species Brassica napus shows diversity in resistance genes. *Funct Integr Genomics,* 14, 643–55.

Drywa, A., Pocwierz-Kotus, A., Dobosz, S., Kent, M.P., Lien, S. & Wenne, R. 2014. Identification of multiple diagnostic SNP loci for differentiation of three salmonid species using SNP-arrays. *Mar Genomics,* 15, 5–6.

Han, J., Swan, D.C., Smith, S.J., Lum, S.H., Sefers, S.E., Unger, E.R. & Tang, Y.W. 2006. Simultaneous amplification and identification of 25 human papillomavirus types with Templex technology. *J Clin Microbiol,* 44, 4157–4162.

Kwon, H.K., Ahn, C.H. & Choi, Y.E. 2009. Molecular authentication of Panax notoginseng by specific AFLP-derived SCAR marker. *Journal of Medicinal Plants Research,* 3, 957-U4001.

Lapegue, S., Harrang, E., Heurtebise, S., Flahauw, E., Donnadieu, C., Gayral, P., Ballenghien, M., Genestout, L., Barbotte, L., Mahla, R., Haffray, P. & Klopp, C. 2014. Development of SNP-genotyping arrays in two shellfish species. *Mol Ecol Resour,* 14, 820–30.

Obidiegwu, J.E., Sanetomo, R., Flath, K., Tacke, E., Hofferbert, H.R., Hofmann, A., Walkemeier, B. & Gebhardt, C. 2015. Genomic architecture of potato resistance to Synchytrium endobioticum disentangled using SSR markers and the 8.3 k SolCAP SNP genotyping array. *BMC Genet,* 16.

Oh, J.J., Park, S., Lee, S.E., Hong, S.K., Lee, S., Lee, H.M., Lee, J.K., Ho, J.N., Yoon, S. & Byun, S.S. 2015. Genome-wide detection of allelic genetic variation to predict biochemical recurrence after radical prostatectomy among prostate cancer patients using an exome SNP chip. *J Cancer Res Clin Oncol.*

Shaw, P.C. & But, P.P.H. 1995. Authentication of Panax Species and Their Adulterants by Random-Primed Polymerase Chain-Reaction. *Planta Medica,* 61, 466–469.

Shim, Y.H., Park, C.D., Kim, D.H., Cho, J.H., Cho, M.H. & Kim, H.J. 2005. Identification of Panax species in the herbal medicine preparations using gradient PCR method. *Biological & Pharmaceutical Bulletin,* 28, 671–676.

Silva-Junior, O.B., Faria, D.A. & Grattapaglia, D. 2015. A flexible multi-species genome-wide 60 K SNP chip developed from pooled resequencing of 240 Eucalyptus tree genomes across 12 species. *New Phytol,* 206, 1527–40.

Bioinformatics and Biomedical Engineering – Chou & Zhou (Eds)
© *2016 Taylor & Francis Group, London, ISBN 978-1-138-02784-8*

Study on the range of motion of laparoscopic instruments with a simulator

K. Lu & C. Song
School of Medical Instrument and Food Engineering, University of Shanghai for Science and Technology, Shanghai, China

C. Wei
School of Mechanical Engineering, University of Shanghai for Science and Technology, Shanghai, China

ABSTRACT: The purposes of this paper were to measure the range of motion of instruments during the laparoscopic operation, and to determine the influential factors. Two tasks (transferring and suturing) were set up in a laparoscopic simulator and three groups (Group A: transferring via two sites; Group B: transferring via one site; and Group C: suturing via two sites) were classified. A tracking system was used to record the instrument position. The inserted length and shift angles of the instruments were then calculated. The results indicated that the range of motion of instruments differed within and between the different groups. The range of motion of instruments during the laparoscopic operation was limited to a small amorphous spatial space. The type of operation, surgical approaches, and left-right hand are important factors to determine the range of motion. This study can be beneficial for motion analysis or instrument design via different surgical approaches.

Keywords: laparoscopic; instrument; range of motion; tracking

1 INTRODUCTION

Laparoscopic surgery, a representative of minimally invasive surgeries, has been widely adopted in clinics because of fewer traumas, less postoperative pain, reduced hospital stay and better cosmetic effect compared with open surgery (Cuschieri et al., 1991). Instruments were inserted into the abdomen through small incisions to complete the operation. A straight rigid instrument through one incision has four degrees of freedom (Breedveld et al., 1999), including translation, rotation and pivoting within the insertion point. Motion analysis of the laparoscopic instrument is one of the objective methods to evaluate surgical skills (Chmarra et al., 2007a, Reiley et al., 2011, Mason et al., 2013). The total path length, velocity, acceleration and movement smoothness of instruments are usually calculated as the objective parameters (Chmarra et al., 2007b, Oropesa et al., 2011, Sanchez-Margallo et al., 2014). However, the actual workspace of the instrument and the relative position to the operation platform have never been thoroughly discussed. Theoretically, the position that the instrument tip can reach inside the abdomen is a cone-shape space (Ma et al., 2010), but the range of motion of the instrument is limited to a certain spatial space during operation, and determined by the inserted length and shift angles of the instrument. In laparoscopic surgery, the overlap of the workspace between different instruments is common. When the insertion points of the instrument are very close to each other, e.g. Single-Incision Laparoscopic Surgery (SILS) (Romanelli and Earle, 2009, Lirici et al., 2011), there might be severe instrument conflicts, which will hamper the surgical operation. Therefore, the range of motion of the laparoscopic instrument and its relative position to the operation simulator can benefit for the motion analysis of the laparoscopic instrument, and may be used as the preliminary proof for an innovative instrument design.

An optical tracking system was used to record the position of instruments during operation in a laparoscopic simulator. Reference points on the simulator were measured and then a coordinate system was established. The inserted length and shift angles of each instrument were calculated. Differences between groups were compared and the influential factors are discussed.

2 MATERIALS AND METHODS

2.1 *Experimental setup*

The system consists of a laparoscopic simulator and a tracking system (Fig. 1). The laparoscopic simulator (SIMIT Scientific Co., Ltd., Shanghai, China) is a multiple functional system for laparoscopic training. Both traditional laparoscopy (LAP) and SILS can be simulated. In order to avoid the shift of the insertion point of instruments, four universal bearings were used as instrument channels instead of the trocars. Two dissectors and one needle holder (SIMIT Scientific Co., Ltd., Shanghai, China) were used for the measurement.

The tracking system (Claron, 2011) is a third generation optical device (MicronTracker®, 3Hx60, Claron Technology Inc., Toronto, Canada), which has been widely used for motion analysis and visual navigation (Kyme et al., 2008, Sanchez-Margallo et al., 2014). The system uses artificial marks identified in the visible spectrum. Two artificial marks were designed and fixed at the instrument handle (Fig. 2). The marks are lightweight and sturdy, and do

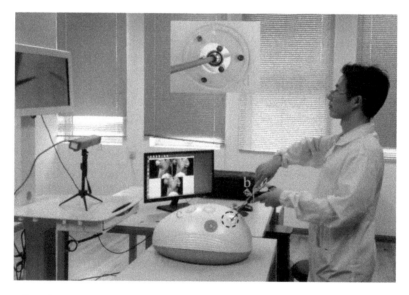

Figure 1. Experimental setup: (a) Tracking system. (b) Instrument with tracking marks. (c) Laparoscopic simulator. (d) Universal bearing for instrument insertion.

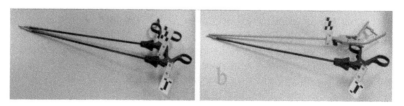

Figure 2. Design and location of marks in each instrument: (a) For the transferring task. (b) For the suturing task.

not interfere with the normal use of the instrument. The position of the mark as well as the instrument tip can be captured and recorded.

2.2 Tasks and groups

Simulated tasks (transferring and suturing) (Fig. 3) were selected and modified based on the tasks included in the Fundamentals of Laparoscopic Surgery program (SAGES/ACS, FLS program, Los Angeles, CA) (Derossis et al., 1998, Peters et al., 2004). In the transferring task, participants were required to transfer four rings from the left columns to the right side and then reversed the procedure to complete the task. In the suturing task, participants were required to do an intracorporeal right-to-left suturing and then complete a half knot with one over-wrapped loop. Three groups of operations were carried out: transferring via LAP (Group A), transferring via SILS (Group B) and suturing via LAP (Group C). The classification of the operations was based on operation type and surgical approach.

2.3 Coordinate system on the laparoscopic simulator

Three reference points A, B and C on the simulator were measured (Fig. 4). Points A and B are insertion points of the instruments, which are the centers of the universal bearings. They are obtained by reading the tip position of a pre-calibrated tool from Claron, while placing the tip of the tool at the bearing center. Point C is the center of the task board (brown area in Fig. 4), which is obtained by reading the position of the instrument tip while placing the tip of the instrument at the center of the task board.

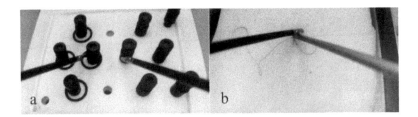

Figure 3.   Operation tasks: (a) Transferring. (b) Suturing.

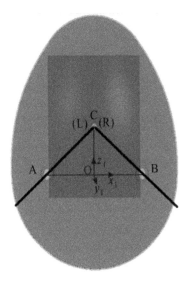

Figure 4.   New coordinate system on the simulator.

327

Suppose that the position of points A, B and C are (x_A, y_A, z_A), (x_B, y_B, z_B) and (x_C, y_C, z_C), respectively. Point O, the middle point of A and B, is defined as the origin of the new coordinate system $Ox_1y_1z_1$. O → B is the positive direction of the x_1 axis, and O → C is approximately the positive direction of the z_1 axis. The unit vector of the three axes can be calculated as follows:

$$\vec{x_1} = \frac{\overrightarrow{OB}}{|\overrightarrow{OB}|}, \; \vec{y_1} = \frac{\overrightarrow{OC} \times \overrightarrow{OB}}{|\overrightarrow{OC} \times \overrightarrow{OB}|}, \; \vec{z_1} = \vec{x_1} \times \vec{y_1} \tag{1}$$

Thus, the position of point P (x_P, y_P, z_P) in $Ox_1y_1z_1$ can be calculated as the dot product of **OP** and the unit vector of the axes:

$$\begin{cases} x_{PN} = (x_P, y_P, z_P) \cdot \vec{x_1} - (x_O, y_O, z_O) \cdot \vec{x_1} \\ y_{PN} = (x_P, y_P, z_P) \cdot \vec{y_1} - (x_O, y_O, z_O) \cdot \vec{y_1} \\ z_{PN} = (x_P, y_P, z_P) \cdot \vec{z_1} - (x_O, y_O, z_O) \cdot \vec{z_1} \end{cases} \tag{2}$$

Suppose that points L and R are the tips of the left and right instruments, respectively (Fig. 4). The inserted lengths l_L and l_R can be calculated by

$$l_L = |\overrightarrow{LA}|, \; l_R = |\overrightarrow{RB}| \tag{3}$$

Shift angles around the x_1 axis can be calculated by

$$\begin{cases} \alpha_L = \arctan(y_{LN}/z_{LN}) \cdot 180/pi \\ \alpha_R = \arctan(y_{RN}/z_{RN}) \cdot 180/pi \end{cases} \tag{4}$$

Also, shift angles around the y_1 axis can be calculated by

$$\begin{cases} \beta_L = \arctan((x_{LN} - x_{AN})/z_{LN}) \cdot 180/pi \\ \beta_R = \arctan((x_{RN} - x_{BN})/z_{RN}) \cdot 180/pi \end{cases} \tag{5}$$

In Formulas 4 and 5, x_{LN} and x_{RN} are the values of points L and R of the x_1 axis in $Ox_1y_1z_1$; y_{LN} and y_{RN} are the values of points L and R of the y_1 axis; z_{LN} and z_{RN} are the values of points L and R of the z_1 axis; and x_{AN} and x_{BN} are the values of points A and B of the x_1 axis.

2.4 Experimental procedures

The measurements started once the tracking system was thermally stable. First, the mark template and the tool tip of each instrument were registered in the tracking system. Then, in each group, the task was repeated 10 times, and positions of the mark and instrument tip were captured by the 'pose recorder' of the tracking system. The measurement of key points A, B and C was after the operation in each group. Finally, the raw data were imported into Matlab, and data processing was carried out according to Formula 1 to 5.

3 RESULTS

In each group, differences between each trial were slight, so all the 10 sets of raw data were put into one reservoir to thoroughly obtain the range of motion. Point cloud (Fig. 5) combined 10 sets of the raw data in each group, while the range of inserted length and shift angles (Table 1) was determined by the minimum and maximum values of the combined data after filtering the noisy data in each group.

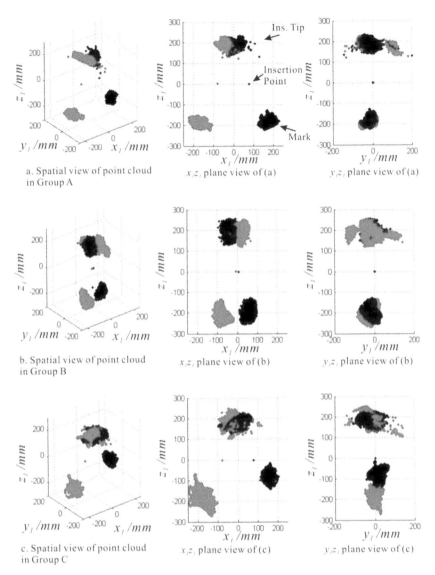

Figure 5.   Point cloud of the mark, instrument tip and the insertion point (the red points represent the left instrument, while the black points represent the right instrument).

The point cloud and the calculated value enabled the comparison of the range of motion within and between the groups. In each group, the inserted lengths of the left instrument and the right instrument are almost the same; the up/down and left/right shift angles are very close between the left and right instruments, except in group B.

Groups A and B are the same task with different approaches. From Figure 5, the first big difference that can be seen is that the cross manner was used in group B. The tip of the left instrument is at the right side, while the tip of the right instrument is at the left side. The second difference is that the range of the inserted length of both instruments and the up/down shift angle of the left instrument in Group B is much larger than those in Group A.

Group A and C are the same approach with different tasks. The first difference is that there is more interference at the instrument tip in Group C than in Group A. The second difference is that the range of the shift angles of the right instrument is approximately 3 deg larger than that of the left instrument in Group C; while in Group A, there is no such obvious difference

Table 1. Range of motion of instruments in each group.

Group	Ins.	Inserted length (mm)	Up/down: range (deg)	Left/right: range (deg)
A	L	173.52~226.92	−17.72~4.36: 22.08	5.57~22.21: 16.64
	R	174.47~226.64	−13.63~6.53: 20.16	−27.40~−7.42: 19.98
B	L	152.61~248.27	−24.48~10.83: 35.31	0.84~16.01: 15.17
	R	153.60~249.75	−15.53~5.23: 20.76	−19.51~−2.65: 16.86
C	L	171.75~234.36	−13.82~6.58: 20.40	4.70~27.63: 22.93
	R	178.49~231.57	−19.41~4.92: 24.33	−34.14~−8.49: 25.65

between the right and left instruments. In addition, the range of motion of instruments in Group C is slightly larger than that in Group A.

4 DISCUSSION

Theoretically, the range of motion that a straight instrument can reach is a large, cone-shaped space. However, from Figure 5, we found that the actual shape of the range of motion was definitely not a cone or large, but limited to a small amorphous space. There are several factors to determine the space.

The first factor is the type of operation. Transferring and suturing were studied in this paper. The range of motion of the instrument in suturing is larger than that in transferring. In addition, tip interference is more frequent in suturing than in transferring. This might be considered as a reason why the FLS identified suturing as an advanced operation and more difficult to learn (Derossis et al., 1998).

The second factor is the surgical approach. Two different approaches were used in LAP and SILS. In LAP, triangulation is easily formed due to the fact that the insertion points for instruments are apart from each other. In SILS, insertion points are very close to each other, resulting in triangulation loss and instrument clashes (Tang et al., 2012). Data from this study clearly showed the cross manner operation in SILS, and the range of motion of instruments in SILS was quite different from that in LAP, especially in the up/down shift angles. In SILS, instruments are usually kept up/down away to avoid instrument clashes.

The third factor is the left-right hand factor. In Groups B and C, there is an apparent difference between the left instrument and the right instrument. The role of the hand in operation is a main reason. In SILS transferring (B), two instruments are very close to each other, so the non-dominant instrument (left) is usually kept aside at the upper or the lower place to avoid conflicts with the dominant one. This is why the range of the up/down shift angle of the left instrument in SILS is much larger than that of the right instrument. In LAP suturing (C), the dominant hand is used to hold the needle holder, while the non-dominant hand is for the grasper. Different functionalities of the instruments result in different ranges of motion. In LAP transferring (A), however, the difference is slight, because there are few instrument conflicts and there is no difference in the functionality of the left and right instruments.

In this study, only one subject was measured. Motion feature (velocity, acceleration and movement smoothness) of the instrument varies according to the subjects' operation habits and skills level. However, the range of motion of the instrument mainly depends on the operation object while the surgical approaches are fixed. Instrument tip will only move around the operation target, so different subjects only have slight influences on the measurement results, and have no influence on the calculation methods.

In conclusion, this study introduced a method to measure the range of motion of laparoscopic instruments during operation. The range of motion of laparoscopic instruments is limited to a small amorphous spatial space. There are several factors to determine the actual range of motion of instruments, including the type of operation, surgical approaches and left-right hand. The study can not only be used for motion analysis of instruments for LAP, but also for motion analysis or instrument design via different surgical approaches.

ACKNOWLEDGMENT

This project was supported by the National Natural Science Foundation of China (Grant No. 51175345) and the Innovation Fund Project for Graduate Student of Shanghai (JWCXSL1301).

REFERENCES

Breedveld, P., Stassen, H.G., Meijer, D.W. & Jakimowicz, J.J. 1999. Manipulation in laparoscopic surgery: overview of impeding effects and supporting aids. *Journal of Laparoendoscopic & Advanced Surgical Techniques. Part A,* 9(6): 469–480.

Chmarra, M.K., Grimbergen, C.A. & Dankelman, J. 2007a. Systems for tracking minimally invasive surgical instruments. *Minimally Invasive Therapy & Allied Technologies,* 16(6): 328–340.

Chmarra, M.K., Kolkman, W., Jansen, F.W., Grimbergen, C.A. & Dankelman, J. 2007b. The influence of experience and camera holding on laparoscopic instrument movements measured with the TrEndo tracking system. *Surgical Endoscopy & Other Interventional Techniques,* 21(11): 2069–2075.

Claron. 2011. http://www.claronav.com/microntracker.php [Online].

Cuschieri, A., Dubois, F., Mouiel, J., Mouret, P., Becker, H., Buess, G., Trede, M. & Troidl, H. 1991. The European experience with laparoscopic cholecystectomy. *American Journal of Surgery,* 161(3): 385–387.

Derossis, A.M., Fried, G.M., Abrahamowicz, M., Sigman, H.H., Barkun, J.S. & Meakins, J.L. 1998. Development of a model for training and evaluation of laparoscopic skills. *American Journal of Surgery,* 175(6): 482–487.

Kyme, A.Z., Zhou, V.W., Meikle, S.R. & Fulton, R.R. 2008. Real-time 3D motion tracking for small animal brain PET. *Physics in Medicine & Biology,* 53(10): 2651–2566.

Lirici, M.M., Califano, A.D., Angelini, P. & Corcione, F. 2011. Laparo-endoscopic single site cholecystectomy versus standard laparoscopic cholecystectomy: results of a pilot randomized trial. *American Journal of Surgery,* 202(1): 45–52.

Ma, R., Wu, D., Yan, Z., Du, Z. & Li, G. Research and development of micro-instrument for laparoscopic minimally invasive surgical robotic system. Robotics and Biomimetics (ROBIO), 2010 IEEE International Conference on, 2010: 1223–1228.

Mason, J.D., Ansell, J., Warren, N. & Torkington, J. 2013. Is motion analysis a valid tool for assessing laparoscopic skill? *Surgical Endoscopy & Other Interventional Techniques,* 27(5): 1468–1477.

Oropesa, I., Sanchez-Gonzalez, P., Lamata, P., Chmarra, M.K., Pagador, J.B., Sanchez-Margallo, J.A., Sanchez-Margallo, F.M. & Gomez, E.J. 2011. Methods and tools for objective assessment of psychomotor skills in laparoscopic surgery. *Journal of Surgical Research,* 171(1): e81–95.

Peters, J.H., Fried, G.M., Swanstrom, L.L., Soper, N.J., Sillin, L.F., Schirmer, B. & Hoffman, K. 2004. Development and validation of a comprehensive program of education and assessment of the basic fundamentals of laparoscopic surgery. *Surgery,* 135(1): 21–27.

Reiley, C.E., Lin, H.C., Yuh, D.D. & Hager, G.D. 2011. Review of methods for objective surgical skill evaluation. *Surgical Endoscopy & Other Interventional Techniques,* 25(2): 356–366.

Romanelli, J.R. & Earle, D.B. 2009. Single-port laparoscopic surgery: an overview. *Surgical Endoscopy & Other Interventional Techniques,* 23(7): 1419–1427.

Sanchez-Margallo, J.A., Sanchez-Margallo, F.M., Pagador Carrasco, J.B., Oropesa Garcia, I., Gomez Aguilera, E.J. & Moreno Del Pozo, J. 2014. Usefulness of an optical tracking system in laparoscopic surgery for motor skills assessment. *Cirugía Española,* 92(6): 421–428.

Tang, B., Hou, S. & Cuschieri, S.A. 2012. Ergonomics of and technologies for single-port laparoscopic surgery. *Minimally Invasive Therapy & Allied Technologies,* 21(1): 46–54.

Biomedical imaging

Bioinformatics and Biomedical Engineering – Chou & Zhou (Eds)
© *2016 Taylor & Francis Group, London, ISBN 978-1-138-02784-8*

The application of photoacoustic tomography in joint tissues

X.C. Zhong
School of Physical Science and Technology, Southwest Jiaotong University Chengdu, China

X.Y. Jing, S.Q. Jing, N.Z. Zhang & J. Rong
School of Physical Electronics, University of Electronic Science and Technology of China, Chengdu, China

ABSTRACT: This paper mainly presents experimental studies in imaging of the joints in vitro and in vivo using our special photoacoustic tomography system. In this pilot study, a chicken claw joint in vitro and a Distal Interphalangeal (DIP) joint from a volunteer were imaged, and a delay-and-sum algorithm was used to reconstruct the two-Dimensional (2D) photoacoustic images. Good reconstructed images have been achieved. In the future, photoacoustic tomography, which combines the merits of pure optical imaging and pure ultrasound imaging, has potential to provide an effective approach of studying the architectures, physiological and pathological properties and metabolisms of joint tissues.

Keywords: photoacoustic tomography; joint; delay-and-sum; two-dimensional

1 INTRODUCTION

When a laser pulse irradiates the biological tissue, a portion of the energy is absorbed by the tissue and converted to heat. A temperature gradient is then produced by the heating based on the energy absorption pattern, and subsequently ultrasonic waves are generated through thermal expansion. This is called photoacoustic effects (Xing, Xiang, 2007; Jin, 2007; Foster, Finch, 1974; Gutfeld, 1980; Bowen, 1981; Olsen, 1982; Lin, Chan, 1984; Oraevsky, Esenaliev, Jacques, Tittel, 1995). Photoacoustic Tomography (PAT) is a new imaging modality based on photoacoustic effects. In recent years, there is an increasing interest in studying it and it has attracted the attentions of researchers from various fields.

Other imaging methods relating different biological parameters have both advantages and limitations, such as X-ray, ultrasound imaging, Positron Emission Tomography (PET), Magnetic Resonance Imaging (MRI), etc. The ultrasound imaging has a relatively high spatial resolution, but limited to image contrast. Traditional X-ray and Positron Emission Tomography (PET) are either ionizing or radioactive, and the expensive cost of MRI prevents its use as a routine screening tool (Jin, 2007). PAT, which is non-invasive and combines advantages of both the high image contrast that results from optical absorption and the high resolution of ultrasound imaging, could be the next successful generation imaging techniques in biomedical application (Xing, Xiang, 2007).

PAT has been successfully applied in imaging vascular structure, brain function and finger joints in vivo in America (Kolkman, Hondebrink, Steenbergen, Mul, 2003; Wang, Pang, Ku, Xie, Stoica, Wang, 2003; Xiao, Tang, Lu, Chen, 2011; Xiao, Tang, Lu, Chen, 2011). Nevertheless, the research of the application of the PAT in joint tissues is still at its preliminary stage in China. So, in this paper, we used the photoacoustic tomography system based on PCI4732 and delay-and-sum algorithm, which are different from those of published articles in bone and joint imaging, to do a pilot PAT study of joint tissues in vitro and in vivo.

2 MATERIALS AND METHODS

2.1 *Photoacoustic tomography system*

The schematic of the photoacoustic experimental setup is shown in Figure 1. A Q-switched Nd:YAG laser (Quantel Brilliant B) was employed to provide 532 nm laser light with a pulse duration of 5 ns and a pulse repetition rate of 10 Hz. The laser beam was expanded by a concave lens and homogenized by a ground glass to provide an incident energy density of ~ 12 mJ/cm², which is far below the safety standard of 22 mJ/cm². The experiment was conducted in a plastic container filled with water, which acted as coupling medium due to its high ultrasonic penetrating ability. The ultrasonic receiver was an unfocused transducer (V323, Panametrics NDT Inc.), with a central frequency of 5 MHz and a diameter of 6 mm. This receiver was immersed in the water and scanned around the Z axis, and its axis was

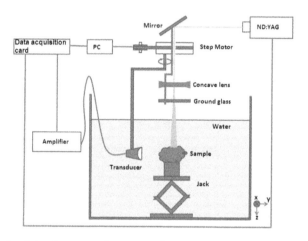

Figure 1. Schematic of the photoacoustic tomography system.

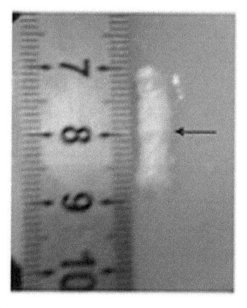

Figure 2. Photograph of the chicken claw joint with skin, flesh, blood vessels and the other tissues removed, The black arrow points to the joint.

336

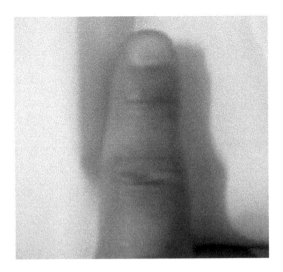

Figure 3.   Photograph of the finger joint. DIP: distal interphalangeal.

aligned with the center of samples. The transducer, driven by the computer and the controller to scan around the sample, detected the photoacoustic signals in the imaging plane at each scanning position. A pulse amplifier received the signals from the transducer and transmitted the amplified signals to a data sampling card (PCI 4732 acquisition card). One set of data at 180 (in vitro sample) or less than 180 (in vivo sample) different positions was taken when the receiver moves over 360 or less than 360 degree. After a series of data were recorded, the photoacoustic image was reconstructed with the delay-and-sum algorithm.

2.2   *The experimental sample in vitro*

In the experiment, a joint, which was cut from a fresh chicken claw and removed its skin and flesh, blood vessels and the other tissues attached to the bones was used as the experimental sample in vitro. To fixed it for imaging conveniently, the joint was embedded in a larger cylindrical background phantom which was simply made of water and agar powder. The photograph of the sample is shown in Figure 2.

2.3   *The experimental sample in vivo*

In the experiment, a volunteer's finger joint was used as the experimental sample in vivo. The photograph of the sample is shown in Figure 3.

3   RESULTS AND DISCUSSION

Existing reconstruction algorithms for photoacoustic tomography are based on the assumption that the acoustic properties in the tissue are homogeneous, biological tissue, however, has heterogeneous acoustic properties, which lead to distortion and blurring of the reconstructed images. Besides, the imperfection existing in the delay-and-sum algorithm results in artifact. That all can be seen from figures above. In the whole study process, firstly, we did a preliminary study on the sample in vitro as shown in Figure 2 with the system in Figure 1. Compared to the physical objects in Figure 2, the reconstructed images in Figure 4 can well correspond with it, from whatever the physical size, the structures and so on. Then, we did further research on joint imaging used the sample of DIP joint in vivo as shown in Figure 3 with the system in Figure 1. Because of a certain angle (greater than zero and less than

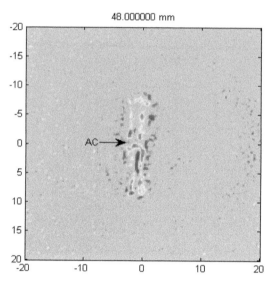

Figure 4. The photoacoustic reconstructed image of the chicken claw joint. The units in both X-coordinate and Y-coordinate are millimeter. AC: articular cavity.

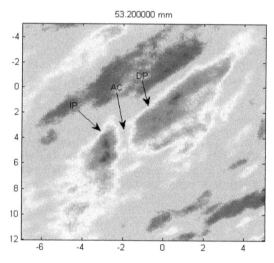

Figure 5. The photoacoustic reconstructed image of the DIP joint. The units in both X-coordinate and Y-coordinate are mm. DP: distal phalanx; IP: intermediate phalanx; AC: articular cavity.

90 degrees) between the finger and transducer, the reconstructed images in Figure 5 can be seen slant. From the reconstructed image in Figure 5, the Distal Phalanx (DP), Intermediate Phalanx (IP) and articular cavity are clearly visible.

4 CONCLUSION

This study represents the first attempt to image the DIP joint in vivo with the special photoacoustic tomography system in Figure 1. The structure of the DIP joint such as, the Distal Phalanx (DP), Intermediate Phalanx (IP) and articular cavity can be clearly imaged. In the future, with the improvement of the reconstructed algorithm and the system, more information of biological tissues will be obtained. And, photoacoustic tomography has the potential to provide an effective approach for joint imaging and further to provide a new diagnosing method for osteoarthritis and other bone joint diseases.

ACKNOWLEDGMENT

This research was partially supported by the International cooperation project of Sichuan Province Science and Technology Agency (2014HH0037 to Rong Jian).

REFERENCES

Bowen, T. 1981. Radiation-induced thermoacoustic soft tissue imaging, Proc. IEEE Ultrason. Symp. 2: 817–822.

Foster, K.R. & Finch, E.D. 1974. Microwave hearing: evidence for thermoacoustic auditory stimulation by pulsed microwaves. Science 185(147): 256–258.

Gutfeld, R.T.V. 1980. Thermoelastic generation of elastic waves for non-destructive testing and medical diagnostics. Ultrasonics 18(4): 175–181.

Jin, X. 2007. Microwave induced thermoacoustic tomography: applications and corrections for the effects of acoustic heterogeneities, Texas A&M University.

Kolkman, R.G.M. Hondebrink, E. Steenbergen, W. & de Mul, F.F.M. 2003. In vivo photoacoustic imaging of blood vessels using an extreme-narrow aperture sensor, IEEE J. Sel. Top. Quant 9: 343–346.

Lin, J.C. & Chan, K.H. 1984. Microwave thermoelastic tissue imaging—system design. IEEE Trans. Microwave Theory Tech 32: 854–860.

Oraevsky, A. Esenaliev, R. Jacques, S. & Tittel, F. 1995. Laser optoacoustic tomography for medical diagnostics—principles, Proc. of SPIE 2676: 22–31.

Olsen, R.G. 1982. Generation of acoustic images from the absorption of pulsed microwave energy. in Acoustic Imaging, edited by J. P. Powers (Plenum Publishing, New York) 53–59.

Sun, Y. Sobel, E.S. & Jiang, H.B. 2011. First assessment of three-dimensional quantitative photoacoustic tomography for in vivo detection of osteoarthritis in the finger joints, Med. Phys 38(7): 4009–4017.

Wang, X. Pang, Y. Ku, G. Xie, X. Stoica, G. & Wang, L.V. 2003. Non-invasive laser induced photoacoustic tomography for structural and functional imaging of the brain in vivo, Nat. Biotechnol 21: 803–806.

Xing, D. & Xiang, L.Z. 2007. Photoacoustic and Thermoacoustic Imaging Application in Cancer Early Detection and Treatment Monitoring. Proc. of SPIE 68260B: 1–8.

Xiao, J.Y. Tang, J.T. Lu, G.Y. & Chen, Z.S. 2011. Functional Photoacoustic Imaging of Osteoarthritis In The Finger Joints, IEEE 299–303.

Bioinformatics and Biomedical Engineering – Chou & Zhou (Eds)
© 2016 Taylor & Francis Group, London, ISBN 978-1-138-02784-8

Frequency response mismatch correction in multichannel time interleaved analog beamformers for ultrasound medical imaging

Amir Zjajo & Rene van Leuken
Circuits and Systems Group, Delft University of Technology, Delft, The Netherlands

ABSTRACT: Time interleaving is one of the most efficient techniques employed in the design of high-speed analog beamformers in ultrasound phased array transducers. However, its implementation introduces several mismatch errors between interleaved subunits, limiting the accuracy of the overall system. In this paper, a method for estimation and correction of frequency response mismatch based on an adaptive equalization is reported. The adaptive algorithm requires only that the input signal is band-limited to the Nyquist frequency for the complete system. The proposed method greatly reduce the computational complexity requirement of sampled signal reconstruction and offers simplified hardware implementation.

1 INTRODUCTION

The ultrasound imaging is frequently employed in medical diagnosis due to its real-time processing capability and less-detrimental effects to the human body in comparison with magnetic resonance imaging, computed tomography, and X-rays scanning. In the ultrasound phased array transducers, normally a beamforming (focusing) operation (Talman, 2003)-(Um, 2014) is performed to enhance the Signal to Noise Ratio (SNR) of the ultrasound image. With the advances in analog delay lines and analog processing elements (e.g. Analog-to-Digital Converters (ADC)), digital processing and digital beamforming was pushed completely into the front-end of a transducers. However, in modern transducers, a large number (e.g., 9216 for a 72×128 array) (Um, 2014) of transducer elements for active signal processing needs to be integrated in the ultrasound probe handle. As a consequence, due to the large number of required ADCs, digital beamforming becomes impractical and analog beamforming is necessary, at least at the front stage of the transducer.

The analog beamformer circuit consists of multiple, time-interleaved, analog sub-beamformers for sequential beamforming of multiple focal points on a scan line. Although time-interleaved principle provides efficient way to increase the conversion rate of sub-beamformers beyond the limit imposed by available technology on the maximum frequency of an overall beamformer, its implementation introduces several mismatch errors between interleaved units (Jenq, 1988), limiting the overall system accuracy. Offset, time and gain mismatches have been analyzed comprehensively (Petraglia, 1991), and in certain cases calibration/correction techniques have been proposed (Zou, 2011). The frequency response mismatch can be avoided through the use of a full-speed front-end sampler (Hsu, 2007). In this case, however, the sampler must be followed by a fast-settling buffer to drive the stages. Similarly, by increasing the bandwidth of interleaved sub analog beamformers, the impact of the frequency response mismatch at the signal frequency becomes lower, although at the cost of a severely reduced SNR.

Recently, iterative frequency response mismatch correction methods (Satarzadeh, 2009)-(Johansson, 2008) have been proposed based on adaptive filters either on each stage or at the system output by employing an inverse Fourier transform (Satarzadeh, 2009) or polynomial time varying filter structures (Johansson, 2008).

In this paper, we demonstrate digital frequency response mismatch correction in time-interleaved analog beamformer as an adaptive equalization process. The estimation method requires only that the input signal is band-limited to the Nyquist frequency for the complete system. The equivalent signal estimation structure can avoid aliasing without over sampling the input signal or operating at full sampling rate. The sparse structure of interleaved signals in the continuous frequency domain is used to replace the continuous reconstruction with a single finite dimensional problem. The implemented method significantly reduce the computational complexity requirement of sampled signal reconstruction and offers effective hardware implementation.

2 CORRECTION ALGORITHM

In the analog, time-interleaved beamformer, there are M identical sampling and sub-beamformers operating in parallel. Each subsystem samples and digitizes the input signal every MT seconds; i.e., the digitizing rate of each subsystem is $1/MT$ samples/s. Although all subsystems operate at the same clock frequency, the sampling clock of subsystem $m+1$ is T seconds behind that of the subsystem m for $m = 0, 1, ..., M-1$. The timing alignment within the required accuracy is obtained by using a master clock to synchronize the different sampling instants. As illustrated in Figure 1, at the back end of these parallel sampling and sub-beamformers is a sequential M:1 multiplexer, which samples the outputs of beamformers stages at a rate of 1 sample/T s. The sub analog beamformer consists of a sample-and-hold circuit, which consists of sampling capacitors and switches and a clock controller. Time domain analog-beamformer output is shown in Figure 2. For M time-interleaved sub-beamformers, each working uniformly with a period of $T_s = MT$ and frequency response offset $\{h_m f_{in}/f_1 = m_i T\}$, the clock generator provides the required M sample clocks for each sub-beamformer stage according to the sampling pattern such that $t_i(n) = (nM + m_i) = (n + m_i/M)T_s$ for $1 \leq i \leq M$. Defining the ith sampling sequence for $1 \leq i \leq M$ as $x_{mi}[n] = x(t = nT = kMT + mT)$, the sequence of $x_{mi}[n]$ is obtained by up-sampling the output of the i-the sub-beamformer with a factor of M and shifting in time with m_i samples. The spectrum of the sampled time-domain signal is then represented as

$$Y_{mi}(e^{j2\pi fT}) = \sum_{n=-\infty}^{+\infty} x_{mi}[n]e^{-j2\pi fT} = \frac{1}{MT}\sum_{k=0}^{M-1} A_k(f)X\left(f - \frac{k}{MT}\right)$$

$$\forall f \in \mathcal{F}_0 = (0, \frac{1}{MT}), 1 \leq i \leq M \tag{1}$$

$$A_k(f) = \sum_{n=0}^{M-1}\left(e^{-jnh_{m,i}f_{in}/f_1}\right)e^{-jkn(2\pi/M)}$$

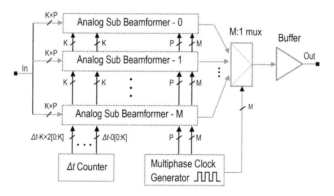

Figure 1. Analog beamformer architecture.

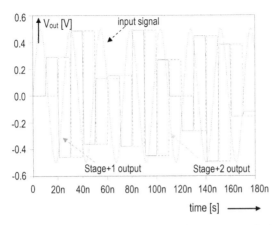

Figure 2.   Time domain analog beamformer sample and hold output. For clarity $M = 3$ is shown.

where $Y_{mi}(e^{j2\pi fT})$ is the Discrete-Time Fourier Transform (DTFT) *of* $x_{mi}[n]$ band-limited to F, h_m for $m = 0, 1, M-1$ is frequency response mismatch encountered at the m-th unit, f_{in} is input frequency and f_1 the unity-gain frequency.

The goal of our reconstruction scheme is to perfectly recover $x(t)$ from the set of sequences of sampled signal $x_{mi}[n], 1 \le i \le M$. We express (1) in a matrix form as

$$y(f) = A(f)s(f), \quad \forall f \in \mathcal{F}_0 \tag{2}$$

where $y(f)$ is a vector of length M whose ith element is $Y_{mi}(e^{j2\pi fT})$,

$$y(f)_{M\times 1} = [Y_1(e^{j2\pi fT}), Y_2(e^{j2\pi fT}), ..., Y_M(e^{j2\pi fT})]^T \tag{3}$$

$A(f)$ is a the $M \times M$ discrete-time Fourier transform matrix containing $A_k(f)$ terms, and the vector $s(f)$ contains M unknowns as (Mishali, 2009)

$$s(f)_{M\times 1} = [X(f), X(f - 1/(MT)), ..., X(f - (M-1)/(MT))]^T, \forall f \in \mathcal{F}_0 \tag{4}$$

A straightforward approach to recover $x(t)$ is to find the sparsest vector $s(f)$ over a dense grid of $\mathcal{F}_0$ and then approximate the solution over the entire continuous interval $\mathcal{F}_0$. However, this discretization strategy cannot guarantee perfect reconstruction. If we only need to reconstruct the distortion amplitudes at the nominal sampling instances $t = (kM + m)T_s$, $m = 0,..., M-1$; $k = 0 ..., -1, 0, 1, ...$ the reconstruction can be simplified. The spectrum given by (2) has M pairs of line spectra, each pairs centered at the fractional of the sampling frequency, such as f_s/M, ..., $(M-1) f_s/M$. Fundamental corresponds to $k = 0$ while $k = 1, ..., N-1$ corresponds to the distortion. The signal amplitude is determined by $A(0)$ while the distortion amplitudes are determined by $A(n), n = 1, ..., M-1$. The spectral index set of the signal is than defined as the set $k = [k_1, k_2, ..., k_q]$ and the reduced signal vector as (Bresler, 2008)

$$z(f) = [X(f - k_1/M), X(f - k_2/M), ..., X(f - k_q/M)]^T \tag{5}$$

that contains only the q sampling instances indexed by the set k. The reduced measurement matrix $A_k(f)$ is derived by choosing the columns of $A_k(f)$ that are indexed by the spectral index k. Assuming $h_m f_{in}/f_1 \ll 1$ and using the first order Taylor expansion, the magnitude of the sidebands components (as illustrated in Figure 3 and Figure 4 (Zjajo, 2010)) can be expressed as

343

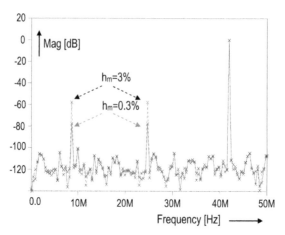

Figure 3. Simulated frequency response mismatch at $h_m = 0.3\%$ and $h_m = 3\%$ for $f_{in} = 41$ MHz, $f_s = 100$ MS/s and $f_l = 350$ MHz.

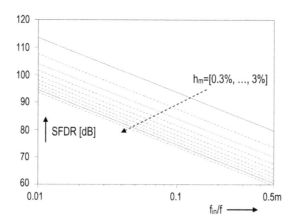

Figure 4. SFDR versus different frequency response ranging from $h_m = 0.3\%$ to 3% for $f_s = 100$ MS/s.

$$\left| A_k(f) = \sum_{n=0}^{M-1} (e^{-jh_m f_{in}/f_1}) e^{-jkn(2\pi/M)} \right| \approx \left| \frac{1}{M} \sum_{n=0}^{M-1} (1 - jh_m f_{in}/f_1) e^{-jkn(2\pi/M)} \right|$$

$$= \begin{cases} \left| \dfrac{f_{in}}{Mf_1} \sum_{n=0}^{M-1} h_m e^{-jkn(2\pi/M)} \right| & \text{for } k \neq 0, \ \pm M \\[3mm] \left| 1 - (jf_{in}/f_1)\left(\dfrac{1}{M}\sum_{n=0}^{M-1} h_m\right) \right| = 1 & \text{for } k = 0, \ \pm M \end{cases}$$

(6)

As a consequence, (2) reduces to

$$y(f) = A(f)z(f), \qquad \forall f \in \mathcal{F}_0 \tag{7}$$

If $A(f)$ has full column rank, the unique solution can be obtained using a left inverse, e.g. the pseudo-inverse of $A(f)$ denoted by $A^\dagger$

$$z(f) = A^\dagger(f)y(f), \qquad \forall f \in \mathcal{F}_0 \tag{8}$$

A time domain solution for the recovery of $x(t)$ than involves filtering of the sequences $x_i[n]$, $i = 1, \ldots M$ to generate $x_{hi}[n]$. The interpolation filter $h[n]$ with cut off frequency at

344

$f_c = f_{max}/M$ filters the sequence $x_i[n]$ that is upsampled with M according to $kMT + mT$, i.e. $x_{hi}[n] = h[n]*x_i[n]$. To recover the desired continuous-time signal $x(t)$ with a standard D/A converter, the reconstruction is than obtained with

$$x(nT) = \sum_{i=1}^{q}\sum_{l=1}^{M}[A^\dagger]_{il}x_{hl}[n]e^{j2\pi k_i n/M} \tag{9}$$

which is the Nyquist-rate sampled version of the desired continuous-time signal $x(t)$. Additionally, advantage of using (9) is that all filters have the same low pass response, offering efficient implementation. An estimation error can be than easily found by several methods from estimation theory (e.g. Gauss-Newton, Levenberg-Marquardt, or gradient search method) by comparing the estimated output $x(n)$ to a desired response.

3   EXPERIMENTAL RESULTS

The test results are shown in Figures 5–7 and Table 1, where Spurious-Free Dynamic Range (SFDR) is employed as a performance matrix. The proposed approach achieves required reconstruction accuracy in less than one thousand samples for gradient search method (Fig. 5).

The iterative correction structure responses almost instantaneously with an improving estimate of mismatch parameters as n increases. To decrease the convergence time, initially a single set of coefficients is utilized and adapted after every decision, as in the conventional

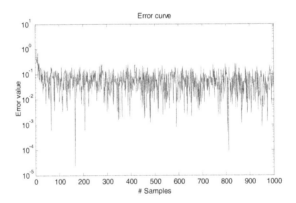

Figure 5.   Mean-square error for one thousand samples.

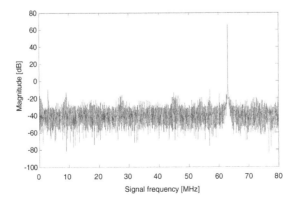

Figure 6.   Spectrum of the time-interleaved beamformer at 160 MS/s.

345

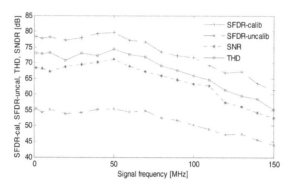

Figure 7.   Measurement result of the time-interleaved beamformer at 160 MS/s.

Table 1.   Example of the calculated and estimated SFDR values.

f_{in} [MHz]	43			57			71		
Δf_1 [MHz]	1	5	10	1	5	10	1	5	10
h_m [%]	0.29	1.42	2.86	0.29	1.42	2.86	0.29	1.42	2.86
Theor. [dB]	78.6	64.7	58.6	76.2	62.3	56.2	74.5	60.4	54.3
Est. [dB]	78.8	64.9	58.9	76.3	62.4	56.4	74.6	60.5	54.4

method, and then after initial convergence, the coefficients are further adapted separately. Validations have been performed for the entire analog beamformer usable signal bandwidth and most probable limitation mechanisms. Four examples are shown in Table 1, where theoretical and estimated values are compared for 11, 43, 57 and 71 MHz input frequencies with unity gain frequency f_1 set at 350 MHz and deviations Δf_1 in the range of 1–10 MHz.

The spectrum of the time-interleaved beamformer sampled at 160 MS/s is illustrated in Figure 6. The spurious harmonics at $f_s/N \pm f_{in}$ due to the frequency response mismatch are held to 72 dB below the fundamental signal. The SNR, SFDR and Total Harmonic Distortion (THD) as a function of input frequency at a sample rate of 160 MS/s are shown in Figure 7. The uncalibrated beamformer SFDR is held below 56 dB, an average loss of 23 dB in comparison with calibrated beamformer. The calibration algorithm (0.5k gates) consumes 6 mW of power. All results are obtained with a 1.2 V supply at room temperature.

4   CONCLUSIONS

Combining time-interleaved analog beamformer with efficient, moderate speed converter result in a high speed ultrasound phased array transducer with low power consumption. Although parallelism of time-interleaved systems enables high-speed operation, mismatch between interleaved units, cause distortion in the processed signal. In this paper, effect of frequency response mismatch is investigated, and estimation method based on an adaptive equalization presented. Obtained results suggest that the proposed method accurately estimate (within 0.5%) and correct mismatch errors greatly enhancing analog beamformer performance (up to 23 dB).

REFERENCES

Bresler, Y. 2008. Spectrum-blind sampling and compressive sensing for continuous-index signals. *Ieee Information Theory and Applications Workshop*, pp. 547–554.

Gurun, G. *et al.*, 2012. An Analog Integrated Circuit Beamformer for High-Frequency Medical Ultrasound Imaging. *IEEE Transactions on Biomedical Circuits and Systems,* vol. 6, pp. 454–467.

Hsu, C.C. *et al.*, 2007. An 11-b 800-MS/s time-interleaved ADC with digital background calibration. *IEEE International Solid-State Circuits Conference Digest of Technical Papers*, pp. 464–465.

Jenq, Y.-C. 1988. Digital spectra of nonuniformly sampled signals: Fundamentals and high-speed waveform digitizers. *IEEE Transactions on Instrumentation and Measurement*, vol. 37, no. 2, pp. 245–251.

Johansson, H. & Lowenborg, P. 2008. A least-squares filter design technique for the compensation of frequency response mismatch errors in time-interleaved A/D converters. *IEEE Transactions on Circuits and Systems-II: Express Briefs,* vol. 55, no. 11, pp. 1154–1158.

Mishali, M. & Eldar, Y. 2009. Blind multiband signal reconstruction: compressed sensing for analog signals. *IEEE Transactions on Signal Processing,* vol. 57, pp. 993–1009.

Petraglia, A. & Mitra, S.K. 1991. Analysis of mismatch effects among A/D converters in a time-interleaved waveform digitizer. *IEEE Transactions on Instrumentation and Measurement*, vol. 40, no. 5, pp. 831–835.

Satarzadeh, P., Levy, B.C., Hurst, P.J. 2009. Adaptive semiblind calibration of bandwidth mismatch for two-channel time-interleaved ADCs. *IEEE Transactions on Circuits and Systems-I: Regular Papers*, vol. 56, no. 9, pp. 2075–2088.

Talman, J.R. *et al.*, 2003. Integrated Circuit for High-Frequency Ultrasound Annular Array. *Proceedings of IEEE Custom Integrated Circuit Conference*, pp. 477–480.

Um, J.-Y. *et al.*, 2012. A single-chip time-interleaved 32-channel analog beamformer for ultrasound medical imaging. *Proceedings of IEEE Asian Solid-State Circuits Conference*, pp. 193–196.

Um, J.-Y. *et al.*, 2014. An analog-digital-hybrid single-chip RX beamformer with non-uniform sampling for 2D-CMUT ultrasound imaging to achieve wide dynamic range of delay and small chip area. *IEEE International Solid-State Circuits Conference Digest of Technical Papers,* pp. 426–427.

Zjajo, A., 2010. Design and Debugging of Multistep Analog to Digital Converters. *Eindhoven University of Technology*, PhD Thesis.

Zou, Y.X., Zhang, S.L., Lim, Y.C., Chen, X. 2011. Timing mismatch compensation in time-interleaved ADCs based on multichannel Lagrange polynomial interpolation. *IEEE Transactions on Instrumentation and Measurement*, vol. 60, no. 4, pp. 1123–1131.

Bioinformatics and Biomedical Engineering – Chou & Zhou (Eds)
© 2016 Taylor & Francis Group, London, ISBN 978-1-138-02784-8

HIFU based photoacoutic tomography

X.C. Zhong
School of Physical Science and Technology, Southwest Jiaotong University, Chengdu, China

W.Z. Qi, S.Q. Jing, N.Z. Zhang & J. Rong
School of Physical Electronics, University of Electronic Science and Technology of China, Chengdu, China

ABSTRACT: The application of photoacoutic tomography is limited by penetration depth, because of the highly optical scattering properties of bio-tissues. To solve this problem, we studied the interaction of HIFU field and photoacoustic signal by processing a series of imaging experiments under same conditions. From these photoacoustic images we come to a conclusion that HIFU field has the ability to improve the penetration depth and SNR of photoacoutic tomography.

Keyword: HIFU; photoacoustic tomography; imaging

1   INTRODUCTION

Photoacoustic tomography, which is a newborn non-invasive medical imaging method combined the high contrast of pure optical imaging and high imaging depth of ultrasound imaging (Wang, Pang, Ku, Xie, Stoica, Wang, 2003), works based on the initial temperature rise caused by bio-tissues absorption of short duration-time (nanosecond magnitude) pulse laser energy. The temperature rise will cause thermal expansion in bio-tissues and thus elastic waves penetrating through bio-tissues, which is known as photoacoustic signals. By detecting these signals around the imaging target and proceeding them with appropriate reconstruction arithmetic, images of laser energy absorption inside bio-tissues can then be obtained (Yang, Xing, Gu, Tan, Zeng, 2005).

The absorption of bio-tissues is mainly determined by the endogenous tissue chromophores such as haemoglobin, melanin, water and lipid (Beard, 2011). Absorption coefficient spectra of different chromophores are different. For wavelengths below 1000 nm, haemoglobin is the most important absorption sources and the absorption of oxyhaemoglobin and deoxyhaemoglobin are also different. The differences in the absorption spectra of HbO2 and HHb can then be exploited to measure blood oxygenation by acquiring images at multiple wavelengths and applying a spectroscopic analysis to obtain functional as well as structural images of the vasculature (Laufer, Cox, Zhang, Bear, 2010).

However, the application of PA imaging is limited by its finite penetration depth due to high scattering of bio-tissues thus for tissues deep beneath the skin PA imaging is hard to be obtained (Wang, Gao, 2014). To improve the penetration depth of PA imaging, HIFU (High Intensity Focused Ultrasound) is involved. HIFU is mainly used as a therapeutic method for tumors for its high intensity focused ultrasound field can lead to a very high temperature (60°C) within its focal area (Zhou, 2011). This high temperature will kill the tumor cells with in target area and does little harm to normal tissues.

In this paper, how HIFU affect PA imaging is investigated through comparison experiment. The methods and experiment setup is described in part 2. Results and discussion is shown in part 3. Part 4 we will draw a conclusion.

2 METHODS AND EXPERIMENT SETUP

Schematic of experiment setup is shown in Figure 1 as below. A ND:YAG 532 nm laser is used as excitation source working at 1 Hz repetition frequency. Laser beam is then reflected by a reflector and expended to irradiate the absorbers, pencil (0.5 mm diameter) inserted in fat block, located in a water tank. A 2.25 MHz central frequency Olympus ultrasound transducer is used as detectors and is motivated by a motorized precision translation stage processing B-scan, which ensures that scanning starting point and step numbers are always the same. Data acquired by transducer is then recorded by a PCI-4732 data acquisition card, and is computed by a linear based delay and sum arithmetic to obtain images. HIFU transducer, working at 25 W, is located aside the fat to ensure the HIFU focal area is between the absorber and ultrasound transducer.

This entire experiment setup is emerged in water to couple the ultrasound impedance. Images of the pencil when HIFU is on and off are obtained to conduct comparison. When HIFU is on, the ultrasound intensity will be so high that photoacoustic signals are covered. Thus we have to turn HIFU on to heat the fat block then turn it off when acquiring PA signals.

To get a priori understanding of HIFU effecting PAT the shape of the focal area, which is shown in Figure 2, simulated by a free of charge MATLAB toolbox named ultrasim, is provided. From this figure, we take HIFU field to be approximately considered as a convex lens. The high ultrasound intensity within the area will cause a high temperature rise which shares the same shape of the focal area.

In fat tissues, ultrasound speed is inverse proportional to temperature. So the convex lens shaped thermal area do works as a convex lens to the photoacoustic signal. The experimental results of infrared thermal imager in Figure 3 verified the simulation

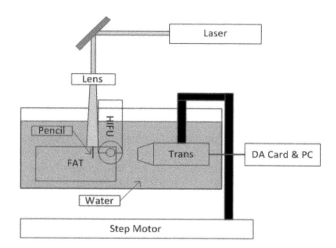

Figure 1. Schematic of experiment setup.

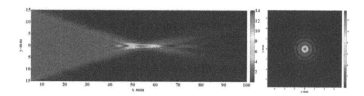

Figure 2. Simulation results of HIFU focal area.

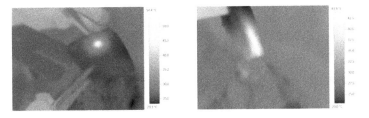

Figure 3. Thermal images of HIFU focal area.

results above. In this experiment, we use HIFU to heat 30 s at 25 W, for higher HIFU power will do greater damage to the fat and less power can't create appropriate thermal lens.

3 RESULTS AND DISCUSSION

3.1 *HIFU acoustic wave propagation in tissue*

We investigate how HIFU effects PA imaging first. 4 pencils are inserted in fat block as imaging target shown in Figure 4 a. When HIFU is off, images of pencils and improfile are shown in Figure 4 b0 and c0. In this image, the point like pencil is shown as a line because the transducer we use is non-focal and has a finite not infinitesimal aperture, but it dose not influence our succedent analysis.

Then we turn HIFU on making it working at 25 W pointing the focal area at point 1, 2, 3, and 4 for 30 s. Then we start PAT B-scan to get images. In order to avoid the influence of previous HIFU heating legions, each time we proceed this experiment, a new fat block and 4 pencils are replaced by another. The shape and size of the fat blocks, from same meat loaf, are nearly the same and multiple experiments are proceeded to reduce random and personal error. Under these conditions, we get Fig b1, b2, b3, b4 and c1, c2, c3, c4 respectively standing for HIFU heating point 1, 2, 3, 4 and improfile on the red line.

Through this figure, we can see that when HIFU is off, the pencil closest to the transducer provide the highest amplitude in the improfile, but the second closer is much more smaller due to acoustic attenuation. when HIFU is on and heating on point 1, 2, 3 and 4 we can see that the amplitude of improfiles of pencil images grow stronger, which means the amplitude of photoacoustic signals is stronger, thus we can come to a conclusion that HIFU thermal effects are able to enhance the penetration depth pf PA imaging, however its capabilities are also limited.

3.2 *HIFU light propagation in tissue*

Second, we investigate HIFU thermal effect of light propagation in tissue. The experiment images are shown in Figure 5 (a1) and (a2). Same as former experiment, new fat block is used after each HIFU heating. When HIFU is off, images of the pencil is obtained in Figure 5 (b1), then HIFU is turned on to heat the area above the pencil at 25 W for 30 s, images shown in Figure 5 (c1).

From these two pictures, no differences but in amplitude is found, which means at this depth, HIFU does little effect in laser penetration. However, when the pencil is overlaid by another 5 mm fat and HIFU is off, images in Figure 5 (b2) shows that laser is hard to penetrate this nearly 1 cm fat for its high optical scattering properties. Then, HIFU is used to heat the same area to obtain another image shown in Figure 5 (c2).

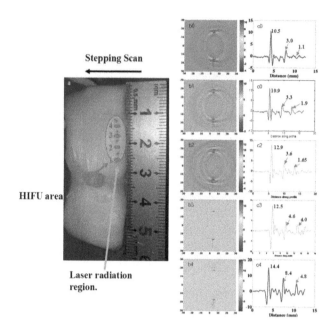

Figure 4. (a) 4 pencils in fat, (b0) HIFU off images, (c0) improfile on red line, (b1) HIFU on heating point 1, (c1) improfile on red line, (b2) HIFU on heating point 2, (c2) improfile on red line, (b3) HIFU on heating point 3, (c3) improfile on red line, (b4) HIFU on heating point 4, (c4) improfile on red line.

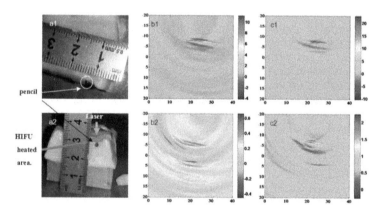

Figure 5. Image of material object and experiment result of a pencil.

4 CONCLUSION

We can come to a conclusion from the results above that the thermal lens generated by HIFU is able to enhance both the amplitude and the penetration depth of PA signals. SNR is also improved when PA signal transmitted through the HIFU thermal area. However, the high intensity and temperature of HIFU will do great harm to normal bio-tissues. As a result HIFU cannot be directly used in vivo PA imaging unless we can find a way to control the temperature and intensity of HIFU within safe levels. HIFU will also cause micro bubbles which is a very complex field to be researched. How these micro bubbles affect PA imaging needs further investigation.

ACKNOWLEDGMENT

This research was partially supported by the International cooperation project of Sichuan Province Science and Technology Agency (2014HH0037 to Rong Jian).

REFERENCES

Beard, P. 2011. Biomedical photoacoustic imaging. Interface Focus 1: 602–631.

Laufer, J. Cox, B. Zhang, E. and Bear, P. 2010. Quantitative determination of chromophore concentrations from 2D photoacoustic images using a nonlinear model-based inversion scheme. Appl. Opt 49(8): 1219–33.

Wang, X.D. Pang, Y.J. Ku, G. Xie, X.Y. Stoica, G. and Wang, L.V. 2003. Noninvasive laser-induced photoacoustic tomography for structural and functional in vivo imaging of the brain, Nat. Biotechnol. Nat. Biotechnol 21: 803–806.

Wang, L.V. and Gao, L. 2014. Photoacoustic microscopy and computed tomography: from bench to bedside. Annual Review of Biomedical Engineering 16: 155–185.

Yang, D.W. Xing, D. Gu, H.M. Tan, Y. and Zeng, L.M. 2005. Fast multielement phase-controlled photoacoustic imaging based on limited-field-filtered back-projection algorithm. Appl. Phys. Lett 87: 19.

Zhou, Y.F. 2011. High intensity focused ultrasound in clinical tumor ablation. World J Clin Oncol 2(1): 8–27.

Bioinformatics and Biomedical Engineering – Chou & Zhou (Eds)
© 2016 Taylor & Francis Group, London, ISBN 978-1-138-02784-8

Analyzer-Based Phase Contrast X-ray Imaging for mouse tissues

H. Li & M. Wang
Medical Science Department, Peking University Health Science Center, Beijing, China

Z. Wang
Radiology Department, Peking University Third Hospital, Beijing, China

S.Q. Luo
School of Biomedical Engineering, Capital Medical University, Beijing, China

ABSTRACT: Analyzer-Based Phase Contrast X-ray Imaging (ABPCXI) can provide high contrast images of biological tissues with exquisite sensitivity to the boundaries between tissues and a reduced tissue radiation dose, compared with conventional absorption radiography. In this study, ABPCXI has been successfully applied to the excised mouse lung, heart and stomach. The anatomical features of these tissues are clearly distinguishable. The results demonstrate that the ABPCXI technique can provide a new source of information related to biological microanatomy.

1 INTRODUCTION

For potential application in medical soft tissue visualization, the phase contrast X-ray imaging modality has attracted considerable attention over the last decade (Castelli et al. 2007, Davis et al. 1995, Donepudi et al. 2012, Kao et al. 2009, Li et al. 2009, Simone et al. 2014, Weitkamp et al. 2002, Wilkins et al. 1996, Zhang & Luo 2011, Zhang et al. 2013, Zhu et al. 2007). The system generates high-resolution images with considerable improvement in contrast compared with conventional radiographic imaging. When monochromatic X-rays traverse weakly absorbed samples, they will be refracted, absorbed and scattered. The interaction of X-ray with the sample can be expressed using the complex refractive index as $n = 1 - \delta + i\beta$, where the real part δ represents the phase shift and the imagery part β corresponds to the absorption of the beam. For biological soft tissue, the real part of the refractive index is three orders of magnitude larger than the imaginary part. This difference in magnitude offers the possibility of a substantial increase in image contrast using the phase contrast imaging over the standard absorption-contrast imaging.

Diffraction Enhanced Imaging (DEI) is a kind of analyzer-based phase contrast X-ray imaging modality. It utilizes the fine angular acceptance of a diffracting analyzer crystal in such a way that it is possible to create several physically unique images. The fine angular acceptance of Bragg diffraction separates out the components of the X-ray beam that have undergone different interactions with the sample. The analyzer crystal's acceptance is described by its Rocking Curve (RC), which is a plot of reflectivity against the angle of incidence of the X-ray beam on the crystal. Different images can be obtained by setting the analyzer on different positions of the rocking curve. Particularly meaningful positions are the peak and Full Width at Half Maximum (FWHM) of the RC. The image acquired at the peak position highlights the so-called extinction contrast, and the image acquired at the FWHM positions maximizes contrast due to refraction. Finally, combining the two FWHM images produces the so-called refraction image, which is different from conventional radiographs because it is entirely the result of refraction and represents a map of the refraction angle at each point on the object plane (Chapman et al. 1997, Zhong et al. 2000). This imaging technique enhances the visibility of tissue borders and helps to visualize low contrast areas that otherwise would be lost. The purpose of this study is to explore the ability of DEI in directly visualizing complex and simple biological soft tissues.

2 MATERIALS AND METHODS

This study was performed in accordance with the guidelines of the National Institutes of Health for the care and use of laboratory animals. Mice were humanely killed by overdose injection of sodium pentobarbital. Then, the lung, heart and stomach were excised and fixed in 10% formalin solution for imaging. During imaging, specimens were cut along the middle center, and each half was used for imaging.

The DEI experiments were performed at the 4 W1 A Biomedical Beamline of the Beijing Synchrotron Radiation Facility (BSRF). The Wiggler source of the beam line 4W1A was 43 m away from the experiment hutch. The incident white synchrotron X-ray beam was first monochromatized by a perfect silicon (111) crystal. The available energy range was 3–22 keV. During the experiment, the X-ray energy was set at 10 keV. The sample was placed between the collimator crystal and the analyzer crystal (i.e. the same as the monochromator crystal). According to the Brag diffraction theory, the analyzer crystal only reflects photons coming from a particular angle, described by its RC. When the highly collimated and monochromatized beams traverse the sample, they will be refracted through very small angles due to the small variations in the refractive index of the sample. Thereafter, the X-ray is refracted by the analyzer crystal. The transmitted beam then strikes the detector, forming an image that is sensitive to the X-ray refraction. The image was detected by a X-ray CCD with a pixel size of approximately 10.9 μm.

During the experiment, the analyzer was first set at the peak position of the rocking curve, and the peak image was obtained. Then, the analyzer was set at each side of the FWHM positions; thereafter, the refraction image was obtained through pixel by pixel subtraction of the two raw images obtained at the FWHM positions of the rocking curve.

3 RESULTS AND DISCUSSION

Figure 1 shows the images of the mouse lung. Figure 1a shows the peak image, and Figure 1b shows the refraction image. The refraction image reveals the boundaries of the major airways and lung lobes, and a speckled pattern is evident across the image of the lung tissue. However, the airway boundaries are almost invisible in the peak image. The image result reveals that much of the contrast arises from phase gradients within the tissue.

Figures 2 and 3, respectively, show the transmitted and diffracted images of the mouse heart and stomach. The peak images of Figures 2a and 3a show a blurred appearance, thus

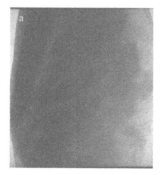

Figure 1. Diffraction-enhanced images of the mouse lung.
a. The peak image. b. The refraction image.

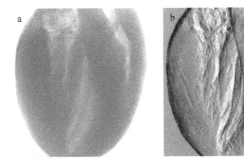

Figure 2.   Diffraction-enhanced images of the mouse heart.
a. The peak image. b. The refraction image.

Figure 3.   Diffraction-enhanced images of the mouse stomach.
a. The peak image. b. The refraction image.

the tissue structure cannot be distinguished. The peak image is similar to a conventional radiograph, but it is free of the undesired scattering that is usually present and reduces the image quality. The inner structure of the heart is composed of cardiac muscle fibers and the stomach is composed of connective tissues, each has a similar density. It is hard to be distinguished by the absorption contrast imaging. The distinct appearance of Figures 2a and 3a comes from the different thicknesses of the sample. The refraction images of Figures 2b and 3b give a clear identification and visibility of the structures with a considerable degree of details present, and the interfaces between soft tissue regions can be clearly distinguished. The refraction image depicts the effect of small-beam deflections owing to refractive index variations in the object. X-ray passing through regions with different phase indices will deviate from their original directions, borders and contours of the microstructures of soft tissues, which will be highlighted by this refractive effect and the image contrast can be greatly enhanced.

Magnetic Resonance Imaging (MRI) is widely used for imaging of soft tissue, but MRI currently does not yield sufficient spatial resolution for visualization of subtle pathologies. Conventional X-ray radiography is incapable of demonstrating any significant details of soft tissues. Here, the results have shown that ABPCXI, using X-rays, produces extremely high-quality images, exhibiting an exquisite detail of soft tissue components. It is particularly meaningful for the early detection of morphological changes of tissues, thus helpful for the early detection of cancer. Furthermore, the resultant increase in contrast is achieved at lower X-ray doses since higher X-ray energies can be used.

Nevertheless, all results presented until now have been obtained using excised specimens from the animal body. It remains to be determined whether the technique can be extended to an *in vivo* study of living animals.

4 CONCLUSIONS

The possibility of DEI to image complex biological specimens, such as the mouse lung, and simple biological specimens, such as the mouse heart and stomach, has been investigated. DEI provided high-resolution images with improved contrast and clear visibility of the anatomical structures, even tissues of the inner heart and stomach with negligible absorption contrast can be clearly visualized, thanks to the edge enhancement. The results show great promise for DEI in characterizing soft tissue structures at high resolution.

ACKNOWLEDGMENTS

The authors would like to thank Professors Zhu P.P., Huang W.X. & Yuan Q.X. of BSRF for their assistance in the preparation of the experimental apparatus and helpful discussion. This research was supported by the Beijing Natural Science Foundation of China. Grant no. 7142084.

REFERENCES

Castelli, E., Arfelli, F., Dreossi, D., et al. 2007. Clinical mammography at the SYRMEP beam line, *Nucl. Instrum. Methods Phys. Res. A* 572(1): 237–240.
Chapman, D., Thomlinson, W., Johnston, R.E., et al. 1997. Diffraction enhanced x-ray imaging, *Physics in Medicine and Biology* 42(11): 2015–2025.
Davis, T.J., Gao, D., Gureyev, T.E., et al. 1995. Phase-contrast imaging of weakly absorbing materials using hard x-rays, *Nature* 373(6515): 595–598.
Donepudi, V.R., Medasani, S., Roberto, C., et al. 2012. Synchrotron-based DEI for bio-imaging and DEI-CT to image phantoms with contrast agents, *Applied Radiation and Isotopes* 70: 1570–1578.
Kao, T., Connor, D., Dilmanian, F.A., et al. 2009. Characterization of diffraction-enhanced imaging contrast in breast cancer, *Physics in Medicine and Biology* 54(10): 3247–3256.
Li, H., Zhang, L., Wang, X.Y., et al. 2009. Investigation of hepatic fibrosis in rats with X-ray diffraction enhanced imaging, *Applied Physics Letters* 94(12): 124101–124103.
Simone, S., Martin, B., Susanne, G., et al. 2014. X-ray phase-contrast tomosynthesis for improved breast tissue discrimination, *European Journal of Radiology* 83: 531–536.
Weitkamp, T., Rau, C., Snigirev, A., et al. 2002. In-line phase contrast in synchrotron-radiation microradiography and tomography, *Proceedings of Spie* 4503: 92–102.
Wilkins, S.W., Gureyev, T.E., Gao, D., et al. 1996. Phase-contrast imaging using polychromatic hard X-rays, *Nature* 384(28): 335–338.
Zhang, L., Luo, S.Q., 2011. Micro Soft Tissues Visualization Based on X-Ray Phase-Contrast Imaging, *The Open Medical Informatics Journal*, 5 (Suppl 1-M2) 19–25.
Zhang, X., Yang, X.R. Chen, Y., et al. 2013. Visualising liver fibrosis by phase-contrast X-ray imaging in common bile duct ligated mice, *Eur Radiol*, 23: 417–423.
Zhong, Z., Thomlinson, W., Chapman, D., et al. 2000. Implementation of diffraction-enhanced imaging experiments: At the NSLS and APS, *Nucl. Instrum. Methods Phys. Res. A*, 450: 556–567.
Zhu, P.P., Wang, J.Y., Yuan, Q.X., et al. 2007. Computed tomography algorithm based on diffraction-enhanced imaging setup, *Applied Physics Letters* 87(26): 264101(1–3).

Biocybernetics and biological effects

Bioinformatics and Biomedical Engineering – Chou & Zhou (Eds)
© 2016 Taylor & Francis Group, London, ISBN 978-1-138-02784-8

Effects of captopril pretreatment on gp130 expression in rats with acute cardiomyocyte injury

Y. Zhou, S.B. Li & Y. Zhang
Department of Histology and Embryology, College of Basic Medicine, Beihua University Jilin, China

X.B. Jing & M. Gu
Department of Cardiology, Affiliated Hospital, Beihua University Jilin, China

ABSTRACT: Effects of the pretreatment with captopril on myocardial apoptosis in myocardial injury and gp130 expression were studied in this experiment. 66 Male Wistar rats were randomly divided into control group, Iso-induced injury group (Isoproterenol—induced injury group) and Cap-pretreated group (Captopril-pretreated group). The myocardial cell apoptosis was examined by TUNEL method, the expression of Bcl-2 and Bax proteins was detected by immunohistochemistry, and the expression of gp130 in the myocardial tissue was measured by Western blot. Compared with that in Iso-induced injury group, the apoptotic index of rats in Cap-pretreated group decreased, the level of Bax and gp130 expression decreased, but the expression of Bcl-2 increased ($P < 0.01$ or < 0.05). Captopril could interfere with the apoptosis by affecting the regulation of apoptosis-related genes and the expression of gp130 in the process of myocardial tissue injury. Captopril may exert its inhibitory effect on the apoptosis both directly by blocking the formation of AngII and indirectly by inhibiting the expression their receptor gp130 in the myocardial injury.

Keywords: captopril; isoproterenol; glycoprotein 130; apoptosis

1 INTRODUCTION

Isoproterenol (Iso) can activate β receptors in the heart to shorten the diastole and increase the wall tension, leading to an imbalance of myocardial oxygen supply and oxygen demand, and further the myocardial tissue injury due to the myocardial ischemia and hypoxia, which is commonly used for the preparation of animal model with ischemic myocardial injury. Angiotensin-Converting Enzyme Inhibitors (ACEI) can reduce the myocardial injury by its ischemic preconditioning and captopril (Cap) a member of the first generation of ACEI, is widely used in clinic. Glycoprotein 130 (gp130) is a tyrosine kinase receptor and can be activated by various cytokines. In this experiment, an animal model of acute myocardial injury were established by giving rats isoproterenol in intraperitoneal injection, and the cardioprotective effect of captopril and whether the cardiomyocyte apoptosis was related with the expression of gp130 were studied, to try to provide an theoretical foundation for the clinical application of ACEI to prevent and reverse the ventricular remodeling.

2 MATERIALS AND METHODS

2.1 *Animal grouping and pretreatment*

66 Healthy male Wistar rats, weighing 175 ± 15 g, (Provided by Experimental Animal Center, Beihua University), were housed in polypropylene cages under controlled temperature at $22 \pm 0.5°C$ and a 12:12-hour light/dark cycle, and on a standard laboratory feed with food

and water. The experimental animals were divided into three groups, namely, control group, Iso-induced injury group and Cap-pretreated group, and there were 22 rats in each group. Rats in Iso-induced injury group were administered Iso (5.0 mg · kg^{-1} · d^{-1}) in intraperitoneal injection successively for two days, rats in the control group were administered the same volume of saline in intraperitoneal injection successively for two days, and rats in Cap-pretreated group were administered Iso (5.0 mg · kg^{-1} · d^{-1}) in intraperitoneal injection successively for two days. On the day before the intraperitoneal injection of Iso, rats in Cap-pretreated group were given captopril (100 mg · kg^{-1} · d^{-1}) intragastrically successively for 3 days, and those in Iso-induced injury group and the control group were given the same volume of normal saline in the same way. At 24 hours after the last administration, the rats were sacrificed and the samples were collected according to the experimental requirements.

2.2 Observation indicators and experimental methods

2.2.1 Analysis of myocardial histopathology

The thorax of rat was opened, the heart was cut apart along the long axis of the heart at the middle of right ventricular free wall, and half of the heart was fixed with 4% paraformaldehyde at room temperature for 24 hours, embedded in paraffin, and prepared into 5 um-thick slices for HE staining. Five non-overlapping horizons in each slice were randomly chosen under a high power light microscope, the number of necrotic myocardial cells and the total number of myocardial cells per field were calculated with a true color pathological image analysis system made by Beijing Aeronautics and Astronautics University, and then the myocardial necrosis area was calculated based on the following formula: myocardial necrosis area = number of necrotic myocardial cells /number of total myocardial cells × 100%.

2.2.2 Detection of cardiomyocyte apoptosis by TUNEL method

The myocardial tissue was fixed with 4% paraformaldehyde for 24 h, which was embedded with paraffin and 5 um-thick slices of it was made. The operation was based on the kit instructions. Under an optical microscope, the myocardial cells whose nuclei showed a specially labeled brown color were taken as positive cells, that is, apoptotic ones, while normal myocardial cells could be stained blue by hematoxylin. Five non-overlapping fields of each slice were randomly selected, and the number of apoptotic myocardial cells and the total number of myocardial cells in each field were counted and recorded for the calculation of apoptotic index of cardiomyocytes (apoptotic index of cardiomyocytes = number of apoptotic myocardial cells/number of total myocardial cells × 100%).

2.2.3 Immunohistochemical detection of Bcl-2 and Bax protein
expression in myocardial cells

The paraffin specimens were stained by immunohistochemistry and DAB method, and the operation steps were according to the kit instructions. After DAB color reagent was added dropwise, the specimens were observed under an optical microscope and cells that there was a brown staining in the cytoplasm were taken as positive ones. Five non-overlapping fields of each slice were randomly selected for the analysis on the total area of positively expressed cardiomyocytes and cardiomyocytes in each field with a pathological image analyzer, and the positive expression rate was calculated (positive expression rate = total area of positively expressed myocardial cells/ total area of cardiomyocytes per field × 100%).

2.2.4 Detection of gp130 expression in myocardial tissue by western blot

The total protein was extracted. The protein concentration of samples was measured by BCA method. Polyacrylamide gel electrophoresis: 6-fold sample buffer used for the denaturation at 100°C for 5 min, the concentration of separation gel was 10%, 50 μg protein was added to each lane. After the sample got into the separation gel electrophoresis, the electrophoresis of it was conducted at 100 V for 1.5 h until it reached to about 1 cm of the lower end of gel; proteins in the sample were transferred at 30 V and 4°C for 12 h onto a PVDF membrane, the membrane was washed 3 times with TBST for 6 min, and 10 ml blocking solution containing the first antibody

(1: 500) was added to the membrane, which was shaken at 37°C in a parallel direction for 1 h; the membrane was washed 3 times with TBST for 5 min, 10 ml blocking solution containing the second antibody (1: 500), which was shaken at 37°C in a parallel direction for 1 h; the membrane was washed 4 times with TBST for 5 min, and then 2 times with TBS for 5 min; after the addition of color reagent and the achievement of the desired color, it was washed twice with deionized water and then dried for the imaging using a digital imager. The density of bands was scanned.

2.3 Drugs and reagents

Isoproterenol: Sigma, USA; Captopril powder: Changzhou Pharmaceutical Factory, China; Apoptosis detection kit: Roche Company, USA; Bcl-2 and Bax rabbit polyclonal rabbit polyclonal antibody: Wuhan Boster Biological Engineering Co; Gp130 rabbit polyclonal antibody: Santa Cruz, USA.

2.4 Statistical analysis

The results belonged to measurement data and were analyzed using SPSS 12.0 statistical software. The data is expressed in mean ± s. The analysis of variance was applied to test differences among multiple samples and q test was used for the comparison between two groups.

3 RESULTS

3.1 Pathological changes of myocardial tissue

No abnormal myocardiocytes were found in the myocardial tissue of rats in the control group; myocardial tissue showed necrosis of various degrees in the Iso-induced injury group, some cardiomyocytes disintegrated and disappeared, and a large number of inflammatory cells could be seen in the myocardial tissue of rats; compared with that in Iso–induced injury group, the myocardial necrosis area in Cap-pretreated group was significantly reduced and there was a significant difference in it between the two groups ($P < 0.01$). The results are shown in Table 1 and Figures 1–3.

Table 1. Myocardial cell apoptotic index and expression of apoptosis-related proteins and gp130 in myocardial tissues of rats ($\bar{x} \pm S$).

Group	No.	Myocardial necrosis area (%)	Apoptotic index (%)	Bcl-2 expression index (%)	Bax expression index (%)	gp130
Control	22	0	0.97 ± 0.35	5.85 ± 0.82	5.98 ± 0.76	281 ± 87
Iso-induced	22	13.31 ± 1.91△	11.85 ± 2.56△	4.38 ± 0.49△	11.85 ± 1.12△	821 ± 128△
Cap-pretreated	22	5.35 ± 0.92*	5.87 ± 1.08*	5.25 ± 0.76*	10.28 ± 1.01**	516 ± 97*

Notes: Compared with those in Iso-induced injury group, * means $P < 0.01$ and ** means $P < 0.05$; Compared with those in control group, △means $P < 0.01$.

Figure 1. Myocardial cells in the control group (HE × 100).

Figure 2. Myocardial cells in the Iso-induced injury group (HE × 100).

Figure 3. Myocardial cells in the Cap-pretreated group (HE × 100).

3.2 Cardiomyocyte apoptosis

According to the color shade and the number of granules in images, the staining intensities were divided into −~+++. Cells with the staining intensity beyond + were considered to be apoptotic ones. As shown in Table 1, apoptotic cells were only occasionally found in the myocardial tissue of rats in the control group; a large number of apoptotic cells could be seen in myocardial tissue of rats in Iso-induced injury group and Cap-pretreated group, which were distributed mostly in the area surrounding the necrotic myocardial tissue, namely, the ischemic area, and some apoptotic cells were also found in areas away from the necrotic myocardium. However, compared with that in Iso-induced injury group, the apoptotic index in Cap-pretreated group was significantly reduced, there was a significant difference ($P < 0.01$) between the two groups. The results are shown in Table 1.

3.3 Bcl-2 and Bax expression

The expression of Bcl-2 and Bax proteins was various in the three groups, and the expression location was approximately consistent with the site of myocardial apoptosis. The results in Table 1 and Figures 4–9 showed that compared with that in the control group, the expression of Bcl-2 protein decreased and that of Bax protein increased in Iso-induced injury group, and the difference between the two groups was significant ($P < 0.01$); compared with that in Iso-induced injury group, the expression of Bcl-2 protein increased and that of Bax protein decreased in Cap-pretreated group, and the difference between the two groups was significant ($P < 0.01$ or $P < 0.05$).

3.4 gp130 expression

Bands presenting an amplified protein with 130 KD molecular weight, gp130, in the myocardial tissue of all rats in the three groups could be found. As shown in Table 1 and Figure 10,

Figure 4. Expression of Bcl-2 protein in the control group (×100).

Figure 5. Expression of Bcl-2 protein in the Iso-induced injury group (×100).

Figure 6. Expression of Bcl-2 protein in the Cap-pretreated group (×100).

Figure 7. Expression of Bax protein in the control group (×100).

Figure 8. Expression of Bax protein in the Iso-induced injury group (×100).

Figure 9. Expression of Bax protein in the Cap-pretreated group (×100).

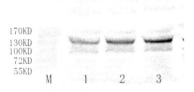

Figure 10.   gp130 protein band of rat myocardium.
M:Molecular marker; 1: Control group; 2: Iso group; 3: Cap pretreatment group.

compared with that in the control group, the level of gp130 expression in Iso-induced injury group was significantly higher ($P < 0.01$); compared with that in Iso induced injury group, the level of gp130 expression in Cap-pretreated group was significantly lower, and there was a significant difference between the two groups ($P < 0.01$).

4   DISCUSSION

This experiment succeeded in producing the animal model of acute myocardial injury in rats, and pathological characteristics are identical to those of previous literature (Manikandan 2002). The phenomenon, in which not only the cumulative injury in tissues and organs after a short and repeated ischemia does not occur, but also the subsequent injury caused by the prolonged ischemia can be oppositely alleviated, is known as Ischemic Preconditioning (IP) (Marongiu et al. 2014). Based on mechanisms of IP, various aspects of IP can be purposefully simulated or the transient ischemia can be improved by some drugs, which can more effectively protect the heart and known as pharmacological preconditioning (Gu et al. 2012). Captopril, a angiotensin converting enzyme inhibitor, has been successfully applied in the establishment of IP animal models (Miki et al. 1996).

Apoptosis is also called programmed cell death, which is is an active death of cells in the process of normal growth or in pathological conditions, and it is regulated by a series of gene. Bcl-2 and Bax respectively play opposite roles, Bcl-2 can inhibit the cell apoptosis, in contrast, Bax can promote the cell apoptosis (Takemura et al. 2004). The results showed that compared with those in the control group, the myocardial necrosis area, the apoptotic index and Bax protein expression in rats in Iso-induced injury group increased significantly, and Bcl-2 protein expression decreased significantly, indicating that Bcl-2 and Bax should be involved in the regulation of the process of myocardial cell apoptosis in the isoproterenol-induced myocardial injury. Compared with that in the control group, the expression of gp130 protein in rats in Iso-induced injury group was also elevated, indicating a responsive elevation of gp130 expression to the myocardial injury induced by isoproterenol in the body. Under physiological conditions, a certain amount of gp130 can be expressed in ventricular myocytes, which can play an important regulatory role in the growth, proliferation and differentiation of cells (Mohammadi et al. 2010). After the myocardial injury, due to hypoxia, the stimulation of angiotensin II and norepinephrine, and the over-activation of IL-6 family cytokines, such as CT-1, IL-6, Leukemia Inhibitory Factor (LIF) (Sano et al. 2000), the expression of gp130 is upregulated, the activated gp130 is dislocated from the cytoplasm to the cell membrane to interact with cytokines, activating Janus tyrosine kinase (JAK)/Signal Transducer and Activator of Transcription (STAT3) signaling pathway to make STAT3 tyrosine phosphorylated to translocate into the nucleus, which can combined with the specific DNA sequences to activate related genes to mediate cardiac hypertrophy, causing ventricular remodeling (El-Adawi et al. 2003).

Compared with those in Iso-induced injury group, the myocardial necrosis area, the apoptosis index and Bax and gp130 protein expressions were significantly reduced, while the expression of Bcl-2 protein significantly increased in rats in Cap-pretreated group, indicating that captopril could interfere with the apoptosis by affecting the regulation of apoptosis-related genes and the expression of gp130 in the process of myocardial tissue injury. It can be speculated from the results that since captopril can block the formation of angiotensin II

(AngII) by inhibiting angiotensin converting enzyme, and AngII can increase the expression of pro-apoptotic gene Bax and decrease the expression of anti-apoptotic genes Bcl-2, that is, leading to the reduction in the ratio of Bcl-2/Bax to induce the apoptosis (Zhou et al. 2009), captopril, on the one hand, may play an inhibitory effect on the apoptosis by blocking the formation of AngII, on the other hand, captopril can also inhibit associated genes that can mediate the myocardial hypertrophy by antagonizing the up-regulated expression of IL-6 family cytokines (CT-1, LIF, IL-6) mediated by angiotensin II and indirectly by inhibiting the expression their receptor gpl30, thereby to inhibit the JAK/STAT3 signaling pathway, delaying the ventricular remodeling caused by the myocardial injury. Moreover, Tsuruda reported that the synergistic effect of gp130 signaling pathway and Endothelin Receptor (ETA) could potentiate the in vitro proliferative effect in dog's fibroblasts (Tsuruda et al. 2002), so that it could be believed that inhibiting gp130 expression should delay the ventricular remodeling by reducing the generation of fibroblasts. However, there has been also some studies suggesting that the activated gp130 can be dislocated from the cytoplasm to the cell membrane, where it can interact with some cytokines to reduce the myocardial apoptosis by inhibiting the transduction of apoptosis signal through the p42/p44 MAPK pathway, playing an role in the protection of myocardial cells (Fahmi et al. 2013).

5 CONCLUSION

Captopril could interfere with the apoptosis by affecting the regulation of apoptosis-related genes and the expression of gp130 in the process of myocardial tissue injury. It can be speculated from the results that on the one hand, captopril may play an inhibitory effect on the apoptosis by blocking the formation of AngII, on the other hand, captopril can also inhibit associated genes that can mediate the myocardial hypertrophy by antagonizing the up-regulated expression of IL-6 family cytokines (CT-1, LIF, IL-6) mediated by angiotensin II and indirectly by inhibiting the expression their receptor gpl30, thereby to inhibit the JAK/STAT3 signaling pathway, delaying the ventricular remodeling caused by the myocardial injury.

REFERENCES

El-Adawi H. et al. 2003. The functional role of the JAK-STAT pathway in post-infarction remodeling. *Cardiovasc Res* 57(1):129–138.

Fahmi A. et al. 2013. p42/p44-MAPK and PI3 K are sufficient for IL-6 family cytokines/gp130 to signal to hypertrophy and survival in cardiomyocytes in the absence of JAK/STAT activation. *Cell Signal* 25(4):898–909.

Gu Ming et al. 2012. Effects and significance of captopril pretreatment on CT-1 mRNA and gp130 mRNA expression in rats with acute myocardial injury. *Chinese Journal of Gerontology* 12(32): 2554–2556.

Manikandan P. 2002. Antioxidant potential of a novel tetrapeptide derivative in isoproterenol-induced myocardial necrosis in rats. *Pharmacology* 65(2):103–109.

Marongiu E. & Crisafulli A. 2014. Cardioprotection acquired through exercise: the role of ischemic preconditioning. *Curr Cardiol Rev* 10(4):336–348.

Miki T. et al. 1996. Captopril potentiates the myocardial infarct size-limiting effect of ischemic preconditioning through bradykinin B2 receptor activation. *J Am Coll Cardiol* 15(6):1616–1622.

Mohammadi Roushandeh A. et al. 2010. Effects of leukemia inhibitory factor on gp130 expression and rate of metaphase II development during in vitro maturation of mouse oocyte. *Iran Biomed J* 14(3):103–107.

Sano M. et al. 2000. Interleukin-6 family of cytokines mediate angiotensin II-induced cardiac hypertrophy in rodent cardiomyocytes. *J Biol Chem* 275(38):29717–29723.

Takemura G. & Fujiwara H. 2004. Role of apoptosis in remodeling after myocardial infarction. *Pharmacol Ther* 104(1):1–16.

Tsuruda T. et al. 2002. Cardiotrophin-1 stimulation of cardiac fibroblast growth: roles for glycoprotein 130/leukemia inhibitory factor receptor and the endothelia type A receptor. *Circ Res* 90(2):128–134.

Zhou Yan et al. 2009. The protective effect of captopril preconditioning on rats of acute myocardial injured and its influence of apoptosis of cardiomyocytes. *Jilin Medical Journal* 30(21):2567–2569.

Bioinformatics and Biomedical Engineering – Chou & Zhou (Eds)
© 2016 Taylor & Francis Group, London, ISBN 978-1-138-02784-8

Effects of sleep restriction and circadian rhythm on human vigilance

H.Q. Yu, Y. Tian, C.H. Wang & S.G. Chen
National Key Laboratory of Human Factors Engineering, China Astronaut Research and Training Center, Beijing, China

J.H. Guo
School of Life Sciences, Sun Yat-sen University, Guangzhou, China

ABSTRACT: Human operators need to keep high vigilance in many tasks. Sleep loss often induces decreased vigilance, while the time of day may modulate the effects of sleep loss on human vigilance. The current study aims to investigate the effects of sleep restriction and circadian rhythm on human vigilance. The Psychomotor Vigilance Task (PVT) metrics are selected as objective measures of human vigilance. The constant routine procedure in circadian rhythm research is adopted in our experimental study: in the days with sleep restriction, participants are restricted to 5 hours of sleep; in the control days, participants are allowed to sleep 8 hours per day. The participants are required to perform the PVT test repeatedly at different times of the day (in the morning, at noon, in the afternoon, in the evening and at midnight). The results show that in the sleep restriction condition, the PVT metrics of the participants are significantly worse. Meanwhile, the vigilance of the participants is shown to be significantly better in the afternoon and in the evening than in the morning and at noon.

1   INTRODUCTION

In tasks such as driving a vehicle and air traffic control, the operators have to keep good vigilance and react quickly and accurately, otherwise accidents may occur. There are many factors that may influence human vigilance, such as sleep, motivation, food and circadian rhythm. Sleep is a very important factor that influences vigilance. If we cannot sleep for sufficient duration, our attention and cognitive performance may be negatively affected during the wake period (Goel et al. 2013). Meanwhile, the time of day may influence human vigilance and cognitive performance as well, and may even modulate the effects of sleep loss on human vigilance and cognitive performance (Van Dongen et al. 2000).

In space flight, cumulative sleep loss is commonly reported by astronauts. Reports have shown that astronauts sleep only about 5 hours at night (Liang et al. 2014, McPhee & Charles 2009). The question whether this extent of sleep loss (5 hours of sleep) causes vigilance change, and whether human vigilance changes at different times of the day under the sleep loss condition have not been clearly addressed in previous research, which is the focus of the present paper.

2   METHODS

2.1   *Experimental design*

In this study, we aim to investigate the effects of fatigue and circadian rhythm on human vigilance. To induce human fatigue, the sleep restriction method is adopted: in the days with sleep restriction, participants are restricted to 5 hours of sleep; in the control days, participants are allowed to sleep 8 hours per day. To investigate the effect of circadian rhythm, participants are required to perform the test repeatedly at different times of the day. In the sleep restriction condition, participants were tested five times per day: once in the morning, once at noon, once in the afternoon, once in the evening and once at midnight. In order not to disturb the participants' sleep (7 hours of night sleep and 1 hour of nap at noon) under the control condition, participants were tested three times on the fourth day: once in the morning, once in the afternoon and once in the evening. Also, four times on the last day: once in the morning, once in the afternoon, once in the evening and once at midnight.

2.2 Participants

Eight male participants with an average age of 23.4 years (range 20 to 28 years) participated in this study. Participants were all right-handed, had a normal sight and hearing, and had a regular daily routine. All participants were free of depression, health or sleep problems. Participants were paid for their participation in the experiment.

2.3 The Psychomotor Vigilance Task (PVT)

The PVT measures human vigilance by recording Response Times (RT) to visual or auditory stimuli that occur at random Inter-Stimulus Intervals (ISI). The PVT test is the most widely used objective measure of human vigilance owing to its high sensitivity to fatigue (Mollicone et al. 2008, Raymann & Van Someren 2007). The PVT test that we used in our experiments is a fully computerized test for a simple visual reaction time. Participants were presented with a visual stimulus (a display of the response time in ms) and were asked to react when they saw the stimulus by pressing the space key on the keyboard. In our test, participants took a 5-minute PVT test, in which they responded for about 40 times. The following PVT outcome metrics were assessed and included in our analyses: (1) number of lapses (Lapses), the trial with response times over 500 ms is defined to be a lapse; (2) mean Response Time (RT_{mean}), the mean response time of all responses excluding lapses and false starts (the trial with a response time less than 100 ms is defined to be a false start); (3) fastest 10% RT ($RT_{10\,fast}$), the mean response time of the fastest 10% of all responses excluding lapses and false starts; (4) slowest 10% RT ($RT_{10\,slow}$), the mean response time of the slowest 10% of all responses excluding lapses and false starts; and (5) 10%-90% RT ($RT_{10\,to\,90}$), the mean response time of the 10% to 90% fastest responses excluding lapses and false starts. Lower RTs and less lapses indicate better vigilance.

2.4 Procedure

The experiment lasted for 5 days. The test arrangements are presented in Table 1. During the 5-day tests, the temperature and illuminance of the test environment were kept at similar levels. Before the 5-day formal experiment, participants were allowed to practice over six times to make sure that they are clear of the test requirements and test procedures, and also to eliminate possible learning effects.

Table 1. Experimental arrangements.

Day num.	Sleep duration	Whether tests are arranged at a particular time of the day*				
		In the morning	At noon	In the afternoon	In the evening	At midnight
1	5 hours	Yes	Yes	Yes	Yes	Yes
2	5 hours	Yes	Yes	Yes	Yes	Yes
3	5 hours	Yes	Yes	Yes	Yes	Yes
4	8 hours	Yes	No	Yes	Yes	No
5	8 hours	Yes	No	Yes	Yes	Yes

*Test in the morning: test performed between 08:30 and 10:00; Test at noon: test performed between 12:00 and 13:30; Test in the afternoon: test performed between 15:30 and 17:00; Test in the evening: test performed between 19:00 and 20:30; Test at midnight: test performed between 22:30 and 24:00.

3 RESULTS AND ANALYSES

Data were analyzed with SPSS for Windows (version 17.0) and are presented as mean ± Standard Error (SE). Statistical significance was set at $p < 0.05$.

3.1 *Correlations between the different PVT metrics*

In this study, five PVT metrics were calculated for analyses: (1) mean Response Time (RT_{mean}); (2) 10%–90% RT ($RT_{10\ to\ 90}$); (3) fastest 10% RT ($RT_{10\ fast}$); (4) slowest 10% RT ($RT_{10\ slow}$) and (5) number of lapses (Lapses). For all the five metrics, lower values indicate better vigilance. As each of the eight participants completed the PVT test 22 times (five times in the first three days, three times in the fourth day and four times in the last day), altogether there were 176 sets of PVT data. The correlation coefficients between the five metrics were calculated based on the 176 sets of PVT data, and are listed in Table 2.

From Table 2, we can see that all the five metrics were significantly correlated; however, the metric "Lapses" seemed to be different from the other four metrics as its correlations with other metrics were relatively low. Upon inspection of the data, we found that "Lapses" occurred very infrequently. On average, the participants had only 0.74 ± 0.31 lapses in the PVT test during the sleep restriction condition (restricted to 5 hours of sleep), and they had even fewer lapses (0.35 ± 0.17) during the control condition (8 hours of sleep). Although the paired t-test showed that the participants' lapses were significantly lower when they got 8 hours of sleep than when slept only for 5 hours ($t(7) = 2.558$, $p = 0.038$), the fact that the 75% of the participants had less than 2 lapses in the PVT test indicated that the metric "Lapses" was not so sensitive to fatigue in our study.

As RT_{mean}, RT_{10to90}, RT_{10fast} and $RT_{10\ slow}$ showed a strong correlation with each other, and all of the metrics were significantly lower when the participants got 8 hours of sleep compared with 5 hours of sleep. To facilitate analysis, RT_{mean} was used for further analysis.

3.2 *PVT metrics under two sleep conditions*

During the first three days of the sleep restriction condition and the last two days of the control condition, the participants completed the PVT test at least three times of the day: in the morning, in the afternoon and in the evening. So, the PVT metric RT_{mean} of the participants for three times of the day were averaged to test the influence of sleep restriction on PVT performance, as shown in Figure 1. The paired t-test showed that the PVT performance was significantly worse in the sleep restriction condition (315.5 ± 9.6) than during the control condition (304.0 ± 8.5) ($t(7) = 4.474$, $p = 0.003$).

3.3 *PVT metrics during different times of the day*

Participants were required to perform the PVT test repeatedly at different times of the day. In the sleep restriction condition, participants were tested five times per day: once in the morning, once at noon, once in the afternoon, once in the evening and once at midnight. They were tested three times on the fourth day (once in the morning, once in the afternoon and once in the evening) and four times on the last day (once in the morning, once in the afternoon, once

Table 2. Correlation coefficients between the five metrics (Pearson, N = 176).

	$RT_{10\ to\ 90}$	$RT_{10\ fast}$	$RT_{10\ slow}$	Lapses
RT_{mean}	0.997**	0.859**	0.910**	0.574**
$RT_{10\ to\ 90}$		0.861**	0.883**	0.574**
$RT_{10\ fast}$			0.665**	0.485**
$RT_{10\ slow}$				0.504**

**Correlation is significant at the 0.01 level (2-tailed).

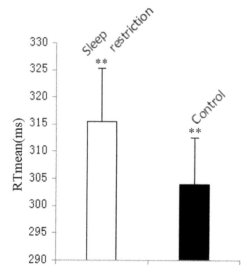

Figure 1.   PVT performance (RT$_{mean}$) under the sleep restriction and control conditions.

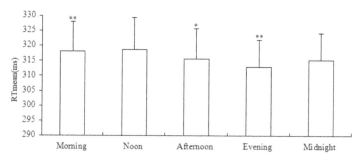

Figure 2.   PVT performance (RT$_{mean}$) of the participants at the five different times of the day under the sleep restriction condition.

in the evening and once at midnight). The PVT performance of the participants at different times of the day was separately analyzed for the sleep restriction and control conditions.

3.3.1   Data under the sleep restriction condition

The PVT performance (RT$_{mean}$) of the participants at the five different times of the day under the sleep restriction condition is shown in Figure 2. Paired t-tests were performed to compare the PVT performance at the five different times of the day (the t-test were performed 10 times), which showed significant differences only between the performance in the evening and that in the morning, and between the performance in the evening and that at noon. The performance in the evening (312.9 ± 8.5) was significantly better than that in the morning (318.1 ± 9.3) (t(7) = 3.628, p = 0.008). The performance in the evening (312.9 ± 8.5) was significantly better than that at noon (318.7 ± 10.0) (t(7) = 2.422, p = 0.046).

 If we consider the time from 08:30 to 13:30 as the "early day" (including the morning time and noon time in this study), and the time from 15:30 to 20:30 as the "late day" (including the afternoon time and evening time in this study), it was found that the participants completed the PVT test twice in both the "early day" and the "late day" during the sleep restriction condition. The PVT performance (RT$_{mean}$) of the participants in the "early day" and the "late day" is shown in Figure 3. The paired t-test showed that the performance in the "early day" (318.4 ± 9.6) was significantly worse than that in the "late day" (314.2 ± 8.9) (t(7) = 2.815, p = 0.026).

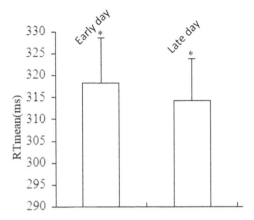

Figure 3. PVT performance in the "early day" (08:30–13:30, left) and the "late day" (15:30–20:30, right) during the sleep restriction condition.

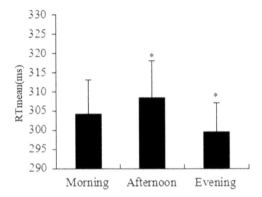

Figure 4. PVT performance (RT_{mean}) of the participants at the three different times of the day under the control condition.

3.3.2 Data under the control condition

The PVT performance (RT_{mean}) of the participants at the three different times of the day under the control condition is shown in Figure 4. Paired t-tests were performed to compare the PVT performance at the three different times of the day (the t-test were performed three times), which showed a significant difference only between the performance in the evening and that in the afternoon. The performance in the evening (299.5 ± 7.1) was significantly better than that in the afternoon (308.4 ± 9.1) ($t(7) = 2.716$, $p = 0.03$).

4 DISCUSSION

4.1 Sensitivity of PVT metrics

In this study, PVT metrics were selected as objective measures of human vigilance. Five metrics reported in previous research were calculated in our study, namely RT_{mean}, $RT_{10\ to\ 90}$, $RT_{10\ fast}$, $RT_{10\ slow}$ and Lapses. The data analysis showed that the metric "Lapses" was not as sensitive to fatigue induced by sleep restriction as the other four metrics, which is consistent with the results reported by Kline (Kline et al. 2010). However, in several other research (Van Dongen et al. 2003, Dorrian et al. 2005), "Lapses" was adopted as a typical PVT metric, indicating vigilance variation. The phenomenon that "Lapses" is sensitive in those studies may be caused by the more severe fatigue induced by experimental manipulation, such as total sleep

371

deprivation for 72 hours or sleep restriction up to 14 consecutive days. Meanwhile, the PVT test in our study lasted for 5 minutes. However, if the test time lasted longer (such as 10 minutes or longer), there may be more lapses and the metric "Lapses" may be more sensitive.

4.2 *Effect of sleep restriction on human vigilance*

PVT performance data of the participants in this study indicated that sleep loss (from 8 hours in the normal condition to 5 hours) can decrease vigilance. However, for the PVT test, which requires sustained attention and simple reaction, the mean reaction time generally increased less than 5 percent, indicating that the decrease in vigilance can be compensated by effort to some extent (Moller et al. 2006). Meanwhile, individual differences can be observed in the effect of sleep loss on vigilance (Van Dongen & Dinges 2005, Frey et al. 2004, Leproult et al. 2003).

4.3 *Effect of circadian rhythm on human vigilance*

In this study, the PVT performance of the participants was shown to be significantly better in the afternoon and in the evening than in the morning and at noon during the sleep restriction condition; during the control condition, the performance of the participants in the evening was significantly better than that in the afternoon. However, Wilhelm (Wilhelm et al. 2001) suggested that vigilance was better in the morning (9:00) and at midnight (23:00), and Kraemer (Kraemer et al. 2000) showed that vigilance peaked at noon. These inconsistent results may result from individual differences or differences among groups. In fact, besides vigilance, other parameters such as discrimination (Bodenhausen 1990) and working memory (West et al. 2002) are also individually dependent. Perhaps peaks and troughs in vigilance are partially contingent upon the individual's habit expressed through preferences in the timing of daily activities, which are dependent partly on their work types. So, some people are consistently at their best in the morning, whereas others are more alert and perform better in the evening (Schmidt et al. 2007). Besides, the circadian rhythm on human vigilance may also be influenced by age (Buysse et al. 2005) and many other factors, according to Roenneberg (Roenneberg et al. 2003), emphasized that sleep time is influenced by genetic disposition, sleep debt accumulated on workdays and light exposure.

5 CONCLUSIONS

The current study focuses on the effects of sleep restriction and circadian rhythm on human vigilance. Five PVT metrics (RT_{mean}, $RT_{10\ to\ 90}$, $RT_{10\ fast}$, $RT_{10\ slow}$ and Lapses) are selected as objective measures of human vigilance. All the metrics are quite sensitive to sleep restriction, except the metric "Lapses", which occurred very infrequently in this study. In the sleep restriction condition, the PVT metrics of the participants were significantly worse, indicating that 5 hours of sleep for three consecutive days will cause a decline in human vigilance. The time-of-day effect on human vigilance was also found in the days of sleep restriction in this study, as the PVT metrics of the participants were significantly better in the afternoon and in the evening ("late day") than in the morning and at noon ("early day").

ACKNOWLEDGMENTS

This study was supported by the Foundation of National Key Laboratory of Human Factors Engineering (No. HF2013-Z-B-02, No. SYFD140051802), the Foundation of Key Laboratory of Science and Technology for National Defense (No. 9140A26070213KG57387), the Feitian Foundation of China Astronaut Research and Training Center (No. FTKY201505), and the National Basic Research Program of China (973 Program, No. 2011CB711000).

REFERENCES

Bodenhausen, G.V. 1990. Stereotypes as judgmental heuristics: Evidence of circadian variations in discrimination. *Psychological Science* 1: 319–322.

Buysse, D.J., Monk, T.H., Carrier, J., & Begley, A. 2005. Circadian patterns of sleep, sleepiness, and performance in older and younger adults. *Sleep* 28: 1365–1376.

Dorrian, J., Rogers, N.L., & Dinges, D.F. 2005. Psychomotor vigilance performance: neurocognitive assay sensitive to sleep loss. In C. Kushida (ed.), *Sleep deprivation: Clinical issues, pharmacology and sleep loss effects*. New York: Marcel Dekker.

Frey, D.J., Badia, P., & Wright, K.P.J. 2004. Inter- and intra-individual variability in performance near the circadian nadir during sleep deprivation. *Journal of Sleep Research* 13: 305–315.

Goel, N., Basner, M., Rao, H., & Dinges, D.F. 2013. Circadian rhythms, sleep deprivation, and human performance. *Progress in molecular biology and translational science* 119: 155–178.

Kraemer, S., Danker-Hopfe, H., Dorn, H., Schmidt, A., Ehlert, I., & Herrmann, W.M. 2000. Time-of-day variations of indicators of attention: Performance, physiologic parameters, and self-assessment of sleepiness. *Biological Psychiatry* 48: 1069–1080.

Kline, C.E., Durstine, J.L., Davis, J.M., Moore, T.A., Devlin, T.M., & Youngstedt, S.D. 2010. Circadian rhythms of psychomotor vigilance, mood, and sleepiness in the ultra-short sleep/wake protocol. *Chronobiology international* 27(1): 161–180.

Liang X., Zhang L., Shen H., et al. 2014. Effects of a 45-day head-down bed rest on the diurnal rhythms of activity, sleep, and heart rate. *Biological Rhythm Research* 45(4): 591–601.

Leproult, R., Colecchia, E.F., Berardi, A.M., Stickgold, R., Kosslyn, S.M., & Van Cauter, E. 2003. Individual differences in subjective and objective alertness during sleep deprivation are stable and unrelated. *American Journal of Physiology: Regulatory, Integrative and Comparative Physiology* 284: R280–R290.

McPhee J.V. & Charles J.B. 2009. *Human health and performance risks of space exploration missions*. Houston: NASA Lyndon B, Johnson Space Center.

Moller, H.J., Devins, G.M., Shen, J., & Shapiro, C.M. 2006. Sleepiness is not the inverse of alertness: Evidence from four sleep disorder patient groups. *Experimental Brain Research* 173: 258–266.

Mollicone, D.J., Van Dongen, H.P., Rogers, N.L., & Dinges, D.F. 2008. Response surface mapping of neurobehavioral performance: Testing the feasibility of split sleep schedules for space operations. *Acta Astronautica* 63(7): 833–840.

Raymann, R.J.E.M., & Van Someren, E.J. 2007. Time-on-task impairment of psychomotor vigilance is affected by mild skin warming and changes with aging and insomnia. *Sleep–New York Then Westchester* 30(1): 96–103.

Roenneberg, T., Wirz-Justice, A., & Merrow, M. 2003. Life between clocks: Daily temporal patterns of human chronotypes. *Journal of Biological Rhythms* 18: 80–90.

Schmidt, C., Collette, F., Cajochen, C., & Peigneux, P. 2007. A time to think: circadian rhythms in human cognition. *Cognitive Neuropsychology* 24(7): 755–789.

Van Dongen, H.P., & Dinges, D.F. 2000. Circadian rhythms in fatigue, alertness, and performance. *Principles and practice of sleep medicine* 20: 391–399.

Van Dongen, H.P., Maislin, G., Mullington, J.M., & Dinges, D.F. 2003. The cumulative cost of additional wakefulness: dose-response effects on neurobehavioral functions and sleep physiology from chronic sleep restriction and total sleep deprivation. *Sleep—New York Then Westchester* 26(2): 117–129.

Van Dongen, H.P., & Dinges, D.F. 2005. Sleep, circadian rhythms, and psychomotor vigilance. *Clinics in Sports Medicine* 24: 237–249.

West, R., Murphy, K.J., Armilio, M.L., Craik, F.I., & Stuss, D.T. 2002. Effects of time of day on age differences in working memory. *Journals of Gerontology Series B: Psychological Sciences and Social Sciences* 57: 3–10.

Wilhelm, B., Giedke, H., Lüdtke, H., Bittner, E., Hofmann, A., & Wilhelm, H. 2001. Daytime variations in central nervous system activation measured by a pupillographic sleepiness test. *Journal of sleep research* 10(1): 1–7.

Bioinformatics and Biomedical Engineering – Chou & Zhou (Eds)
© 2016 Taylor & Francis Group, London, ISBN 978-1-138-02784-8

The impact of environmental factors on the survival status of *Rana*

M.H. Duan, X.H. Li, A.J. Jiang, Y.Y. Pei & D.D. Guan
Changchun University of Chinese Medicine, Jilin, China

Y.X. Sun, H. Qiang & Y. Yue
Jilin University, Jilin, China

Y.P. Li
Changchun University of Chinese Medicine, Jilin, China

ABSTRACT: Chinese *Rana* (QIU Bao-hong et al. 2010) frogs have an important medicinal and economic value. Therefore, the survival and growth of *Rana* is important in China. In this review, we examine and discuss the impact of environmental factors on the survival status of *Rana*, with the aim of promoting environmental protection of *Rana* species.

1   INTRODUCTION

Rana (Amphibia: Anura: Ranidae), a unique species of frog (Yu Dan et al. 1992), is a secondary protected animal in China. The species of *Rana* is *temporaria*. Excessive capture, human intervention and environmental pollution, among other factors, have led to the classification of *Rana* as a vulnerable (V) species due to dwindling resources.

   In this review, we aimed to promote the protection of *Rana* as a species of medicinal value by highlighting the changes in wildlife resources and the impact of environmental disruption on the survival status of *Rana*. We further hope to raise the awareness of the vulnerable status of *Rana* to improve contributions toward the protection of this endangered species.

2   THE INFLUENCE OF NATURAL ENVIRONMENT ON *RANA*

2.1   *Effects of forest types on the survival of Rana*

Rana mainly lives in secondary deciduous broadleaf weed vegetation in low mountain regions (Wang Chuang et al. 2012) and the Qiuling zone. The varying environment of broadleaf forests, containing trees, shrubs and grasses, provides an optimal habitat for *Rana*. Studies on *Rana* have shown that broadleaf trees; Hu, lespedeza, and Rose Hill hazel shrubs (Chen Bing et al. 2011); ferns; and *Artemisia* were also prevalent in optimal habitats for *Rana*.

2.2   *Effects of different types of forests on frog populations*

In Canada's boreal mixedwood forests (Paul D. Klawinski et al. 2014), different surrounding mountain landscapes were used to compare vegetation structures, plants and animals at different distances from a small lake in the forest. The most important feature of this habitat is the large lakeside forest coverage, with many tall, large aspens. This differs from upland forests. However, the absolute differences may be quite small, and no studies have shown whether there is greater structural diversity in this habitat. Despite this, the richness and diversity of herbs are lower than those in lakeside forests and lowlands. Two Anura (wood frog [*R. sylvatica*] and northern toad) were more abundant in the lake and the surrounding forest (100 m) than in higher grounds at 400–1200 m above sea level. However, in non-forested riparian zones (W. Chris Funk et al. 2003), their numbers were not very different. For amphibians, this was also a common habitat (especially for juvenile *Rana*), particularly in late winter, probably because the upland areas provide useful locations for hibernation and breeding. Small mammals (e.g. redback vole [brown gapperi] and deer mice [*Peromyscus maniculatus*]) were also common in this area, as well as meadow voles (voles), meadow jumping mouse (Americas Lin jerboa, genus *Hudsonicus*) and shrews. Species were also abundant in areas with 50 m of forest to more than 600 m of open water. In the case of insectivorous birds, the abundance and richness of songbirds in these regions are high (Linda Dupuis et al. 1997) because of the abundance of prey near lakes. Overall, their results do not strongly support our hypothesis on the ecology of the riparian forest lakefront property. We encourage forest management agencies to rethink and replan an appropriate geographic scope of protection to suit a wide range of conservation objectives.

2.3 Impact of pollution from the Fukushima nuclear power plant leak on Rana populations

Amphibians are the key to the food chain (Teruhiko et al Takahara. 2015) in the forest. When examining the radioactive contamination of tailless amphibians, understanding the transfer of radioactive cesium and understanding the forest ecosystem from low to high nutrient levels is important. Takahara investigated the levels of radioactive cesium (134Cs and 137Cs) in the forest floor at approximately 2.5 years after the Fukushima Nuclear Power Plant (FNPP) accident (Qi-Ya Zhang et al. 1999), and determined the effects of radioactivity in captured *R. tagoi tagoi*. Radioactive cesium accumulation was observed in air and garbage, but did not vary in frogs based on the distance from the FNPP. The body weights and lengths of frog did not differ according to the levels of radioactivity. Their findings suggested that existing food items (Marina Paolucci et al. 2003) may lead to individually significant differences in pollution. Therefore, it is necessary to continue the monitoring of terrestrial and aquatic amphibians by analyzing the degree of contamination and radioactive cesium, as well as the forest food web transfer mechanism near the FNPP.

3 THE INFLUENCE OF HUMAN ENVIRONMENT ON *RANA*

3.1 Population analysis

Population decline in California red-legged frog populations (Jonathan Q. Richmond et al. 2014) has affected all species within the distribution areas. However, the results obtained for species in northern and southern regions differed, confusing many geneticists. In a laboratory study, Richmond used microsatellite and mitochondrial DNA data to compare species from the northern region (i.e. Sierra Nevada, NV), such as *R. draytonii* (harp frog), with those of coastal areas (i.e. San Francisco, with an abundant, long history of inhabitation by frogs), and performed analyses (Elly S.W. Ngan et al. 1999) of the population genetic structure and diversity. It was shown that frogs within the Sierra Nevada exhibited lower genetic diversity and increased frequency of species differences than those within the Gulf region. This trend appears to be true for all of California. Indeed, lower allelic variation was observed in frogs from the Gulf region compared with those from the Sierra Nevada mountains. Additional analyses of geographical differences showed that there is great similarity in haplotype mtDNAs in frogs from the northern region of Nevada and from the southern coastal region. These data have made it difficult to determine which populations face the most urgent threat. Therefore, analysis of the genetic reasons for this population decline (Tibor Hartel et al. 2009) is necessary. No evidence has supported contemporary Nevada gene transfer (*R. draytonii*) between groups of organisms. Thus, we propose that appropriate species management policies should aim to support the maintenance and creation of typical ponds in local habitats to facilitate the reproduction of the species.

3.2 Effects of road traffic on frog populations

Road traffic and loss of forest roads (Felix Eigenbrod et al. 2008) have serious negative effects on tailless amphibians. However, the relative importance of road traffic itself on frog populations is poorly understood because the landscape of forest cover is usually associated with road and traffic density. In order to assess the impact of traffic and major forest cover, Eigenbrod selected 36 ponds near Ottawa, Canada in four locations: low forest/low traffic; low forest/high traffic (Miguel Yanes et al. 1995); high forest/low traffic; and high forest/high traffic. All ponds in our survey were first examined in 2005 and resurveyed in 2006. For 23 frog species, a strong negative correlation between species richness and traffic density was observed (partial R2 = 0.34, P < 0.001), while a positive correlation was observed between species richness and forest cover (part R2 = 0.10, P > 0.05). Moreover, variations in the effects of forest cover and traffic density were observed for different species (Martin A. Schoaepfer et al. 2002). These results indicated that the negative effects of traffic on toad populations in northeastern North America were similar to the negative impacts of deforestation.

4 THE INFLUENCE OF ECOLOGICAL ENVIRONMENT ON *RANA*

4.1 Impact of the fungal pathogen Batrachochytrium dendrobatidis (bd) on Rana

Infection with BD can lead to the fungal disease chytridiomycosis (Jonah Piovia-Scott et al. 2011), which is associated with the decline and extinction of many mountain amphibians. To better understand the distribution and prevalence of BD in *Rana* frogs, Scott focused on frogs within the Klamath Mountains in northwestern California. The waterfall frog (*R. cascadae*) is one of the most common amphibians in these mountains, and was used for the BD experiment, in which it exhibited high mortality. Other parts of California have recently seen a dramatic decline in the number of *R. cascadae* because of BD infection. Between 1999 and 2002, Scott surveyed 112 sites in Klamath Mountains and described the distribution of BD. They also assessed changes in the *R. cascadae* distribution and the drivers for BD infection. BD was widely distributed in Klamath Mountains; in some locations, up to 64% of frogs were infected. Moreover, 79% of *R. cascadae* were infected with BD in some locations. These results indicated that BD infection did not cause a sharp decline in *R. cascadae* populations within Klamath Mountains. Interestingly, subadult *R. cascadae* had a higher prevalence of BD than other age groups (M. Nasser et al. 2011) (subadult: 36%, adults: 25%, metamorphs: 4% and larvae: 1%). The highest BD infection rates occurred in the areas of high altitude, which may indicate that living in high-altitude areas may be associated with a higher risk of chytridiomycosis. In addition, three other common amphibian species have also been shown to be positive for BD infection (M. Nasser et al. 1997): the Pacific chorus frog (*Pseudacris regilla*); the western toad (*Anaxyrus boris*); and rough-skin newt (*Taricha particles*).

4.2 Effects of frog density on growth

In the complex life cycle of the species, density (Jon Loman et al. 2009) adjustment can be applied at any growth stage. However, there are few studies examining adult densities in terrestrial habitats. Therefore, Loman studied the effects of four densities of adult frogs on growth. Four 30 × 30 m experimental habitats of wet grass were used. During early summer, *R. arvalis* and *Rana* frogs, which exhibit similar ecological competition and breeding migration, were organized into two high-density plots and two low-density plots. In early autumn (Jon Loman et al. 2002), the frogs were caught, and summer growth data were recorded. The growth of *R. arvalis* was increased in low-density plots compared with that in higher-density plots, while no changes were observed for *R. temporaria*. Lincoln believed that summer density would not affect the growth of any species; however, the results of this analysis showed that the growth of *R. arvalis* could be affected by a suitable growing environment and density regulation. Moreover, the significant differences in *R. arvalis* growth, without changes in *R. temporaria* response to changes in density, suggest that *R. arvalis* dominates over

R. temporaria in habitats with strong competitors (Domenico Sanzo et al. 2006). Moreover, as the density of frogs did not affect other status indicators, changes in the growth rate may actually be an adaptive trait in *R. arvalis*. This study suggested that regulation of density may be associated with different summer habitats of frogs.

4.3 *Effects of environmental pollution on the endocrine systems of Xenopus laevis and Rana*

The effects of sewage on endocrine system function (C, Bogi, J. et al. 2003) in frogs were examined in *Xenopus laevis* and *Rana* frogs. Larvae developed to complete the process of metamorphosis in the southern Bavaria region, which was exposed to sewage treatment plant effluent. The river (Würm) sample stored as reference was found to contain a mixture of sewage and other pollutants at varying concentrations. Industrial wastewater unexpectedly affected (Werner Kloas et al. 2002) benchmarks in development, including sex differentiation, as shown by histological analysis in all treatment groups; there was a particularly high probability of disruption in female animals. For example, the sex ratio between species (in the semi-field study) showed reduced proportions of female frogs and a relatively strong correlation between sewage and disruption of the sex ratio (from 1:12 to 1:2). *Xenopus* liver biomarkers were used to measure the changes in gene expression by the semi-quantitative reverse transcription polymerase chain reaction. A significant increase in vitellogenin mRNA levels was observed in female juveniles when estrogen was accumulated in the effluent; however, no differences in the expression levels of androgen and estrogen receptor genes were observed. Moreover, the results of estrogen-induced (Robin P. Ertl et al. 1998) biomarkers demonstrated that water contained higher levels of estrogen receptor agonists at the end of the experiment, with the ability to displace [3H]estradiol from estrogen receptors.

5 CONCLUSION

As China's rare medicinal materials, especially of Changbai mountains, *Rana* has a high medicinal value. The improvement of the natural environment and the human environment could provide a better environment for the survival and breeding of *Rana*, improve rare medicinal material collection and enhance the medicinal value of *Rana*, which will benefit the people.

REFERENCES

Chen Bing. & Lei Zhen-huan. & Yuan Tao. 2011. Analysis of forest frog breeding practice of wetland ecological restoration. Science and Technology Innovation Herald 29: 137–138.
C, Bogi, J. & Schwaiger. & H. Ferling. & U. Mallow. & C. Steineck. 2003. Endocrine effects of environmental pollution on Xenopus laevis and *Rana* temporaria. Environmental Research 93(2): 195–201.
Chris Funk. W. & L. Scott Mills. 2003. Potential causes of population declines in forest fragments in an Amazonian frog. Biological Conservation 111(2): 205–214.
Domenico Sanzo. & StephenJ. Hecnar. 2006. Effects of road de-icing salt (NaCl) on larval wood frogs (*Rana* sylvatica). Environmental Pollution 140(2): 247–256.
Elly S.W. Ngan. & Lillian S.N. Chow. & Dicky L.Y. Tse. 1999. Functional studies of a glucagon receptor isolated from frog *Rana* tigrina rugulosa: implications on the molecular evolution of glucagon receptors in vertebrates. FEEBS Letters 457(3): 499–504.
Felix Eigenbrod. & Stephen J. Jecnar. & Lenore Fahrig. 2008. The relative effects of road traffic and forest cover on anuran populations. Biological Conservation 141(1): 35–46.
Jonah Piovia-Scott. & Karen L. Pope. & Sharon P. 2011. Lawler. Factors related to the distribution and prevalence of the fungal pathogen Batrachochytrium dendrobatidis in *Rana* cascadae and other amphibians in the Klamath Mountains. Biological Conservation 144(12): 2913–2921.
Jon Loman. & Bjorn Lardner. 2009. Density dependent growth in adult brown frogs *Rana* arvalis and *Rana* temporaria-A field experiment. Acta Oecologica 35(6): 824–830.

Jon Loman. 2002. *Rana* temporaria metamorph production and population dynamics in the field-Effects of tadpole density, predation and pond drying. Journal for Nature 10(2): 95–107.

Jonathan Q. Richmond. & Adam R. Backlin. & Patricia J. 2014. Tatarian. Population declines lead to replicate patterns of internal range structure at the tips of the distribution of the California red-legged frog (*Rana* draytonii). Biological Conservation 172: 128–137.

Linda Dupuis. & Doug Steventon. 1999. Riparian management and the tailed frog in northern coastal forests. Forest Ecology and Management 124(1): 35–43.

Lowcock. L.A. & T.F. Sharbel, J. Bonin. & M. Ouellet. 1997. Flow cytometric assay for in vivo genotoxic effects of pesticides in Green frogs (*Rana* clamitans). Aquatic Toxicology 38(4): 241–255.

Marina Paolucci. 2003. An androgen receptor in the brain of the green frog *Rana* esculenta. Life Sciences 73(3): 265–274.

Miguel Yanes. & Jose M. Velasco. & Francisco Suarez. 1995. Permeability of roads and railways to vertebrates: The importance of culverts. Biological Conservation 71(3): 217–222.

Martin A. Schoaepfer. & Michael C. Runge. & Paul W. 2002. Sherman. Ecological and evolutionary traps. Trends in Ecology & Evolution 17(10): 474–480.

Nasser. M. & Gerardo R. Melgar. & David L. Longworth. 1997. Incidence and Risk of Developing Fungal Prosthetic Valve Endocarditis after Nosocomial Candidemia. The American Journal of Medicine 103(1): 25–32.

Paul D. Klawinski. & Ben Dalton. & Aaron B. Shiels. 2014. Coqui frog populations are negatively affected by canopy opening but not detritus deposition following an experimental hurricane in a tropical rainforest. Forest Ecology and Management 332(15): 118–123.

Qi-Ya Zhang. & Zhen-Qiu Li. & Jian-Fang Gui. 1999. Studies on morphogenesis and cellular interactions of *Rana* grylio virus in an infected fish cell line. Aquaculture 175(3–4): 185–197.

Qiu Bao-hong. & Piao Long-guo. & Zheng Ren-jiu. 2010. Status of population of *rana* chensinensis in Changbai Mountain Natural Reserve and Measures for population propagation. Journal of BEIHU Unibersity 11(3): 259–262.

Robin P. Ertl. & Gray W. Winston. 1998. The microsomal mixed function oxidase system of amphibians and reptiles: components, activities and induction. Comparative Biochemistry and Physiology Part C: Pharmacology, Toxicology and Endocrinology 121(1–3): 85–105.

Teruhiko Takahara. & Satoru Endo. & Momo Takada. 2015. Radiocesium accumulation in the anuran frog, *Rana* tagoi tagoi, in forest ecosystems after the Fukushima Nuclear Power Plant accident. Environmental Pollution 199: 89–94.

Tibor Hartel. & Szilard Nemes. & Dan Cogalniceanu. 2009. Pond and landscape determinants of *Rana* dalmatina population sizes in a Romanian rural langscaape. Acta Oecologica 35(1): 53–59.

Werner Kloas. 2002. Amphibians as a model for the study of endocrine disruptors. International Review of Cytology 216: 1–57.

Wang Chuang. & Tong Qing. & Cui Li-yong. 2012. Research progress of the effects of environmental factors on the Northeast Forest Frog. Jiangsu Agricultural Sciences 40(2): 190–192.

Yu Dan. & Zhao Ke-zun. & Li Jun-tao. 1992. The breeding ecology and resource recovery of Chinese forest frog. Journal of Northeast Forestry University 20(5): 89–93.

Bioinformatics and Biomedical Engineering – Chou & Zhou (Eds)
© 2016 Taylor & Francis Group, London, ISBN 978-1-138-02784-8

Balance role of dopamine system on exciting and inhibitory involved in motor function

J.L. Cheng, X.W. Wang & L.J. Hou
Exercise Physiology Laboratory, PE and Sports Science Department, Beijing Normal University, Beijing, China

ABSTRACT: Dopamine is a neurotransmitter of the catecholamine and phenethylamine families, and it plays important roles in the human brain and body. Major dopamine pathways are parts of the reward pathway which is also involved in motor functions regulation. Several important diseases of the nervous system are associated with dysfunctions of the dopamine system. This paper will discuss the balance role of dopamine system on exciting and inhibitory involved in motor function.

Keywords: dopamine system; motor function; dopamine cell activity; dyskinesia

1 INTRODUCTION

In the brain, dopamine functions as a neurotransmitter—a chemical released by nerve cells to send signals to other nerve cells. Dopamine is manufactured in nerve cell bodies located within the Ventral Tegmental Area (VTA) and is released in the nucleus accumbens and the prefrontal cortex (Keeler 2014). Dopamine system mainly includes dopaminergic project, and they are neural pathways in the brain that transmit dopamine neurotransmitter from one region of the brain to the other (Fig. 1) (Usher 2002). Dopaminergic nerve cell bodies in such areas as the substantia nigra tend to be pigmented due to the presence of the black pigment melanin. There are eight dopaminergic pathways.

Dopamine releasing midbrain neurons are crucial for fundamental and complex brain functions, such as voluntary movement and goal-directed behavior, as well as cognition, emotion, reward, motivation, working memory, associative learning and decision making (Dragicevic 2015).

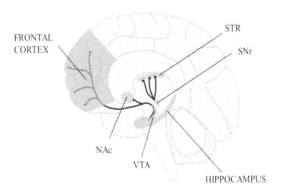

Figure 1. Dopamine System in Basal Ganglion (STR: striatum, SNr: Substantia Nigra pars reticulata, NAc: Nucleus Accumbens, VTA: Ventral Tegmental Area).

2   DOPAMINECIRCUIT AND REGULATION ROLE IN MOVEMENT

D1 and D2 pathway are two described pathways for transmission of signals through the dopamine pathway. The basic loop circuit includes an excitatory Glu projection from the cortex to the striatum and then inhibitory GABA striatal projection to the internal Globus Pallidus (GPi). GABA neurons in GPi project to targets in the thalamus and brainstem.

2.1   *D1 pathway and D2 pathway of dopamine system*

There are two described pathways for transmission of signals through the dopamine system, D1 pathway and D2 pathway. The D1 pathway operates relatively independent of the other. Binding of dopamine to the D1DR can be considered such that dopamine can act to stimulate actions through the D1 pathway. However, this is only effective in the presence of appropriate cortical excitation. In contrast, the D2 pathway operates as a feed forward and feedback hub, allowing activation to spread laterally through the basal ganglia. In the both pathways, striatum is a highly complex structure coordinating motor and cognitive functions. Striatum neurons can be divided into two populations. The first population is projects directly from the striatum to the substantial nigra pars reticulate (SNr) and the internal segment of the globus pallidus, which is referred as the D1 pathway, since it directly projects to the basal ganglia output nuclei. The second population also projects to the SNr and GPi, but indirectly via several intermediate relays in the external part of the Globus Pallidus (GPe) and the Sub-Thalamic Nucleus (STN) and is referred as the D2 pathway. In general, D1 pathway increases the activation level (excitatory) via binding of dopamine to the D1DR, but exerts inhibitory influences via the D2 pathway which is of the form capable of mediating such competition processes (Fig. 2). On the contrary, the new 'Prepare And Select' (PAS) model suggests that both the D1 and D2 pathways carry excitatory signals. During action selection, the D1 pathway prepares a set of possible responses, whereas the D2 pathway provides a more refined selection (Surmeier 2007, Barros 2004, Howe 2013, Foley 2008).

2.2   *Characteristics of dopamine cell activity*

Dopamine release is typically divided into two modes: tonic and phasic (Fig. 3). Tonic activity refers to the continuous output of the system. In contrast, phasic activity refers to the shorter term burst activity of the dopaminergic cells (Calabresi 2007, Dreyer 2010, Overton 2014, Fiorillo 2013, Bromberg-Martin 2010a, Frank 2004).

Characteristics of dopamine cell activity is very different from other non-dopamine neurons in the brain, and it has two basic types of activity pattern: single spike firing and burst firing.

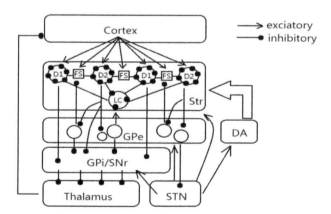

Figure 2.   D1and D2 pathways in dopamine system (D1: D1DR-expressing MSN, D2: D2DR-expressing MSN, FS: Fast-Spiking interneuron, LC: Large Cholinergic interneuron).

Figure 3. Characteristics of dopamine cell activity (tonic irregular and phasic burst activity).

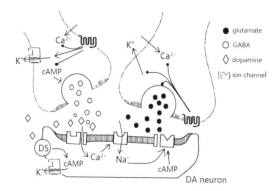

Figure 4. Relationship among the glutamatergic neurons, GABAergic neurons and dopamine neurons. (GABA neurons are inhibited via G protein-mediated inhibition of Ca2+ and activation of K+ conductances, and GABA release is decreased via inhibition of cAMP-dependent facilitation of transmitter release. This leads to disinhibition of dopamine neurons).

In vivo, electrophysiological recording suggested that spontaneous activities of dopamine neurons containing those two basic types. However, the proportions of two basic types are very different in different dopamine neurons, when given sensory stimuli or drug; the firing pattern will change even in the same dopamine neurons (Fig. 4). These phenomena indicate that inter-transform of the single spike firing and burst firing may mean they changed output of the postsynaptic information. There have been numerous demonstrations that dopamine neurons in a range of species are responsive to sensory stimuli. Under anesthesia, responses to non-noxious and noxious sensory stimuli are usually tonic in nature, although long duration changes in activity have been reported in the awake preparation as well. However, in the awake preparation, short-latency, phasic changes in activity are most common. In general, dopamine cell activity frequency less than 10 Hz. Depending on intracellular recording, action potentials of dopaminergic neurons contain four parts: slow depolarization potential, initial segmental potential, somato-dendritic potential and after-hypolarization (Gomes 2002).

3   REGULATION ROLE OF DOPAMINE IN RELATIVITY DISEASES

Accordingly, dysfunctions of the dopamine midbrain system are associated with neurological and psychiatric diseases, such as schizophrenia, addiction, Attention Deficit Hyperactivity Disorder (ADHD) and Parkinson's Disease (PD), the second most common neurodegenerative disease (Frank 2004). The majority of dopamine midbrain neurons are located in two overlapping nuclei, the Substantia Nigra (SN, A9) and the Ventral Tegmental Area (VTA, A10). So-called mesostriatal SN dopamine neurons, projecting to the dorsal striatum, are particularly important for voluntary movement control as well as for motor learning, habit formation, goal-directed actions election and exploration (Strausfeld 2013, Connolly 2014, Bromberg-Martin 2010b).

3.1   Dopamine and parkinson diseases

In human patients, the typical motor symptoms of PD (rigor, tremor, bradykinesia/akinesia) occur late with in the neurodegenerative process, when a significant number of nigrostriatal axons (an early feature in PD) as well as SN dopamine neuron somata are already lost. Early in the course of the disease, the most obvious symptoms are movement-related; these include trembling of limbs, stiffness of movement and difficulty with walking and gait. In advanced stages, it progresses to dementia and depression (Bromberg-Martin 2010b). Other symptoms include sensory, sleep and emotional problems and eventually death. Consequently, there is still no cure or causal neuroprotective therapy available, but currently established PD therapies attempt to compensate the progressive loss of striatal dopamine by administering its precursor L-DOPA, and/or dopamine D2-receptor agonists. L-DOPA is converted into dopamine in the dopaminergic neurons by dopamine decarboxylase unfortunately, however, involuntary movements, or 'dyskinesias', represent a dramatic complication of long-term treatment with this pharmacological compound in the vast majority of patients with PD. In the 6-OHDA model of PD, LTP is lost. Nevertheless, in animals sustaining unilateral 6-OHDA lesions of the substantia nigra, corticostriatal LTP can be restored by chronic treatment with the dopamine precursor l-DOPA (Picconi 2003, Dallérac 2015).

3.2 Dopamine and Huntington Diseases

Huntington's Disease (HD) is a late-onset fatal neurodegenerative disorder caused by a CAG trinucleotide repeat expansion in the gene coding for the protein Huntington and is characterised by progressive motor, psychiatric and cognitive decline. Dopaminergic neuronal dysfunction is a key early event in HD disease progression. The initial increase in dopamine release appears to be related to a loss of SK3 channel function, a protein containing a poly glutamine tract. Research shows HD due to the projection neurons of the striatum is damaged, especially the indirect pathway of striatum-pallidum GABAergic neurons is dying. Studies demonstrated that normal synaptic function in HD could be restored by application of dopamine receptor agonists, suggesting that changes in the release or bioavailability of dopamine may be a contributing factor to the disease process (Adopaminem 2004).

The dicarbanion form of 3-NP, which is a suicide inhibitor of succinate dehydrogenase (respiratory complex II), a functional member of both the Krebs cycle and the aerobic respiratory chain, can irreversibly inhibit the active site of the enzyme, leading to a depletion of intracellular ATP, membrane depolarization and NMDA toxicity through the relief of a voltage-dependent $Mg2+$ block. At corticostriatal synapses, 3-NP can induce a metabotropic glutamate type 1 receptor–Protein Kinase C (PKC) pathway-dependent LTP of NMDA-mediated corticostriatal transmission. All these data suggest that the involvement of the D2 receptor–PKA pathway is a crucial requirement for the generation of this form of pathological synaptic plasticity (Howe 2013).

4 CONCLUSIONS

Dopamine releasing midbrain neurons are crucial for fundamental and complex brain functions, such as voluntary movement and goal-directed behavior, as well as cognition, emotion, reward, motivation, working memory, associative learning and decision making. D1 and D2 pathways are two described pathways for transmission of signals through the dopamine pathway. The basic loop circuit includes an excitatory Glu projection from the cortex to the striatum and then inhibitory GABA striatal projection to the internal Globus Pallidum (GPi). Due to the unbalance between the two pathways of disease are associated with movement disorders, such as Parkinson Diseases and Huntington Diseases.

ACKNOWLEDGMENT

This research was supported by National Natural Science Foundation of China (31171138/31401018).

REFERENCES

Adopaminem, G.C. & Bernardo, L.S. 2004. State-dependent calcium signaling in dendritic spines of striatal medium spiny neurons. Neuron 44(3):483–493.

Barros, R.C. & Branco, L.G. & Carnio, E.C. 2004. Evidence for thermoregulation by dopamine D1 and D2 receptors in the anteroventral preoptic region during normoxia and hypoxia. Brain Res 1030(2):165–171.

Bromberg-Martin, E.S. & Matsumoto, M. & Hikosaka, O. 2010a. Distinct tonic and phasic anticipatory activity in lateral habenula and dopamine neurons. Neuron 67: 144–155.

Bromberg-Martin, E.S. & Matsumoto, M. & Hikosaka, O. 2010b. Dopamine in motivational control: rewarding, aversive, and alerting. Neuron 68:815–834.

Calabresi, P. & Picconi, B. & Tozzi, A. et al. 2007. Dopamine-mediated regulation of corticostriatal synaptic plasticity. Neuroscience 30(5):211–9.

Connolly, B.S. & Lang, A.E. 2014. Pharmacological treatment of Parkinson disease: a review. American Medical Association 311(16):1670–1683.

Dallérac, G.M. & Levasseur, G. & Vatsavayai, S.C. 2015. Dysfunctional dopaminergic neuron in mouse models of Huntington's disease: A Role for SK3 Channels. Neurodegenerative Dis15 (2).

Dragicevic, E. & Schiemann, J. & Liss, B. 2015. Dopamine midbrain neurons in health and Parkinson's disease: emerging roles of voltage-gated calcium channels and ATP-sensitive potassium channels. Neuroscience 284(22):798–814.

Dreyer, J.K. & Herrik, K.F. & Berg, R.W. et al. 2010. Influence of phasic and tonic dopamine release on receptor activation. Neuroscience 30(42):14273–14283.

Fiorillo, C.D. & Yun, S.R. & Song, M.R. 2013. Diversity and homogeneity in responses of midbrain dopamine neurons. Neuroscience 33:4693–4709.

Foley, T.E. & Fleshner, M. 2008. Neuroplasticity of dopamine circuits after exercise: implications for central fatigue. Neuromolecular Med 10(2):67–80.

Frank, M.J. & Seeberger, L.C. & O'Reilly, R.C. 2004. By carrot or by stick: cognitive reinforcement learning in Parkinsonism. Science 306(5703):1940–1943.

Gomes, P. & Soares-Dopamine-Silva, P. 2002. D2-like receptor-mediated inhibition of Na+2-K+2-ATPase activity is dependent on the opening of K+ channels. Am J Physiological Renal Physiological 283(1):114–123.

Howe, M.W. & Tierney, P.L. & Sandberg, S.G. et al. 2013. Prolonged dopamine signals in striatum signals proximity and value of distant rewards. Nature 500:575–579.

Keeler, J.F. & Pretsell, D.O. & Robbins, T.W. 2014. Functional implications of dopamine D1 vs. D2 receptors: A 'prepare and select' model of the striatal direct vs. indirect pathways. Neuroscience 282:156–175.

Overton, P.G. & Vautrelle, N. & Redgrave, P. 2014. Sensory regulation of dopaminergic cell activity: Phenomenology, circuitry and function. Neuroscience 282:1–12.

Picconi, B. & Centonze, D. & Håkansson, K. et al. 2003. Loss of bidirectional striatal synaptic plasticity in L-DOPA-induced dyskinesia. Nat Neuroscience 6(5):501–6.

Surmeier, D.J. & Dopaminey, M. & Jun, D. et al. 2007. D1 and D2 dopamine-receptor modulation of striatal glutamatergic signaling in striatal medium spiny neuron. Neuroscience 30(5):228–235.

Strausfeld, N.J. & Hirth, F. 2013. Deep homology of arthropod central complex and vertebrate basal ganglia. Science 340(6129):157–161.

Usher, Z. 2002. Hick's law in a stochastic race model with speed-accuracy tradeoff. Math Psychology 46(6):704–715.

Bioinformatics and Biomedical Engineering – Chou & Zhou (Eds)
© *2016 Taylor & Francis Group, London, ISBN 978-1-138-02784-8*

Pemetrexed and carboplatin chemotherapy combined with whole brain radiotherapy for non-small cell lung cancer with brain metastasis

H.B. Li[†], Y.F. Yun, X.F. Zhou & J.D. Luo
Department of Radiotherapy, Changzhou Tumor Hospital, Soochow University, Changzhou, China

H. Zhu[†]
Department of Radiation Oncology, Minhang Branch of Cancer Hospital of Fudan University, Shanghai, China

L.Y. Zhou[†]
Department of Oncology, Huaian No. 2 Hospital, Huaian, China

Q. An
Department of Radiation Oncology, Cancer Hospital of Jiangsu Province, China

ABSTRACT: *Objective:* To evaluate the safety and efficacy of pemetrexed and carboplatin chemotherapy combined with whole brain radiation therapy for non-small cell lung cancer with brain metastasis.

Methods: A retrospective analysis of 45 cases of non-small cell lung cancer with brain metastasis treated with combined pemetrexed and carboplatin chemotherapy plus whole brain radiotherapy. The primary endpoints were disease progression and the completion of chemotherapy given after whole brain radiotherapy.

Results: All patients with intracranial metastases had a Disease Control Rate (DCR) of 82.2%, with the rate of those with systemic lesions being 46.6%, which was significantly different from that of the intracranial metastases (P = 0.006). The median OS in the whole group of patients was 12.2 months (95% CI, 10.5–13.8 months), while the median PFS was 5.3 months (95% CI, 4.2–6.1 months). No serious adverse reactions were observed.

Conclusion: Pemetrexed plus carboplatin chemotherapy combined with whole brain radiotherapy for non-small cell lung cancer brain metastasis is effective and associated with few adverse reactions.

Keywords: pemetrexed; non-small cell lung cancer; brain metastases; chemotherapy; whole brain radiotherapy

1   INTRODUCTION

We herein examined the impact of pemetrexed and carboplatin chemotherapy combined with whole brain radiotherapy for the treatment of non-small cell lung cancer with brain metastasis. The incidence of Brain Metastases (BM) in patients with Non-Small Cell Lung Cancer (NSCLC) is 10% at the time of diagnosis and 40% during the course of the disease. These BM are generally unresectable and directly cause mortality up to 30%–50% of patients (Barlesi F, et al., 2013, Meng FL, et al., 2013). Whole-Brain Radiation Therapy (WBRT) is currently the standard treatment for non-small cell lung cancer patients with

---

[†]These three authors contributed equally to this work.

brain metastases (Lind JS, et al., 2011) but is contraindicated for patients who undergo surgery for single or multiple lesions, or who have active systemic disease or are in poor general condition. Whole-brain radiotherapy can improves local tumor control, reduces the brain metastases, and relieves symptoms associated with the central nervous system. Previous studies have shown that the survival of patients after radiotherapy or surgery may be extended to three to five months (JS, et al., 2011, De Ruysscher D, et al., 2012), but the prognosis of patients following whole brain radiotherapy is affected by many factors, and whole brain radiotherapy has only been shown to increase the survival by one month (Hodgson DC, et al., 2013).

Recent studies have shown that combination chemotherapy plays a role in the treatment of non-small cell lung cancer with brain metastasis treatment. Pemetrexed is a novel multi-targeted anti-folate cytotoxic drug with high efficiency and low toxicity. It is being applied more frequently for clinical applications, and as a fourth-generation chemotherapy drug, has been approved for the treatment of non-squamous non-small cell lung cancer in the United States. Several studies have shown that pemetrexed-based protocols are useful for brain metastases (Scagliotti GV, et al., 2008, Shepherd FA, et al., 2001). The present study retrospectively analyzed the effects of pemetrexed and carboplatin chemotherapy combined with whole brain radiotherapy for patients with non-small cell lung cancer with brain metastasis to determine whether it is suitable for clinical treatment.

2   MATERIALS AND METHODS

2.1   *General information*

A total of 45 patients with cytologically- or histologically-confirmed non-small cell lung cancer treated from March 2011 to March 2013 at our hospital were enrolled in the study. All of the patients had a physical status (performance status, PS) score of 0 to 2, underwent MRI or CT scans that showed measurable lesions that could be assessed for short-term efficacy, were not suitable for surgical treatment, normal biochemical, blood and heart function tests and had not received whole brain radiotherapy before. All patients provided informed consent.

There were 25 males (55.6%) and 20 females (44.4%) aged 35 to 76 years old, with a median age of 57 years old. Forty-two patients had adenocarcinoma (93.3%), two cases did not show a specific type (4.4%), and one case had mixed glandular scales cancer (2.2%). Nineteen patients were smokers (42.2%), while 26 were non-smokers (57.8%). Newly diagnosed brain metastases were found in 35 cases (77.8%), 10 of whom received treatment for the brain metastases (22.2%). Twelve patients had single brain metastasis (26.7%) and 33 had multiple brain metastases (73.3%); 32 of the former received pemetrexed chemotherapy. Thirteen patients received pemetrexed after whole brain radiotherapy (71.1%), and 13 received pemetrexed without whole brain radiotherapy (28.9%). Pemetrexed was given as the first-line treatment in 19 cases (42.2%), and as the second-line or later treatment in 26 patients (57.8%). The clinical and pathological data of the 45 patients are shown in Table 1.

2.2   *Methods*

2.2.1   *Method of administration*

Eligible patients undergoing their first treatment with pemetrexed receive intramuscular injections of vitamin B12 at 1000 µg/second, repeated every nine weeks; followed by oral folic acid at 400 µg/day during treatment and continuing for 21 days after the last administration, and also receive oral dexamethasone at 8 mg/day, divided into two doses, on the day of and the day after pemetrexed treatment. Pemetrexed is given at 500 mg/m^2 intravenously over a period of 10 min; the dose of carboplatin is AUC = 5, and it was infused intravenously on day 1, and the treatment was repeated every three weeks. The efficacy was assessed after every two cycles, and if effective, we continued to use the original program for another two to four cycles. When routinely given before chemotherapy, serotonin receptor

Table 1. The clinicopathological characteristics of 45 non-small cell lung cancer patients with brain metastasis.

Variable	Case, n (%)
Age in years	
Median (range)	57 (35–76)
Gender	
Female	20 (44.4)
Male	25 (55.6)
Histological type	
Adenocarcinoma	42 (93.3)
Others	3 (6.7)
Smoker	
Yes	19 (42.2)
No	26 (57.8)
Number of metastases	
Single	12 (26.7)
Multiple	33 (73.3)
Initial brain metastasis	
Yes	35 (77.8)
No	10 (22.2)
First-line chemotherapy	
Yes	19 (42.2)
No	26 (57.8)

antagonists prevented vomiting, and if grade II or higher myelosuppression occurred, we administered recombinant human granulocyte colony-stimulating factor (G-CSF), while if adverse reactions higher than grade III occurred, the next cycle of pemetrexed and carboplatin was reduced by 25%. After four to six cycles, or in cases with disease progression or unacceptable toxicity, the patients received standard whole brain radiation therapy (Dt 30 Gy/10 times/14 days).

2.2.2 Response evaluation criteria

The evaluation criteria were judged by the RECIST version 1.0, and were considered to be a Complete Remission (CR), Partial Response (PR), Stable Disease (SD) or Progressive Disease (PD). The total efficiency (overall response rate, ORR) was considered to be CR + PR, and the Disease Control Rate (DCR) was calculated as CR + PR + SD. The Overall Survival (OS) and Progression-Free Survival (PFS) were calculated after the initiation administration of pemetrexed.

2.2.3 Classification of adverse reactions

Adverse reactions were graded according to the U.S. NCI toxicity evaluation criteria (CTCAE version 3).

2.3 Statistical methods

The SPSS17.0 software program was used for the statistical analyses. A Pearson χ^2 test or Fisher's exact test was used to compare different factors, such as differences in efficiency and the disease control rate, while the survival analysis was performed using Kaplan-Meier survival curves, and Log-rank univariate and multivariate Cox regression analyses were performed to assess the differences in survival. A value of $P < 0.05$ was considered to be statistically significant.

3 RESULTS

3.1 Statistical results

Each patient completed at least two cycles of treatment, and the whole group completed a total of 157 treatment cycles of joint programs, with a median of four cycles completed. There were three patients with a PS score of 2, and four patients developed grade IV neutropenia, and subsequent treatment was 25% dose reduction for these cases. Of the 45 evaluable patients, the efficacy in all patients with intracranial metastases was the following: a PR in 17 patients (37.8%), SD in 20 (44.4%), and PD in eight patients (17.8%). The ORR was 37.8% and the DCR was 82.2%. The overall efficacy for the systemic lesions was a PR in five patients (11.1%), SD in 16 (35.6%), and PD in 24 patients (53.3%). The ORR was 35.6% and the DCR was 46.6%. The differences in the DCR for brain lesions and systemic metastases was statistically significant (P = 0.006), while the ORR was not significant different between the two (P > 0.05). In the patients with intracranial metastases who did not receive chemotherapy before whole brain radiotherapy, the ORR was 46.2% and the DCR was 84.6%; the ORR and DCR for systemic lesions were 7.6% and 61.5%, respectively. Fisher's exact test was used to compare the impact of age, gender, smoking status, and the number of brain metastases on the clinical outcomes of radiotherapy (P > 0.05, Table 1).

3.2 Comparison of the OS and PFS

Using a Log-rank univariate analysis, the progression-free survival of the 45 patients with NSCLC was analyzed, and Kaplan-Meier survival curves were plotted. The findings showed that all patients had a median OS of 12.24 months (95% CI, 10.59–13.89 months). The median OS of patients with intracranial metastases was 13.70 months, while that of the patients with systemic lesions was 11.38 months, and the difference was statistically significant (P = 0.018, Fig. 1). In all patients, the median PFS was 5.3 months (95% CI, 4.21–6.10 months). The PFS of patients with intracranial metastases was 6.22 months, with the median PFS was 3.97 months in patients with systemic lesions. The difference between these groups was statistically significant (P = 0.026, Fig. 2). The median Overall Survival (OS) for patients with cerebral and extracerebral disease (intracranial and systemic metastases) (P = 0.018).

3.3 Adverse reactions in patients

Most of the patients experienced mild adverse reactions, mainly bone marrow suppression, fatigue and gastrointestinal reactions (Table 2).

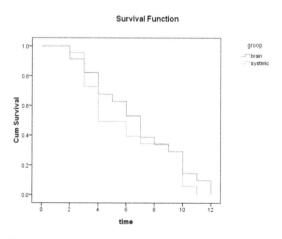

Figure 1.   The median Overall Survival (OS) for cerebral and extracerebral diseases (P = 0.018).

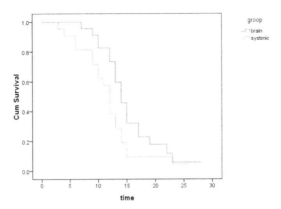

Figure 2. The median Progression-Free Survival (PFS) for patients with cerebral and extracerebral disease (intracranial and systemic metastases) (P = 0.026).

Table 2. Adverse events associated with pemetrexed treatment.

Adverse event	n (%)
Leukopenia	2 (4.4)
Neutropenia	5 (11.1)
Febrile neutropenia	0 (0)
Anemia	3 (6.7)
Thrombopenia	0 (0)
Nausea/vomiting	4 (8.9)
Fatigue	1 (2.2)
Rash	0 (0)

4 DISCUSSION

Currently, the treatments used for non-small cell lung cancer brain metastases include whole-brain radiotherapy, stereotactic radiotherapy, surgery, targeted therapy, and chemotherapy. Previous studies have reported that chemotherapy led to efficacy rates for non-small cell foci of 35% to 50% (Scagliotti GV, et al., 2008. Shepherd FA, et al., 2001). Recently, Galetta et al. (Galetta et al., 2011) conducted a multi-center Phase II clinical study, and all patients given three cycles of platinum-based chemotherapy prior to whole brain radiotherapy showed a DCR of 60% after two cycles in the brain lesions, while the systemic lesions had a DCR of 48%, indicating that the response rate of intracranial metastatic lesions was significantly higher than that of the systemic lesions located in other parts of the body, which is consistent with the results of this study.

In recent years, pemetrexed, a new multi-target anti-folate agent, has shown efficacy in the treatment of non-small cell lung cancer. Grønberg et al. (Grønberg et al., 2009) published a study comparing pemetrexed and gemcitabine with carboplatin as the first-line treatment for advanced NSCLC, and both groups showed a similar OS (7.3 months vs 7.0 months), while the hematological toxicity was significantly lower in the former. Chiappori et al. (Chiappori et al., 2010) evaluated 39 patients who received pemetrexed as the second-line treatment for non-small cell lung cancer with brain metastases, and showed a good short-term effect, no serious adverse reactions, indicating that pemetrexed is effective for non-small cell lung cancer brain metastases and is well tolerated.

Our present study demonstrated a longer median OS of 12.24 months (95% CI, 10.59–13.89 months), while the median PFS was 5.3 months (95% CI, 4.21–6.10 months). The longer survival time was probably due to several reasons. First, in this trial, most of the patients had adenocarcinoma histology, while the study by Guo et al. (Guo et al., 2012) examined patients with other histological types of non-small cell lung cancer. Second, the traditional view is that chemotherapy is not a standard solution for brain tumor metastases, as most chemotherapy drugs do not pass through the blood brain barrier. However, in recent years, there has been increasing evidence that the blood-brain barrier damage induced by radiotherapy may allow the penetration of chemotherapy drugs. The results of several studies have shown that sequential chemotherapy after radiotherapy can improve the survival of patients (Sperduto PW, et al., 2013, Gerstner ER et al., 2007). Gerstner and others showed that, compared with primary brain tumors, brain metastases secreted lower levels of P-glycoprotein, which is involved in the exchange of important materials across the blood-brain barrier. The lower P-glycoprotein expression may lead to increased blood brain barrier permeability in these cases. It has been reported in preclinical studies (Zhao R, et al., 2004) that when pemetrexed is combined with radiotherapy, it shows a synergistic effect, and the physiological pH value decreased with the intake of folic acid, which likely enhances the tumor cells' sensitivity to pemetrexed.

Recently, a number of Phase I and Phase II clinical trials have demonstrated that pemetrexed has low toxicity in patients when used in combination with whole brain radiation therapy. In the present study, grade 3 hematological or non-hematological toxicity was found in 34.2% and 46.2% of the patients, respectively. Our findings regarding adverse reactions were generally mild, mainly bone marrow suppression, fatigue and gastrointestinal reactions.

In conclusion, our present study shows that pemetrexed and carboplatin chemotherapy combined with whole brain radiotherapy is effective against non-small cell lung cancer brain metastases and is associated with good tolerability. However, our study was a retrospective study, so randomized controlled trials are needed to confirm the efficacy and safety of the regimen. In addition, performing genotyping for the methylenetetrahydrofolate reductase and thymidylate synthetase genes may help to clarify the mechanisms of action of pemetrexed and identify patients who would most benefit from the treatment.

ACKNOWLEDGEMENTS

This work is supported by the National Natural Science Foundation of China (81402518), Jiangsu Provincial Special Program of Medical Science (BL2012046), Changzhou Scientific Program (ZD201315;CY20130017; CE20135050; CJ20140050; 2014260).

REFERENCES

Barlesi F, Khobta N, Tallet A, et al. 2013. Management of brain metastases for lung cancer patients. Bull Cancer, 100 (3): 303–308.

Chiappori A, Bepler G, Barlesi F, et al, 2010. Phase II, double-blinded, randomized study of enzastaurin plus pemetrexed as second-line therapy in patients with advanced non-small cell lung cancer. Thorac Oncol. 5 (3): 369–375.

De Ruysscher D, Wanders R, van Baardwijk A, et al. 2012. Radical treatment of non-small-cell lung cancer patients with synchronous oligometastases: Long-term results of a prospective phase II trial (Nct01282450). Thorac Oncol, 7 (10): 1547–1555.

Galetta D, Gebbia V, Silvestris N, et al. 2011. Cisplatin, fotemustine and whole-brain radiotherapy in non-small cell lung cancer patients with asymptomatic brain metastases: a multicenter phase II study of the Gruppo Oncologico Italia Meridionale (GOIM 2603). Lung Cancer 72 (1): 59–63.

Grønberg BH, Bremnes RM, Fløtten O, et al. 2009. Phase III study by the Norwegian lung cancer study group: pemetrexed plus carboplatin compared with gemcitabine plus carboplatin as first-line chemotherapy in advanced non-small-cell lung cancer. Clin Oncol. 27 (19): 3217–3224.

Guo S, Reddy CA, Chao ST, et al. 2012. Impact of non-small cell lung cancer histology on survival predicted from the graded prognostic assessment for patients with brain metastases Lung Cancer, 77 (2): 389–393.

Gerstner ER, Fine RL. 2007. Increased permeability of the blood-brain barrier to chemotherapy in metastatic brain tumors: establishing a treatment paradigm. Clin Oncol, 25 (16): 2306–12.

Hodgson DC, Charpentier AM, Cigsar C, et al. 2013. A multi-institutional study of factors influencing the use of stereotactic radiosurgery for brain metastases. Radiat Oncol Biol Phys, 85 (2): 335–340.

Lind JS, Lagerwaard FJ, Smit EF, et al. 2011. Time for reappraisal of extracranial treatment options Synchronous brain metastases from nonsmall cell lung cancer. Cancer, 117 (3): 597–605.

Meng FL, Zhou QH, Zhang LL, et al. 2013. Antineoplastic therapy combined with whole brain radiation compared with whole brain radiation alone for brain metastases:. A systematic review and meta-analysis. Eur Rev Med Pharmacol Sci,. 17 (6): 777–787.

Scagliotti GV, Parikh P, von Pawel J, et al. 2008. Phase III study comparing cisplatin plus gemcitabine with cisplatin plus pemetrexed in chemotherapy-naive patients with advanced-stage non-small-cell lung cancer. Clin Oncol, 26 (21): 3543–3551.

Shepherd FA, Dancey J, Arnold A, et al. 2001. Phase II study of pemetrexed disodium, a multitargeted antifolate, and cisplatin as first-line therapy in patients with advanced nonsmall cell lung carcinoma: a study of the National Cancer Institute of Canada Clinical Trials Group. Cancer, 92 (3): 595–600.

Sperduto PW, Wang M, Robins HI, et al. 2013. A phase 3 trial of whole brain radiation therapy and stereotactic radiosurgery alone versus WBRT and SRS with temozolomide or erlotinib for non-small cell lung cancer and 1 to 3 brain metastases: Radiation Therapy Oncology Group 0320. Radiat Oncol Biol Phys, 85 (5): 1312–1318.

Zhao R, Gao F, Hanscom M. 2004. A prominent low-pH methotrexate transport activity in human solid tumors: contribution to the preservation of methotrexate pharmacologic activity in HeLa cells lacking the reduced folate carrier. Clin Cancer Res, 10 (2): 718–727.

Bioinformatics and Biomedical Engineering – Chou & Zhou (Eds)
© 2016 Taylor & Francis Group, London, ISBN 978-1-138-02784-8

Metabolites analysis of functional lactic acid bacteria strain *Lactobacillus paracasei* HD1.7

R.P. Du
Key Laboratory of Microbiology, Life Science Department, Heilongjiang University, Harbin, China

D. Zhao
Key Laboratory of Microbiology, Life Science Department, Heilongjiang University, Harbin, China
Engineering Research Center of Agricultural Microbiology Technology, Ministry of Education,
Heilongjiang University, Harbin, China

X.Y. Wang, Q. Wang & J.P. Ge
Key Laboratory of Microbiology, Life Science Department, Heilongjiang University, Harbin, China

ABSTRACT: *Lactobacillus paracasei* HD1.7 (CCTCCM 205015), which was isolated from fermentation liquid of Chinese fermented cabbage, was capable of producing bacteriocin Paracin1.7 and lactate. These two products suggested *L.paracasei* HD1.7 great potential in industrial use. In order to get a better understanding and explore furtherer application of *L.paracasei* HD1.7, this research conducted a comprehensive investigation on metabolic characteristics of this strain in MRS liquid medium. The growth curve was established and pH value was detected. Several metabolites and key enzymes were assayed, including lactate, acetate, pyruvate, fructose-1,6-diphcsphate, pyruvate kinase and fructose-1,6-diphcsphate kinase. The depletion of glucose and production of mannitol were also investigated. The results derived from this research would set a foundation for strain *L.paracasei* HD1.7 in theoretical study and practical application.

Keywords: *Lactobacillus paracasei* HD1.7; fermentation characteristics; metabolites

1 INTRODUCTION

Lactic Acid Bacteria (LAB) are of great importance in biotechnology industries such as food fermentation and brewing. They are widely used as starters for manufacturing food, such as cheese, yoghurt, butter, and sauerkraut by producing lactic acid, alcohol and other metabolites, which improve the taste and afford nutrients (Tripathi et al. 2012). Many kinds of LAB, such as *L.plantarum, L.brevis, Leuconostoc mesenteroides* and *Pediococcus acidilactici*, are reported to be used safely in the different stages of fermentation (Jeong et al. 2013, Douillard et al. 2014). Furthermore, LAB are closely associated with the normal function and healthy of intestinal tract (Ng'ong'ola-Manani et al. 2014). Content changes in metabolites during glucose fermentation were detected for LAB metabolism in the production of various metabolites which contribute to various flavors and tastes of food (Mozzi et al. 2013).

In fermented food industry, such as free sugars (glucose and fructose), organic acids (lactic acid and acetic acid), and other flavor compounds (mannitol and amino acids) are important determinants of taste and flavor, and represent more direct and collective phenotypic outcomes resulting from microbial activities in food fermentation (Jung et al. 2011). Flavoring compounds in food fermentation were investigated at a particular time of homolactic fermentation, but metabolite changes during the entire homolactic fermentation period have never been explored. Therefore, barely anything is known about the metabolites of LAB as starter cultures during food fermentation so far (Jung et al. 2012). Spontaneous fermentation using various naturally-occurring bacteria usually results in variations in the sensory qualities of fermented food, which makes it difficult to produce commercial food with uniform quality. Therefore, the use of a starter culture has been considered for the production of standardized fermented food with uniform high quality (Ge et al. 2014).

L.paracasei HD1.7 was isolated from fermentation liquid of Chinese cabbage, which is one of the most popular traditional fermented vegetables in Northeastern China. Strain *L.paracasei* HD1.7 produces a bacteriocin called Paracin1.7, which is similar to many other bacteriocins produced by LAB, has antimicrobial activity. It is known that *L.paracasei* is homofermentative LAB that produce lactate, acetate from sugars (Ge et al. 2009). Studies of not only microbial communities but also metabolite changes are needed in order to understand the relationships between microbial populations and metabolites in food fermentation since a rational approach to the control of the microbial community is currently almost impossible (Yang et al. 2014, Azaza et al. 2015). The investigation of metabolites produced by *L.paracasei* HD1.7 would lead to a better understanding of the LAB's role during food fermentation and the flavor formation mechanism.

2 MATERIAL AND METHODS

2.1 *Bacterial strains and culture media*

Strain *L.paracasei* HD1.7 was maintained and cultured in MRS medium. The strain grown in MRS broth were inoculated (1.0%) in 50 ml of fresh MRS broth at 30 °C with shaking (170 rpm) for 18 h until the cell density reached 10^8 CFU/mL. The inoculum cultures were inoculated into a 500 ml flask containing 300 ml medium. Culture temperature and agitation speed were maintained at 30 °C and 170 rpm, respectively. MRS broth contained soya peptone, 10 g/L; beef extract, 10 g/L; yeast extract, 5 g/L; glucose, 20 g/L; K_2HPO_4, 2 g/L; Na_2SO_3, 0.1 g/L; sodium acetate, 5 g/L; $MgSO_4 \cdot 7H_2O$, 0.2 g/L; $MnSO_4$, 0.05 g/L; ammonium citrate, 0.4 g/L; Tween 80, 1 mL. The initial pH value was adjusted to 5.5.

2.2 *Measurement of the pH and viable cell numbers*

The pH value was directly measured using pH meter (Delta 320A, Mettler Toledo, Switzerland). The numbers of microorganisms in samples were determined by turbidimetric method.

2.3 *Key enzyme and metabolite assay*

The mycelial extract obtained prior from cell extraction and partial enzyme purification was analyzed for the activities of the enzymes of interest including Pyruvate Kinase (PK) and fructose-1,6-diphcsphate kinase (PFK) during cultivation of the living of *L.paracasei* HD1.7. The activity of PK and PFK were determined using PK assay kit and PFK assay kit (NY2 and NY3, Suzhou Comin Biotechnology Co., Ltd., China). The pyruvate and 1,6-diphosphate fructose was measured based on Nigam (1962) and Zhenning and Wutong (1993)'s method, respectively.

2.4 *Characteristics measurement of lactic fermentation*

The *L.paracasei* HD1.7 metabolites were analyzed in triplicate using HPLC (Shimadzu LC-10 ATvp) using a HPX-87H column (300 mm 67.8 mm, Aminex HPX-87H ion exclusion column) at 65 °C with a refractive index detector (RID-10A). The mannitol was measured based on Demain (1977)'s method. L-8800 analyzer (Hitaichi, Japan) was used for determining concentration of amino acids.

3 RESULTS AND DISCUSSION

3.1 *The pH profiles and numbers of strain during fermentation*

The pH profiles of the fermentation broth over the 108 h-fermentation process were similar to those of LAB fermentation (Patel et al. 2014). As was shown in Figure 1, in this study, the pH

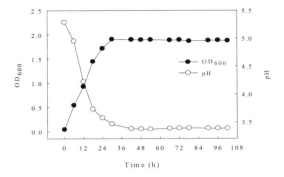

Figure 1.   Changes in pH and total bacterial cells during fermentation.

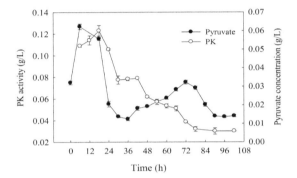

Figure 2.   Changes in PK and pyruvate during fermentation.

values were about 5.3–5.4 during early stage of the fermentation, which quickly decreased to approximately 3.6 in 24 h of fermentation. After 30 h of fermentation, the pH value became relatively stable until the end of fermentation, reached approximately 3.3. Similar conclusion have been reported by Patel and others that the addition of Lactobacillus strains as starter cultures resulted in lower acidity in fermented Chinese cabbage (Patel et al. 2014).

L.paracasei HD1.7 utilized glucose from MRS culture-medium and producted of large organic acid, resulting in the pH value decreased during the fermentation until glucose depletion. The results demonstrated that the use of L.paracasei HD1.7 made the fermentation liquid more acidic and tasted more refreshing. The increased bacterial abundance was inversely correlated with a decreased in pH value during the early fermentation period (0–24 h). The average initial bacterial cell number in the liquid was approximately 2.5×10^3 CFU/ml, which reached the peak value of about 8.2×10^8 CFU/ml at 30 h. After 30 h of fermentation, bacterial abundance remained stable. The OD_{600} analysis also showed that the abundances of bacteria and glucose content were negatively correlated during early stage of the fermentation (Fig. 1), (Fig. 4). The increased bacterial cells due to glucose can provide energy for the cell during the early fermentation period indicating that the glucose could improve the effect of the LAB growth (Chang et al. 2011).

3.2   *Enzyme assay during homolactic fermentation*

From the fermentation kinetics shown in Figure 2 and Figure 3, it is clear that regardless of the presence of the regulators, the activities of PK and PFK were increased with time, the pyrurate concentration was also increased with the prolonging of the fermentation time, but the fructose-1,6-diphcspahte maintain a high level. During the early fermentation period (0–18 h), the enzyme activities were increased then started to drop with the growth of bacteria. The L.paracasei HD1.7 attempted to reverse the direction of the PK and PFK

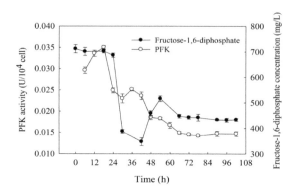

Figure 3. Changes in PFK and fructose-1,6-diphcsphate during fermentation.

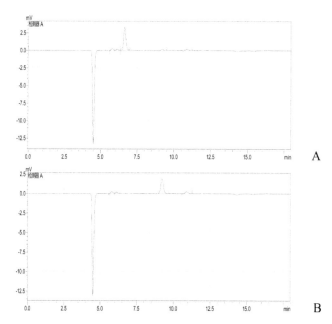

Figure 4. The peak spectrum of metabolites during *Lb.paracasei* HD1.7 fermentation with HPLC. A: 6 h, B: 102 h.

reaction to acetaldehyde instead of the end product ethanol. This eventually led to the large lactate production. Accordingly, with the cell growth, the enzyme activity first increased and later decreased, its metabolites corresponding positive correlation. This result is consistent with the metabolic growth of homolactic fermentation LAB (Nariya et al. 2003, Callejón et al. 2014). Therefore, during the fermentation of LAB, monitoring of lactic acid fermentation process will help us understanding of the accumulation of lactic acid.

3.3 *Content changes in metabolites related to lactate-production during lactate fermentation*

The peak spectrum and content changes of metabolites during the entire fermentation were illustrated in Figure 4, Figure 5. The levels of lactate and acetate, which were major products of homolactic fermentation increased inversely with the decrease of free sugars as fermentation progressed (García-Ruiz et al. 2014). The concentrations of lactate, acetate, mannitol and glycerol were very low during the early fermentation period (0–6 h). Then the concentrations of free sugars declined rapidly, despite the increase of the lactate, mannitol, and acetate.

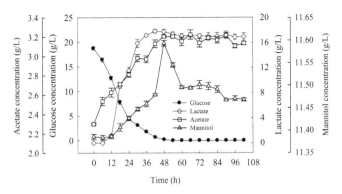

Figure 5.   Changes in metabolites during *Lb.paracasei* HD1.7 fermentation.

Table 1.   Terminal content and variance analysis of free amino acids in homolactic fermentation.

Amino acid	Concentration ± SE (g/L)
Aspartic acid	0.0546 ± 0.0001
Threonine	0.1572 ± 0.0001
Serine	0.1305 ± 0.0001
Glutamic acid	0.2541 ± 0.0002
Glycine	0.6651 ± 0.0004
Alanine	0.3246 ± 0.0001
Cystine	0.0366 ± 0.0006
Valine	0.2326 ± 0.0001
Methionine	0.0734 ± 0.0001
Isoleucine	0.1644 ± 0.0003
Leucine	0.3568 ± 0.0004
Tyrosine	0.2049 ± 0.0001
Phenylalanine	0.2821 ± 0.0002
Histidine	0.0709 ± 0.0001
Arginine	0.6645 ± 0.0001
Proline	0.1027 ± 0.0005
Lysine	0.2677 ± 0.0001
Total	4.0427 ± 0.0004

After the maximum levels of lactate, mannitol and acetate production were reached, the lactate and acetate concentrations were relatively constant until the end of fermentation. Meanwhile, the mannitol concentration decreased gradually after 48 h. At the same time, the production of glycerol, succinic and ethanol were not observed. Many LAB inhibit the growth of yeasts due to the production of organic acids or antimicrobial compounds (Azaza et al. 2015). In this research, *L.paracasei* HD1.7 also had antagonistic effects on yeasts growth. Mannitol, a six-carbon polyol produced by the reduction of fructose by LAB, increased during fermenta-tion and the profile was inversely correlated with the levels of fructose (Jeong et al. 2013). Its concentrations increased as fermentation progressed in the early period of fermentation and keep stable at the end of fermentation. The results indicated that the use of *L.paracasei* HD1.7 as a starter increased the production of mannitol and improve the flavor of fermented Chinese cabbage. It was helpful to produce fermented Chinese cabbage with more nutritive value.

3.4   *Terminal content of amino acids in fermentation*

Amino acids are important for determining flavors and tastes (Kim et al. 2001). Seventeen amino acids including cystine, methionine, aspartic acid, threonine, serine, glutamic acid, proline, glycine, alanine, valine, isoleucine, leucine, tyrosine, phenylalanine, histidine, lysine, and arginine were identified in 48 h of fermentation. The total amino acids were 4.0427 ± 0.0004 g/L. Glycine and arginine were significantly higher (0.6651 ± 0.0004 g/L and 0.6645 ± 0.0001 g/L). Glycine, glutamic acid, alanine, leucine, arginine, phenylalanine, and lysine were considered the main amino acids (>0.25 g per L sample) throughout fermentation. Amino acids play important roles in aroma and taste development in food as they are involved in Maillard reactions and Strecker degradation. Glutamic acid was reported as the most abundant amino acid in Chinese fermented cabbage fermentation paste during ripening and storage (Juhász et al. 2015). For instance, alanine is a precursor for a key flavor compound of fermentation food that intensifies the sweetness and taste of the food.

4 CONCLUSIONS

L.paracasei HD1.7 as starter cultures could improve the characteristics and flavor of fermented Chinese cabbage. The results derived from this research would link of relationships among microbial communities, metabolites, and sensory characteristics are necessary in food, which would provide useful guidelines for producing fermented food featuring both uniform and high qualities.

ACKNOWLEDGEMENTS

This research was financially supported by National Nature Science Youth Foundation of China (31300355), Foundation of Harbin Municipal Science and Technology Bureau (2014RFQXJ101), Master Degree Candidate Research Innovation Fund of Heilongjiang University (YJSCX2015-090HLJU) and National Nature Science Foundation of China (31270534).

AUTHOR CONTRIBUTIONS

Renpeng Du and Dan Zhao (first author): these two authors contributed equally to this work, including acquisition of funding, literature research, experimental studies, manuscript preparation.

Xiaoyu Wang: data collection and data analysis.

Qi Wang: data collection and data analysis.

Jingping Ge (corresponding author): acquisition of funding, study concepts, study design, final version approval.

REFERENCES

Azaza, M.S., Khiari, N., Dhraief, M.N., Dhrauef, M.N., Aloui, N., Kraïem, M. & Elfeki, A. 2015. Growth performance, oxidative stress indices and hepatic carbohydrate metabolic enzymes activities of juvenile Nile tilapia, *Oreochromis niloticus* L., in response to dietary starch to protein ratios. *Aquaculture Research* 46: 14–27.

Chang, J.Y. & Chang, H.C. 2011. Growth inhibition of foodborne pathogens by kimchi prepared with bacteriocin-producing starter culture. *Journal of Food Science* 76(2011): M72–M78.

Callejón, S., Sendra, R. Ferrer, S. & Pardo, I. 2014. Identification of a novel enzymatic activity from lactic acid bacteria able to degrade biogenic amines in wine. *Applied Microbiology and Biotechnology* 98(2014): 185–198.

Douillard, F.D. & de Vos W.M. 2014. Functional genomics of lactic acid bacteria: from food to health. *Microbial Cell Factories* 13(Suppl 1): 1–21.

Ge, J.P., Ping, W.X., Song, G., Du, C.M., Ling, H.Z., Sun, X. & Gao, Y. 2009. Paracin 1–7, a bacteriocin produced by *Lactobacillus paracasei* HD1–7 isolated from Chinese cabbage sauerkraut, a traditional Chinese fermented vegetable food. *Acta Microbiol Sin* 49(4): 609–616.

Ge, J.P., Fang, B.Z., Wang, Y., Song, G. & Ping, W.X. 2014. *Bacillus subtilis* enhances production of Paracin1.7, a bacteriocin produced by *Lactobacillus paracasei* HD1–7, isolated from Chinese fermented cabbage. *Annals of Microbiology* 64: 1735–1743.

García-Ruiz, A., de Llano, D.G. & Esteban-Fernandez, A. 2014. Assessment of probiotic properties in lactic acid bacteria isolated from wine. *Food Microbiology* 44(2014): 220–225.

Jung, J.Y., Lee, S.H. & Kim, J.M. 2011. Metagenomic analysis of kimchi, a traditional Korean fermented food. *Applied and Environmental Microbiology* 77: 2264–2274.

Jung, J.Y., Lee, S.H., Lee, H.J., Seo H-Y., Park W-S. & Jeon, C.K. 2012. Effects of *Leuconostoc mesenteroides* starter cultures on microbial communities and metabolites during kimchi fermentation. *International Journal of Food Microbiology* 153: 378–387.

Jeong, S.H., Lee, H.J., Jung, J.Y., Lee, S.H., Park W-S., Seo H-Y. & Jeon, C.K. 2013. Effects of red pepper powder on microbial communities and metabolites during kimchi fermentation. *International Journal of Food Microbiology* 160(2013): 252–259.

Juhász, Z., Boldizsár, Á. & Nagy, T. 2015. Pleiotropic effect of chromosome 5 A and the *mvp* mutation on the metabolite profile during cold acclimation and the vegetative/generative transition in wheat. *BMC Plant Biology* 15(57): 1–12.

Kim, J.H. & Sohn, K.H. 2001. Flavor compounds of dongchimi soup by different fermentation temperature and salt concentration. *Food Science and Biotechnology* 10: 236–240.

Mozzi, F., Ortiz M, E. & Bleckwedel, J. 2013. Metabolomics as a tool for the comprehensive understanding of fermented and functional foods with lactic acid bacteria. *Food Research International* 54(2013): 1152–1161.

Nariya, H. & Inouye, S. 2003. An effective sporulation of *Myxococcus Xanthus* requires glycogen consumption via Pkn4-activated 6-phosphofructokinase. *Molecular Microbiology* 49(2): 517–528.

Ng'ong'ola-Manani T.A, Østlie H.M. & Mwangwela, A.M. 2014. Metabolite changes during natural and lactic acid bacteria fermentations in pastes of soybeans and soybean-maize blends. *Food Science & Nutrition* 2(6): 768–785.

Patel, A., Praiapati, J.B., Holst, O. & Ljungh, A. 2014. Determining probiotic potential of exopolysaccharide producing lactic acid bacteria isolated from vegetables and traditional Indian fermented food products. *Food Bioscience* 5(2014): 27–33.

Tripathi, P., Beaussarta, A., Andre, G., Rolain, T., Lebeer, S., Vanderleyden, J. & Hols, P. 2012. Towards a nanoscale view of lactic acid bacteria. *Micron* 43(2012): 1323–1330.

Yang, J., Ji, Y., Park, H., Lee, J., Park, S., Yeo, S., Shin, H. & Holzapfel, W.H. 2014. Selection of functional lactic acid bacteria as starter cultures for the fermentation of Korean leek (*Allium tuberosum* Rottler ex Sprengel.). *International Journal of Food Microbiology* 191(2014): 164–171.

Bioinformatics and Biomedical Engineering – Chou & Zhou (Eds)
© 2016 Taylor & Francis Group, London, ISBN 978-1-138-02784-8

4-aminopyridine inhibits cell proliferation and affects anti-tumor activities of cell cycle-specific drugs in human breast cancer cells

Q. Ru, X. Tian, J.S. Liu & C.Y. Li
Wuhan Institutes of Biomedical Sciences, Jianghan University, Wuhan, China

ABSTRACT: Evidence is growing that potassium channels play important roles in the development and growth of human cancer, including breast cancer. In this study, we investigated the effect of 4-aminopyridine (4-AP), an inhibitor of voltage-gated potassium channels, on the growth of human breast cancer MCF-7 cells, with an attempt to know whether pre-arresting tumor cells at a certain phase could affect anti-proliferative activities of cell cycle-specific drugs, and found that 4-AP significantly inhibited the cell proliferation and induced apoptosis in a dose-dependent manner. 4-AP also significantly blocked the cell cycle in G1 phase at 1 mmol/l, and induced mainly S phase arrest at 5 and 10 mmol/l. Further studies exhibited strong synergistic action between 4-AP (5 mmol/l) and S phase-specific drug 5-FU, and antagonistic action between 4-AP (5 mmol/l) and G2/M phase-specific drug paclitaxel, indicating that cell cycle parameters should be considered in the combination therapy involving cell cycle-specific agents.

Keywords: potassium channels; human breast cancer cells; proliferation; cell cycle

1 INTRODUCTION

Accumulating evidence has proved that a variety of K^+ channels, including voltage-gated K^+ (Kv) channels (Kunzelmann, 2005), calcium-activated K^+ (K_{Ca}) channels (Asanuma et al., 2010) and inward rectifier K^+ (Kir) channels (Huang et al., 2009) are over-expressed in tumorous tissues compared to their healthy counterparts, and the channels play important roles in the regulation of migration, proliferation, cell cycle progression and apoptosis of tumor cells (Jang et al., 2009, Blackiston et al., 2009). Moreover, a significant increase in K^+ channel expression has been found to be associated with tumorigenesis, suggesting that it is possible to use these proteins as transformation markers and to reduce tumor growth by selectively inhibiting the channel activities (Ouadid-Ahidouch et al., 2008).

Breast cancer is one of the most common malignancies and the most common causes of cancer-related deaths in women worldwide (Ait-Mohamed et al., 2011). Studies have shown that a number of K^+ channels, such as Kv, Kir, and K_{Ca} channels, are over-expressed in breast cancer cells, and exhibit oncogenic potential (Brevet et al., 2008, Ouadid-Ahidouch and Ahidouch, 2008). Similar to other epithelial tumors (Kunzelmann, 2005), Kv channels have been found to be correlated to the proliferation of breast cancer cells. For instance, α-dendrotoxin (α-DTX), a blocker of Kv1.1, inhibits breast cancer MCF-7 cell proliferation in a dose-dependent manner (Ouadid-Ahidouch et al., 2000).

Although numerous reports suggest that Kv channels play important roles in the proliferation of human breast cancer cells (Ouadid-Ahidouch et al., 2000, Ouadid-Ahidouch, 2004), the underlying mechanism remains poorly understood. The aim of the present study was to investigate the effect of 4-aminopyridine (4-AP), a nucleotide analog that blocks Kv channels, on the growth of human breast cancer MCF-7 cells and the underlying mechanism. The effects of 4-AP alone or in combination with low-dose cell cycle-specific chemotherapeutic agents on the growth of MCF-7 cells were also evaluated not only to further validate the effect of 4-AP on cell cycle, but also to know whether pre-arresting tumor cells at a certain phase could affect anti-proliferative activities of cell cycle-specific drugs.

2 MATERIALS AND METHODS

2.1 Chemicals and drug preparations

4-aminopyridine (4-AP), 5-fluorouracil (5-FU), 3-(4,5-dimethylthiazol-2-yl)-2,5-diphenyltetrazolium bromide (MTT), paclitaxel and Propidium Iodide (PI) were products of the Sigma Chemical Corp (St Louis, MO, USA). Dulbecco's Modified Eagle Medium (DMEM) and Fetal Bovine Serum (FBS) were obtained from Life Technologies (Carlsbad, CA, USA). Annexin V-fuorescein isothiocyanate/propidium iodide (Annexin V-FITC/PI) apoptosis detection kit was procured from Antgene Biotech (Wuhan, China). RNase A was bought from Fermentas International Inc. (Burlington, Ontario, Canada). All other chemicals were of standard analytical grade.

2.2 Cell culture

Human breast cancer MCF-7 cells were purchased from American Type Culture Collection (ATCC, VA, USA) and grown in DMEM supplemented with 10% FBS and 100 U penicillin/streptomycin in 5% CO_2 at 37 °C.

2.3 MTT proliferation assay

Briefly, cells were seeded at 5000 cells/well into a 96-well plate and incubated overnight. After drug treatment for different time lengths, 20 µl of MTT solution (5 mg/ml) was added to each well and the samples were incubated for another 4 h. Subsequently, the supernatant was removed and cells were dissolved in 150 µl DMSO. Finally, absorbance at 570 nm was measured by using a 96-well microplate reader.

2.4 Colony formation assay

Briefly, cells were treated with 4-AP for 24 h and then cells (500 cells/well) were re-plated into 6-well plates under normal conditions. After 1-week incubation, the colonies were washed with $1 \times$ PBS and stained with 0.1% trypan blue in 50% ethanol. Colonies containing 50 or more cells were counted as viable clonogenic cells.

2.5 Hoechst 33342/PI staining assay

Cells were plated at an initial density of 1×10^5 cells/well into a 24-well plate and incubated for 24 h. After treated with different concentrations of 4-AP for 48 h, the cells were washed gently with PBS twice. After stained with Hoechst 33342 (5 µg/ml) at 4°C for 10 minutes, cells were stained with PI (15 µg/ml) in the dark for 10 minutes at 4°C and washed gently with PBS twice. Cells fluorescently stained blue or red were examined under a fluorescence microscope.

2.6 Annexin V-FITC/PI apoptosis assay

Cells were double-stained by using an Annexin V-FITC/PI apoptosis detection kit. The detailed experimental steps were conducted following manufactory's directions. Cells were analyzed on a flow cytometer within 1 h.

2.7 Cell cycle assay

Following treatment, cells were harvested into cold PBS at different time points, fixed in ice-cold 70% ethanol, and stored at 4°C overnight for subsequent cell cycle analysis. Fixed cells were washed with PBS twice and re-suspended in 1 ml staining reagents (100 µg/ml RNase A and 50 µg/ml PI). Samples were incubated in the dark for 30 minutes, and the distribution of cells in the various phases of the cell cycle was measured by flow cytometry.

2.8   *Statistical analysis*

Coeffients of Drug Interaction (CDI) were calculated as follows: CDI = AB/(A·B), where A and B are the survival values with single agents and AB is the observed values of the two-drug combination (Chen et al., 2005). The effects of two drugs were considered to be antagonistic when CDI > 1.0, additive when CDI = 1.0, synergistic when CDI < 1.0, and significantly synergistic when CDI < 0.7 (Chen et al., 2005).

Data were expressed as mean ± standard error $(\bar{x} \pm SE)$. Statistical significance was assessed by using analysis of variance (ANOVA). *P* values of less than 0.05 were considered to be statistically significant.

3   RESULTS AND DISCUSSION

3.1   *Effect of 4-AP on the cell growth of MCF-7 cells*

As shown in Figure 1a, 4-AP significantly inhibited the cell proliferation of MCF-7 cells in a dose- and time-dependent manner ($P < 0.01$). In addition to MTT assay, we also tested the effect of 4-AP on colony formation and found that treatment with 4-AP at 10 mmol/l effectively inhibited colony formation of MCF-7 cells (Fig. 1b). These results indicate that 4-AP was able to inhibit the proliferation of human breast cancer MCF-7 cells.

To determine whether the reduced cell viability by 4-AP was related to apoptotic cell death, the effect of 4-AP on apoptosis of MCF-7 cells was studied by Hoechst 33342/PI staining and Annexin V-FITC/PI staining. Hoechst 33342 can readily pass through all cell membranes, while PI can only penetrate into late-stage apoptotic or necrotic cells, the positive staining with PI was indicative of late-stage apoptotic or necrotic cells. As shown in Figure 2a, these morphological findings suggest that 4-AP was able to induce apoptosis or necrosis in MCF-7 cells. The apoptotic rates of MCF-7 cells treated with 1, 5, and 10 mmol/l 4-AP were 8.7%, 13.5% and 78.4%, respectively (Fig. 2b). However, the effect of 4-AP on necrosis was slight. At 10 mmol/l, 4-AP only induced necrosis in approximately 1.2% of MCF-7 cells, indicating that 4-AP could induce apoptosis in MCF-7 cells in a dose-dependent manner, without causing obvious necrosis.

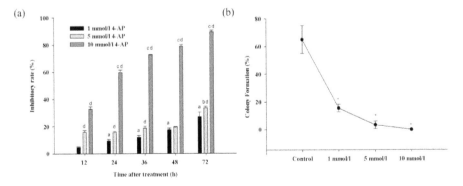

Figure 1.   Effect of 4-AP on the growth of human breast cancer MCF-7 cells. (a) MTT proliferation assay. [a]$P < 0.01$, as compared with group treated with 1 mmol/l 4-AP for 12 h. [b]$P < 0.01$, as compared with group treated with 5 mmol/l 4-AP for 12 h. [c]$P < 0.01$, as compared with group treated with 10 mmol/l 4-AP for 12 h. [d]$P < 0.01$, as compared with group treated with 1 mmol/l 4-AP at same time period. (b) Colony growth assay. *$P < 0.01$, as compared with control group.

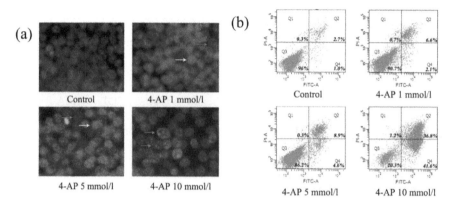

Figure 2. Induction of cell apoptosis by 4-AP in MCF-7 cells. (a) MCF-7 cells were treated with 4-AP for 48 h at the indicated concentrations and then stained with Hoechst 33342/PI (×400). The white arrows indicate early-stage apoptotic cells and red arrows show the late-stage apoptotic cells. (b) MCF-7 cells were treated with 4-AP for 48 h at the indicated concentrations.

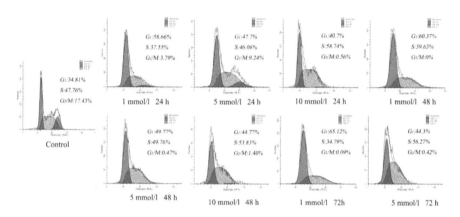

Figure 3. Cell cycle analysis of MCF-7 cells treated with 4-AP.

Compared to control cells, cell cycle arrest at G_1 phase was significantly induced by 4-AP at 1 mmol/l in a time-dependent manner (Fig. 3). Interestingly, a greater accumulation of cells in S phase was also observed in cells treated with 5 and 10 mmol/l 4-AP. These results indicate that 4-AP had diverse effects on the cell cycle events in MCF-7 cells at different concentrations.

3.2 Effect of 4-AP in combination with cell cycle-specific anti-cancer drugs on MCF-7 cells

The lengthened G_1 phase and/or S phase of MCF-7 cells caused by 4-AP suggest that these cells should be more sensitive to G_1 or S phase-specific chemotherapeutic agents. MTT assay was used to evaluate the effects of 4-AP in combination with two different cell cycle specific anti-cancer drugs in MCF-7 cells, including S phase-specific drug 5-FU and G_2/M phase-specific drug paclitaxel.

In drug combination experiments, MCF-7 cells were pre-treated with 4-AP for 16 h and co-administrated with 4-AP and each of the two cell cycle-specific agents for additional 48 h respectively. As shown in Figure 4 (a, c), the combinations of 4-AP and 5-FU or paclitaxel exerted stronger effects than each drug alone. The CDI values for 4-AP (1 mmol/l) combined with 5-FU (2 μmol/L and 16 μmol/l) were greater than 1.0 and the CDI values for 4-AP (1 mmol/l) combined with 5-FU (4 μmol/L and 8 μmol/l) were close to 1.0 ($P > 0.05$),

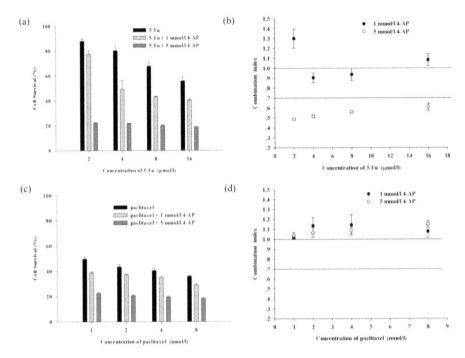

Figure 4. Cytotoxicity of 4-AP in combination with anti-cancer drugs in MCF-7 cells. CDI > 1.0, antagonistic effect; CDI < 1.0, synergistic effect; CDI < 0.7, significant synergistic effect.

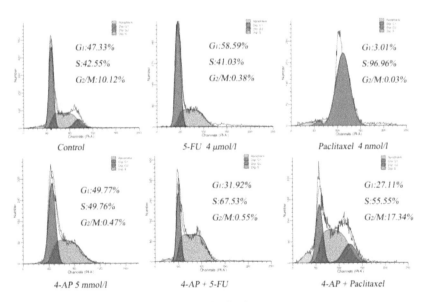

Figure 5. Flow cytometric analysis of cell cycle distribution.

suggesting an antagonistic action between 4-AP (1 mmol/l) and 5-FU was involved (Fig. 4b). Interestingly, when 4-AP concentration was increased to 5 mmol/l, the CDI values for 4-AP combined with 5-FU (2, 4, 8 and 16 µmol/l) were less than 0.7, indicating a strong synergistic action (Fig. 4b). Figure 4d illustrates that the CDI values for 4-AP at both 1 and 5 mmol/l combined with paclitaxel (1, 2, 4 and 8 nmol/l) were greater than 1.0, suggesting that an antagonistic action between 4-AP and paclitaxel was involved.

To further verify the synergistic and antagonistic actions between 5 mmol/l 4-AP with 5-FU and 5 mmol/l 4-AP with paclitaxel were related to cell cycle, we used flow cytometry to examine the cell cycle progression of MCF-7 cells treated with 5-FU or paclitaxel in the presence or absence of 4-AP. The results were depicted in Figure 5. Compared to the cells treated with 5-FU only, the number of cells arrested at the S phase was significantly increased when the cells were pre-treated and co-administrated with 5 mmol/l 4-AP. Figure 5 also shows that paclitaxel induced apparent arrest at G_2/M phase, but this paclitaxel-induced arrest was clearly blocked when the cells were pre-treated and co-administrated with 5 mmol/l 4-AP, indicating that 4-AP antagonized the cytotoxicity of paclitaxel by preventing tumor cells from entering the G_2/M phase.

4 CONCLUSION

In summary, our study demonstrates that 4-AP significantly inhibits the proliferation and induces cell apoptosis of MCF-7 cells in a dose-dependent manner. Its diverse effects on the cell cycle events at different concentrations suggest that, apart from working on channel activity, 4-AP may exert other biological effects. Since protracted S phase leads to enhanced cytotoxicity to S-phase-targeted agents, cell cycle parameters should be taken to account in the combination therapy involving the cell cycle-specific anti-cancer drugs.

ACKNOWLEDGEMENTS

This work was supported by Natural Science Foundation of China (81302203).

REFERENCES

Ait-mohamed, O. 2011. Acetonic extract of Buxus sempervirens induces cell cycle arrest, apoptosis and autophagy in breast cancer cells. *PLoS One* 6: e24537.

Asanuma, M. 2010. Calcium-activated potassium channels BK and IK1 are functionally expressed in human gliomas but do not regulate cell proliferation. *PLoS One* 5: e12304.

Blackiston, D.J. 2009. Bioelectric controls of cell proliferation: ion channels, membrane voltage and the cell cycle. *Cell Cycle* 8: 3519–28.

Brevet, M. 2008. Expression of K+ channels in normal and cancerous human breast. *Histology and Histopathology* 23: 965–72.

Chen, S.Z. 2005. HERG K+ channel expression-related chemosensitivity in cancer cells and its modulation by erythromycin. *Cancer Chemotherapy and Pharmacology* 56: 212–20.

Huang, L. 2009. ATP-sensitive potassium channels control glioma cells proliferation by regulating ERK activity. *Carcinogenesis* 30: 737–744.

Jang, S.H. 2009. Silencing of Kv4.1 potassium channels inhibits cell proliferation of tumorigenic human mammary epithelial cells. *Biochemical and Biophysical Research Communications* 384: 180–6.

Kunzelmann, K. 2005. Ion channels and cancer. *Journal of Membrane Biology* 205: 159–73.

Ouadid-ahidouch, H. 2004. Functional and molecular identification of intermediate-conductance Ca2+-activated K+ channels in breast cancer cells: association with cell cycle progression. *AJP: Cell Physiology* 287: C125–C134.

Ouadid-ahidouch, H. 2008. K+ channel expression in human breast cancer cells: involvement in cell cycle regulation and carcinogenesis. *Journal of Membrane Biology* 221: 1–6.

Ouadid-ahidouch, H. 2000. KV1.1 K(+) channels identification in human breast carcinoma cells: involvement in cell proliferation. *Biochemical and Biophysical Research Communications* 278: 272–7.

Biostatistics and biometry

Bioinformatics and Biomedical Engineering – Chou & Zhou (Eds)
© *2016 Taylor & Francis Group, London, ISBN 978-1-138-02784-8*

Epidemiological characteristics of outpatients in a Grade 3, Class A general hospital

Y.M. Li, Y.J. Tan, F. Wu, C. Zheng, N.Y. Sun & K.Y. Wang
Chengdu Military General Hospital, Chengdu, Sichuan Province, China

ABSTRACT: The present study aims to provide a basis for outpatient management strategy through a comprehensive understanding of epidemiologic features of outpatients in a general hospital. Between 2010 and 2013, data were collected from the hospital information system of a Grade 3, Class A general hospital in Chengdu city, Sichuan province. The annual indicators, such as demographics and expense category, were collected and analyzed through hospital management statistical analysis methods. The total number of outpatient visits during the statistical interval was 3,036,300, with a mean annual growth rate of 24.07%, and with 44.65% males and 55.35% females. The male-female ratio was 0.81:1, the mean patient age was 40.24 ± 19.30 years, the proportion of Medicare patients was 18.27%, and the medical-surgical department ratio was 1.35:1. Compared with surgical departments, the average annual growth rate, male-female ratio, mean patient age and proportion of insured patients in medical departments were higher.

1 INTRODUCTION

A comprehensive and systematic understanding of epidemiologic features of outpatients could provide a basis for outpatient management strategy and ultimately achieve refine management. It is extremely useful for the scientific setting up of out-patient consulting room, and for improving the utilization of scarce human, financial and material resources (Yuan et al. 2005, Xiong 2002.). In recent years, some outpatient-related investigations have been carried out by domestic scholars. However, most of the researches focus mainly on patient satisfaction, disease spectrum, population characteristics in specific departments/with specific diseases (Zhang et al. 2011, Liu et al. 2010, Pan et al. 2011) and there is still a lack of epidemiologic data of outpatients from large-scale general hospitals. In order to obtain objective information that would facilitate the planned expansion of the outpatient building of a grade 3, class A (first-class) general hospital in Chengdu city, Sichuan province, the present study was designed to collect the data of outpatients during the last four years. Annual Indicators, such as demographics, expense category, medical-surgical Dept. ratio and growth rate were analyzed through hospital management statistical analysis methods.

2 MATERIALS AND METHODS

2.1 Data source

Our study adopted the retrospective survey research design. The outpatients of Chengdu Military General Hospital in Chengdu, Sichuan Province, China were investigated. Data were retrieved from the outpatient medical records of the No. 1 Military Medical Project, which was the Hospital Information system of Chengdu Military General Hospital, between January 1, 2010 and December 31, 2013. Information concerning the gender, age, department, expense category and registration time etc. was collected. The number of outpatients during the statistical interval was 3,036,300.

2.2 Statistic method

The data were statistically analyzed by PASW 18.0 (Li et al. 2010) and Microsoft Office Excel 2007. The enumeration data were presented by frequency and percentage, and the measurement data by mean ± standard deviation ($\bar{x} \pm sd$). Patients were divided into 7 age groups: ≤10, 11-, 21-, 31-, 41-, 51-, ≥ 61 years old. Medicare category: free medical service, provincial and municipal medical insurance, county medical insurance, urban residents medical insurance, new rural cooperative medical care and commercial insurance etc.

3 RESULTS

3.1 General information concerning the outpatients

As shown in Table 1, the cumulative total number of outpatient visits during the statistical interval was 3,036,300, with a mean annual growth rate of 24.07%. The patients seen during these visits included 1,355,700 males (44.65%) and 1,680,600 females (55.35%). The male-female ratio presented a descend trend, with a total ratio of 0.81:1; The mean age of all the patients was 40.24 ± 19.30 years, with 1.60 years increased over this four-year period, presenting a rising trend. From the perspective of age group, 31–40 age group and 41–50 age group accounted for the highest rate in 2010 and 2011, respectively. The proportion of ≥ 61 age group has increased gradually, while the rest with little change. In addition, Medicare patients accounted for 18.27%.

3.2 Comparison between medical and surgical outpatients

As shown in Table 2, the average annual growth rate of medical outpatients (26.73%) was higher than that of surgical outpatients, and the medical-surgical Dept. ratio presented a rising trend, with an average of 1.35:1. By comparing the data between "corresponding" medical-surgical Dept. (gastroenterology Dept. and general surgery Dept., neurology Dept. and neurosurgery Dept., cardiology Dept. and cardio-thoracic surgery Dept., nephrology

Table 1. General information concerning the outpatients of the hospital (ten thousand (%)).

Group	Year				
	2010	2011	2012	2013	Total
Total	52.20	67.36	84.32	99.74	303.63
Gender					
Male	24.05 (46.07)	30.91 (45.89)	37.81 (44.84)	42.79 (42.90)	135.57 (44.65)
Female	28.15 (53.93)	36.45 (54.11)	46.51 (55.16)	56.95 (57.10)	168.06 (55.35)
Ratio	0.85:1	0.85:1	0.81:1	0.75:1	0.81:1
Age	39.30 ± 19.22	39.60 ± 19.29	40.55 ± 19.35	40.91 ± 19.27	40.24 ± 19.30
Age group					
≤10	4.59 (8.79)	5.91 (8.77)	7.16 (8.50)	8.05 (8.07)	25.71 (8.47)
11-	3.70 (7.08)	4.36 (6.47)	4.96 (5.89)	5.60 (5.61)	18.61 (6.13)
21-	8.11 (15.53)	10.94 (16.24)	13.23 (15.69)	16.09 (16.13)	48.37 (15.93)
31-	11.21 (21.48)	13.47 (20.00)	15.22 (18.05)	17.2 (17.24)	57.11 (18.81)
41-	9.73 (18.64)	13.59 (20.17)	18.21 (21.59)	22.58 (22.64)	64.11 (21.11)
51-	7.30 (13.98)	8.99 (13.34)	12.32 (14.61)	14.13 (14.16)	42.73 (14.07)
≥61	7.57 (14.50)	10.11 (15.01)	13.22 (15.68)	16.10 (16.14)	47.00 (15.48)
Payment					
Medicare	8.43 (16.15)	12.10 (17.96)	16.20 (19.21)	18.76 (18.80)	55.49 (18.27)
Self-pay	43.77 (83.85)	55.26 (82.04)	68.12 (80.79)	80.98 (81.20)	248.14 (81.73)

Table 2. Comparison between medical and surgical outpatient visits to the hospital (ten thousand).

Group	Year				Total	Average annual growth rate (%)
	2010	2011	2012	2013		
Total	52.20	67.36	84.32	99.74	303.63	24.07
Internal medicine Dept.	29.07	37.57	48.64	59.17	174.44	26.73
Surgical Dept.	23.13	29.79	35.68	40.58	129.19	20.60
Ratio*	1.26:1	1.26:1	1.36:1	1.46:1	1.35:1	
Gastroenterology Dept.	5.76	7.72	9.38	10.61	33.46	22.62
General surgery Dept.	1.86	2.51	2.86	3.08	10.31	18.28
Ratio*	3.09:1	3.07:1	3.28:1	3.45:1	3.24:1	
Neurology Dept.	3.29	3.97	5.02	5.25	17.51	16.89
Neurosurgery Dept.	0.63	0.71	0.82	0.93	3.08	13.76
Ratio*	5.23:1	5.60:1	6.14:1	5.67:1	5.69:1	
Cardiology Dept.	2.03	2.63	3.46	3.95	12.07	24.79
Cardio-thoracic Dept.	0.61	0.78	0.83	0.99	3.22	17.47
Ratio*	3.31:1	3.38:1	4.16:1	3.97:1	3.75:1	
Nephrology Dept.	1.63	2.12	3.10	3.60	10.45	30.07
Urologic surgery Dept.	1.92	2.51	3.00	3.41	10.84	21.19
Ratio*	0.85:1	0.85:1	1.03:1	1.05:1	0.96:1	

*The outpatient visits ratio of Internal medicine Dept. and Surgical Dept.

Dept. and urologic surgery Dept.), it was found that the average annual growth rate of medical outpatients was higher than that of surgical outpatients, with a medical-surgical Dept. ratio of 3.24: 1, 5.69: 1, 3.75: 1, 0.96: 1, respectively. In addition, the medical-surgical Dept. ratio presented a rising trend. As shown in Table 3, compared with gastroenterology Dept., the patient age and proportion of Medicare patients in general surgery Dept. were higher; the patient age in neurology Dept. was higher than that in neurosurgery Dept., while the proportion of Medicare patients was the opposite; the patient age in cardiology Dept. was higher than that in cardio-thoracic surgery Dept., while the proportion of Medicare patients was the opposite; compared with urologic surgery Dept., the patient age and proportion of Medicare patients in nephrology Dept. were higher.

3.3   General information of outpatients in major clinical departments

As shown in Table 3, the male-female ratio of medical patients (0.92:1) was higher than that of surgical patients (0.67:1), the mean age of medical patients (41.23 ± 20.43) was higher than that of surgical patients (38.92 ± 17.58), and the proportion of Medicare patients of medical patients (22.57%) was higher than that of surgical patients (12.47%). Internal medicine Dept.: It was found that a greater proportional difference between the male and female (more male than female) appears in traditional Chinese medicine Dept., nephrology Dept., endocrinology Dept., oncology Dept., respiratory medicine Dept., and gastroenterology Dept.; the mean age was highest in cardiology Dept., oncology Dept. and endocrinology Dept.; the proportion of Medicare patients was lowest in dermatology Dept., rehabilitation Dept. and neurology Dept.; the average annual growth rate was highest in oncology Dept., nephrology Dept. and hematology Dept.. Surgical Dept.: It was found that a greater proportional difference between the male and female appears in urologic surgery Dept., burn and plastic surgery Dept. and stomatology Dept.; the mean age was highest in general surgery Dept., orthopedics Dept. and urologic surgery Dept.; the proportion of Medicare patients was lowest in emergency Dept., otorhinolaryngological Dept. and obstetrics and gynecology Dept.; the average annual growth rate was highest in obstetrics and gynecology Dept., ophthalmology Dept. and stomatology Dept.

Table 3. General information of outpatients in major clinical Dept. of the hospital (ten thousand).

System/Dept.	N	Male-female ratio	Age (years old)	proportion of insured patients (%)	Average annual growth rate (%)
Internal medicine Dept.	174.44	0.92:1	41.23 ± 20.43	22.57	26.73
Gastroenterology Dept.	33.46	1.12:1	45.58 ± 14.78	15.90	22.62
Neurology Dept.	17.51	0.73:1	47.09 ± 16.32	14.07	16.89
Dermatology Dept.	13.10	1.00:1	33.24 ± 17.91	5.98	18.88
Cardiology Dept.	11.87	0.93:1	55.63 ± 16.05	30.82	22.70
Respiratory medicine Dept.	11.59	1.22:1	47.14 ± 17.04	15.09	23.42
Nephrology Dept.	10.45	0.68:1	45.66 ± 15.81	41.11	30.07
Endocrinology Dept.	8.00	0.70:1	48.13 ± 15.79	20.48	20.24
Traditional Chinese medicine Dept.	6.53	0.47:1	45.21 ± 15.51	23.76	22.33
Rehabilitation Dept.	3.36	0.76:1	44.23 ± 14.60	6.51	13.58
Hematology Dept.	3.06	0.74:1	42.92 ± 17.82	33.86	28.86
Oncology Dept.	2.95	1.32:1	55.35 ± 14.36	28.61	45.88
Surgical Dept.	129.19	0.67:1	38.92 ± 17.58	12.47	20.60
Obstetrics and gynecology Dept.	26.90	–	33.40 ± 10.75	7.88	25.49
Otorhinolaryngological Dept.	16.29	1.02:1	36.54 ± 17.51	7.72	20.36
Emergency Dept.	15.96	1.43:1	38.86 ± 19.60	7.31	12.42
Stomatology Dept.	11.99	0.77:1	40.92 ± 21.04	22.89	21.41
Urologic surgery Dept.	10.84	3.29:1	41.13 ± 17.42	13.30	21.19
Burn and plastic surgery Dept.	10.78	0.32:1	38.91 ± 15.65	11.29	21.25
General surgery Dept.	10.31	0.91:1	47.41 ± 16.23	17.24	18.28
Orthopedics Dept.	9.90	0.98:1	42.87 ± 17.26	9.47	19.98
Ophthalmology Dept.	9.70	0.96:1	39.98 ± 20.76	19.09	23.65
Cardio-thoracic surgery Dept.	3.22	1.22:1	39.87 ± 20.88	32.33	17.47
Neurosurgery Dept.	3.08	1.27:1	38.08 ± 19.37	16.60	13.76

4 DISCUSSION

Big data possess the characteristics of 4 Vs: Volume, Velocity, Variety and Value, which refers to adopting all of the data rather than randomly selected data (sample survey), as Viktor Mayer-Schönberger and Kenneth Cukier pointed out profoundly in their book Big Data. Hospital information system has been applied in Grade 3, Class A general hospitals across China for nearly 20 years, resulting in a large accumulation of data information involving clinical service, treatment and nursing, with the same characteristics as big data. Using the concept of big data for reference, the present study collected the medical records of outpatients from the hospital information system during the last four years in order to provide a basis for outpatient management strategy. In this study, we analyzed general information of outpatients of a Grade 3, Class A general hospital by means of epidemiological methods and statistical methods used in hospital management. The main achievements are as follows:

4.1 *There was an obvious variation in male-female ratio and patient age of outpatients*

The male-female ratio has declined from 0.85:1 to 0.75:1 over the past 4 years, with the possible causes as follows: 1. In modern society, women are more health-conscious, resulting in more outpatient visits of women than men; 2. The proportion of elderly female patients are rising because of the increasing aging population as well as the higher average life in females than in males; 3. Departments with a higher female visit rate (e.g. obstetrics and gynecology department, traditional Chinese medicine department) are developing faster, and thus attracting a large number of female patients. The mean age of outpatients has increased by 1.60 years over the past 4 years, and the proportion of patients aged

41–50 years has reached first place. On the one hand, this indicates the problem of aging population. On the other hand, this suggests that the gerontology department in this hospital develops fairly well. Yu Qilin etc. reported that the male-female ratio of outpatients in Navy General Hospital of Chinese PLA was 0.97:1 (Yu et al. 2003), Liu Yuanqiang etc. reported that the male-female ratio of outpatients in primary medical organizations of Gansu province was 0.81:1(Liu et al. 2010), Yao min etc. reported that the male-female ratio of outpatients in Peking University School of Stomatology was 0.76:1 (Yao et al. 2009). However, the above results, which are different from what we got from the present study, are calculated from sample surveys, as a result, they are not representatives of the overall conditions of outpatients. Hospital managers should give attention to variations in male-female ratio and patient age.

4.2 *There was a great difference between medical and surgical outpatients in growth rate and general information*

The average annual growth rate of medical outpatients was higher than that of surgical outpatients, and there was a great difference between medical and surgical outpatients in male-female ratio, mean patient age and type and composition of Medicare. The implementation of the new rural co-operative medical system and the medical insurance system for urban residents resulted in a rapid increase in the number of emergency and outpatient visits over the country as a whole. For example, compared with the year 2010, the total number of emergency and outpatient visits in 2011 rose by 7.67% nationwide, with increases of 5.27% in western areas of China, and 7.07% in Sichuan Province. The national increase in emergency and outpatient visits to general hospitals was 11.08%. The hospital in the present study is a Grade 3, Class A general hospital with a mean annual increase in outpatient visits of 24.07%, higher than that reported nationally, and higher than the values for western areas of China, and Sichuan Province. Hospital managers should give attention to variations in growth rate between medical and surgical outpatients.

4.3 *Make the out-patient consulting rooms scientifically designed and make the attempt to establish integrated consulting rooms*

The final goal of data statistic analysis is to support management policy (Li 2012). Based on the results obtained, hospital managers should take gender differences into consideration when allocating doctors to departments with a great difference in male-female ratio (e.g. obstetrics and gynecology department, urologic surgery department) in order to protect patient privacy and provide better service. Besides, departments with target patients at an advanced age (e.g. cardiology department, general surgery department) should be allocated with consulting room on the lower floor in order to facilitate elderly patients. When making a plan for departments with growing numbers of outpatients, hospital managers should reserve space for future expiation and for the purpose of allocating additional consulting rooms and medical staffs at the appropriate time. Now, clinical disciplines (subdiscipline) are highly developed in large general hospitals, resulting in more and more clinical department settings and thus making patients do not know which department to visit. In addition, this study also found that there were differences between "corresponding" medical-surgical outpatients in growth rate and general information. As a result, hospital managers should make the attempt to establish integrated consulting rooms under the guidance of Academy Combination and Discipline Integration so as to facilitate patients and promote harmonious development of medical and surgical departments. At present, the hospital in the present study has established integrated consulting rooms in the fields of cardiovascular disease, pain and tumor, which have shown initial effect.

In conclusion, using the concept of big data for reference, we could provide a basis for refined outpatient management strategy through a comprehensive investigation of general information of outpatients by means of collecting clinical information from hospital information system and adopting statistical methods used in hospital management.

ACKNOWLEDGMENTS

This work was supported by the 53 of China postdoctoral science foundation projects (No. 2013M532123) and the Chengdu Military General Hospital first funded research talents (No. 2013YG-B021).

REFERENCES

Li Z. 2012. *Information department manual of Xi Jing.* Xian: Publishing House of The Fourth Military Medical University.

Li Z.H. & Luo P. 2010. *PASW/SPSS statistic analysis coursebook (3rd ED.).* Beijing: Publishing House of Electronics Industry.

Liu Y.Q, Zhang Z.N., Ma G.G, etc. 2010. Hospital visit status of outpatients for basic medical institutions in Gansu province. *Chinese Journal of Public Health,* 26(9):1192–1193.

Pan L.Y. & Lin X.l. 2011. Investigation and analysis on outpatients' demand in grade 3, class A general hospitals. *International Journal of Nursing* 30(2):240–242.

Xiong J.Q. 2002. *Hospital management and medical statistics.* Beijing: People's Medical Publishing House.

Yao M., Luo Y., He Q.N. 2009. Investigation and analysis on 1000 outpatients. *Chinese Journal of Hospital Statistics* 16(3):269–270.

Yu Q.L. & Liu H.Y. 2003. Investigation and analysis of the kinds of diseases afflicting outpatients. *Chinese Journal of Hospital Administration* 19(7):435–437.

Yuan J.C., Gao X.W., Shi Y.Q., etc. 2005. Survey and analysis of distribution of outpatients registration and visiting time. *J Fourth Mil Med Univ* 26(1):83–85.

Zhang N, Zhang C.Y., Zhou H.Q. 2011. Investigation and analysis on outpatients' demand in large general hospitals. *Chinese Hospitals* 15(11):18–19.

Bioinformatics and Biomedical Engineering – Chou & Zhou (Eds)
© *2016 Taylor & Francis Group, London, ISBN 978-1-138-02784-8*

Methane estimation of food waste in Chinese household and environmental benefits from an energy perspective

S. Ding, G.B. Song & S.S. Zhang
Key Laboratory of Industrial Ecology and Environmental Engineering (China Ministry of Education), School of Environmental Science and Technology, Dalian University of Technology, Dalian, China

ABSTRACT: As the necessity material to the survival of humans, food is the carrier of nutrition, water resources, land resources and greenhouse gas. However, resource shortage along with food waste analysis of the environmental impact of food waste is of great significance. Based on CHNS survey data, the amount of food waste in nine representative Chinese provinces was calculated. On the basis of the existing domestic research, the theoretical amount of greenhouse gas emissions from food waste after complete fermentation was estimated, and the environmental benefit of food waste recycling was evaluated. The results showed that the theoretical CH_4 production per kilogram food waste after complete fermentation was 0.68 m^3, which could save an equivalent amount of 8.45×106 tyr^{-1} of standard coal. In seven years, the total average annual CO_2 emission reduction was 2.25×10^7 tyr^{-1}, and for SO_2, it was 2.53×10^5 tyr^{-1}.

Keywords: food waste; climate change; methane; bio-gas; China

1 INTRODUCTION

In the 21st century, the most significant environmental challenge faced by human is global climate change, which is characterized by climate warming. The impact of greenhouse gas emissions on the global climate change has attracted wide attention throughout the world. Methane is one of the most important greenhouse gases, the content of which is much lower than CO_2 in the atmosphere, while its global warming potential is 21 times as much as CO_2. The IPCC report indicated that the municipal waste landfills were the main source for the discharge of CH_4. Some research scholars have shown that there are 22 to 36 million tons of CH_4 emissions from solid waste landfills in the world each year (Bogner, 1997). However, food waste is a double-edged sword, representing both the contribution to global warming and the opportunity to benefit the environment through energy production from landfill.

On the one hand, food waste is the important component of urban solid waste. As the most basic material in people's life, food is the carrier of nutrition, water resources, land resources and greenhouse gas emissions. With the rapid growth of population, the demand for food is increasing significantly, while the food waste phenomenon is becoming even serious. According to the Food and Agriculture Organization report (FAO, 2011), one-third of the world's food (close to 1.3 billion tons) is lost or wasted. Main ingredients of food waste are meat, vegetables, fruits and staple food, which contain a large amount of proteins, lipids and carbohydrates that are suitable for biodegradation. For a long time, food waste is generally shipped to landfill facilities, the process of which will release a great deal of CH_4. Along with the increasing food waste, the release of CH_4 from the landfill process has reached to a degree that cannot be ignored.

On the other hand, as the country with a large amount of coal consumption, China's coal consumption in 2014 was 2.48×10^9 t. Coal is the main cause of air pollution in China, and

80% of carbon dioxide emissions and 85% of sulfur dioxide emissions are from burning coal according to the statistics.

As food consumers, people should bear the environmental responsibility of resource consumption and greenhouse emissions that come along with food consumption, especially the shortage of resources and environmental degradation. In this paper, we quantitatively analyzed the environmental impact of food waste, calculated the methane release quantity during the landfill process and the reduced CO_2 and SO_2 emissions by converting methane into standard coal consumption. Thus, we may draw consumers' attention to the potential environmental impact of food waste, in order to help them adjust their daily diet and behaviors to achieve sustainable development.

2 MATERIALS AND METHODS

2.1 Data sources of the household kitchen garbage

Data are collected by the China Health and Nutrition Survey (CHNS). This study was first established by the Principal Investigator of the University of North Carolina (UNC), Chapel Hill, aiming at discussing the influence of economic and social changes in a large country. The CHNS database contains the survey data of intra-household food consumption and waste in the 9 provinces of China from 1989 to 2009. Due to the fragmentary of the 1989 data, this work aims at researching the survey data from 1991 to 2009.

2.2 The sources of the demographic data

The demographic data from 1991 to 2008 are cited from the *Statistical data assembly of China in the recent 60 years* and the 2009 data are derived from the 2010 China statistical data. Due to the differences in statistical caliber (e.g. Resident population and Registered population), the data of Hunan, Hubei, and Henan provinces are combined based on the data from 2010 China statistical data of each province. The population data of these provinces are provided in Table 1.

2.3 Determination of the formula of the household kitchen garbage

The main component of the household kitchen garbage is protein, starch and fat. The elemental contents of C, H and O are 46.00%, 5.86% and 31.50% (Liu et al., 2005, Li et al., 2009), respectively, meaning that the three elements account for about 83.36% of all the household kitchen garbage. The molar ratio of the elements is calculated by the normalization method according to the molecular weight of the elements. The results are presented in Table 2, from which we can obtain the molecular formula of the household kitchen garbage as approximately $C_{35}H_{53}O_{18}$.

2.4 Methane calculation produced from food waste

The gas production rate of the household kitchen garbage directly relates to the content of C, H, O and the fermentation process. The article assumes that the gas production rate of the

Table 1. The population of each province from 1990 to 2009 (million).

Year	Liaoning	Heilongjiang	Jiangsu	Shandong	Henan	Hubei	Hunan	Guangxi	Guizhou
1991	3938.5	3575.0	6843.7	8570.0	8763.0	5512.3	6166.3	6527.0	3314.6
1993	3982.9	3640.0	6967.2	8642.0	8946.0	5653.5	6245.6	6936.7	3408.7
1997	4077.1	3751.0	7147.9	8785.0	9243.0	5872.6	6465.0	7779.7	3605.8
2000	4135.3	3807.0	7327.2	8997.0	9488.0	5960.0	6562.1	8650.0	3755.7
2004	4172.8	3816.8	7432.5	9180.0	9717.0	6016.1	6697.7	9110.7	3903.7
2006	4271.0	3823.0	7549.5	9309.0	9820.0	6050.0	6768.1	9304.0	3757.2
2009	4319.0	3826.0	7725.0	9470.3	9967.0	6141.9	6900.2	9638.0	3798.0

Table 2. Elemental contents of C, H and O from kitchen garbage and approximate molar ratios.

Element	C	H	O
Elemental contents (%)	46.00	5.86	31.50
Molar ratio	3.83	5.86	1.97
Normalized molar ratio	35	53	18

household kitchen garbage in the fermentation process is the theoretical value, which can be calculated according to the Buswell-Mueller formula:

$$C_nH_aO_b + \left(n - \frac{1}{4}a - \frac{1}{2}b\right)H_2O \rightarrow \left(\frac{1}{2}n - \frac{1}{8}a + \frac{1}{4}b\right)CO_2 + \left(\frac{1}{2}n + \frac{1}{8}a - \frac{1}{4}b\right)CH_4 \qquad (1)$$

Based on the molecular formula ($C_{35}H_{53}O_{18}$) of the household kitchen garbage, this formula can be written as follows:

$$C_{35}H_{53}O_{18} + 12.75H_2O \rightarrow 19.63CH_4 + 15.38CO_2 \qquad (2)$$

In the case of complete fermentation, 0.34 kg methane could be produced per kilogram household kitchen garbage, and the theoretical gas production of methane is 0.44 m³ kg⁻¹ at the standard condition of 0°C along with 1.01×10^5 Pa.

3 RESULTS AND DISCUSSION

3.1 Food waste in household in China

The overall quantity of food waste per capita in China's 9 provinces is 16 kgyr⁻¹. Of these, the province with the largest amount of waste is Hubei, whose quantity of food waste per capita in seven years is up to 38.6 kgyr⁻¹; the province with the least amount of waste is Heilongjiang, whose quantity of food waste per capita in seven years is only 13.8 kgyr⁻¹.

3.2 Food waste estimation of spatial pattern of methane food waste

There are obvious differences between the provinces in relation to the amount of CH_4 released by completely fermented food waste. Figure 1 intuitively shows the amount of CH_4 released by completely fermented food waste in each year among the different provinces. In 7 years, CH_4 emission of Hubei Province is 7.57×10^5 tons per year and exceeds that of the other six provinces. This phenomenon is due to its large amount of food waste per capita as well as the large population base.

3.3 Potential environmental benefits of methane from food waste

China is rich in resources, but deficient in per capita occupation. If CH_4 released by food waste is used rationally, it can become a useful supplement to our energy security. If the volume of CH_4 accounts for 65% of biogas, then the biogas volume that 1 kg of food waste produces is 0.68 m³.

Biogas can completely substitute fuels such as coal, firewood and straw used in daily life. In terms of the amount, assuming that the main purpose of the biogas is for cooking, and taking the effective calorific value as the transformation basis, then the amount of raw coal that could be saved by using biogas can be calculated as follows:

$$M_C = \frac{V_G \times CV_G \times TE_G}{CV_C \times TE_C} \qquad (3)$$

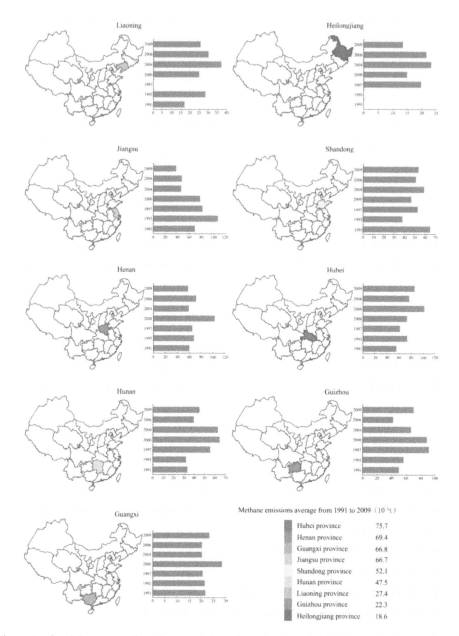

Figure 1. Spatial patterns of methane emissions from the household kitchen of nine provinces in China during 1991 to 2009.

where M_C refers to the amount of raw coal replaced by household biogas; V_G refers to the amount of household biogas; CV_G refers to net calorific value of household biogas, whose average value is 2.09×10^4 kJm^{-3}; TE_G refers to the thermal efficiency of household biogas stove, whose value is 60% (Wu et al., 2007); CVC refers to the average net calorific value of standard coal, which can be estimated as 2.09×10^4 kJkg^{-1}; and TE_{AE} refers to the thermal efficiency of raw coal boiler, whose value is 35% (Liu, 2009).

Consider that 1 kg raw coal can be converted to 0.71 kg standard coal, while 2.66 kg carbon dioxide and 0.03 kg sulfur dioxide are emitted per kilogram of standard coal combustion.

Table 3. Reduced emission of each province in seven years (10^4 tyr^{-1}).

Emission reductions	Liaoning	Heilongjiang	Jiangsu	Shandong	Guizhou
CO_2	137.95	93.80	335.70	262.51	112.29
SO_2	1.56	1.06	3.79	2.96	1.27
Emission reductions	Henan	Hubei	Hunan	Guangxi	Average
CO_2	349.12	380.84	238.96	336.06	4847.33
SO_2	3.94	4.30	2.70	3.79	54.67

The reduced per year carbon dioxide and sulfur dioxide emissions due to the use of CH_4 within 7 years are summarized in Table 3. Also, it can be seen from the table that the average reduced emissions of CO_2 and SO_2 nationwide are 4.85×10^7 tyr^{-1} and 5.67×10^5 tyr^{-1}, respectively.

Biogas energy has many environmental benefits from both the consumption perspective and the production perspective. The coal mining process will destroy the mine ecological environment and the acid mine drainage will cause water pollutions; however, the problems could be overcome by the process of transforming food waste to CH_4. It will have positive effects on the environment and effectively purify the living environment.

4 CONCLUSION

In nine provinces across the north and south of China, food waste quantity per capita differs widely. The province with the least waste is Heilongjiang Province, with the largest amount of waste per capita being 13.8 kg yr^{-1} in 7 years. The province with the largest waste is Hubei Province, with the largest amount of waste per capita being 38.6 kg yr^{-1} in 7 years. When food waste was disposed by using the biogas fermentation technology, the theoretical amount of methane production is 0.44 m^3 kg^{-1} in the standard condition after complete fermentation. The province with the largest theoretical gas production is Hubei Province. If all of the food waste from the nine provinces in seven years is anaerobically fermented, it can save an equivalent amount of 8.44×10^6 tons of standard coal per year. In seven years, the emission of CO_2 and SO_2 could be reduced to 2.25×10^7 tyr^{-1} and 2.53×10^5 tyr^{-1}, respectively.

Food waste causes great issues on the aspect of greenhouse gas emissions and potential resource wastage. Methane, however, can serve as an alternative energy source to substitute coal consumption for the purpose of reducing global warming and other pollutant emissions, such as SO_2 in China, where the environment is highly challenged by the commitment to mitigate climate change in the next few decades. We therefore expect that a great opportunity would be available if a policy of garbage classification is executed to divide the carbon-rich organic food waste from urban construction waste, and to feed the biodegradable waste to landfill around the city. Thus, China will achieve two things at one stroke by producing more methane for energy purpose and conserving natural resource embedded in food waste.

ACKNOWLEDGMENTS

This work was supported by the Fundamental Research Funds for the Central Universities of China under Grant DUT14LAB17.

REFERENCES

Bogner, J.E., Spokas, K.A. Burton, E.A. Kinetics of Methane Oxidation in a Landfill CoverSoil: Temporal Variations, a Whole-Landfill Oxidation Experiment, and Modeling of Net CH₄ Emissions [J]. Environ. Sci. Technol., 1997, 31: 2504–2514.

China Health and Nutrition Survey; www.cpc.unc.edu/projects/china.

Food and Agriculture Organization Website; www.fao.org/home/en/.

Li, D., Sun, Y.M. Yuan, Z.H., et al. Methane production by anaerobic co-digestion of food waste and waste paper [J]. Acta Scientiae Circumstantiae, 2009, 29(3): 577–583.

Liu, H.Y., Wang J.H. Zhao, D.G. Study of anaerobic digestion treatment technology for food waste and swill [J]. Energy Technology, 2005, 26(4): 150–154.

Liu, Y.Z. Environmental benefits evaluation of using methane to reduce greenhouse gas [J]. Journal of Yangtze University, 2009, 6(1): 81–84.

Wu, L.F., Deng, S.M. Liao, G.C., et al. Economic evaluation of rural methane project based on CDM [J]. Jiangxi Energy, 2007, (3): 41–43.

Bioinformatics and Biomedical Engineering – Chou & Zhou (Eds)
© 2016 Taylor & Francis Group, London, ISBN 978-1-138-02784-8

A method of Chemiluminescence Enzyme Immunoassay for Zearalenone

K.H. Li, L.X. Zhu, W. Meng & R.R. Liu
School of Life Science, Jiangxi Science and Technology Normal University, Jiangxi, Nanchang, China

ABSTRACT: Zearalenone (ZEN) is a deleterious mycotoxin towards human and animals. An immunoassay, Chemiluminescence Enzyme Immunoassay (CLEIA) is established for ZEN based on the luminescence properties of the ZEN. In the experimentation, antigen coated concentration, antibody dilution proportion, the concentration of Horseradish Peroxidase (HRP)-conjugated anti-antibody, PH value and ionic strength of the buffer, and concentration of different organic solvents were optimized to obtain the best reaction condition. The method shows followed results, the 50% Inhibitory Concentration (IC_{50}) of CLEIA is 0.143 ng/ml. With adding standards to rice meal 30 ng~100 ng, the recovery ratio ranged 80%~95.4%, and the Relative Standard Deviation (RSD) ranged 1.88%~5.89%; with adding standards to corn flour 10 ng~30 ng, the recovery ratio ranged 93.0%~103.7%, and the RSD value ranged 2.97%~4.41%. In conclusion, CLEIA is an appropriate and applicable method for detection of ZEN.

1 INTRODUCTION

Zearalenone (ZEN) what also called F2 toxin is a Resorcylic Acid Lactone (RAL) mycotoxin without steroidal structure secreted by *F. graminearum*, *C. gibberella* or *F. tricinctum* (Popiel, D 2014a, b, Fink-Gremmels, J & Malekinejad 2007). ZEN is a water-fast white crystal, but in alkaline water solution, methanol, acetonitrile, ethyl acetate, acetone ZEN is diffluent, and ZEN is semisoluble in sherwood oil. Figure 1 shows the constitutional formula of ZEN.

ZEN was first extracted from moldy corn by Stob and his team in 1962 (Wang, Y.K. et al. 2013). It is proved that ZEN causes harm to humans mainly through contaminating corn, wheat and other cereals. Many studies have shown a variety of hazards towards human and animals, and the reproductive toxicity is the most serious harm because of its estrogen-like effects. Other hazards such as immunotoxicity, hepatotoxicity, cytotoxicity, genotoxicity and induced carcinogenicity were also reported (Jarvis, B.B. & Miller, J.D. 2005).

In recent years, food safety issues are received more and more attention. As a food contaminant, ZEN gathered the food experts' attention. At the same time, varies of detection methods were established for ZEN (Sun, Y.N. et al. 2014). Chemiluminescence Enzyme Immunoassay (Maiolini, E. et al. 2013) has many advantages such as high sensitivity, good specificity, simple and rapid. That makes it become a good measure for detection of traceable ZEN.

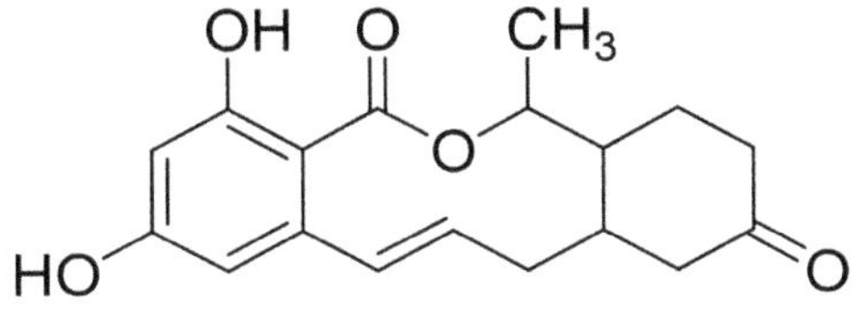

Figure 1.   The constitutional formula of ZEN.

Through this study, conditions of CLEIA (Wang, Y.K. et al. 2012) for detection of ZEN could be optimum, and could give a better ratio of recovery adding experiments for samples. It is supposed to provide a believable and responsible data for further study of ZEN.

2 MATERIALS AND METHODS

2.1 Reagents and instrument

Mouse anti-zearalenone monoclonal antibody. Horseradish peroxidase labeled goat anti-mouse anti-antibody immunoglobulins and hapten were purchased from Sigma. Methanol (HPLC grade) was obtained from Sigma. Phosphate-Buffered Saline (PBS). Luminoskan ascent, Wellwash versa (Thermo Scientific, USA). 96-well white polystyrene plates (Costar, USA). 5804R High-speed Refrigerated Centrifuge (Eppendorf, Germany).

2.2 Experimental details

The whole experiments were performed in 96-well white polystyrene high-binding microplates for chemiluminescence. ZEN-BSA antigen (120 μl per well) incubated 2 h at 37°C lucifuge moisturize for coating. And the uncoated sites were blocked by adding 330 μl per well of 5% skim milk powder, incubated 3 h at 37°C lucifuge moisturize. The third step of adding anti-body and the fourth step of adding Horseradish Peroxidase (HRP)-conjugated anti-antibody take the same incubated condition as the former except for time (40 min). Finally, adding the intermixture (100 μl per well) of chemiluminescent substrate A and B (A: B = 1:1).

2.3 Optimized antigen coated concentration and antibody dilution proportion

The optimal concentrations of coated-antigen and antibody for the chemiluminescent assay were first carried by checkerboard titration using the twice antigen coated concentration range between 0.125 μg/ml~4 μg/ml and doubling dilution for antibody concentration range 1:4000~1:2560000.

2.4 Optimized Horseradish Peroxidase (HRP)-conjugated anti-antibody concentration

The optimal concentrations of Horseradish Peroxidase (HRP)-conjugated anti-antibody for the chemiluminescent assay was also performed by checkerboard titration. Antigen coated concentration used the optimized one in first step. Antibody dilution proportion also ranged from 1:4000 to 1:2560000 and performed by doubling dilution. The concentrations of Horseradish Peroxidase (HRP)-conjugated anti-antibody are 1:2500, 1:5000, 1:7500, and 1:10000, respectively.

2.5 Optimized PH value, ionic strength of the buffer and concentration of organic solvents

In this three experiments, the PH value of the buffer ranges from 5.5 to 9.0, and the ionic strength of the buffer ranges from 0 mol/l to 2.0 mol/l, and the concentration of organic solvents (methanol-water) ranges from 5% to 30% all using the same concentration range, respectively, and other conditions stay same.

2.6 Calibration curves and the standard addition recovery

The calibration curves were performed in the 0.02–5 ng/ml range. And the association rates were normalized between 0% and 100% according to the expression (Eq. 1):

$$\%B/B_0 = 100(E_X/E_0) \tag{1}$$

where E_x is the chemiluminescent emission value when the concentration of ZEN-hapten is × ng/ml. E_0 is the emission value of a blank control.

The standard addition recovery experiments were performed based on rice meal and corn flour. Adding 30 ng, 50 ng and 100 ng standard of ZEN into rice meal, respectively. And then using methanol-water (9:1) as the extraction solvent, vibrating 10 minutes, centrifuging 15 minutes at 8000 r/min for extraction of ZEN. Finally, the respective concentration was thinned by methanol-water (1:9). And adding 10 ng, 30 ng standard of ZEN into corn flour, respectively, and other steps for extraction were the same as the former one.

3 RESULT AND DISCUSSION

3.1 *The result of optimized experiment*

In the optimized experiment, the experiment condition of a later one used the result of the former experiments. But the orthogonal test was unperformed, so it was irresponsible that the combination of all optimized condition also is the best condition for immunochemical method.

Figure 2 (a) and Figure 2 (b) show the effects for Horseradish Peroxidase (HRP)-conjugated anti-antibody dilution proportion and ionic strength, respectively. Especially, the influence for PH value of buffer towards luminescence value showed in Figure 2 (c) is ruleless. The effect of the concentration of different organic solvents is expressed by changing of IC_{50} showed in Figure 2 (d).

As the luminescence value, blank value, and IC_{50} are considered, chose 0.5 ug/ml, 1:200000, 1:5000, 0 mol/ml, 7.0, 10% methanol-water as the antigen coated concentration, antibody dilution proportion, Horseradish Peroxidase (HRP)-conjugated anti-antibody dilution proportion, ionic strength of the buffer, PH value of the buffer, organic solvents concentration, respectively.

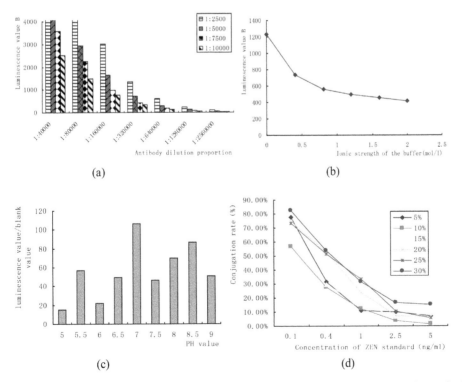

(a)   (b)

(c)   (d)

Figure 2.   Effect of Horseradish Peroxidase (HRP)-conjugated anti-antibody dilution proportion (a), ionic strength (b), PH value (c), organic solvents concentration (d) on the ZEN immunoassay based on CLEIA.

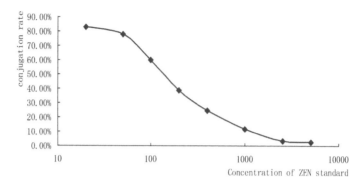

Figure 3.   CLEIA calibration curves for ZEN immunoassay.

Table 1.   The recovery ratio and RSD value for rice meal and corn flour.

Sample	Adding standard (ng)	Recovery ratio (%)	RSD value (%)
	30	91.2	3.20
Rice meal	50	95.4	1.88
	100	80.0	5.89
Corn flour	10	103.7	2.97
	30	93.0	4.41

3.2   *The results of calibration curves and the standard addition recovery experiments*

Calibration curves. Figure 1 shows the calibration curves for ZEN immunoassay based on CLEIA. From the calibration curves, we can calculate that the IC_{50} is 0.143 ng/ml. The IC_{50} value is lower and the method has good sensitivity.

The result of standard addition recovery experiments. In this procedure, there is interference from matrix what can not be ignored, so take action of dilution to eliminate the interference. The recovery ratio and RSD value for rice meal and corn flour are listed in Table 1.

From Table 1 we can see that the recovery ratio of the adding experiments are acceptable and the all RSD values are lower than 10%. It is prove that the CLEIA method has a good precision.

4   CONCLUSIONS

A new, simple and fast CLEIA has been established to be a very useful method for a semi-quantitative determination of ZEN. The method has showed good sensitivity and precision for ZEN detection and it is suitable for detection of trace ZEN in food. The IC_{50} is 0.143 ng/ml. The recovery ratio all ranged in 80%~103.7% and the RSD value ranged all ranged in 1.88%~5.89%. Compared with another CLEIA (Wang, Y.K. et al. 2012) for ZEN detection in corn, the method mentioned above has the lower IC_{50} value. It is prove that the method is more sensitive.

What's more, compared with the method of HPLC and GC, CLEIA avoids the use of abundant organic solvent and large-scale instrument. The significant reduction in the time per analysis and the organic solvent makes CLEIA become a promising method for secure detection of ZEN.

ACKNOWLEDGEMENT

This work was financially supported by grants from Natural Science Foundation of Jiangxi Province (No. 20122BAB214006, 20132BAB214004) and Science Foundation of Jiangxi Educational Bureau (No. GJJ13573).

REFERENCES

Fink-Gremmels, J. & Malekinejad. 2007. H. Clinical effects and biochemical mechanisms associated with exposure to the mycoestrogen zearalenone. *Animal Feed Science & Technology*. 137:326–341.

Jarvis, B.B. & Miller, J.D. 2005. Mycotoxins as harmful indoor air contaminants. *Applied Microbiology & Biotechnology*. 66(4):367–372.

Maiolini, E. et al. 2013. Bisphenol A determination in baby bottles by chemiluminescence enzyme-linked immunosorbent assay, lateral flow immunoassay and liquid chromatography tandem mass spectrometry. *Analyst*. 139(1):318–324

Popiel, D. 2014. Zearalenone lactonohydrolase activity in Hypocreales and its evolutionary relationships within the epoxide hydrolase subset of a/b-hydrolases. *BMC Microbiology*, 14(4):485–488.

Sun, Y.N. et al. 2014. Development of an Immunochromatographic Strip Test for the Rapid Detection of Zearalenone in Corn. *J Agric Food Chem*, 62:11116–11121.

Wang, Y.K. et al. 2012. Detection of Zearalenone by Chemiluminescence Immunoassay in Corn Samples. *Advances in Chemical Engineering II*. 550–553:1911–1914.

Wang, Y.K. et al. 2013. Novel chemiluminescence immunoassay for the determination of zearalenone in food samples using gold nanoparticles labeled with streptavidin-horseradish peroxidase. *J Agric Food Chem*. 61(18):4250–4256.

Bioinformatics and Biomedical Engineering – Chou & Zhou (Eds)
© 2016 Taylor & Francis Group, London, ISBN 978-1-138-02784-8

Estimation of blood glucose noninvasively using near infrared spectroscopy

R. Peng, D.X. Guo, Y.Z. Shang, S.S. Yong & X.A. Wang
The Key Laboratory of Integrated Microsystems, Peking University Shenzhen Graduate School, Shenzhen, Guangdong Province, China

ABSTRACT: This paper presents a unique method for non-invasive estimation of blood glucose concentration using near infrared spectroscopy. The method has been performed using transmission photoplethysmograph. We introduce support vector machine as a nonlinear regression method to construct calibration model which use eigenvalues of photoplethysmograph and blood glucose concentration as input parameters. A new modeling method is proposed which combines the classifier and predictor with regression analysis based on support vector machine for mutual calibration. Our results show that the mutual calibration model enables a crucial improvement over single regression model. For the glucose estimation, the correlation coefficient of determination for reference versus prediction increases by 16.61% and root mean square error decreases by 10.69%.

1 INTRODUCTION

According to International Diabetes Federation report, the estimated diabetes prevalence for 2013 is 382 million and it is expected to affect 592 million people by 2035. Poor management of diabetes can lead to serious health problems such as cardiovascular diseases, damage of blood vessels, stroke, blindness, chronic kidney failure, nervous system diseases and early death (Coster et al. 2000). In order to maintain Blood Glucose Concentration (BGC), frequent detection is an essential part of diabetic management. In current clinical practice, the blood glucose has been measured by pricking the patient's finger or vein to extract a small quantity of blood. Although most blood glucose monitor is minimally invasive, patients still suffer from the inevitable pain and infection several times a day (Kottmann et al. 2012, Guo et al. 2012). Thus, a number of techniques employed for blood glucose noninvasive measurement have been proposed, such as reverse iontophoresis, thermal emission spectroscopy, absorbance spectroscopy (including Near-Infrared (NIR) spectroscopy Mid-Infrared (MIR) spectroscopy, Raman spectroscopy and so on (Li et al. 2014).

NIR spectroscopy has become one of the most promising techniques for blood glucose noninvasive measurement. The concept is to irradiate a particular wavelength of infrared light on a vascular region of the body and obtain the transmission Photoplethysmograph (PPG). The key to NIR spectroscopy is to model the relationship between eigenvalues of PPG and BGC. The most important linear calibration method for spectroscopic data analysis is Partial Least Squares Regression (PLSR) (Thissen et al. 2004a). The main problem in PLS methodology is that the spectrum property relationship is assumed to be linear. However, the assumption of linearity may fail under the influence of fluctuations in process and system variables, such as changes in temperature and physiological glucose dynamics. Hence, nonlinear modeling techniques are the significant requirements for building robust calibration model since such modeling techniques have the potential of reflecting intrinsic nonlinearities that can be found in natural multivariable systems. Support Vector Machine (SVM) (Suykens 1999, Van Gestel et al. 2004) might be regarded as the perfect candidate for spectral regression purposes. Since its initial formulation by Vapnik (Vapnik 1995, Cortes & Vapnik 1995), SVM has been used extensively for classification problems. Moreover, SVM has been extended to develop non-linear regression models capable of quantitative prediction. For example, SVM calibration models have exhibited excellent potential for NIR absorption-based concentration prediction in chemical mixtures, even when the acquired spectra are nonlinearly affected by temperature fluctuations (Thissen et al. 2004b).

In this paper, we propose a method for non-invasive estimation of BGC using NIR spectroscopy. The method uses eigenvalues of PPG and a calibration model is constructed between PPG and BGC based on SVM. This paper presents a new modeling method that a multi-class classifier for different range of blood glucose is proposed with regression model for mutual calibration when predicting BGC. We compare the difference of performance on estimation of blood glucose level using the mutual calibration model or single regression model and our experimental results show that the mutual calibration model enables a significant improvement for the glucose estimation.

2 THEORETICAL ANALYSIS

According to Beer-lambert's law, the intensity of incident light I_0 and transmission light I can be given as follow:

$$I = I_0 \exp(-\varepsilon c d) \tag{1}$$

where ε = absorbancy of material; c = concentration of material; and d = thickness of material. When taking account of the transformation coefficient Z and light-absorbing substance is split into tissue and blood, the Beer-lambert's equation can also be given as follow:

$$V = ZI_0 \exp(-\varepsilon_a c_a d_a - \varepsilon_t c_t d_t) \tag{2}$$

where a represents blood and t represents tissue of human body. The intensity of transmission light depends on not only the change of blood volume, but also the change of the blood components' concentration over a long period of time. Hence, equation can be explained as follows:

$$V = V_0 \exp(-\varepsilon_a c_a d_a) \tag{3}$$

$$\ln V = \ln V_0 - \varepsilon_a c_a d_a = k_1 - k_2 c_a d_a \tag{4}$$

where k_1 represents the effect of the intensity of incident light and transformation coefficient. k_2 represents the effect of the wavelength of incident light. c_a represents the effect of the

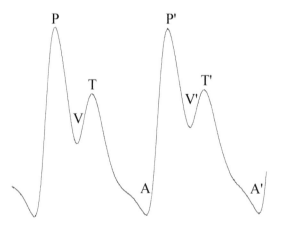

Figure 1. The obtained PPG signal.

Table 1. Eigenvalues of PPG signal.

Number	Name of eigenvalues
1	Amplitude of main peak h_P
2	Amplitude of main wave trough h_A
3	Amplitude of secondary peak h_T
4	Amplitude of secondary wave trough h_V
5	Time of adjacent main peaks t_{PP}
6	Time between main peak and main trough t_{PA}
7	Time between main peak and secondary peak t_{PT}
8	Time between main peak and secondary wave trough t_{PV}

blood components' concentration. In the current research, different wavelengths of infrared have been applied and corresponding PPG signals are obtained. And wavelengths of 850 nm, 875 nm, 940 nm, and 1050 nm have been chosen because the cost of the components is less but glucose almost has the same absorption. Meanwhile, four different infrared wavelengths can also provide more eigenvalues.

The obtained PPG signal is shown in Figure 1. There are four key points over a period, main peak P, secondary peak T, main wave trough A and secondary wave trough V. While the PPG signal has eight eigenvalues. And these eigenvalues can be given in Table 1.

3 MATERIAL AND METHODS

3.1 Experimental protocols

Experimental studies are conducted based on independently designed system to accomplish the following objectives: (1) investigate the relationship between PPG and BGC and (2) verify the effectiveness of the mutual calibration model compared to single regression model.

The details of experiment can be described as follows: 5 healthy adult volunteers were recruited for this experiment (22 to 25 years old; all males). PPG signals of each volunteer were collected for 8–10 times, including 3 times before meal and 5–7 times after meal. Measurements were performed every 10 minutes, each lasting 3 minutes. PPG signals collected in the first two minutes were used to construct the calibration model, while those in last one minute were treated as test data. In addition, finger prick measurements were performed immediately after the PPG collection to get the reference values of BGC using an ACCU-CHEK glucose analyzer. During the experiment, each subject was asked to sit comfortably in a chair and environment temperature was controlled at about 25°C.

3.2 Data treatment and model construction

The data treatment and model construction procedure are shown in Figure 2. The obtained PPG is denoted as $S_{PPG}(t)$ and L_{PPG} is denoted as length of signal. A vector $S^{(i)}_{Windows}(t)$ (i means cycle index) is extracted from $S_{PPG}(t)$ (window length, $L_{Windows} = 5s$). Then mean values of features shown in Table 1 have been extracted and put into a vector:

$$X^{(i)} = \begin{bmatrix} h_P & h_A & h_T & h_V & t_{PP'} & t_{PA} & t_{PT} & t_{PV} \end{bmatrix}^T \tag{5}$$

Since wavelengths of 850 nm, 875 nm, 940 nm, and 1050 nm have been applied in our study, PPG signals and features can be denoted as $S_{PPG}(t)_j$ and $X^{(i)}_j$ ($j = 1, 2, 3, 4$ represent 850 nm, 875 nm, 940 nm and 1050 nm, respectively). Meanwhile, $X^{(i)}_F$ is shown as:

$$X^{(i)}_F = \left[\left(X^{(i)}_1\right)^T \quad \left(X^{(i)}_2\right)^T \quad \left(X^{(i)}_3\right)^T \quad \left(X^{(i)}_4\right)^T \right]_{32\times1} \tag{6}$$

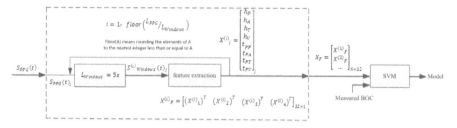

Figure 2.   Construction procedure of calibration model using SVM.

Table 2.   The class labels for multi-class classifier.

Range of BGC (mmol/L)	Labels
$4 \le$ BGC < 5	0
$5 \le$ BGC < 6	1
$6 \le$ BGC < 7	2
$7 \le$ BGC < 8	3
BGC ≥ 8	4

Finally, a eigenvalue array is constructed:

$$X_F = \begin{bmatrix} X^{(1)}_F \\ X^{(2)}_F \\ ... \end{bmatrix}_{N \times 32} \tag{7}$$

where, $N = floor\ (L_{PPG}/L_{Windows})$ (floor (A) means rounding the element of A to the nearest integer less than or equal to A).

After that, we take the eigenvalue array X_F and the measured BGC as input parameters to construct the SVM model. In order to improve the accuracy of the glucose estimation, this paper presents a new modeling method that a multi-class classifier for different range of blood glucose is constructed. The class labels of the training data for classifier are given in Table 2.

Meanwhile, a regression model is also constructed using the same training data. When predicting BGC, the outputs of multi-class classifier and regression model are combined for mutual calibration. In order to implement the procedure, the software package Libsvm (version 2.3.1) and MATLAB (version R2012a) are used in this paper.

4   RESULTS AND DISCUSSION

46 sets of data for the PPG signals of different wavelengths and the measured BGC over the range of 4.1–9.0 mmol/l (73.8–162 mg/dl) were obtained and used as the training data sets to construct the calibration models, and 43 sets of data for test of models.

The performance of calibration model based on the mutual calibration model or single regression model is evaluated according to Root Mean Squares Error (RMSE), correlation coefficients (R^2) and Table 3 illustrates the prediction parameters based on different modeling method.

When compared with single regression model, classification accuracy of the mutual calibration model increases by 13.95%. Meanwhile, R^2 of the mutual calibration model increases by 16.61% and RMSE decreases by 10.69%.

Table 3. Prediction parameters with the mutual calibration model and single regression model.

Parameters	Single regression model	Mutual calibration model
Accuracy of classification	63.52%	77.47%
R^2	0.6328	0.7379
RMSE	0.7163	0.6397

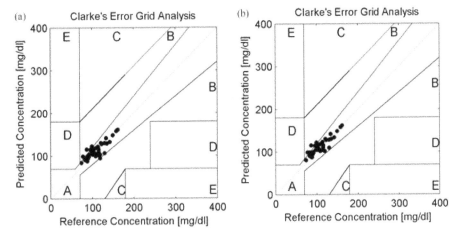

Figure 3. Blood glucose predictions of different modeling methods shown on the Clarke error grid: (a) single regression model; (b) mutual calibration model.

The estimated BGC versus measured BGC using single (a) regression model and (b) mutual calibration model are shown in Figure 3 plotted on the Clarke error grid (Clarke et al. 1987), which is a widely used method for evaluating the clinical usefulness of glucose predictions. Predictions in zones A and B are considered acceptable, and predictions in zones C, D, and E are potentially dangerous if used in clinical judgment. As can be seen in Figure 3(a), all data points using single regression model are within clinically acceptable regions: the region A and B (region A: 88.37%, region B: 11.63%: A + B = 100%). When compared to single regression model, data points in region A using mutual calibration model increase to 90.70%, and 9.3% of points are in region B (A + B = 100%). Therefore, it might be suggested reasonably that the mutual calibration model enables a significant improvement over the single regression model. A probable reason is that the mutual calibration model can better reflect the relationship between PPG signals and BGC compared with single regression model.

5 CONCLUSION

This paper investigates a technique for non-invasive estimation of BGC using NIR spectroscopy. PPG signals are obtained based on independently designed system and SVM is used as a nonlinear modeling method to model the relationship between PPG signals and BGC. A new modeling method is proposed which is combined the classifier with regression analysis based on SVM for mutual calibration. Our experimental results show that PPG signal is a promising object of study for non-invasive estimation of BGC. Meanwhile, the mutual calibration model presented in paper enables a significant improvement over single regression model.

REFERENCES

Clarke, W.L., Cox, D., Gonder-Frederick, L.A., Carter, W. & Pohl, S.L. 1987. Evaluating clinical accuracy of systems for self-monitoring of blood glucose. *Diabetes Care* 10: 622–628.

Cortes, C. & Vapnik, V. 1995. Support-vector networks. *Machine Learning* 20: 273–297.

Coster, S., Gulliford, M.C., Seed, P.T., Powrie, J.K. & Swaminathan, R. 2000. Monitoring blood glucose control in diabetes mellitus: A systematic review. *Health Technol Assess* 4: i–iv, 1–93.

Guo, X., Mandelis, A. & Zinman, B. 2012. Noninvasive glucose detection in human skin using wavelength modulated differential laser photothermal radiometry. *Biomed Opt Express* 3: 3012–21.

Kottmann, J., Rey, J.M., Luginbuhl, J., Reichmann, E. & Sigrist, M.W. 2012. Glucose sensing in human epidermis using mid-infrared photoacoustic detection. *Biomed Opt Express* 3: 667–80.

Li, Z., Li, G., Yan, W.-J. & Lin, L. 2014. Classification of diabetes and measurement of blood glucose concentration noninvasively using near infrared spectroscopy. *Infrared Physics & Technology* 67: 574–582.

Suykens, J.A.K. 1999. Least squares support vector machine classifiers. *Neural Processing Letters* 9: 293–300.

Thissen, U., Pepers, M., Üstün, B., Melssen, W.J. & Buydens, L.M.C. 2004a. Comparing support vector machines to pls for spectral regression applications. *Chemometrics and Intelligent Laboratory Systems* 73: 169–179.

Thissen, U., Üstün, B., Melssen, W.J. & Buydens, L.M.C. 2004b. Multivariate calibration with least-squares support vector machines. *Analytical Chemistry* 76: 3099–3105.

Van Gestel, T., Suykens, J.A.K., Baesens, B., Viaene, S., Vanthienen, J., Dedene, G., De Moor, B. & Vandewalle, J. 2004. Benchmarking least squares support vector machine classifiers. *Machine learning* 54: 5–32.

Vapnik, V.N. 1995. The nature of statistical learning theory. *Neural Networks IEEE Transactions on* 10: 988–999.

Restrictive factors of the measurement accuracy to estimate Blood Pressure with Pulse wave Transit Time

X.M. Chen, Y.B. Li, Y. Zhang & N. Deng
Institute of Microelectronics, Tsinghua University, Beijing, China

ABSTRACT: Pulse wave Transit Time (PTT) is a promising approach to achieve a continuous and non-invasive measurement of Blood Pressure (BP). However, compared with invasive and cuff-based methods, the PTT-based method is not accurate enough for some circumstances. In this paper, its major restrictive factors are investigated. According to the experimental results and analysis, the breathing rhythm and movement of subjects have an adverse impact on the detection of the pulse wave. In order to obtain better signals, noise should be eliminated and the time delay resulting from filters should be considered. The validity of PTT detection is a dominating factor of accuracy as well. Furthermore, the measured PTT is not the real PTT from the physiological perspective, which is mainly due to the acquisition difficulty of the pulse wave from the main artery. Finally, the influence of the regulatory mechanism is also analyzed.

1 INTRODUCTION

Blood Pressure (BP) is one of the most important vital signs for healthcare and reflects the cardiovascular status of people (Parati, G. et al. 1995). Direct measurement of BP is accurate but invasive and only limited to certain clinical situations. Indirect methods based on the Pulse wave Transit Time (PTT) have attracted the public's attention because of their continuous and non-invasive characteristics (Fiala, J. et al. 2010; Joseph, J et al. 2009). However, some factors affect the accuracy of the BP measurement.

This paper intensively studies the major restrictive factors of the PTT-based method. First, the relationship between the PTT and BP is introduced. Then, the acquisition system and experimental methods are described. Based on the experimental results, we summarize the effects of these factors on the BP measurement, and give reasonable explanations and appropriate solutions.

2 RELATIONSHIP BETWEEN PTT AND BP

From the definition of PTT (Geddes, L.A. et al. 1981), it is inversely proportional to Pulse Wave Velocity (PWV). PWV has been confirmed to be a kind of reflection of BP (Gribbin, B. et al. 1976). In 1808, Thomas Young pointed out the quantitative description of PWV for an incompressible fluid in a flexible tube:

$$c = \sqrt{\frac{Eh}{\rho D}} \tag{1}$$

where c is the velocity of the pulse wave; E is Young's modulus of the blood vessel; h is the thickness of the arterial wall; ρ is the density of the blood; and D is the diameter of the artery.

After Thomas Young, a large amount of ideas have been brought forward to calibrate this relationship (Chen, W. 2010; Chen, Y. et al. 2009; Wang, B. et al. 2000). In order to eliminate the difficulties of measuring the thickness of the arterial wall and Young's modulus, Bramble and Hill modified the formula as follows:

$$c = \sqrt{\frac{V \partial P}{\rho \partial V}} \tag{2}$$

where V is the blood volume and P is the BP of that pulse. Because the density of the blood can be assumed to be constant, the expression can be simplified as follows:

$$\Delta P \propto c^2 \left(\frac{\Delta V}{V} \right) \tag{3}$$

Namely

$$BP \propto \frac{L^2}{PTT^2} \left(\frac{\Delta V}{V} \right) \tag{4}$$

where L is the distance of the transit. In summary, the BP and the PTT are inversely related.

3   INTRODUCTION OF THE ACQUISITION SYSTEM

In order to get the PTT, the Electrocardiogram (ECG) and the pulse wave should be collected. The ECG signal is detected by two electrodes attached to the left and right arms, respectively. Besides, the pulse wave signal can be acquired by a pressure sensor or a photoelectric sensor at the peripheral artery, such as the fingertips and earlobes.

The collected analog signals are amplified and filtered to reduce the noise. They are observed with an oscilloscope before being sampled by an A/D converter. The digital processing system is built on the platform of Freescale K60 MCU, which is based on ARM Cortex-M4. After AD conversion, useful data is sent to the computer via the UART series port, and finally coped with MATLAB software to process and display. The sample rate of AD is 500 samples/second, and the serial transmission baud rate is 115200 bps.

The real BP is measured with the Omron HEM-1020 BP Monitor, which is cuff-based and can display the Systolic BP (SBP), Diastolic BP (DBP) and Heart Rate (HR).

4   METHOD OF THE EXPERIMENT

A total of five healthy youths aged 23 to 26 years were selected for the study. They are required to be seated still and relaxed during the procedure.

Experiments are conducted at two states: static rest state and exercise recovery state. In the static rest state, subjects should keep quiet and be calm before and during the test, which can reflect their real normal BP. The exercise recovery state is the process of recovery just after exercise.

First, subjects are at the static rest state. The PTT and the real BP are obtained simultaneously. Since the acquisition of the real BP is not continuous and lasts for about 20 seconds, the mean value of the PTT within this period of time is regarded as the corresponding value. Overall, 10–15 groups of values are recorded for each subject.

Second, in the exercise recovery state, subjects perform vigorous exercise for about 30 minutes to remarkably change the BP. Thereafter, the PTT and the real BP are recorded for 15–25 groups.

5 EXPERIMENTAL RESULTS

The collected ECG, Photoplethysmography (PPG) and pressure pulse wave signals can be processed by Matlab, as shown in Figure 1. The left three images are raw signals and the right three images are their power spectrum. The raw signals are shown with sampling points as the horizontal axis and the amplitude derived from the signals' voltage as the vertical axis.

Figure 2 shows the comparison between the original signals and the filtered signals of PPG (the signal with a relatively smaller amplitude) and pressure pulse wave. The filtered signals become smooth and easy to deal with. The horizontal axis represents time and the vertical axis indicates the voltage of the signal.

The illustration of PTT is displayed in Figure 3. The horizontal axis refers to sampling points (500 points equal to 1 s) and the vertical axis refers to the voltage of the signal. The

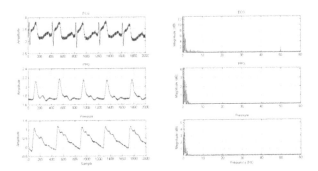

Figure 1. The original signal and its power spectrum.

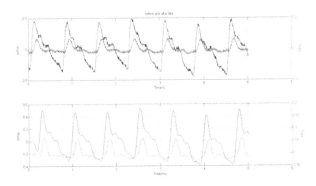

Figure 2. The filtered signals are displayed below the raw signals.

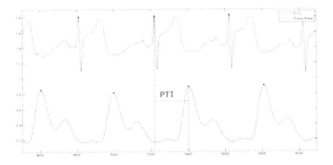

Figure 3. Illustration of PTT.

437

PTT is the time interval that the pulse wave propagates from the central artery to the peripheral artery in a cardiac cycle. Usually, the R-peak of ECG represents the beginning of the ventricular systole and the peak of the peripheral pulse wave represents the arrival of the signal.

As shown in Figure 4, the PTT, SBP, DBP and HR are acquired by the method mentioned above. The horizontal axis indicates the group number, and the vertical axis describes their values. From Figure 4, PTT is found to be about 400 ms, SBP ranges from 115 to 135 mmHg, DBP ranges from 65 to 80 mmHg, and HR ranges from 60 to 120beat/min. Besides, groups 1 to 11 are in the static rest state, and groups 12 to 26 are in the exercise recovery state.

Figure 5 shows the pulse waves of different testing states plotted with the overlay of their beginning and normalization of the amplitude from 0 to 1. Part I represents the pulse waves at the static rest state (groups 1–11 in Fig. 4). Part II represents the pulse waves just after the exercise (groups 12–20 in Fig. 4). Part III contains the pulse waves when the BP is already back to normal (groups 20–26 in Fig. 4).

As shown in Figure 5, the cardiac cycle of part I, part II and part III is about 950 ms (63 beat/min), 500 ms (120 beat/min) and 750 ms (80 beat/min).

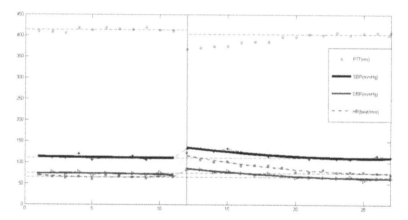

Figure 4.   PTT and its BP measured from the static rest state to the exercise recovery state.

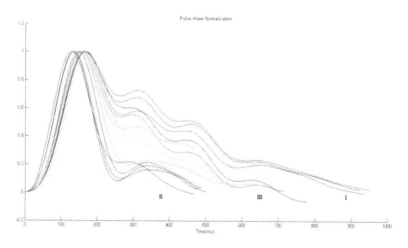

Figure 5.   Comparison of pulse waves from the static rest state to the exercise recovery state.

6   DISCUSSION AND ANALYSIS

6.1   *The body movement affects the accuracy of measurement*

The sensors, which are used to detect pulse signals, are generally placed at the fingertips, toes, radial artery and earlobes. During the experiment, the slight movement and the breathing rhythm of subjects can cause the baseline drift, i.e. the pulse waves will swing back and forth, or, even worse, may damage the signals. In other words, the signals cannot be recognized by naked eyes or programs. In order to obtain the high-quality signals, subjects need to keep quiet, be relaxed and motionless during the procedure.

6.2   *Bad Signal-to-Noise Ratio (SNR) and delay from filters cause measurement problem*

The signals are exported by the sensors in small voltage, which have low-frequency characteristics as shown in Figure 1 and Figure 2. For the purpose of acquiring useful information and suppressing random noisy, the low-pass filter and the 50 Hz notch filter are necessary. Furthermore, from the algorithm's aspect, the method of wavelet transform could be adopted. However, the filter would cause time delay when compared between the original and filtered signals, as shown in Figure 2. As we rely on the transit time, the delay must be counted and considered seriously.

6.3   *The method of obtaining the PTT is also a restrictive factor of accuracy*

As the peripheral pulse wave is the superposition of the forward wave and the reflected wave, its shape is not totally the same as the aortic pressure, but the information on peak is still valuable. So, the peak is commonly acceptable to gain the PTT. In other words, the more the accurate peaks are detected, the more reliable the PTT is. Therefore, it is beneficial to normalize the vertical axis, sharp the signal and optimize the peak extraction algorithm.

6.4   *PTT acquired by ECG and PPG is not strictly equal to the real PTT*

As shown in Figure 6, the PTT acquired by the measurement is PTT_m; however, the real PTT is from the peak of the aortic pressure to the peak of the peripheral pulse wave. PTT_m

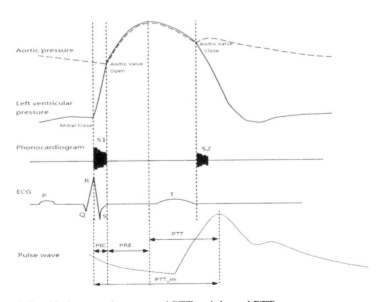

Figure 6.   Relationship between the measured PTT and the real PTT.

contains the Period of Isovolumic Contraction (PIC), the Period of Rapid Ejection (PRE) and the real PTT. Some researchers regard PIC and PRE as a whole, which is called the Pre-Ejection Period (PEP) (Zhang, G. et al. 2011). As shown in Figure 4, PTT_m is about 400 ms; however, PIC is about 50 ms and PRE is about 100 ms according to a previous study (Yao, T. 2013).

When the aortic pressure rises or myocardial contractility declines, the time of PIC will be prolonged. It has also been reported that PRE is associated with BP (Yao, T. 2013; Bers, DM. 2001). So, the existence of PIC and PRE affects the accuracy of the model between the BP and the PTT. For further research, this part should be reconsidered.

6.5 Complicated regulation mechanism of the cardiovascular system affects the experimental result

In order to study the relationship between the BP and the PTT, the exercise-recovery process is implemented. During exercise, both BP and HR are increased by neuromodulation, humoral regulation and autoregulation of the cardiovascular system. However, some hormones produced by the regulation can be quickly metabolized, but some are not. During the recovery process. Some functions of the regulation affect the BP alone, some affect the HR alone, and the factors affecting both BP and HR also exist. It is therefore complex and coupled. As a result, the recovery of BP is faster than HR (see Fig. 4).

In Figure 5, comparing part I with part II, with the rising of BP and HR, both the ascent and descent of the pulse wave are affected, but the impact on the descent is much obvious. Comparing I with part III, the BP remains the same, but the HR is different, and the ascent of pulse waves nearly coincides with each other, but the descent of pulse waves has a distinct difference.

These aforementioned factors cause adverse influences to the experimental result. Under ideal conditions, the PTT before and after exercise should be identical at the same BP level. However, from Figure 4, it can be seen that the PTT after exercise is a bit shorter than before. The reasonable explanation of this phenomenon is that vascular compliance and peripheral resistance have been changed by the regulation of the body, which is detrimental to setting up the relationship between the BP and the PTT.

7 CONCLUSION

Some major factors that affect the accuracy of the BP measurement based on the PTT are studied in this paper. The origin, physiology explanation, seriousness and proper solution of each factor are intensively discussed with experiments. Considering the restrictive factors is helpful for enriching and improving the theory of PTT's application. Further research lies in the better reorganization of these factors, and efforts are made to avoid these disadvantages and reduce the impact of these factors gradually to improve the accuracy of the measurement.

ACKNOWLEDGMENT

This work was funded by the R&D Programs of Tsinghua University (Grant No. 2012z01001).

REFERENCES

Bers, D.M. 2001. *Excitation-Contraction Coupling and Cardiac Contractile Force.* 2nd edition Dordrecht. Netherlands: Kluwer Academic Press Publishers.
Chen, W. 2000. Continuous Estimation Of Systolic Blood Pressure Using The Pulse Arrival Time And Intermittent Calibration. *Medical & Biological Engineering & Computing* 38(5):569–574.
Chen, Y., Wen, C., Tao, G., Bi, M. & Li, G. 2009. Continuous and Noninvasive Blood Pressure Measurement: A Novel Modeling Methodology of the Relationship Between Blood Pressure and Pulse Wave Velocity. *Annals of Biomedical Engineering* 37(11):2222–2233.

Fiala, J., Bingger, P., Foerster, K., Heilmann, C., Beyersdorf, F., Zappe, H. & Seifert, A. 2010. Implantable sensor for blood pressure determination via pulse transit time. *Sensors IEEE*: 1226–1229.

Geddes, L.A., Voelz, M.H., Babbs, C.F., Bourland, J.D. & Tacker, W. 1981. Pulse transit time as an indicator of arterial blood pressure. *Psychophysiology* 18(1):71–74.

Gribbin, B., Steptoe, A. & Sleight, P. 1976. Pulse Wave Velocity as a Measure of Blood Pressure Change. *Psychophysiology* 13(1):86–90.

Joseph, J., Jayashankar, V. & Kumar, V.J. 2009. A Pc Based System For Non-Invasive Measurement Of Carotid Artery Compliance. *Instrumentation and Measurement Technology Conference, 2009. I2MTC '09. IEEE*: 680–685.

Parati, G., Saul, J.P., Rienzo, M.D. & Mancia, G. 1995. Spectral analysis of blood pressure and heart rate variability in evaluating cardiovascular regulation. A critical appraisal. *Hypertension* 25(6):1276–1286.

Wang, B., Yang Y. & Xiang J. 2000. A Noninvasive Method for Radial Pulse-Wave Velocity and the Determinants of Pulse-Wave Velocity. *Journal of Biomedical Engineering* 17(2):179–182.

Yao, T. 2013. *Physiology*. 2nd edition. Beijing: People's Medical Publishing House.

Zhang, G., Gao, M., Xu, D., Olivier, N.B. & Mukkamala, R. 2011. Pulse arrival time is not an adequate surrogate for pulse transit time as a marker of blood pressure. *Journal of Applied Physiology* 111(6):1681–1686.

Rehabilitation engineering

Bioinformatics and Biomedical Engineering – Chou & Zhou (Eds)
© 2016 Taylor & Francis Group, London, ISBN 978-1-138-02784-8

The research on motion recognition based on EMG of residual thigh

T.Y. Zhang
National Research Center for Rehabilitation Technical Aids, Beijing, China
School of Biological Science and Medical Engineering, Beihang University, Beijing, China

ABSTRACT: Movement pattern recognition is the key to control the intelligent lower limb prostheses. In this paper, surface Electromyographic (EMG) signals from six muscles of the thigh amputee stump were collected. After wavelet de-noised, all the starting and ending time of the effective action was determined by calculating the wavelength of the EMG signal in real time. A variety of time-domain and frequency domain features of the EMG signals were extracted, three movement pattern were recognized based on the Support Vector Machine (SVM) including flat walking, up stairs and down stairs, and the efficiency of identification was improved by feature optimizing. Experimental results show that, the three movement patterns can be classified online by EMG signals from different subjects using the method in this paper, the recognition rate was above 95%, so that just using the stump EMG to recognize the movement intention was proved to be feasible.

Keywords: motion recognition; EMG; residual thigh; SVM

1 INTRODUCTION

The EMG and movement have a strong correlation. Using the EMG signals to recognize the amputee's motion intent has become an important development direction for intelligent prostheses control. Currently, the surface EMG signals mainly applied to control the upper extremity prostheses (Gao et al. 2011 & Gini et al. 2012). Because of the particularity and complexity for the lower limb movement, the research on myoelectric prosthetic leg is still in the exploratory stage (Rubana et al. 2013). Therefore, the movement pattern recognition of the lower limb based on EMG has become the core of the lower limb prostheses research. Currently, many studies are based on healthy people, and the real-time recognition failed to achieve satisfactory results. For example, She & Luo et al. (2010) who used plantar pressure, knee angle and surface EMG of healthy people to classify each gait phases of flat walking, up and down the stairs. Huang (2009 & 2011) used 11 groups of foot pressure and EMG from amputees' hip and thigh muscles to identify seven kinds of movement patterns, such as flat walking, cross obstacles, stairs and turned, and made better results that was worthy learn. However, the studies above were all achieved recognition using EMG and motion information, using only the EMG for real-time action classification has not yet been solved. To solve these problems, in this paper, the EMG signals from thigh amputee's stump muscles were collected and action segments were divided online. After feature extraction, three motion patterns were recognized based on SVM, so that to solve the problem of lower limb action online recognition.

2 EXPERIMENT DESIGN

2.1 Subjects

20 patients with thigh amputation were selected as experiment subjects, including 15 males and 5 females. The age distribution was 8 to 59 years old, and mostly was 20 to 30 years old.

The left and right amputations were each ten cases, and the time of subject worn prostheses was 0.5 to 28 years. Before the experiment, the purpose of this study and the experiment steps were all explained to each subject, and consented to take the form of voluntary participation.

2.2 Instrument

The Biometrics electromyography and motion analysis system was used. 6 channels were used to synchronously collect the EMG signals from 6 muscles of the amputee stump, including Rectus Femoris (RF), Vastus Lateralis (VL), Tensor Fascia Lata (TFL), Biceps Femoris (BF), Semitendinosus (ST) and Gluteus Maximus (GM). The sampling frequency was 1000Hz.

2.3 Content and procedure

The basic information of all subjects was recorded and the corresponding parts of the subjects' skin were cleaned before the experiment. The electrodes were required to paste on suitable locations. The subjects were required to complete the maximal voluntary contraction of the muscle according the specific action, and the EMG signals were recorded.

In the experiment, the subjects were required to complete movements in accordance with the action essentials. The required action includes: flat walking, up stairs and down stairs. The flat walking action was 10 steps for a group, a total of six groups, and the interval between two groups was 30 seconds. The up and down stair action was 6 steps for a group, a total of 10 groups. The interval between up stair and down stair was 10 second. 1 to 2 minutes rest was allowed after completion of each group.

3 DATA SEGMENTATION

The original EMG signals contain various noises, so before the signal feature extraction, wavelet de-noising was done firstly. Figure 1 shows the comparison of EMG signals before and after wavelet de-noising.

Due to each action is continuous in the online identification process, the most important before feature extraction and action recognition is the dividing the EMG signals according the action, looking for the start and end of each single actions, and remove the signal which corresponds no action, so that to reduce the workload of the signal analysis and improve the

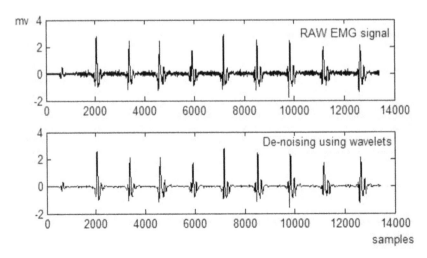

Figure 1.   The EMG signals before and after wavelet de-noising.

446

analysis accuracy (Li et al. 2013). Generally, the EMG magnitude and muscle strength is a positive correlation (Dideriksen et al. 2010). Before the movement, the muscle in a relaxed state, and the surface EMG signal is weak; with the intensification of the movement, the EMG signal amplitude increases. Therefore, a certain threshold was set to the judge action start by the amplitude of EMG, as follows:

The EMG signal was been scanned using a moving window with 100 data points, and made the Wavelength (WL) of data within the window as a feature to determine whether has an action. Wavelength reflects the combined effect of the signal amplitude, frequency and duration, can reflect the complexity of EMG signal waveform. The wavelength is defined as Equation 1.

$$WL = \sum_{k=1}^{N-1} |x_{k+1} - x_k| \tag{1}$$

where x_k = current EMG; x_{k+1} = next EMG and N = data legth.

When the wavelength value exceeds the set threshold, the current window flag was 1, otherwise marked 0.

$$A = \begin{cases} 1 & WL \geq S \\ 0 & WL < S \end{cases} \tag{2}$$

where S = threshold; WL = wavelength and A = window flag.

If the flag of the current window is $A(k)$, then the previous window is $A(k-1)$, the next window is $A(k + 1)$, and so on. In the process of moving the window, when $A(k-1) = 0$, $A(k) = 1$, $A(k + 1) = 1$ and $A(k + 2) = 1$, the start point of the current window was marked as the start point for the action. Calculation process showed in Figure 2.

Figure 3 is the EMG signal of TFL when subject walking on flat. The horizontal axis is data points, and the vertical axis is the EMG signal amplitude. The broken lines showed the calculated start points of each myoelectric action. Seen from Figure 3, the calculated start points consistent with the fact ones, proved the validity of this calculation method.

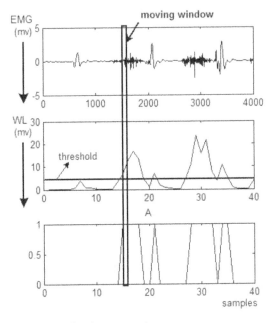

Figure 2.   The calculation process of action start point.

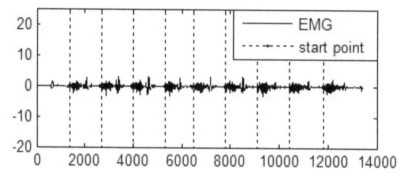

Figure 3.　The calculation result of action start point.

4　FEATURE EXTRACTION AND ACTION RECOGNITION

4.1　Action recognition for the same subject

40 groups of EMG signals were randomly selected from one subject's data in once experiment for the motion recognition, normalized by its maximum voluntary contraction value of each muscle, and then the signal features were extracted, including integrated EMG (IEMG), Root Mean Square (RMS), Standard Deviation (STD), Wavelength (WL), the average frequency (MPF) and the Median Frequency (MF).

For each action, 30 sets of data were selected as training data, and 10 sets of data as the verification data. The Support Vector Machine (SVM) was used for training, and the training process with parameter optimization. The correct recognition result of the flat walking, up stairs and down stairs was 100% (30/30).

Randomly selected two different times of experimental data of the same subject, experiment time was one week apart, then mixed two experimental data together, extracted feature and trained in the same way. Similarly, 30 sets of data were selected as training data, and 10 sets of data as the verification data. The correct rate of recognition was 93.33% (28/30). When 50 sets of data were selected as training data for each action, and 20 sets of data as the verification data, the correct rate was 98.33% (59/60).

The experiment results confirmed that even for the same subject, different experimental conditions (such as: electrode paste location, physical fatigue condition, etc.) would have some impact on the accuracy of EMG recognition (Tkach et al. 2010 & Rainoldi et al. 2004). However, by increasing the number of training data, the differences under different conditions would be eliminated as far as possible, and the recognition accuracy could be improved.

4.2　Action recognition for different subjects

There are some differences between EMG characteristics of different subjects. To test the action recognition method in this paper whether versatile for different subjects, experimental data from several different subjects were randomly selected and mixed together. Extracted feature and trained was same as the previous way. Likewise, 50 sets of data were selected as training data for each action, and 20 sets of data as the verification data, the correct rate was 93.33% (56/60).

The result showed that due to the different characteristics of the subjects, EMG differences, the recognition rate of the mixed data was lower than data of the single subject. But overall, the recognition rate can reach 90%, proved that the identification method is universal for data of different subjects and can be used to identify EMG action from different subjects. By increasing the number of training data, the accuracy of recognition is expected to improve further.

Table 1. Correlation matrix between six EMG features.

Correlation	IEMG	RMS	STD	WL	MPF	MF
IEMG	1.000	0.891	0.744	0.659	0.171	0.087
RMS	0.891	1.000	0.965	0.557	0.341	0.155
STD	0.744	0.965	1.000	0.466	0.417	0.189
WL	0.659	0.557	0.466	1.000	0.213	0.263
MPF	0.171	0.341	0.417	0.213	1.000	0.889
MF	0.087	0.155	0.189	0.263	0.889	1.000

Table 2. Recognition results compare between different feature amounts.

	Six feature	Three feature
Same subject		
Single data	100% (30/30)	100% (30/30)
Mixed data	93.33% (28/30)	93.33% (28/30)
	98.33% (59/60)	96.67% (58/60)
Different subjects	93.33% (56/60)	95% (57/60)

4.3 *Feature optimization*

Though a good recognition result has reached using the features in the front pattern recognition, the time for training and recognition was too long because of vast features and training data. In the actual application, due to the limit of processor speed, the timeliness of recognition, convenience of EMG acquisition and other requirements, the satisfactory recognition results should be achieved using as few EMG features as possible. Therefore, how to extract the most effective feature is a focus of solving the problem.

In this paper, the SPSS software was used to calculate the correlation between the six kinds of features in the previous section. The result showed in Table 1.

As can be seen from the result, the highest correlation between RMS and STD reached 0.965, while the relatively high correlation between the RMS and IEMG close to 0.9, so the RMS feature may be selected to using for the EMG recognition. In addition, the correlation between MPF and MF is also relatively high, close to 0.9, so one of them could be choose as an effective feature.

Eventually, RMS, WL and MPF were used for pattern recognition, the recognition results and the results of all six feature vectors were compared in Table 2.

Table 2 showed that, after analysing the correlation between the six features, reducing some strongly related features, the recognition accuracy of same subject's data was not significantly reduced, while the accuracy of different subjects' data improved. These illustrated that after reducing the feature amount in this way, not only the training efficiency improved, but also excluded the possibility of redundant features, and the recognition rate reached the expectations.

5 CONCLUSIONS

By collecting surface EMG signals from amputees' thigh stumps, using moving window to scan and capture the effective signal, the method of feature extraction and motion classification was researched, simultaneously discussed the feature optimization methods. Recognition accuracy for data among different subjects reached more than 95%, so that the feasibility of lower limb amputee motion recognition relying solely on the stump surface EMG was preliminary verified. In follow-up studies, the EMG features will be further optimized to minimize the number of muscles used for recognition and increase the types of motions to be classified.

ACKNOWLEDGEMENTS

The study funded by the Fundamental Research Projects of NRRA.

REFERENCES

Dideriksen J.L., Farina D. & Enoka R.M. 2010. Influence of fatigue on the simulated relation between the amplitude of the surface electromyogram and muscle force. *Philos Trans R Soc A—Math Phys Eng Sci* 368:2765–2781.

Gao Y.Y., Meng M., Luo Z.Z. & She Q.S. 2011. Multi-Mode and Gait Phase Recognition of Lower Limb Prosthesis Based on Multi-Source Motion Information. *Chinese Journal of Sensors and Actuators* 24 (11):1574–1578.

Gao Y.Y., She Q.S., Meng M. & Luo Z.Z. 2010. Recognition method based on multi-information fusion for gait patterns of above-knee prosthesis. *Chinese Journal of Scientific Instrument* 12:2682–2688.

Gini. G., Arvetti. M., Somlai. I. & Folgheraiter. M. 2012. Acquisition and analysis of EMG signals to recognize multiple hand movements for prosthetic applications. *Applied Bionics and Biomechanics* 9(2):145.

Huang H., Kuiken T. & Lipschutz R. 2009. A Strategy for Identifying Locomotion Modes using Surface Electromyography. *IEEE Trans. on Biomedical Engineering* 56(1):65–73.

Huang H., Zhang F., Hargrove L.J., Dou Z., Rogers D.R. & Kevin B. 2011. Englehart. Continuous Locomotion-Mode Identification for Prosthetic Legs Based on Neuromuscular–Mechanical Fusion. *IEEE Transactions on Biomedical Engineering* 58(10):2867–2875.

Li L., Wang J.H. & Gu S.S. 2013. Improved Automatic Segmentation Method of sEMG Based on Signals Energy Value. *Computer Science* 40(6A):188–190.

Rainoldi A., Melchiorri G. & Caruso I. 2004. A method for positioning electrodes during surface EMG recordings in lower limb muscles. *Journal of Neuroscience Methods* 134:37–43.

Rubana H.C. Mamun B.I. 2013. Surface Electromyography Signal Processing and Classification Techniques. *Sensors* 13:12431–12466.

She Q.S., Luo Z.Z., Meng M. & Xu P. 2010. Multiple kernel learning SVM-based EMG pattern classification for lower limb control. *International Conference on Control, Automation, Robotics and Vision* 2109–2113.

Tkach D., Huang H. & Kuiken T. 2010. Study of stability of time-domain features for electromyographic pattern recognition. *Neuroeng Rehabil* 7:21.

Bioinformatics and Biomedical Engineering – Chou & Zhou (Eds)
© 2016 Taylor & Francis Group, London, ISBN 978-1-138-02784-8

Author index